Project Mercury

NASA's First Manned Space Programme

Springer
London
Berlin
Heidelberg
New York
Barcelona
Hong Kong
Milan
Paris
Santa Clara
Singapore
Tokyo

John Catchpole

Project Mercury

NASA's First Manned Space Programme

John Catchpole
Basingstoke
Hampshire
UK

SPRINGER–PRAXIS BOOKS IN ASTRONOMY AND SPACE SCIENCES
SUBJECT *ADVISORY EDITOR*: John Mason B.Sc., Ph.D.

ISBN 1-85233-406-1 Springer-Verlag Berlin Heidelberg New York

British Library Cataloguing-in-Publication Data
Catchpole, John
Project Mercury: NASA's first manned space programme. –
(Springer-Praxis books in astronomy and space sciences)
1. Project Mercury (U.S.) – History
I. Title
629.4′5

ISBN 1-85233-406-1

Library of Congress Cataloging-in-Publication Data
Catchpole, John, 1957–
Project Mercury: NASA's first manned space programme / John Catchpole.
p. cm. – (Springer-Praxis books in astronomy and space sciences)
Includes bibliographical references and index.
ISBN 1-85233-406-1 (alk. paper)
1. Project Mercury (U.S.)–History. 2 Manned space flight–United States. I. Title. II. Series.

TL789.8.U6 M4523 2001
629.45′4′0973–dc21 2001020835

Printed by MPG Books Ltd, Bodmin, Cornwall, UK

Copy editing: Alex Whyte
Cover design: Jim Wilkie
Typesetting: BookEns Ltd, Royston, Herts., UK

Printed on acid-free paper supplied by Precision Publishing Papers Ltd, UK

This book is dedicated to all of the
civilian and military personnel
who worked on
Project Mercury

Contents

List of figures

List of tables

Author's preface

The formation of the world's first Communist State, in November 1917, led to a rift between the new Union of Soviet Socialist Republics on one side and, on the other, America and her European Allies with their democratically elected governments and free market. For the next two decades the mistrust between two of the largest nations on Earth grew stronger with each passing year.

When Adolf Hitler's National Socialist Party rose to power in Germany the Soviet Union signed a non-aggression pact with the new power. This allowed Soviet Premier Joseph Stalin to concentrate on the modernisation and industrialisation of his vast nation while also allowing Hitler to prepare for war in Western Europe with no Soviet intervention against Germany's Eastern border. In 1939, Britain and her Allies declared war on Germany. Following the collapse of Western Europe (with the exception of Britain), Hitler ignored the non-aggression pact and invaded the Soviet Union. The Americans and Soviets temporarily put 25 years of political enmity behind them in order to cooperate with Britain and each other to bring about the downfall of Nazi Germany.

With the end of hostilities in Europe, that continent was divided along the lines where the two great armies met as they had crushed Germany from opposite sides. America, Britain and France had fought together to liberate Western Europe and return all but Germany to the control of their own national governments. In the East, the Soviet Union held on defiantly to the countries their forces had 'liberated'. Those countries formed an occupied buffer zone between the USSR and the West, from where invaders had marched East on so many occasions in the past. Puppet governments sympathetic to Moscow were eventually installed in those countries.

Germany was divided into four Occupation Zones with one of the four major Allies occupying each of the zones. Berlin, Germany's capital city, was within the Soviet Occupation Zone but was similarly divided into four sectors and the Western Allies were assured free access through the Soviet Zone in order to carry out their occupation duties in Berlin.

When America financed the rebuilding of Western Europe the Soviets did likewise in the East, but on a much smaller scale. When the Western nations formed the North Atlantic Treaty Organisation the Soviets formed the Eastern Bloc. In time the occupation forces were removed from the Western part of Germany and the German nation was encouraged to reform its industrial base at the heart of Europe's economy. On that occasion the Soviets did not reciprocate. Rather, they kept the Eastern part of Germany deliberately weak, so that it would remain reliant on the Soviet Union and not able to rise up against it.

Originally, there were few limitations on crossing between occupation zones within Berlin and many people lived in one zone but worked in another. This freedom of movement presented its own problems as thousands of Germans moved out of the Soviet Zone, leaving that zone short of skilled labour. The Soviets made several attempts to persuade the Western powers to remove their troops and hand the entire city over to Soviet control, but these were unsuccessful and Western military forces in Berlin remained a matter of contention in Moscow throughout the post-war period. In 1961, to prevent any further movement of labour, the Soviets physically divided the city and constructed the Berlin Wall.

The 20 years following Victory in Europe Day were a period of world-wide political change. They were a time when the old Imperial powers – Britain, France and Austria – were coming to terms with the loss of their empires. Following the collapse of those empires a number of nations underwent Communist uprisings and installed Communist governments independently of the Soviet Union. In other countries independence brought bloody civil war, as happened in India and Africa. The division of Palestine and the formation of a Jewish homeland introduced a new conflict zone in the Middle East. At the same time, America and the Soviet Union were still learning their roles as 'the world's policemen'. The Soviet Union supported Communist revolutionaries around the world while America supported many of the regimes under attack by those revolutionaries. Both nations sought to attract the non-aligned, newly independent countries.

At the end of the World War II, German military science and engineering had been widely sought by all of the Allied nations. Among the trophies were Germany's ballistic missile programme and the engineers behind it, and the Americans and Soviets both took German rocket technology and engineers back to their own countries. This resulted in a race to develop a missile capable of carrying nuclear warheads directly from either of these countries to the other. Having developed such missiles, both nations sought a way of demonstrating their capability without using them for their primary purpose. The race to place the first artificial satellite into orbit was one result of that quest. While both nations developed scientific satellites, both were also working on parallel programmes to perfect a reconnaissance satellite, to spy on the military capabilities of their adversaries. With the completion of the satellite race a new round of competition began – a race to place the first man into space. Project Mercury was America's competitor in this second race. The Soviet contender was the Vostok programme.

Project Mercury was the product of two very different kinds of warfare. The technology that made it possible to launch a man into space, support him in the vacuum beyond Earth's atmosphere, bring him back to Earth and recover him from the vast Atlantic Ocean was largely developed from the military technology and research available at the end of World War II. The political decisions that brought about the race to place a man in space were products of the post-1945 world order. Without the ideological posturing of a Capitalist America and a Communist Soviet Union during the early years of the Cold War, Project Mercury would not have happened as it did.

Exactly how and why Project Mercury came about is a convoluted story. Rather than follow the complicated chronology of events as they occurred, this text splits the story into a number of self-contained sections. The sections are further broken down into loosely related chapters. It is here that the chronological histories of individual elements in the story are recorded. The history of each element stands alone, making each one easier to understand in its own right. When the stories of these elements are taken together they weave the intricate story that is Project Mercury.

Upon reading the accounts of individual flights within Project Mercury the modern reader might feel that very little happened when compared to the complicated construction tasks carried out by today's Shuttle crews as they build the International Space Station in Earth orbit. In one sense you would be right. The Mercury astronauts remained strapped in their couches at all times and barely had room to fully extend their arms. But the modern reader must not forget that at the end of the 1950s and the beginning of the 1960s just the idea of strapping a man on top of a ballistic missile and sending him to a region high above the majority of Earth's sensible atmosphere was nothing short of audacious. The media coverage at the time made the Mercury astronauts out to be courageous beyond measure, almost superhuman. But the courage shown by the Mercury astronauts was nothing special. It was the same courage that was shown by their fathers and grandfathers in two world wars. It was the courage of all patriotic young men and women called to the service of their country at a time of national need.

The seven Flight A astronauts were only the visible representation of a very large group of people who had dreamed of manned space flight and then devoted their lives to turn those dreams into reality. These other people were often referred to as the nameless and faceless thousands who worked to make space flight possible. Such terminology does each and every one of them a great injustice. Everyone who worked on Project Mercury had a name and a face. They were known to their work colleagues and to the management of their individual companies. Their families knew that they were devoting long hours to the attempt to make an American pilot the first man in space and were proud that they were doing so. A few of these people have since become well known within the space programme, so much so that even the general public know their names, but most have never been recognised for their efforts. I cannot name everyone concerned, nor can I show their faces to the world, but as you read about the exploits of the few that are known, remember that this book also tells the story of those that history has decided shall remain anonymous.

Even as Project Mercury was being undertaken the military and political rivalry between America and the Soviet Union continued. Communist expansionist policies and the West's continued presence in West Berlin brought the world to the brink of war on more than one occasion. Meanwhile, America was undergoing a period of intense domestic upheaval as Negroes demonstrated for their Civil Rights and white segregationists fought to maintain the status quo.

In this volume I have not sought to isolate my account Project Mercury from the international and domestic chaos that surrounded it. Even so, this volume is about Project Mercury and therefore, after a brief history of American–Communist relations I have kept my account of the issues surrounding the fledgling space programme to brief diary-like entries and then only in the chapter giving accounts of the Project Mercury flight programme. This is in keeping with my belief that the original manned space programme was just one more field of competition in the Cold War.

Acknowledgements

Many of the sources that I have used in compiling this manuscript were sent to me some years ago. The names of Jim Grimwood and his secretary Sally, in the History Office at the Lyndon B. Johnson Space Center, Houston, Texas, and that of Lee Seagesser in the History Office at NASA Headquarters, Washington DC, will be known to many older space enthusiasts. I want to express my thanks for their past assistance and friendship to all three of them once again, even though they are now retired. Today the long wait for replies to my letters has been replaced by the instant communications of e-mail and the Internet, and my gratitude also goes to the current generation of NASA Public Affairs officers for answering my queries with their usual polite efficiency. All figures in this volume are credited to NASA unless otherwise acknowledged.

I have experienced difficulty in contacting the copyright holders for a number of contemporary books that I have used as sources in my text. I have nevertheless used these sources and have given full listings of those copyright holders in my bibliography in the hope that this is sufficient.

Recognition must also be given to the officials of the British Interplanetary Society, who have encouraged my writing over the years in their excellent magazine *Spaceflight* and have allowed me to use illustrations of the Society's 'Megaroc' proposals from the late 1940s in this volume. The BIS is the longest surviving of the early Rocket and Space Flight Societies that were responsible for popularising the concept of space flight long before it was technically possible. For longer than I care to remember *Spaceflight* has been the only magazine in Britain dedicated solely to Astronautics. When I was younger it was almost the only magazine covering the subject at all and almost my only source of contemporary information.

David Harland has written many excellent books for Praxis Springer. I was introduced to him (by e-mail at least) when we were both writing articles for Geoffrey Lindop the editor of the now defunct British magazine *Modern Astronomer*. It was David who made me aware that Praxis Springer wanted to publish a book on Project Mercury and encouraged me to become its author. At Praxis, Clive Horwood has been an unending source of encouragement throughout the project and I want to take this opportunity to recognise the role played by Geoff, David and Clive in bringing this volume to fruition.

Finally, I wish to say, *'Thank-you'* to my parents, the numerous family members, close friends and a select group of fondly remembered teachers, who have encouraged my love of writing over the years. Special thanks go to my long-suffering wife Sue, who has supported my interests in both space flight and writing these past 25 years, even though she does not share those interests herself.

John Catchpole

List of acronyms

AASS	Automatic Abort Sensing System (Mercury Redstone)
ABL	Allengany Ballistics Laboratory
ABMA	Army Ballistic Missile Agency
AEC	Atomic Energy Commission
AFB	Air Force Base
AFRC	Ames Flight Research Centre
AMR	Atlantic Missile Range
ARDC	Air Force Research and Development Command
ARPA	Advance Research Projects Agency
ASCS	Automatic Stabilisation and Control System
ASFWC	Armed Services Special Weapons Centre
ASIS	Abort Sensing and Implementation System (Mercury Atlas)
BECO	Booster Engine Cut-Off (Mercury Atlas)
BJ	Big Joe
BMD	Ballistic Missile Division (USAF)
CapCom	Capsule Communicator
CapSep	Capsule (Spacecraft) Separation
CIA	Central Intelligence Agency
CM	Command Module (Apollo spacecraft)
Convair	Consolidated Vultee Aircraft Corporation
CRV	Crew Return Vehicle
CSM	Command and Service Modules (Apollo Spacecraft)
DoD	Department of Defence
ECS	Environmental Control System
ELS	Earth Landing System (parachutes)
EOR	Earth Orbit Rendezvous
EST	Eastern Standard Time
ET	External Tank
FBW	Fly by Wire (Control System)
GALCIT	Guggenheim Aeronautical Laboratory of the California Institute of technology

GEC	General Electric Corporation
GET	Ground Elapsed Time (see $T+$)
GSE	Ground Support System
GSFC	Robert H. Goddard Space Flight Centre
HQ	Headquarters
ICBM	Intercontinental Ballistic Missile
IFEP	In-flight Experiment Panel
IRBM	Intermediate Range Ballistic Missile
ISS	International Space Station
JATO	Jet Assisted Take-Off
JLRMPG	Joint Long Range Missile Proving Ground (Cape Canaveral)
JPL	Jet Propulsion Laboratory
LA	Launch Area (Wallops Island)
LaFRC	Langley Flight Research Centre
LC	Launch Complex (Cape Canaveral)
LEM	Lunar Excursion Module. (shortened to LM)
LES	Launch Escape System
LFRC	Lewis Flight Research Centre
LOR	Lunar Orbit Rendezvous
LOS	Loss of Signal
LSD	Landing Ship Dock
LSS	Life Support System
MA	Mercury Atlas
MCC	Mercury Control Centre
MISS	Man In Space Soonest
MIT	Massachusetts Institute of technology
MPCS	Manual Proportional Control System
MSC	Manned Spacecraft Centre
MSEP	Mercury Scientific Experiment Panel
MSFC	George C. Marshall Space Flight Centre
NAA	North American Aviation Corporation
NACA	National Advisory Council for Aeronautics
NASA	National Aeronautics and Space Administration
NATC	Naval Air Test Centre, Patuxent River
NATO	North Atlantic Treaty Organisation
OMS	Orbital Manoeuvring System
ORDCIT	Ordnance at California Institute of technology
PARD	Pilotless Aircraft Research Division
PMR	Pacific Missile Range
PSAC	Presidential Scientific Advisory Committee
RCA	Radio Corporation of America
RCI	Reaction Motors Incorporated
RCS	Reaction Control System
Scout	Solid Controlled Orbital Test System
SECO	Sustainer Engine Cut-Off (Mercury Atlas)

SM	Service Module (Apollo Spacecraft)
SRB	Solid Rocket Booster
SSME	Space Shuttle Main Engine
STG	Space Task Group
STL	Space Technology Laboratory
T	Time of lift-off (T– time remaining to lift-off; T+ time after lift-off)
UDMH	Unsymmetrical Dimethyl Hydrazine
USAAF	United States Army Air Force
USAF	United States Air Force
USMC	United States Marine Corps
USN	United States Navy
WSMR	White Sands Missile Range
WSPG	White Sands Proving Ground
WWTN	World-Wide Tracking Network
X	Experimental Aircraft (USA)

Part 1: GENESIS

1

A war of ideals

1.1 THE PARTING OF THE WAYS

In 1914 Europe exploded into war as the Austro-Hungarian Empire attempted to crush nationalist feelings in occupied Serbia. The Austro-Hungarian battle plan called for them to defeat the Western European nations as soon as possible before they were faced with a war on two fronts if the Russians declared war on them from the East. To that end the Austro-Hungarian forces, predominantly German, invaded Luxembourg, France and Belgium. Belgium's alliance with Britain brought that island nation into the war, while Russia subsequently invaded Hungary. The major European powers had empires that stretched around the world and the war was carried to the furthest reaches of those empires. Historians call the resulting conflict by various names, including World War I, The Great War, and The War To End All Wars.

When the front in Western Europe became bogged down in trench warfare Britain, France and Germany developed the military aircraft as a means of trying to break the deadlock on the ground. Not wishing his nation to be left behind by the advances in European military aviation, American President Woodrow Wilson formed the National Advisory Committee on Aeronautics (NACA). NACA was charged with finding engineering solutions to the problems raised by advanced military aviation. No one knew it at the time, but America had taken the first step towards Project Mercury.

In April 1917 America entered World War I, allying itself to those European countries fighting against the Austro-Hungarian aggressors. The influx of vast numbers of fresh, well-equipped soldiers on the side of the Allies turned the war against the Austro-Hungarians.

Russia had entered the war in 1914, but her forces were ill equipped compared to those of the Western European nations. Her army suffered a huge defeat in the summer of 1917 and her soldiers were discontented with their conditions and their government. In that year the Russian people rose up and overthrew the Royal Family and the associated aristocracy that had ruled the country with feudal harshness for hundreds of years. Tsar Nicholas II abdicated on 16 March and Russia became a Republic under a provisional government. When the provisional government failed, the Bolshevik Party assumed power in Russia on 7 November. (This was 25 October on the old Russian Calendar, thus the Bolsheviks call their seizure of power the 'October Revolution'.) The Bolsheviks established the world's first Communist regime under the leadership of Vladimir Lenin. It was based on the writings of Karl Marx.

During the revolution in Russia, Britain and a number of other European countries sent soldiers to fight against the Bolshevik forces. When Japan invaded Vladivostok America also sent forces, to observe the Japanese. The Bolsheviks murdered the entire Russian Royal Family in July 1917, and the American government broke off all diplomatic relations with Russia one month later. President Wilson had taken the second step towards Project Mercury.

In Germany, four years of Allied blockade had taken their toll. The civilian population was fed up with the privations of war and the German armed services were slowly bleeding to death on the battlefield. World War I came to an end on 11 November 1918, with the surrender of the Austro-Hungarian forces around the world and the signing of the Armistice document. President Wilson's 14-point plan for peace in Europe failed when the victorious European countries demanded crippling reparation payments from Germany through the Versailles Treaty and refused to allow her to maintain any large standing military forces. The German people felt cheated and oppressed by the terms of the Treaty and some even demonstrated their support for the Russian Bolshevik revolution on the streets of German towns and cities.

1.2 THIRTY YEARS OF PEACE

On 5 January 1919 a new political party had been formed in Germany. A few years later the National Socialist Party elected as their Chairman an Austrian who had fought in the war and reached the rank of Lance Corporal in the Bavarian army. His name was Adolf Hitler.

In the Middle East the country of Palestine was formed as a way of thanking the Arabs in that region for the assistance they had given to the British during the war. Minor wars continued around the world as countries sought to settle their differences with military power.

The world embraced peace and the austerity of war slowly gave way to the 'Roaring Twenties,' but life was only 'roaring' if you were lucky enough to be extremely rich. Across Europe poverty and unemployment were rife and Trade Unions with Socialist sympathies led numerous strikes for better pay and working conditions. In Italy and Germany, Fascism was on the rise as Bolsheviks and Jews were blamed for those countries' difficulties. In Germany hyperinflation meant that the national currency was to all intents and purposes worthless. Meanwhile, a modernised and industrialised Japan began seeking to enlarge its Empire in the Pacific Basin.

America turned its back on Europe and began a policy of Isolationism.

Prohibition brought gang warfare onto the streets of many cities. In the southern states the rise of the Ku Klux Klan, which had been formed in 1915, meant that millions of American Negroes lived in fear of white supremacists, and in August 1925 the Ku Klux Klan marched openly on the streets of the nation's capital.

With the assistance of NACA, military aviation advanced faster in America than in any other nation. In the latter half of the decade the American stock market seemed unstoppable and people with the money to do so invested heavily. Many European banks lent money to traders on the American stock market, which reached an all time high in September 1929 before it crashed in October, making thousands of Americans destitute.

The 'Wall Street Crash' of 1929 left much of America and Western Europe in a state of economic depression, with previously unseen numbers of unemployed people who were unable to purchase even the most basic needs. In November 1932 American President Franklin D. Roosevelt was elected for the first of two terms in a landslide victory. During his election campaign he had promised to cut government spending and balance the budget. Within months he passed legislation through Congress that would begin the American economic recovery. The American people had taken the third step towards Project Mercury.

In the following year Hitler was elected to power in Germany. Anti-Jewish legislation was passed and the German people readily accepted the new leader's talk of an Arian Master Race. The press was censored, all political opposition banned, and 'anti-German' books were burned in the streets. In 1934 the offices of President and Chancellor were combined and Hitler assumed the new all-powerful post of Führer and promptly set about secretly re-arming Germany, in direct contravention of the Versailles Treaty. German troops re-occupied the Rhineland in May 1936 and, on 1 November 1936, German and Italy formed an Axis. When Spain fell to civil war both Germany and Italy sent military forces to support the Nationalist cause. Russia supplied forces to the Republican side and left-wing supporters from around the world formed International Brigades to fight in Spain.

In 1938 Germany annexed its old ally Austria, and Europe once again began the rapid slide into war. On 23 August 1939 Germany and the Soviet Union signed a non-aggression pact. Germany employed the same war plan as it had in 1914 – destroy the Western European countries first. The pact with Russia assured that Germany would only be fighting on one front.

1.3 A SECOND 'EUROPEAN' WORLD WAR

When Britain declared war against Germany in September 1939, Western Europe was unprepared for the fast mobile ground forces supported by aircraft and tanks that Germany had developed. Germany's new Blitzkrieg (Lightning War) tactics were successful and German forces were looking across the English Channel at Britain within eight weeks. European forces retreated to Britain. In the summer of 1940, the Royal Air Force destroyed Germany's hopes of invading Britain and the stage was set for the long fight back against German occupation of Western

Europe. Resistance groups across the continent were organised and supplied from Britain.

Unable to advance further in the West, Hitler sent his forces into the Soviet Union in an attempt to capture the oil fields there, and to provide land for German expansionist policies. The fighting in the Soviet Union was among the bitterest of the war as two diametrically opposed political regimes fought for supremacy. Terrible atrocities were performed on both sides.

In July 1941 a British report suggested that it might be possible to construct a nuclear weapon that would be far more powerful than even the most effective high explosive of the time. In America President Truman was encouraged to commence the development of such a weapon and established a programme called the 'Manhattan District'. America had taken a further step towards Project Mercury.

On 7 December 1941 Japanese carrier-borne aircraft attacked the American fleet as it lay at anchor in Pearl Harbour, Hawaii. The unprovoked attack brought America into the war on the side of the Western European alliance. While Britain and America declared war on Japan the following day, America declared war on Germany and Italy on 11 December. British Prime Minister Winston Churchill and American's President Roosevelt agreed that, while both countries would defend their territories against Japanese aggression, they would employ a 'Europe first' policy in their application of the war.

With America's industrial might once more drawn into a European war against German aggression the war slowly but surely turned against Hitler's forces. The Royal Air Force began bombing Germany by night, while the United States Army Air Force (USAAF) bombed the country by day. In order to keep American bomber crews alive at altitudes above 3,048 metres oxygen-breathing systems and even partial and full pressure suits were developed. American industry had made another advance towards Project Mercury.

Convoys of merchant ships with naval escorts braved German, Italian and Japanese submarines and aircraft to carry the material of American military might across the oceans of the world. Britain was sustained by such convoys, not only at home but also in far-flung reaches of her empire. The Royal Navy also escorted convoys of American and British material to the Soviet Union, where they supported the Communist fight against the German advance.

In September 1942 the US Army constructed a new facility at Los Alamos, New Mexico, where the American and British scientists of the Manhattan District would develop the world's first nuclear weapon. To provide the material for the weapon the world's first nuclear reactor was built in Chicago in December 1942.

Following an Allied invasion of their country, Italy surrendered on 8 September 1943, having overthrown their Fascist government. German forces in Italy continued to fight and the war carried on in Italy, although Italian military forces took no further part. Slowly, over the next two years, the war turned against the Axis powers and in favour of the Allies. In January 1944 the German siege of Leningrad in Russia collapsed after a two-year struggle. Thereafter, Soviet forces began to advance against the German invaders, pushing them back across the vast tracts of land they had occupied since 1941. Meanwhile, Allied forces in Britain made preparations to open a second front against Germany. On 6 June 1944 the Allied forces invaded occupied Western Europe, landing at a number of beaches in Normandy, France.

German research into jet and rocket-propelled aircraft was first seen on Allied reconnaissance photographs in 1942. As a result the USAAF charged NACA to develop an aircraft capable of flying at Mach 1. In 1944 work began on two piloted rocket-propelled aircraft designed to break the 'Sound Barrier'. America had taken another step towards Project Mercury.

As the Allies pushed East across Western Europe and the Soviets pushed West across Russia and Eastern Europe, the Germans bought a new, technically advanced weapon into play for the first time. On 8 September 1944 the first A-4 ballistic missile was fired from occupied Holland, against London. The missiles came too late to change the outcome of the war, but they were sufficiently important that the American military intelligence service was instructed to return the parts for 100 A-4 missiles to America, where they could be studied. American forces in Europe were also told to locate the German engineers responsible for developing the missiles and any documentation associated with their development. Meanwhile American engineers were put to work developing America's answer to the A-4, an all-American ballistic missile to be called Hermes. Two further steps had been taken towards Project Mercury.

American, British, and Russian leaders met in Yalta on 12 February 1945 and decided between them how post-war Europe would be divided. Germany would be split in to four sectors: American, British, French and Soviet occupation forces would each oversee one sector. Berlin, Germany's capital city, lay within the Soviet sector but it was agreed that it would be similarly split into four zones, each to be overseen by one of the victorious Allies. American, British and French troops would be given access through the Soviet sector of Germany to access their zones in Berlin. A further step had been taken towards Project Mercury. (See Fig. 1.)

On 20 April 1945 Soviet troops reached the outskirts of Berlin and three days later American and Soviet troops met for the first time in the ruins of Hitler's Germany. Berlin fell to the Soviets on 2 May and Germany surrendered unconditionally on 7 May.

The core of the group of German engineers at the centre of A-4 missile development surrendered to the Americans. American forces removed parts for 100 A-4 missiles from the underground factory at Nordhausen in the Soviet sector, before it was turned over to Soviet forces. Likewise several tonnes of A-4 documentation were removed from a mineshaft in the British sector. The German engineers were removed to America along with their missiles and documents. America had taken another step towards Project Mercury.

At a place now called Trinity Site in Alamogordo, New Mexico, the Americans exploded the world's first nuclear weapon on 16 July 1945, and advanced once more towards Project Mercury.

The war in the Pacific and the Far East had been long and hard. While the Americans island hopped across the Pacific, British Empire forces fought the Japanese on the mainland. The Japanese forces fought to the last man, preferring to die rather than surrender and bring dishonour to their Emperor, their country and

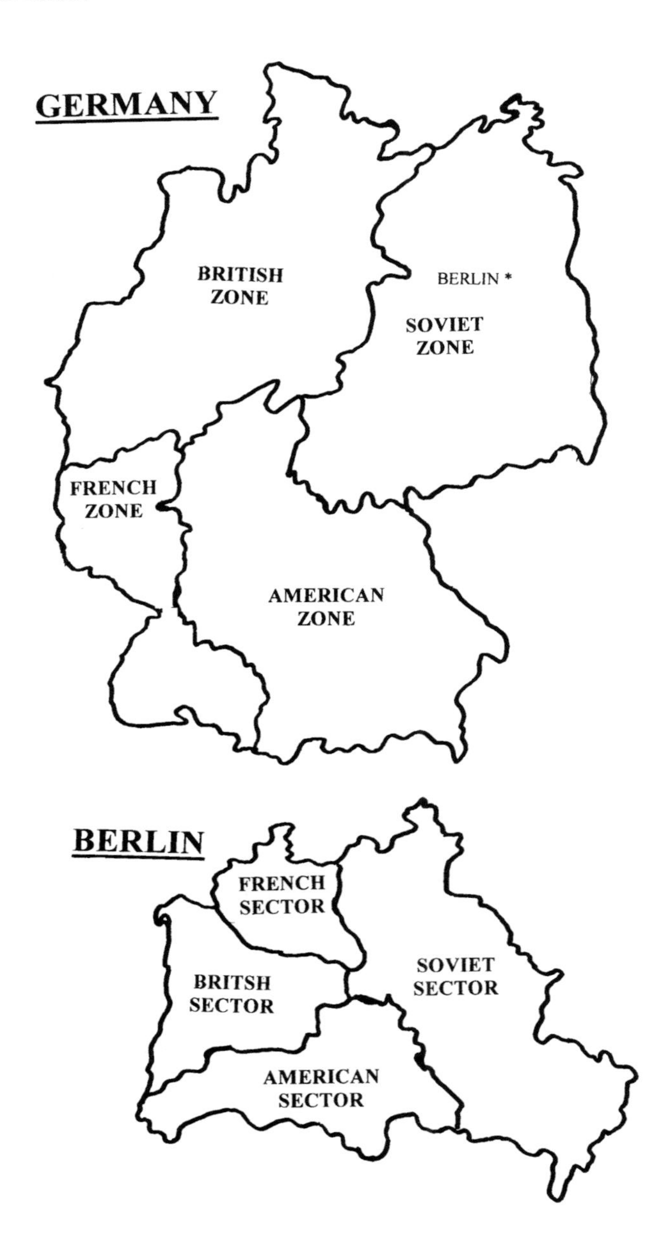

Fig. 1. At the end of World War II Germany was divided into four occupation zones, each policed by one of the four Allied powers. Berlin, deep inside the Soviet zone, was similarly divided. When the three Western powers re-unified their zones the Soviet Union refused to follow suit and both Germany and Berlin were divided into East and West. In 1961 the Soviets constructed the Berlin Wall to physically divide the city in an attempt to stem the flow of skilled workers to the Western half of the city.

their families. The prospect of having to invade the Japanese homeland was too terrible to contemplate, although all three Allied leaders had agreed plans to do so.

On 6 August 1945 an American bomber dropped the first nuclear weapon on the Japanese city of Hiroshima. When the Japanese Emperor refused to surrender Russia declared war on Japan on 9 August. A second American nuclear weapon was dropped on the city of Nagasaki on 9 August and Japan surrendered five days later. World War II was at an end but a new 'Cold War' was soon to begin.

In the aftermath of the Holocaust – the Germans' attempt to eradicate the Jewish race – the victorious Allies agreed to form a Jewish Homeland. To do this Palestine was divided and Israel was formed in one part. The racial tensions formed by the division of the Palestinian Homeland formed after World War I and the ancient religious tensions between the Arabs and Jews (stretching back to the times of the Ancient Roman Empire) led to the Middle Eastern conflicts that continue to this day.

After the war the British gave liberation to India, which slipped into an extremely bloody civil war between religious factions. A similar thing happened when the British moved out of Africa and independent African nation states began fighting each other. Other Western powers attempted to use military force to hang on to their pre-World War II empires. On most occasions this led to protracted wars of independence and defeat for the colonial power.

The world order had changed almost beyond recognition, but a new form of empire had been forged. In the West, America's empire was formed around free government and Capitalist economics. In the East, the Soviet empire was based on military occupation, and central control of government and economic policies from Moscow.

1.4 AMERICA: THE FIRST MODERN SUPERPOWER

When America entered World War II Britain and the remains of her European Allies had already been fighting the Axis powers for three years, albeit with considerable amounts of material purchased from America. As the war progressed America found herself in the fortunate position where she was out of range of Axis bombers. As a result, the nation's entire national production could be turned to war material without fear of destruction from the air. No other nation involved in the war had that security.

American political will and her military might was vital to victory in both Europe and the Pacific theatres of war. When World War II ended America was the only nation to possess nuclear weapons. Although British scientists had worked on the Manhattan District and helped to develop the nuclear weapons dropped on Japan, a post-war political decision barred the American President from sharing nuclear weapons with anyone, including the British.

With most of Europe devastated by war and the old Colonial nations almost bankrupted by the fight, the Western European nations looked to America for protection against Soviet expansion in Central Europe. As early as 1946 President

Truman, who had come to the post in April 1945 when President Roosevelt died, was advised that the Soviet Union represented a threat to Europe, America and the world. It was a point of view that many Americans would share over the next 40 years.

As financial difficulties forced Britain to withdraw from Greece and Turkey, America had little choice but to undertake security of the region, in order to prevent Soviet expansion. President Truman informed congress that America could not return to a policy of Isolationism as it had in 1919, but must stand fast and support Western Europe. This time American politicians gave their full support to the establishment of a United Nations (UN) – a forum for political discussion aimed at preventing future wars.

In 1945 a severe winter blanketed a devastated Europe with snow and was followed by an equally severe drought that led to crop failures in 1946. America's answer to the dire situation in Western Europe was the Marshall Plan, economic and material aid to rebuild the infrastructure of the individual European countries.

In 1947 America mounted a huge overt and covert operation to ensure that the Communist Party did not come to power in Italy. The Democratic Party's victory was also helped by the fact that a Communist Italy would not be able to partake in the Marshall Plan.

On 24 June 1948 the Soviets stopped all traffic travelling in and out of Berlin. The part of the city that was occupied by Western Allies was under blockade and American and European planes began the huge task of supplying the city by air. When negotiation failed to end the blockade, America flew six B-29 Superfortress bombers to Britain, the implication being that they contained nuclear weapons for use against the Soviet Union. The bombers carried no nuclear weapons, but the Soviet leaders believed that they did, and the blockade of Berlin ended in May 1949. The previous month America, Canada and 10 Western European countries had formed the North Atlantic Treaty Organisation (NATO) and agreed to support any member state that was attacked by the Soviet Union.

Between 1946 and 1948 American scientists tested 23 nuclear devices in the Bikini Atoll. Research was slowly making these devices smaller and lighter, while increasing their destructive output. At home the nation was gripped by the actions of the House Un-American Activities Committee (the McCarthy trials) and later the first trials of spies accused of selling nuclear weapon secrets to the Soviet Union.

In August 1949 the Soviet Union surprised American experts by exploding their first nuclear weapon. The Soviet advance led President Truman to approve the development of a thermonuclear weapon, on 31 January 1950. The first such technology was exploded at Eniwetok Atoll on 1 November 1952, but the 'Mike' thermonuclear test system was massive and was not a practical weapon.

UN and Communist forces fought each other for the first time during the Korean War, which is briefly described later in this section

In November 1952 Dwight D. Eisenhower, the Allied Supreme Commander in Europe during World War II, was elected President. His Vice President was Richard M. Nixon.

In 1954 John Foster Dulles, President Eisenhower's Secretary of State, proposed the concept of 'Massive Retaliation'. This called for any Soviet nuclear attack, even if it was only one missile or bomb, to be met by an all-out American nuclear retaliation.

In May 1955 West Germany became a free state and all occupation forces were withdrawn; only defensive NATO troops remained. West Germany was allowed to re-arm and was admitted to NATO; the new state was named the German Federal Republic. The Soviet reply was to establish the Eastern Bloc, the Communist equivalent of NATO.

In July 1955 the American, British, French and Russian leaders met in Geneva to discuss the future relationship between Russia and the Western powers. President Eisenhower suggested a policy of 'open skies', whereby both sides would be given freedom to over-fly and photograph each other's territory. Nikita Khrushchev, the new Soviet leader, refused the idea. The talks came to nothing and on 19 July 1956 American Central Intelligence Agency pilots began flights over the Soviet Union in the new Lockheed U-2 high altitude reconnaissance aircraft. The flights gave the American intelligence services first-hand information on the Soviet Union's true military strength.

At the 1955 May Day Parade in Red Square, Moscow, the Soviets unveiled a new bomber. In order to make it look as though there were many more of the aircraft than there actually were, they flew round in large circles and passed over the parade several times. American intelligence was fooled, as they were when a U-2 reconnaissance aircraft returned a photograph of several of the new bombers lined up on a military airfield. The Americans assumed that the bomber was operational across the Soviet Union, when in fact the photograph showed the entire operational quota of that model of bomber in existence at that time. Similar misunderstandings led to a great over-accounting of Soviet military strength. The myth of a 'bomber gap' in the Soviet's favour was born within American military intelligence circles.

In October 1956, while America's eyes were on the Soviet reaction to a popular uprising in Hungary, the British, French and Israelis launched an invasion to 'liberate' the recently nationalised Suez Canal in Egypt. The action brought condemnation from around the world, including Washington. Many saw the invasion as the final exercise of Imperial power. It failed when Washington made it clear that it did not have America's backing.

At home America had serious domestic troubles. The Negro population were beginning to seek their Civil Rights under the country's Constitution. In September 1957 the Federal District Court demanded that nine Negro students should be enrolled at the previously segregated Central High School in Little Rock, Arkansas. Local white segregationists, including the Governor of Arkansas, were against the move. Governor Faubus used the Arkansas National Guard to bar the Negro students from the school. President Eisenhower sent 1,000 paratroopers to Arkansas and Federalised the National Guard, taking control of the unit away from the Governor. The students were enrolled, but the newspapers were full of photographs of American soldiers pointing rifles with bayonets at the white population of a southern American town.

1.5 COMMUNISM RISING

Throughout the first four decades of the twentieth century there were numerous calls for national independence from the countries within the Western European empires. These uprisings were almost always met with military strength. In the twenty years after World War II most Western European nations gave independence to their colonies. Sometimes it was given peacefully, but on many occasions it came only at the end of a long fight against the colonial power in question. In several cases independence was followed by an equally long and bitter civil war as rival factions fought for control of the newly independent nations.

Having learnt the lessons of World War I, Germany's reparation payments were less crippling after World War II. Although the nation was divided, the Western Allies' view of Germany's position after the war was different this time. The UN was established and enjoyed both American and Soviet support. Meanwhile, around the world, Communism was on the march. The following section will give the reader a very brief history of how the Communist struggle affected American political policy in the late 1950s and early 1960s. The expansion of world Communism and the American reaction to it is pivotal to the story of how and why Project Mercury transpired.

1.6 RUSSIA

When Lenin died of a stroke in 1924, he was replaced by Joseph Stalin as Chairman of the Soviet Communist Party. Stalin was a despot, holding on to power through the fear that his people had for of his secret police force. Millions of Soviet people fell victim to Stalin's purges and the nation's Gulag prison system. Stalin forced industrialisation upon his country within a very short time. He also enforced the concept of huge collective farms on his people along with absolute control from Moscow on all aspects of Soviet life. When Germany invaded Russia in 1941 Stalin personally oversaw the defence of his vast nation. He instructed that important factories be removed from the path of the advancing Germans and rebuilt behind the Ural Mountains. When Moscow was besieged, Stalin refused to leave. He also stationed his secret police around the outskirts of the city to kill anyone trying to desert the beleaguered city.

Finding themselves in a war against a common foe, the Western European nations accepted Communist Russia as an ally. America and Britain both provided Stalin's armed forces with vital material to allow them to carry the fight to the invaders. Having turned the tide of the war, the Soviet Red Army pushed the Germans all the way back to Berlin and beyond. What they called 'The Great Patriotic War' cost the Soviet population over 20 million dead, far more than any other nation involved in the conflict.

As the war drew to a close in 1945 Stalin attended the Yalta Conference with American President Roosevelt and British Prime Minister Winston Churchill. These three men divided post-war Europe between the victors. In addition to half of

Germany and Austria, the countries of Poland, Hungary, Romania, Yugoslavia and Bulgaria were under Soviet occupation when the war ended in May 1945. Stalin insisted on holding any territory occupied by Soviet troops at the end of the war as the occupied Eastern European countries would serve as a buffer zone to prevent any future invasion of the Soviet Union from the West. In these countries governments were installed that were pro-Moscow. Relations with Russia's wartime allies were strained by this move and the Soviet position was viewed in the West as a threat to Western Europe. In a 1946 speech made while visiting America, Winston Churchill summed up the situation by saying, '...*an iron curtain has descended across the continent*'. The American press criticised him, though history would prove that he spoke the truth.

Stalin originally refused to pull his troops out of Persia (Iran), having jointly occupied it with Britain during World War II to prevent its oil falling into German hands. When the British withdrew, the Soviets remained, and the Soviet troops were only removed when faced with strong Western opposition in the UN. It was the first Soviet test of how far they could push the West before provoking military action.

The major problem in the Soviet Union's relationship with the West was Germany. Russia wanted to keep Germany in a political and economical position where it would never again be able to engulf Europe in war. On the other hand, American politicians, realising that Germany's industrial base was vital to Europe's economic recovery, wanted to see Germany re-unified to enable it to contribute to the Western European economy and an international free market. Therein lay the heart of 50 years of Cold War.

When Turkey refused Soviet ships access through the Dardanelle Straits in 1946, Soviet and American armed forces faced each other for the first time. The Soviets admitted defeat and withdrew their request for access. On the other hand, when Czechoslovakia and Poland expressed an interest in applying for Marshall Aid, Stalin refused to allow them. As a result, America turned its back on Czechoslovakia and it became part of Soviet Eastern Europe.

In June 1948 the Soviets blockaded West Berlin, but the Western Allies supplied the city by air. The blockade of Berlin ended in May 1949 when the Soviets relented, and Germany effectively became two countries as a result. During the blockade a split also occurred between Moscow and the Communist government in Yugoslavia, which came to power in 1945 without Soviet assistance. Moscow placed an economic blockade on Yugoslavia and that country was forced to seek aid from America.

In 1945 Stalin had set up a programme to develop a Soviet nuclear weapon. Klaus Fuchs, one of the British scientists working with the Manhattan District, had given the Soviets full details of the weapon dropped on Nagasaki. With this and similar espionage the Soviets tested their first nuclear weapon in 1949, several years earlier than the Americans had expected. The bomb was similar in design to the weapon in Fuchs's notebooks – the weapon dropped on Nakasaki. The balance of power was shifting.

When the Korean War broke out in 1951 Stalin withdrew most of the Soviet advisers from North Korea and encouraged Mao Zedung's Chinese People's Army to carry the Communist ideals into war against the UN forces. The fact that Stalin made them pay for the Soviet military material caused bad feeling between the two Communist leaders, but at the time the rift had not been recognised in the West.

In March 1953 Stalin died and there was brief talk in Moscow of a re-unified but neutral Germany in the centre of Europe. Stalin, the leader who had dragged the Soviet Union into the twentieth century and had led the defence against the German invasion, was discredited as the Soviet people were told of some of the atrocities that had been performed against the Soviet population on his orders.

On 12 August 1953 the Soviet Union exploded its first thermonuclear weapon, just nine months after the first America thermonuclear explosion.

After a two-year power struggle Nikita Khrushchev was secure in the position of leader of the Soviet Union, just in time for the Geneva talks in July 1955. In the same year Soviet troops were removed from Austria in return for that nation's neutrality in central Europe. A strike in Poland gave rise to a popular revolt against Soviet rule in June 1956. Although Khrushchev initially put troops on the streets he withdrew them when he realised the strength of feeling in the country. The Soviet leader met the Polish people's demand for a change of first secretary in the Polish Communist Party and what had become a national uprising subsided. Hungary was the next Eastern European country to stand up to Moscow, in October 1956, but this time Khrushchev was ruthless. He put tanks on the streets of Budapest, where they fought to crush the uprising. When the Soviets regained control, in November, 35,000 protesters were arrested and the Hungarian leader who was in power at the time of the uprising was later executed by the Soviets.

At a site called Baikonur, in the Kazakhstan desert, the Soviet Union launched a ballistic missile on 15 May 1957. It was the first of eight such launches between May and August, resulting in the Soviets declaring that they had developed the world's first Intercontinental Ballistic Missile (ICBM). Designated the R-7, the missile had been developed in just seven years by a design bureau under the charge of Sergei Korolev. Its role was to launch a thermonuclear warhead against the American mainland. The R-7 was huge and powerful, and by the end of the year it would put fear into the minds of millions of people in the Western hemisphere.

Post-World War II Berlin had proved an embarrassment to the Soviets. Since 1946 thousands of East Berlin's professional and young people had made their way to the West. In November 1958 Khrushchev formulated a plan to bring Berlin back to the forefront of European politics. Western leaders would be given six months to sign a treaty that recognised the two halves of the divided Germany as two separate states. If the Western powers failed to sign the agreement then the Soviet Union would sign a separate peace agreement with the German Democratic Republic (GDR), East Germany. If that happened there would no longer be a state of war between the Soviet Union and the GDR, Berlin would be part of an independent country and the Western Allies would be forced to remove their defensive forces from West Berlin. Berlin would be re-unified under Communist rule and the West would be removed from the centre of East Germany. As Khrushchev put it at the time: '*Berlin is the testicles of the West ... When I want to make the West scream, I squeeze on Berlin.*'

1.7 CHINA

Mao Zedung began the Communist fight against Chiang Kai-Shek's Chinese Nationalist government in the early 1920s. Throughout much of the same period China was also fighting a war against the occupying forces of Japan and the fight was long and hard on both fronts. In 1935 the Communist troops were forced to break out of a Nationalist blockade and walked thousands of kilometres through some of China's most mountainous and inhospitable country – an incident that became known as the 'Long March'.

In 1945, with the end of World War II, America began supporting the Chinese Nationalists while the Soviet Union supported the Communist forces. When the Japanese surrendered to the Americans in 1945, following the dropping of the first nuclear weapons, Mao Zedung resumed the internal ideological war against the Nationalist government.

Chiang Kai-Shek was elected President of China on 29 March 1948. In the post-war years America quickly realised the vast commitment required to support Chiang Kai-Shek's government as inflation and corruption weakened its support at home. The Chinese President resigned on 21 January 1949, and with no American support for the Nationalists, Mao Zedung's troops succeeded in forcing Chiang Kai-Shek and his supporters to flee to Taiwan. The Nationalist retreat began on 16 July 1949 and on 1 October 1949 Mao Zedung renamed his vast country the People's Republic of China. In America the Soviet Union and China were seen as one monolithic Communist block. That view was incorrect almost from day one, but the differences of opinion between the two vast communist states were not immediately visible.

Mao Zedung began a two-month visit to Moscow, where he met with Joseph Stalin in December 1949. He agreed to be led by Moscow in the development of the Communist Chinese State and on 15 February 1950 the two Communist leaders signed the Sino-Soviet Treaty.

On 25 June 1950 North Korean forces moved into South Korea and almost occupied the entire southern half of the country. As the UN forces fought their way back up the Korean peninsula they crossed the 38th Parallel and continued north. Not wanting to face America in Korea, Stalin withdrew Soviet forces from the country and let Mao Zedung take responsibility for North Korean Communism. Chinese and UN forces met in combat for the first time in October. After one month of combat against huge numbers of Chinese forces in the north, the UN advance was halted in November 1950 and the UN forces were pushed back to the south once more. The Chinese advance crossed the 38th Parallel, but was not held until January 1951. The American Commander, Douglas MacArthur, talked openly of UN forces invading China and was dismissed for doing so. On the other hand, the newly elected President Eisenhower threatened to use America's nuclear weapons against Beijing. In April 1951 the Chinese began a new offensive and after initial advances the offensive was held, both sides digging trenches to secure their positions. Armistice talks commenced in October and dragged on for 18 months. Finally, with the Armistice signed, the fighting ended on 27 June 1953 but Korea remained divided along the 38th Parallel.

In Moscow Stalin had died in March 1951 and it took Nikita Khrushchev two years to secure his hold on power. When Khrushchev denounced Stalin in 1956, Mao Zedung began to distance his country from Moscow's leadership.

1.8 KOREA

During the Sino-Japanese war the Japanese occupied Korea, which had been an independent country until that time. The country was split along the 38th Parallel in 1945. Soviet forces replaced the occupying Japanese north of the line, while American forces did the same south of the border.

By 1949 elections had taken place in the south and Soviet forces had withdrawn from the north, which was left with a Communist puppet government. In the same year political power in China fell to the Communist forces of Mao Zedung. China and the Soviet Union signed the Sino-Soviet Treaty and American leaders began to fear a monolithic Communist power in the Eastern hemisphere. At the same time the Soviet Union exploded its first nuclear weapon, breaking America's monopoly in that area.

In 1950 North Korea's Communist leader sought permission to invade South Korea and re-unite the country, and Stalin agreed to supply military material and advisers to support the action. The invasion began on 25 June in Korea, and the UN agreed to support South Korea against the Communist aggression. As the Communist forces flowed south they devastated the UN forces confronting them, but the UN troops stood fast along the Pusan Perimeter, where the Communist advance was finally halted. The UN staged an amphibious landing at Inchon Harbour, behind Communist lines, on 15 September 1950. The Northern attack collapsed and Stalin withdrew the Soviet military advisers. All Soviet material had then to be purchased by the Chinese and relations between the two huge Communist countries were strained from the beginning. On 1 October UN forces crossed the 38th Parallel and entered North Korea. Eighteen days later they captured Pyongyang, the Northern capital. The race for the Chinese border began. Determined not to engage Soviet troops directly against American forces, Stalin instructed Mao Zedung that the Chinese should go to the aid of the North Koreans. On 25 October Chinese and UN forces met in combat for the first time. The main Chinese attack began on 26 November 1950 and once more the combined UN and South Korean forces were pushed below the 38th Parallel. Both Inchon and Seoul fell to the Communists. UN forces attacked for the third time in April and had considerable success, reaching Seoul, but once again they were pushed back to the 38th Parallel.

American General MacArthur pushed for an invasion of China and the bombing of Chinese cities. President Truman sacked his military commander in Korea and replaced him with General Ridgeway. Peace talks commenced in July 1951, but came to nothing.

Ground forces now faced each other from trenches while American and Soviet jet fighters, the latter with both Chinese and Soviet pilots, met in combat for the first time. By October peace negotiations were underway, but they went on for 18 months and the fighting continued. When Dwight D. Eisenhower became President of America in November 1952, he threatened to use nuclear weapons against China. After further negotiations a cease-fire began on 27 July 1953. The first war between the two opposing ideologies had simply petered out and Korea remained dissected along the 38th Parallel.

1.9 CUBA

On 26 July 1953, Fidel Castro began a revolution against the government of Fulgencio Batista in Cuba but was defeated, arrested and exiled. He returned to Cuba in 1956 and, with the aid of Ernesto 'Che' Guevara, organised a guerrilla war against Batista's rule. On 29 October 1957 Batista suspended the Cuban Constitution. On 5 April 1959, Castro declared 'Total War' on Batista's government. After six years of war Batista fled to Dominica, on 1 January 1959. The following day a new Nationalist government was established with Dr. Manuel Urrutia at its head. Castro was named as Commander in Chief of the armed forces. One week later Castro led the victory parade into Havana. The new government was immediately recognised by America, Britain, France and the Soviet Union.

On 4 June 1959 Urrutia nationalised the sugar mills in Cuba that belonged to American companies. As a result President Eisenhower placed a trade embargo on Cuban sugar, the primary Cuban export. Faced with the loss of their primary market, the Cuban government was pleased when the Soviet Union offered to purchase the country's sugar output. What was originally a Cuban Nationalist government had been forced into the arms of the Soviet Union.

1.10 SOUTH-EAST ASIA

Ho Chi Minh formed the Indo-Chinese Communist Party in 1924, while working among Vietnamese exiles in China. He was expelled from China by the Nationalist government, but returned in 1938 to help train Mao Zedung's army. Vietnam itself was part of French Indo-China, until the Japanese occupied it in 1941. Ho Chi Minh returned to Vietnam and led the Viet Minh against the new occupiers.

When that war ended the Viet Minh established a Communist government in North Vietnam with Ho Chi Minh as President. South Vietnam was 'liberated' from the Japanese by the British, who used their Japanese prisoners as policemen throughout the period before the country was given back to the French.

The Viet Minh continued the fight to re-unify the two parts of the country. They fought a guerrilla war against the French, which came to a close in 1954 after the French were defeated at Dien Bien Phu. The French withdrew and the country was divided along the 17th Parallel. The Communists ruled in the North, while the South would hold elections in 1956.

In America, President Eisenhower believed that French Indo-China was the key to South-East Asia. If Vietnam fell to Communism the countries around it would also fall, perhaps even as far as India in one direction and Australia in the other, and he was not prepared to let that happen.

In the South, Ngo Dinh Diem, a Vietnamese Catholic, established a Nationalist government. America supported Diem's government, even when he abolished free elections in 1956 and established a regime that was almost as strict as that in the North. American military advisers were sent to South Vietnam, although they were only allowed to operate in Saigon, the Southern capital. It was the beginning of American involvement in South-East Asia.

2

Dreaming of space

2.1 ROBERT GODDARD

America's most prominent first generation rocket pioneer was Robert Goddard, who was born in Worchester, Massachusetts, on 5 October 1882. He was a sickly child and missed a lot of state schooling as a result. Despite that, at age 17 he was making notes on the prospect of using rocket vehicles to carry scientific instruments to extreme altitudes within the atmosphere. Largely self-taught, he graduated from Worchester Polytechnic Institute in 1908 and took a position there as a Physics lecturer while he completed his post-graduate course at Clark University, Worchester. Goddard obtained his Doctorate of Physics in 1911, and took a position as a lecturer in the subject at the university the following year.

While at Clark University, he began practical experiments into the subject of rocket propulsion. He hand built rocket motors and endeavoured to improve the solid propellants in use at the time. Using one of his own small solid propellant motors he became the first person to prove categorically that a rocket motor performs more efficiently in a vacuum than it does in air. Through his years of research he made meticulous notes and took care to patent any new inventions. In 1914 he was awarded his first patents, for combustion chambers, a liquid propellant delivery system and for the muti-stage rocket principle.

By 1916 his personal funds had almost run out. In an attempt to obtain sponsorship he submitted a short paper to the Smithsonian Institution, *A Method of Reaching Extreme Altitudes*, which detailed the technical requirements for a rocket to reach a series of incremental altitudes. It stated that a multi-stage rocket (called a step-rocket in 1916) would be required to overcome Earth's gravitational attraction. In a minor piece, at the end of the paper, he suggested that one day a rocket might be able to reach the Moon. An appendix quoted his research, which had convinced him that liquid propellants would produce far greater thrust than the solid propellants then in use. On 5 January 1917 the Smithsonian Institution replied, releasing $5,000 to Goddard.

That was the year that America entered World War I and Goddard began work on rockets for the government. He suggested the development of two rockets and began work at Clark University. In time the Army moved him to Pasadena, California, where he continued his work. One of the projects he worked on, small rockets fired from a hollow tube, is recognised as the forerunner of the Bazooka anti-tank weapon used by Allied troops in World War II. The project was demonstrated to the US Army Air Force (USAAF) over 6–8 November 1918, but when the Armistice was signed three days later, the US military lost their interest in rocket weapons.

Goddard returned to Clark University and to his non-military rocket studies. In 1918 he received his $5,000 research grant from the Smithsonian Institution. In the same year Ester Kisk, 19 years his junior, began work at the University. Goddard married her in 1924. Ester Goddard was a keen photographer and recorded the long years of trial that lay ahead. The rocket pioneer also gathered a small team of engineers around him to help him with his work.

In 1919 the Smithsonian Institution released *A Method of Reaching Extreme Altitudes* for public consumption. The press of the day ignored most of the paper, but they pounced on his suggestion that a rocket might one-day fly to the Moon. Goddard was mocked in print and the shy pioneer never recovered from the incident. Not one to seek publicity for himself, or his work, he now shunned it completely and became reluctant to discuss his research with all but a very few confidants. His rocket work continued with the test stand firing of solid propellant rocket motors then he redirected his research towards liquid propellants in September 1921. The following year he made his first firing of a rocket motor that used both a liquid oxidiser and a liquid fuel. One year later he had developed a pump fed motor that burned liquid oxygen and petrol. In 1925 Goddard moved his research to a farm in Auburn, Massachusetts, where he developed and tested a new motor that burned liquid oxygen and petrol, fed to the combustion chamber by nitrogen pressure.

On 16 March 1926 the world's first flight of a liquid-propelled rocket took place. The vehicle used an engine that was similar in design to the1925 model, but employed oxygen rather than nitrogen pressurisation. It employed an open-frame construction, with the rocket motor at the top, directly above the propellant tanks. By today's standards it is barely recognisable as a rocket. Goddard sent the Smithsonian Institution written details of the 2.5-second flight that had reached an altitude of 28.6 m and landed in a snow-covered cabbage patch 57 m down range from the launch frame. He did not inform the press of his achievement. Repairing the rocket after the flight, he launched it a second time on 3 April. (See Fig. 2.)

In May Goddard's research led him to redesign his rocket, placing the rocket motor at the base of the vehicle. This allowed the propellant tanks to be placed on top of each other above the motor. He later developed a new rocket four times the size of the previous one, but it never flew. He then developed a smaller vehicle, which also had the rocket motor at the base, and this smaller rocket flew for the first time on 26 December 1928.

The 1920s saw a growing interest in rockets in America and across Europe. Numerous amateur societies grew up to promote their development. Goddard was invited to join two such societies in America, and although he refused he carried on private correspondence with a select number of rocket engineers in both America and Europe.

Fig. 2. Robert Goddard poses alongside the world's first liquid-propelled rocket as it rests within its launch frame. This vehicle was successfully launched on 16 March 1926. Note the lack of external casing and the rocket motor mounted above the propellant tanks. (NASA)

Following the second launch of Goddard's new rocket, on 17 July 1929, the local residents dispatched the fire brigade to the farm. The brigade was not required, but Goddard's work received more bad publicity. Once again, the press missed the whole point of the launch, which had carried a camera that was rigged to photograph a barometer when the recovery parachute deployed. It was the first rocket-launched scientific experiment, an attempt to gain first-hand information about the atmosphere at an altitude that no aircraft could reach at that time. The press coverage of the day's events made no mention of the experiment, but the Smithsonian Institution convinced the US Army to allow Goddard to conduct static rocket motor firings on the artillery range at Camp Devens, Massachusetts.

During the same year Goddard's work came to the attention of Charles Lindbergh, the aviation pioneer, who had been the first man to fly solo across the Atlantic Ocean. Lindbergh was a personal friend of Daniel Guggenheim, a millionaire and a keen follower of advances in aviation. The aviator brought Goddard's work to the attention of the Guggenheim Foundation and arranged for the Foundation to assign him a $50,000 two-year grant. Goddard moved to Hell Pond, New Mexico, where he continued his work. He chose the location partly because of its isolated nature and partly because the dry air would be good for his recurring ill health.

On 30 December 1930 he launched a new rocket, the inner workings of which were enclosed in an aerodynamic casing. Four fins at the base of that casing assisted in maintaining a controlled flight, and small gyroscopically controlled vanes acted in the exhaust plume to provide a limited amount of attitude control. Meanwhile, Goddard's correspondence with other rocket pioneers had kept him aware, but not abreast, of rocket research in Germany.

As Europe threatened to descend into war yet again, Goddard moved his research to Eden Valley, Roswell, New Mexico. He had overcome the problem of in-flight stability by the development of a gyroscopic stabiliser, which he patented. The unit applied corrective steering through a series of vanes acting in the rocket motor's exhaust plume. He had also developed a parachute recovery system, which reduced landing loads and allowed him to re-use many rocket parts. Testing of the new rocket continued until July 1932. At that time the money he had received from the Guggenheim Foundation expired and in September he returned to his job as a Physics lecturer at Clark University. With a limited grant from the Smithsonian Institution he was able to continue his experiments with static liquid-propelled rocket motors.

In September 1933 he received a second grant from the Guggenheim Foundation and spent it developing reliable fuel pumps, gyroscopes and rocket motors. In September 1934 he returned to Roswell, New Mexico, to continue his flight testing. With continuous grants from the Guggenheim Foundation he would continue his work for the next seven years (see Table 1).

By 1939 Goddard's rockets were as sophisticated as the A-4 then in development in Nazi Germany, but the German rocket would not fly for a further three years.

Table 1: Robert Goddard's work programme, September 1934 to November 1939

Test series	Start date	Test type	Finish date
A	Sep. 1934	Flight tests	Oct. 1935
K	Nov. 1935	Static tests	May 1936
L(A)	May 1936	Flight tests	Nov. 1936
L(B)	May 1936	Flight tests	May 1937
L(C)	Jul. 1937	Flight tests	Aug. 1938
Pumps	Jan. 1939	Static tests	Feb. 1939
Generators	Mar. 1939	Static tests	Apr. 1939
Turbines	Jul. 1939	Static tests	Aug. 1939
Solid propellant	Nov. 1939	Static and flight tests	Oct. 1941

Following the Japanese attack on Pearl Harbour America became involved in World War II. Goddard made a presentation to representatives of the US Navy and US Army on the potential of the rocket in military service. The Army was unreceptive, but the Navy was already working on Jet Assisted Take Off (JATO) devices to help heavily laden aircraft take off from the relatively short flight decks fitted on aircraft carriers. Goddard and his team were awarded a contract to develop JATO rockets for both services. The group moved to the Naval Engineering Experiment Station, Annapolis, Maryland, in September 1941, and over the next three years they continued to experiment with liquid-fuelled JATO rockets.

Robert Goddard died on 10 August 1945, still unrecognised in his own country. He left numerous patents covering all areas of rocket technology. His family registered a number of new patents, to secure his heritage, bringing the total to 214. In 1969, following the first manned Moon landing, the American government paid Goddard's estate $1 million for the use of his patents.

Until the day he died Goddard maintained that the German engineers that had developed the A-4 ballistic missile had used his ideas, without acknowledging his precedent. This was not true. The German Army spent the money to develop and test their own rocket technology and, although they may have studied Goddard's published papers, they advanced at their own pace.

Although Goddard was the first rocket pioneer in America and made considerable advances – many equal to the vast A-4 programme in Germany – his work was hindered by his near-paranoid avoidance of publicity and his refusal to share his work with anyone else, lest they steal it. The concept of the solo rocket pioneer died with Robert Goddard in 1945. There is no direct link between Goddard's work and the huge missile and space launch vehicle programmes that followed World War II.

2.2 THE AMERICAN ROCKET SOCIETY

On 21 March 1930 the American Interplanetary Society (AIS) was formed in the New York apartment of G. Edward Pendray. The twelve original members were science fiction writers for a magazine called *Science Wonder Stories*, and shared a desire to see rocket propulsion and space flight become a reality.

David Lasser was the senior editor of *Science Wonder Stories* and is most often quoted as the moving force behind the setting up of the American Interplanetary Society. He was a trained engineer and the Society's first President. In 1931 Lasser wrote one of the earliest factual American books to try to explain the principles of how space flight would one day be achieved. A year later Pendray gave a lecture on an early American television station on the same subject.

The Society's *Rocket 1* was completed in January 1932. Based on a German design, the rocket motor was at the top of the vehicle and burned liquid oxygen and gasoline. Work to prepare the vehicle for launch began in August 1932 and resulted in an extensive redesign and redesignation as *Rocket 2*. The new rocket was static fired for the first time on a farm outside Stockton, New Jersey, on 12 November

1932. *Rocket 2* was launched from Great Kills, Staten Island, New York, on 14 May 1933 and reached an altitude of 76 m. The flight ended when the liquid oxygen tank failed. The Society's members spent the rest of the year working on three further rockets, but made no further launch attempts. On 6 April 1934 the AIS changed its name to the American Rocket Society (ARS). On 9 September 1934 *Rocket 4* was launched from Great Kills and reached an altitude of 116 m.

After the launch of *Rocket 4* the ARS members decided to concentrate their efforts on developing better rocket motors and firing them in static test stands. The Society would make no more free launches. This decision was made partly due to lack of funds and partly because the ARS members realised that it was possible to obtain more data from repeated static firings than from free flights. The first test firing was made on 21 April 1935, in a suburb of New York City. Over the following months five different rocket motors were tested. The ARS completed two further series of static firings, in August and October 1935.

2.3 AGGREGATE 4

During the 1920s and 1930s rocket enthusiasts formed themselves into numerous groups across Europe, including the Soviet Union. These groups consisted of keen amateurs and professional engineers with an interest in rocket propulsion. In Germany one such group, the Verein für Raumschiffahrt (VfR), of which Wiley Ley was a founder member, came to the notice of the German Army. Following a number of visits to VfR launches, in 1932, Captain Walter Dornberger invited Wernher von Braun, a young VfR member, to join his Army rocket research group at Kummersdorf, outside Berlin. Von Braun accepted the Army's offer of employment.

At Kummersdorf the German Army, assisted by von Braun, developed a liquid-propelled rocket motor for use in the Aggregate 1 (A-1) test missile. The A-1, which was ready for testing by late 1933, looked like a large bullet, with four fins at the base surrounding its rocket motor. Tests showed that the rocket's centre of gravity was too far forward and it never flew. The A-2 was similar in design, except that it contained an early gyroscopic stabilisation system. Two identical A-2s were launched from the island of Borkum, in the North Sea, in December 1934. Both launches were successful.

In 1935 Hitler introduced compulsory military service and the new German Army was renamed the Wehrmacht. The Germany Air Force was called the Luftwaffe.

Work at Kummersdorf continued until 1936, when the Commander in Chief of the Wermacht toured the site. After watching the static firing of three rocket motors as part of the tour, Dornberger approached the subject of funding for a much larger rocket research establishment. His C-in-C agreed to fund such an establishment, and the search began to find a suitable location. Von Braun's mother is credited with suggesting the Peenemünde peninsula in Pomerania, which was the site selected. The cost of developing the Peenemünde site was shared with the Luftwaffe. While the Wermacht used Peenemünde to develop their ballistic missile, the Luftwaffe used the site to develop a pilotless, ramjet-powered flying bomb – a predecessor of the modern cruise missile.

Work on the larger and more powerful A-3 missile continued at Kummersdorf while Peenemünde was being built. In 1937 four A-3s were launched from the island of Greifswalder Oie, off the Peenemünde coast. All four vehicles became unstable in flight and the test-flight programme was cancelled.

With the designation A-4 already assigned to the operational ballistic missile, work began immediately on a research programme for an A-5 prototype. In 1938 four A-5s, minus the guidance system being developed for the vehicle, were launched from Greifswalder Oie. The first two complete A-5s were launched from the island in October 1939 and were both successful.

After a vertical launch the rocket's guidance system rotated the pitch gyroscope away from vertical in the direction of the required trajectory. The change in direction was fed to motors controlling the position of carbon vanes acting in the rocket motor's exhaust plume. By altering the position of the vanes the exhaust plume was redirected and the rocket then moved in reaction to the new thrust vector. During 1939–41, 25 of the A-5 rockets were launched from Peenemünde.

Meanwhile, World War II had begun. Using new tactics, German forces had captured most of Western Europe in just a few weeks. Britain stood alone, and became the centre of resistance against joint German and Italian aggression around the world. In the East, Japan was also attempting to expand its empire. In the summer of 1940 the Royal Air Force had caused the cancellation of Germany's planned invasion of Britain. Hitler then turned his forces East and invaded the Soviet Union in June 1941. Despite selling arms to Britain, America took no part in the war until the Japanese Air Force attacked the US fleet at Pearl Harbour, in December 1941.

Work on the A-4 had continued alongside flight-tests of the A-5. The first two attempts to launch an A-4 from Test Stand VII, Peenemünde, failed on 13 June and 16 August 1942. The third launch, on 3 October 1942, was the first missile to fly the full length of the A-4's ballistic trajectory. Dornberger and von Braun began seeking support for the A-4 as a weapon system and received orders that they should plan towards mass production. In February 1942 Albert Speer took over as Germany's Minister of Armaments and Munitions and assigned Gerhard Degenkolb to oversee the A-4's mass production. Degenkolb had earned his reputation when he modernised the German locomotion production industry by introducing standardised parts. A-4 production was to be established at Peenemünde and two external factories. At Peenemünde flight tests continued and gave rise to a series of design changes. The A-4, however, would never be mass produced in anywhere near the numbers demanded by the Nazi government.

The A-4 was the world's first practical application of a large, modern rocket. It was a single-stage vehicle that carried a 975-kg high explosive warhead in its pointed nose. The remainder of the vehicle housed the rocket required to place that warhead on its intended target. The missile, with warhead, stood 14 m tall, 1.68 m across the body and 3.57 m across the fins. The rocket motor was mounted at the base of the vehicle, below the liquid oxygen (oxidiser) and ethyl-alcohol/water (fuel) tanks,

Fig. 3. The German A-4 ballistic missile was the first practical application of a large liquid-propelled rocket. Launched from German-occupied Europe its principal targets were London, England, and the Belgian port of Antwerp, although small numbers were also used against other targets. This example is one of many A-4 missiles taken to America at the end of World War II. (US Army)

which were stacked one on top of the other. The missile's guidance system was placed in a compartment above the fuel tank and below the warhead.

At launch the A-4 stood on the base of its fins, on a small square launch table. Steam drove the turbopumps that drove the propellants into the combustion chamber, where they were mixed, ignited and burned. Exhaust gases were compressed as they passed through the throat of the rocket motor and then expanded rapidly in the extension nozzle. As the rocket rose off the launch table the guidance system controlled its attitude via aerodynamics surface in the corners of the four fins and through the movement of four carbon vanes acting in the exhaust plume. (See Fig. 3.)

After 4 seconds of vertical flight the pitch gyroscope began to move from the horizontal to the vertical, causing the guidance system to move the vanes in the exhaust plume and thereby directing the missile towards its target. A radio signal shut the engine down after 50 seconds of powered flight, but the missile continued to climb under the momentum already built up. The missile reached an apogee of 96 km, just 4 km short of the modern definition of the beginning of space. Gravitational attraction then pulled it back towards the ground along a ballistic arc. As it re-entered the thick lower atmosphere the missile was subjected to atmospheric heating and many early flight tests ended when the missile broke up in a condition known as an airburst. Insulating the propellant tanks provided some protection against airburst on later, operational missiles. The four fins led to the missile adopting a nose down, vertical descent in the final stages of its flight. Travelling at Mach 3, no one in the target area heard the missile arrive until it exploded. Many witnesses would later report hearing a sonic boom and a dull roar after the explosion – the sound of the missile breaking the sound barrier and its final descent through the atmosphere.

In April 1943 Hienrich Himmler, the head of the SS, visited Peenemünde. It was an attempt on his part to obtain information on the A-4 programme. It was Himmler's intention that the SS would take over the A-4 programme, one of the largest projects in Nazi Germany. Von Braun was an honorary officer in the SS and Himmler's visit to Peenemünde was the one occasion when he was known to wear his SS uniform in public. Two A-4s were launched for Himmler; the first failed, but the second was successful and flew its full range.

On 26 May 1943 two A-4s were launched from Test Stand VII, Peenemünde, in an organised demonstration for a number of military and civilian VIPs. Twelve days later Dornberger and von Braun made a presentation on the A-4 to Hitler, at his headquarters in Rastenburg. They showed him the film of the successful third launch and demonstrated how the A-4 worked using models. Hitler assigned the country's highest priority to the A-4 programme, giving it the same right to vital war materials as the tanks and aircraft destined for Europe and Russia.

At that time an underground chemical store at Konstein, near Nordhausen, was identified as a potential location for A-4 production. Prisoners from the SS-run concentration camps were brought in to clear out the chemical storage facilities and extend the existing network of tunnels.

On the night of 17 August 1943, the Royal Air Force bombed Peenemünde. The majority of the bombs fell on the housing estate and the prisoner-of-war camp. Only one senior German rocket engineer was killed, but several of the East European prisoners who had smuggled information out and allowed the Allies to learn of the A-4's existence died while locked in their prison huts. After the raid work continued at Peenemünde, but much of the damage was left unrepaired in order to give the appearance that the site had been abandoned.

Shortly afterwards the two external factories where the A-4 was to have been produced were also bombed. This led to all future A-4 production work being concentrated at Nordhausen. The SS provided the prison workforce to man the new underground facilities and also provided security at the site. The prisoners were kept inside the underground facility while they extended its tunnel system and installed the machinery necessary to manufacture the A-4. Meanwhile an external concentration camp – Dora – was constructed by a second group of prisoners.

When Dora was finished all prisoners were housed within its heavily guarded fences. A prisoner's life at Nordhausen was brutal and short. Coming from the Nazi's concentration camp system, the prisoners were expendable. Discipline was kept by means of malnourishment, poor medical conditions, beatings and public hangings. The senior Wehrmacht and A-4 engineering staff accepted the intense suffering of the Russian, French, Polish and even German political prisoners as the price to be paid if the rockets were to be mass produced.

A letter survives (signed by von Braun) informing the Commandant of Dora that he had visited a neighbouring concentration camp and requested that the Commandant there redirect suitably qualified prisoners to Dora. There they would replace those that had been worked to death producing the A-4. In all, some 22,000 people would die in Dora and its associated sub-camps.

At the same time all A-4 flight-tests were moved from Peenemünde to an old SS training camp at Blizna, in Poland. The SS provided security at the site and SS Brigadier (later General) Hans Kammler was put in charge of the A-4 launch crews.

The first A-4 missiles launched from Holland fell on London, England, on 8 September 1944. These did not come as a total surprise as intelligence had been reaching London since November 1939 suggesting that the Germans were developing large liquid-propelled rockets. In 1944 the parts of three captured A-4 rockets had been flown to Farnborough in Hampshire, England, where they had been reconstructed to produce a nearly complete missile. Meanwhile, following the 20 July attempt to kill Hitler, Himmler's SS took control of the A-4 programme.

Between 8 September 1944 and 27 March 1945 1,346 A-4s were launched against London. By that date the liberation of Europe by the Allies was already underway following the amphibious landings on the French Channel coast on 6 June 1944. The advancing Allied forces pushed the A-4 launch teams back and they were finally recalled to Germany, from where the missile could no longer reach the south of England. The A-4 was also used against the port of Antwerp in Belgium, through which much of the Allies' war material was shipped into Europe following the invasion.

2.4 PAPERCLIPS AND CLASSIFIED MILITARY FILES

It has been stated in many publications that the A-4 was an ineffectual weapon. That may be true, but its effect on Allied leaders in the West and East was to change the shape of the future of warfare. All of the major Allied nations placed scientists and engineers with their advancing forces as they liberated Europe. Their job was to identify new German technology and the people responsible for developing it, and to transfer them out of war-torn Germany. The Americans and Russians were particularly interested in capturing the German rocket engineers. Dornberger and von Braun led their team away from the advancing Soviet Red Army and finally surrendered to the Americans, whom they hoped would fund a new programme of rocket development and even space exploration.

American intelligence officers rounded up those Germans that they felt would be of use to their country after the war and censored their official files, even going to the extent of rewriting some files, to remove entries that would bar them from entering America. The files of the Germans in question were identified by an inconspicuous paperclip. In this way von Braun and many of his rocket engineers were offered contracts to work in America, while the American occupying forces looked after their families in Germany. The parts of 100 A-4 missiles were removed from Nordhausen, before it was handed over to the Russians (both Peenemünde and Nordhausen lay in the Russian Occupation Zone, as identified at the Yalta

Conference). The Americans also removed many tonnes of records and documentation from Peenemünde that had been hidden in the British Occupation Zone.

The Germans had hidden them in a mineshaft and dynamited the entrance to protect the vital documents from the chaos that filled Germany in the final days of the war.

The missiles were transported across Europe to Antwerp and then to America, where they were taken to the newly activated White Sands Proving Ground (WSPG), and the German engineers followed shortly afterwards. On 20 September 1945 von Braun and six other German rocket engineers were transported to America and housed at Fort Strong, in Boston Harbour. Over the next four months 185 Peenemünde engineers were transported to New Mexico and housed in unused hospital buildings at Fort Bliss, part of the WSPG complex. At Fort Bliss the US Army had established the Sub-Office Rocket of the Ordnance Department. The Germans were under the charge of Major James Hamill and for six months they assisted the American Army and its contractor, the General Electric Company (GEC), in understanding how the A-4 worked and how it was prepared and launched. However, they soon became frustrated that they were not tasked with developing new missiles.

Dornberger did not move to America with von Braun and the other 'Paperclippers'. He was originally handed over to the British authorities, who wanted to try him as a war criminal for his part in the A-4 attacks on London. After two years without trial he was released and travelled to America, where he took a job with Bell Aircraft Corporation.

2.5 REACTION MOTORS INCORPORATED

James Wyld of the American Rocket Society was one of the founder members of Reaction Motors Incorporated (RMI), which was formed on 16 December 1941. In 1938 Wyld had developed a regeneratively cooled rocket motor based on a series of static firings carried out by members of the ARS in 1935. The US Navy purchased the 444.8 newton (N) thrust motor.

RMI's second rocket motor produced 4448 N and was designed as a JATO motor for the US Navy's aircraft. The company's third motor produced a thrust of 15,123 N and was constructed under a second US Navy contract.

When the USAAF awarded the contract for the development of a rocket aircraft designed to fly faster than Mach 1 to Bell Aircraft Corporation, Buffalo, New York, in March 1945, the contract for the four-chamber rocket motor went to RMI. The engine they developed was the XLR-11 and the company relocated to New Jersey in 1946. In September 1955 RMI also received a contract to develop the XLR-99 rocket motor for the X-15 hypersonic research aircraft. RMI was also responsible for developing the propulsion system for Viking, the first all-American liquid-propelled sounding rocket, flown out of WSPG in the late 1940s and early 1950s. This was the first American rocket to employ a gimballed motor for in-flight attitude control. On 30 April 1958, RCI merged with the Thiokol Chemical Corporation.

3

Preparing for war

The Ancient Chinese were the first people to deploy black powder rockets as military weapons. Their example was followed by many other nations over the years, until the rocket's capability was overtaken by that that of contemporary artillery. In the first half of the twentieth century the German Army developed and deployed the A-4 ballistic missile. The A-4's deployment during World War II was the first time that a large 'modern' rocket, burning liquid propellants, had been deployed as a weapon of war. When the war in Europe ended many of the German rocket engineers were taken to America with contracts to work for the American Army.

During World War II, American and European scientists working on the Manhattan District developed two nuclear bombs with different ignition processes. The US Army Air Force (USAAF) dropped the first operational nuclear bomb on Hiroshima, destroying the city with just one bomb. When the second nuclear bomb destroyed the city of Nagasaki a few days later, Japan surrendered and World War II came to an end in August 1945. At that time the ballistic missile and the nuclear bomb were at the leading edge of military technology.

Although the A-4 was of questionable value as a weapon, the combination of the ballistic missile and the nuclear bomb would change the face of superpower warfare. From the 1950s to the present day the nuclear arsenal of the Soviet bloc and those of America, Britain and France have faced each other with the potential to destroy all life on Earth. The missiles developed to deploy nuclear warheads across intercontinental distances also gave mankind the capability to leave Earth and explore space.

3.1 JET PROPULSION LABORATORY

The Guggenheim Aeronautical Laboratory of the California Institute of Technology (GALCIT) was formed in 1936. GALCIT was headed by Theodore von Karman, a Hungarian immigrant, and originally consisted of a small group of graduate students from the California Institute of Technology (CALTECH). The leader of that group was Frank Malina.

The original GALCIT group also contained William Bollay, John Whiteside Parsons and Edward Forman. The latter two had corresponded with Ley in Germany and were aware of the work being carried out by the VfR. In 1936 Apollo Smith and Hsue Shen Tsien joined the group. Following a series of accidents involving chemicals at CALTECH, the GALCIT group gained the name the 'Suicide Squad'. GALCIT static fired its first liquid propellant rocket motor on 29 October 1936.

In August 1938 Consolidated Vulture asked CALTECH to research the possibility of using rocket motors to assist heavily laden seaplanes in taking off. Von Karman tasked GALCIT to study the concept. In December 1938 Malian's report was submitted to the National Academy of Sciences (NAS) Committee on Army Air Corps Research. The NAS awarded GALCIT $1,000 to prepare a report on their Jet Assisted Take Off (JATO) concept. On 1 July 1939 the NAS awarded GALCIT a $10,000 contract to develop the JATO idea under the Army Air Corp Jet Propulsion Research Group. The contract was extended by a further allocation of $22,000 on 28 May 1940 and the work moved in to Phase II. While carrying out this work the GALCIT group moved from CALTECH to a location outside Pasadena, where they established a new research centre.

The first flight of a GALCIT JATO-assisted USAAF aircraft occurred at March Field, Riverside, California, on 12 August 1941. It used 12 solid propellant JATO units developed by Parsons. In April 1942 the first American aircraft was assisted into the air by a liquid propellant JATO system developed by new GALCIT member Martin Summerfield. On 19 March 1942 GALCIT set up the Aero Jet Engineering Corporation to market their two JATO motors.

In late 1943 GALCIT engineers were asked to review British Intelligence information regarding the development of a large liquid-propelled ballistic missile, the A-4, by Nazi Germany. Malina and Stein's reports were sufficiently convincing that USAAF Colonel W. Joiner recommended that America commence development of its own ballistic missile.

The Ordnance Department of the US Army Rocket Development Branch tasked GALCIT with the development of an American long-range ballistic missile in January 1944. The missile was to throw a 450-kg payload downrange between 120 and 161 km. Ordnance at California Institute of Technology (ORDCIT) was formed to develop the missile. The proposal for the new missile was submitted to the Army on 28 February 1944 and ORDCIT were instructed to develop it in June. This programme resulted in the Private solid propellant missile. At about that time, GALCIT was renamed the Jet Propulsion Laboratory (JPL).

ORDCIT's engineers developed the Corporal A missiles, which burned solid propellant. Liquid propellant models were designated Corporal E and F. As Allied military forces swept across German-occupied Europe, members of ORDCIT went with them, inspecting captured German missile technology. Hsue Shen Tsien was present during von Braun's initial interrogation by US military personnel.

ORDCIT was also developing the WAC Corporal sounding rocket, designed to carry instruments into the upper atmosphere for research purposes, and the Douglas Aircraft Corporation was contracted to produce the missile. The first WAC

Corporal was launched on 26 September 1945, and nine more were launched before the end of October. The WAC Corporal programme led to the Aerobee sounding rocket. Aerojet Engineering Corporation developed Aerobee with assistance from the Naval Research Laboratory.

As early as 1944 the GALCIT engineers instructed their lawyer to seek prospective buyers or companies that would be prepared to provide financial and material support for Aerojet. The General Tire and Rubber Company, Akron, Ohio, purchased 50 per cent of the company's shares for $75,000 in 1947. With the shares they had purchased separately from individual shareholders, the Ohio company gained the controlling interest in the company, which was renamed Aerojet General Corporation and taken out of GALCIT's hands.

3.2 HERMES

The US Army established Project Hermes on 15 November 1944. On 20 November 1944 the Army Ordnance Department contracted the General Electric Company (GEC) to cooperate with them in Project Hermes, the study and development of an American missile, to counteract the A-4 then under development in Germany. Like ORDCIT, GEC also sent representatives into Europe with American forces in order to study German missile technology. The American Army captured the Nordhausen underground factory on 11 April 1945. Plans were put in place for the parts for 100 A-4 missiles and the German project's technical documentation to be transported to WSPG, New Mexico, which was activated by the US Army Ordnance Department on 9 July 1945. German rocket engineers were rounded up, interrogated and some were subsequently offered contracts to travel to America and commence work developing missiles for the US Army. The engineers were transported to Fort Bliss, part of the WSPG complex, where they would teach American engineers to prepare and launch the A-4. The Americans employed the captured A-4s to gain first-hand experience with large missiles while carrying out scientific experiments in the upper atmosphere. GEC was contracted to launch the majority of the A-4s flown out of WSPG and the programme was assigned as part of Hermes. (See Fig. 4.)

As the American military services each developed their first new missiles, which were capable of flying longer distances than the A-4, a new test range was required. On 9 July 1946 the search began for a location at which to construct a new Joint Long-Range Missile Proving Ground (JLRMPG). Two sites were short-listed: El Centro, California, and the Banana River, Florida. When one A-4, launched out of WSPG, landed in Mexico negotiations regarding the El Centro site were stillborn. As a result, the JLRMPG was built at Cape Canaveral, Florida. The Department of Defence (DoD) was formed on 26 July 1947, and the US Air Force (USAF) became a military service separate from the US Army on the same date. The USAF took over the Banana River Naval Air Station in September 1948, and work began on clearing the surrounding area. JLRMPG was activated on 1 October 1949, and the first launch from the new site took place on 29 July 1950.

Throughout the post-war period various American companies began the

Fig. 4. Test Flight 1 (TF 1) the first A-4 missile to be launched in America by an all Army launch team. The flight set an altitude for the A-4 missile. (Arnold Crouch)

development of rocket motors, ballistic missiles and cruise missiles for the military forces, and sounding rockets for upper atmosphere research. The first was GALCIT's Corporal, which was developed in both a guided and unguided 'Without Any Control' (WAC) version. In 1946, Malina and Somerfield proposed mounting a WAC Corporal on top of an A-4, to produce the first two-stage rocket to be flight tested. The programme was called Project Bumper, and eight Bumper launches were made between 13 May 1948 and 29 July 50. On 24 February 1949 a Bumper–WAC combination was launched on a vertical trajectory out of WSPG. The WAC Corporal second stage reached an altitude of 390 km and became the first object to enter space. A separate subprogramme, called Blossom, carried a series of early experiments carrying live monkeys into the stratosphere in pressurised containers. None of the monkeys used in the Blossom experiments survived. Four died when the parachutes attached to their pressurised containers failed to open. The fifth landed safely, but perished in the desert heat before it could be located and recovered.

The Hermes missile development programme went through many different stages as well as the A-4 launches. As several of the concepts involved drew heavily on research that originated at Peenemünde during World War II, the Hermes studies led to the development of new propulsion systems and radio guidance for use on future missiles. In June 1946, GEC began a feasibility study for a missile designated Hermes CI, and this was the starting point in the development trail that would lead to the Mercury Redstone launch vehicle.

The first large all-American liquid-propelled rocket was the Viking Sounding Rocket. This used the first gimballed rocket motor for in-flight steering. Reaction Motors Incorporated developed Viking's gimballed rocket motor. Fourteen Vikings were launched, with the first flying on 3 May 1949 and the last on 1 May 1957. Viking proved too expensive so, when the supply of A-4s and Vikings ran out at WSPG, America's future sounding rocket programme became based around cheaper solid propellant vehicles.

3.3 REDSTONE

On 1 June 1949 the Redstone Arsenal, Huntsville, Alabama, was placed on stand-by and, on 28 October, the Army Ordnance Research and Development Division, Sub-Office (Rocket) was transferred from Fort Bliss, New Mexico, to Redstone Arsenal. The German Peenemünde engineers were among those transferred. Their leader, Dr Wernher von Braun, was named Technical Director of the new establishment. The unit was redesignated Ordnance Guided Missile Centre on 15 April 1950.

On 10 July, the Office of the Chief of Ordnance directed the engineers at Redstone Arsenal to study the technical requirements of a tactical nuclear missile with a range of 804.6 km to be used in support of Army battlefield operations. The German and American engineers were tasked with defining a missile with a tactical range of between 240.5 and 805 km with a throw weight between 680 and 1,360 kg. The following day GEC's Project Hermes contract was altered to allow the transfer of all Hermes CI work to Redstone Arsenal, and the transfer took place on 10 July 1951. Work continued under a series of designations, including Hermes CI, Major and XSSM A-14.

The XSSM A-14 was little more than a second-generation A-4 and had many similarities with the German missile. The new missile was constructed in three parts (propulsion unit/propellant tanks/warhead and instrument unit) and designed to be a mobile weapon, to be launched in the field. For launch it stood on its own fins, on a small square table. It was erected on its launch table using an A-frame. In flight, four carbon vanes acted in the rocket exhaust to provide in-flight steering by deflecting the exhaust plume. On each fin an aerodynamic surface acted within the air. Both the aerodynamic surfaces and the carbon vanes acted under the control of the ST-80 guidance unit installed in the missile and developed by Ford Instrument Company, Long Island, New York. When this guidance unit was not ready for the earliest Redstone missiles, original Peenemünde blueprints were used to manufacture the A-4 guidance units that were installed in the early missiles. The single rocket motor burned liquid oxygen and a mixture of 75 per cent alcohol/25 per cent water, while decaying hydrogen peroxide produced steam, which drove the turbines that drove the fuel pumps.

XSSM A-14 also carried several innovations. It employed moncoque construction, whereby the walls of the propellant tanks also acted as the exterior wall of the missile. This was also the first missile to use inertial guidance. The propulsion unit was a single North American Aviation A-6 rocket motor, which gave a range of 321.8 km, almost the same as the A-4. After engine shutdown the warhead and instrument unit separated from the body of the missile, under the control of its own guidance unit the warhead had a predicted accuracy of 300 m. This overcame the problem of warhead loss as the result of airburst that had plagued the German missile programme. When the Korean War started in 1951, the XSSM A-14 was cleared for development and production. The new missile was named Redstone, after its design bureau, on 8 April 1952. (See Fig. 5.)

The German engineers had wanted to establish the Redstone missile production line at Redstone Arsenal, but the Office of the Chief of Ordnance rejected the idea on 1 April 1952. Rather, Redstone Arsenal would develop the missile and build the first 12 research and development models. Thereafter, a contractor would be engaged to mass produce the new weapon. The Chrysler Corporation received the first contract for production of the Redstone Intermediate Range Ballistic Missile (IRBM) on 15

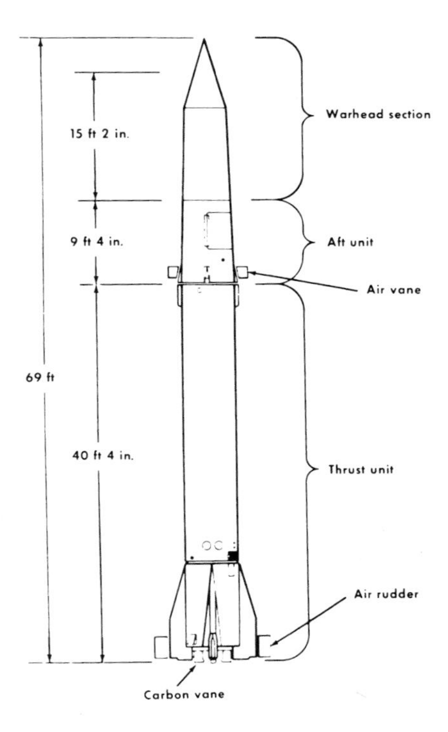

Fig. 5. Redstone Missile. (NASA)

June 1955. When Chrysler was too slow to find facilities for their production line, Redstone Arsenal also produced missiles 18–29.

On 20 August 1953 the first research and development (R&D) Redstone was fired from Launch Complex 2 Eastern Test Range. It failed. Like the German A-4, the third missile made the first successful flight. Despite the early failures, Redstone quickly gained a reputation for reliability. When the 38th and last R&D missile was launched in November 1958, 27 missiles had been successfully launched. After that date 24 operational missiles were deployed with the US Army, each capable of carrying a W-39 tactical nuclear warhead.

3.4 ATLAS

The USAAF called for proposals for a series of four missiles on 31 October 1945. The Consolidated Vultee Corporation (Convair) designed an ICBM. On 22 April 1946 Convair began work on ten R&D rounds to prove the concept of an ICBM, under a contract to the newly independent USAF. The first American ICBM was called RTV A-2, or Hi Roc. Many sources record the missile by its contract number: MX-774. The missile employed four Thiokol rocket motors to provide 8,896 N of thrust at launch. The rocket motors were gimballed to provide in-flight attitude control. Hi Roc had a similar appearance to the German A-4 but it employed a detachable warhead and monocoque design. The walls of the propellant tanks were so thin that they had to be internally pressurised to make the missile sufficiently rigid to support its own weight.

Project Hi Roc was cancelled by the USAAF on 1 July 1947, but sufficient funding remained available for three R&D rounds to be developed and launched. The first launch took place from WSPG on 13 July 1948. The other two followed, on 27 September and 2 December 1948. All three launches were only partially successful

Convair continued to fund its own in-house missile studies, without government backing, for two more years and in November 1948 the first 1.5 stage design for an ICBM came off of the drawing boards at General Dynamics, a division of Convair. One rocket motor was attached to the base of a sustainer stage and four others were mounted around it in a detachable skirt. At lift-off all five motors were to be ignited. After passing through the thickest portion of the atmosphere the four outer motors would be shut down and the detachable skirt would be jettisoned, taking the four rocket motors with it. The single remaining motor would continue to push the nuclear warhead to the desired altitude and velocity.

On 16 January 1951 the USAF Air Research and Development Command (ARDC) requested Convair to advise them on the relative merits of long-range cruise missiles and ICBMs and decide which the service should develop. Convair's report favoured the ICBM and the company supplied the USAF with the latest details of its own studies of a seven-motor ICBM called Atlas. The USAF accepted Convair's design and began a development programme designated MX-1593. Convair were instructed to develop Atlas with a range of 10,187 km. In December 1952 a USAF Scientific Advisory Board reviewed the project and approved its continuation.

In 1953 the Atlas concept changed to a five-rocket motor configuration with the four-first stage motors being arranged around the central Sustainer motor in a square. This allowed for the four-first stage motors to be jettisoned while the single Sustainer motor continued to fire.

At that time American military leaders learned of the Soviet Union's R-7 ICBM programme. As a result, the Pentagon established the Strategic Missiles Evaluation Committee under the code name 'Teapot'. The Teapot Committee was tasked with evaluating America's ballistic missile and thermonuclear warhead programmes. The Committee's report, released in February 1954, stated that America had wasted five years, during which, according to intelligence reports, the Soviets had tried to develop their own ballistic missiles resulting in the huge R-7 ICBM. The report recommended the redesign of the Atlas ICBM to take a smaller and more effective thermonuclear warhead. It also stated that the USAF and President Eisenhower's government had to put in place the funding and management programme to ensure that the ICBM received the priority it demanded.

The missile's design changed again in October 1954, when two Booster motors were placed one on either side of a single Sustainer motor. On 16 December 1954 Convair and USAF agreed that the October design would be the missile that would become America's first ICBM. Approval was given for Atlas to be developed in January 1955. The USAF designated to programme WS 107A-1.

Atlas's design called for a monocoque stainless steel skin that was so thin that it would collapse under its own weight if it were not internally pressurised. The design dispensed with the heavy internal structures used on other missiles of that time and allowed Atlas to obtain the maximum range from its propellant supply.

A single set of propellant tanks fed liquid oxygen and kerosene to three main propulsion units. The original Sustainer was 18.2 m tall and 3 m in diameter. Two Booster rockets and their associated plumbing were mounted in a booster skirt that mated mechanically to the base of the Sustainer thrust ring. When in position the two Booster motors were on either side of the single Sustainer motor. Four sets of vernier attitude control motors were spaced at 90 degrees around the Sustainer, just above the booster skirt mating line. Alternate sets faced forward and to the rear.

At the lift-off of an Atlas missile the rearward facing verniers ignited first and burned at 4,472 N throughout the powered phase. The three main propulsion units then ignited and built up to mainstage before the vehicle was released. At T + 145 s the two Booster motors were shut down and the booster skirt was jettisoned. The single Sustainer motor continued to fire until the pre-programmed velocity was reached, at which time it was also shut down. The vernier motors then fired in solo mode to provide final attitude pointing before the warhead was separated. At separation the two forward-facing verniers fired to push the inert Sustainer away from the warhead and thereby provide positive separation in the microgravity environment. The nuclear warhead re-entered the lower atmosphere and was protected by its heat shield. Guided by its own navigation system it fell to Earth with the intention of impacting the Soviet homeland. The unprotected Sustainer would re-enter the lower atmosphere and burn up.

When the production of the Sustainer motors fell behind schedule the decision was made to flight-test Atlas with just the two Booster motors. The Atlas A was flown from the JLRMPG in this configuration, with a dummy warhead. The first launch took place on 11 June 1957, but the Range Safety officer had to destroy the vehicle when one rocket motor lost thrust and the round became unstable in flight. The second launch, on 25 September 1957, also suffered an in-flight propulsion failure. On 17 December 1957 an Atlas A flew its programmed 965.5-km range. The last Atlas A flew on 3 June 1958, bringing to an end a flight-test series of eight vehicles, of which only half were successful.

Atlas B included an active Sustainer motor and flew for the first time on 19 July 1958. The first flight was a failure. On 18 December 1958 the Sustainer from Atlas 10B was placed into orbit and a taped Christmas message from President Eisenhower was broadcast to the world below. This was Project Score, America's answer to Sputnik II. The Atlas Sustainer re-entered the atmosphere and burned up on 21 January 1959. An Atlas B flew 4,350 km downrange on 2 August 1959. On 9 August 1959 the USAF declared the Atlas ICBM to be operational.

Atlas C carried the General Electric Mark II warhead, with copper heat-sink re-entry protection, and improved guidance equipment. The next model, Atlas D, was capable of carrying the Mark III warhead, or an alternative warhead employing Avco's new ablative re-entry protection.

3.5 SURVIVING RE-ENTRY

While developing the A-4 the German Army lost many rounds to airburst. The A-4 launched vertically and quickly passed beyond the upper limit of the troposphere, the lowest band of thick air in the Earth's atmosphere. When Earth's gravitational attraction overcame the missile's forward velocity it began a long, ballistic fall back to the planet's surface. Some missiles encountered the troposphere and their propellant tanks exploded – an airburst – destroying the missile. Because the V-2's warhead remained permanently attached to the missile, it too was lost when the airburst occurred. Both Atlas and Redstone overcame the airburst problem by detaching the warhead from the missile while it was above the troposphere. If the now useless missile suffered an airburst when it re-entered the troposphere it was no longer of any consequence. Employing its own guidance package, the warhead passed back through the troposphere to strike its target.

Energy cannot be lost, it can only be transferred to another form of energy. As the warhead re-entered the atmosphere the friction caused by contact with the gases in the atmosphere acted against it, slowing it down. The kinetic energy lost was transferred to intense heat, which would vaporise the warhead unless it was protected in some way. In the early 1950s two forms of heat protection were available to warhead designers. Heat-sink protection required the warhead to be surrounded by a shield manufactured from highly conductive metals. The heat-sink shield had to be thick enough to hold the heat built up during the descent, without allowing it to reach the physical structure of the warhead. Adequate protection was 'purchased' with weight, and additional launch weight required a more powerful rocket motor to launch the warhead in the first place. The alternative method of heat protection required the warhead to be surrounded with ablative material. When heated the material ablated, by turning from a solid to a gas and carried the heat of re-entry away with it. The ablative heat shield also had to be thick enough to allow material to vaporise and carry away the heat throughout re-entry, while ensuring that sufficient material remained to protect the warhead.

The earliest ICBM warheads were designed as cones, the point of the cone being the leading edge to push through the atmospheric gases both during exit and re-entry. Subsequent wind tunnel tests predicted that such a conical warhead's heat protection would burn in an irregular fashion, which might lead to the warhead's trajectory being affected and the target being missed.

On 18 June 1952, NACA engineer Harold Allen, working at Ames with Alfred Eggers, realised that during re-entry heat was produced in two places: that produced within the bow wave next to the re-entry vehicle could be transferred to that vehicle, while that produced outside the bow wave would largely bypass the re-entry vehicle. Allen worked out an equation that showed how a wide, blunt, high-drag leading face on the warhead would produce the majority of heat outside of the bow wave. Even so, heat reaching many thousands of degrees would still be produced within the bow wave and would reach the heat shield protecting the warhead. Allen and Egger's radical theory demanded that the conical warheads being planned for America's ICBMs should re-enter the atmosphere wide end first. Only time and experience would show the superiority of the ablative heat shield over the heavy heat-sink type.

3.6 JUPITER C RE-ENTRY TESTS

In 1955 the US Army proposed Project Orbiter, a programme to launch an artificial satellite using an Army IRBM. Although the project was refused national funding the Army continued to develop the idea under the disguise of an IRBM warhead re-entry test programme, designed to test Allen and Egger's theory. The Jupiter C launch vehicle was developed at Redstone Arsenal, consisting of a Redstone IRBM with elongated propellant tanks, and a change of fuel from alcohol/water to unsymmetrical dimethyl hydrazine (UDMH). The Jupiter C was then fitted with a second stage consisting of 11 Recruit solid propellant rocket motors. A third stage consisting of 11 Recruit and three Baby Sergeant solid rocket motors completed the new launch vehicle. The new missile produced a cut-off velocity similar to that of the proposed Atlas ICBM.

On 1 February 1956 responsibility for producing the Redstone and Jupiter C missiles was transferred to the Army Ballistic Missile Agency (ABMA). This represented little more that the reorganisation and renaming of the design bureau at Redstone Arsenal. The first Redstone missile constructed by Chrysler Corporation was launched on 19 July 1956. Twelve warhead re-entry tests were planned under the Jupiter C programme. Only three were flown, on September 1956, and May and August 1957. The third dummy warhead was recovered after flying the full range of an IRBM mission. Von Braun referred publicly to the third warhead as '*The first object recovered from space*'. With the successful completion of the third flight the warhead re-entry tests stopped, but the Army maintained two Jupiter C missiles with upper stages at Redstone Arsenal, in case they were given the opportunity to revive Project Orbiter.

Meanwhile the Pilot Aircraft Research Division (PARD) of the National Advisory Committee for Aeronautics (NACA) was launching sounding rockets out of Wallops Island, Virginia, to test various heat shield materials. The USAF was also testing ablative materials to replace the heavy copper re-entry vehicles that were to be fitted to the early Atlas ICBMs.

3.7 AMERICA'S COLD WAR MISSILE DEVELOPMENT

During the final years of World War II the US Navy, US Army and USAAF all competed for control of their country's new missile programmes, and each service began its own programme of missile development. In 1944 the USAAF was given jurisdiction over winged cruise missiles, while the Army Ordnance Department was assigned control of ballistic missiles. This changed two years later when the USAAF was given control of all missile projects. The Department of Defence (DoD) was formed in July 1947 and the USAF was formed as a separate service at the same time. One year later missile development was split between the USAF, the Navy and the Army depending on each service's needs.

In 1950 the Korean War began and American Secretary of Defence Louis Johnson set up the Special Interdepartmental Guided Missiles Board, which attempted to minimise the confusion that surrounded missile development in the military. At that time the German Peenemünde engineers were in the process of moving from Fort Bliss to Redstone Arsenal and were assigned development of the Hermes C-1 missile, which would ultimately be renamed Redstone. Meanwhile, the USAF had assigned contract MX-1593 to Convair to develop the Atlas ICBM.

Having been denied American nuclear secrets after the end of World War II, despite providing scientists for the Manhattan District, Britain developed its own nuclear bomb. In July 1952 British military leaders visited America to propose a strategy whereby nuclear weapons became the leading portion of Western defence policy. The Americans were initially suspicious of Britain's motives, but the plan was ultimately accepted. To back-up their plan, the British exploded their first nuclear device on 3 October 1952. Using technical information supplied by Convair, Britain developed her own Atlas style missile, called Blue Streak. Britain's Labour Party cancelled Blue Streak as a cost-cutting measure and then negotiated with America to either purchase American missiles, or have American missile squadrons stationed in Britain.

President Dwight D. Eisenhower was elected to office on 4 November 1952. He had served as Supreme Allied Commander during the liberation of Europe and was a well-respected war hero. In October of that year a National Security Council paper had identified the Soviet Union as a potential threat to American security and called for '*emphasis on the capability for inflicting massive retaliatory damage by offensive striking power*'. Eisenhower's Secretary of State for Defence, John Dulles, accepted the concept of 'Massive Retaliation'. Meanwhile, the development of the thermo-nuclear weapon meant that warheads were becoming more effective, while shrinking in size. The prospect of making them small and light enough to be launched by a ballistic missile was becoming a reality.

With all three military services requesting additional budgetary funds to develop their own ballistic missiles, Eisenhower established the Killian Committee to review Soviet military capabilities and the threat they posed to America and her Allies. The report, published in late 1955, told how the Soviets had deployed missiles in Eastern Europe and placed Western Europe and Britain under threat of nuclear attack.

3.7.1 Thor

The USAF began planning its own IRBM, to be called Thor, which crossed the perceived boundaries between USAF missile requirements and those of the US Army and US Navy. This caused consternation among the Chiefs of Staff in Washington and resulted in IRBM development being split between the three services. The USAF would develop Thor, while the Army would develop a second liquid-propelled IRBM called Jupiter. Meanwhile, the Army would cooperate with the Navy on the development of a ship-launched version of Jupiter.

Douglas Aircraft Corporation was awarded contract WS-315A for the development of Thor on 27 December 1955. The USAF IRBM was 19.5 m long and 3.2 m in diameter. Its single Rocketdyne MB-1 motor produced 667,000 N of thrust at launch. It would carry a General Electric Mark II nuclear warhead with heat-sink re-entry protection. Thor's first launch attempt failed on 25 January 1957. Four more attempts also failed before the fifth missile made the first successful flight on 20 September 1957. Thor became operational on 5 November 1958 and the first overseas squadron was established in Britain. The missile also served as a launch vehicle for testing warhead re-entry characteristics. In time it would be withdrawn from service and developed into the Delta space launch vehicle.

3.7.2 Jupiter

The Army's Jupiter IRBM was assigned to von Braun's engineers at Redstone Arsenal. The final design, released in February 1956, was a missile 17.6 m long and 2.6 m across. The Redstone missile would be used to flight-test a number of Jupiter systems, giving rise to the Redstone configurations designated Jupiter A and Jupiter C. Meanwhile, on 26 November 1956 the Army was restricted to developing missiles with a range of 321 km. Any missile with a greater range became the USAF's responsibility. The first Jupiter flew on 1 March 1957 and was only partially successful. The first fully successful flight took place on 31 May 1957. Jupiter IRBMs were stationed in Italy and Turkey between 1958 and 1963.

From the Jupiter project's outset the Navy was reluctant to consider the deployment of liquid-propelled missiles on operational warships, due partly to the results of 'Project Pushover', carried out at WSPG. In that experiment a German A-4 had been launched from the simulated flight deck of an aircraft carrier. At the moment of launch two legs of the A-4's launch table had been dynamited, causing the missile to fall onto the simulated flight deck and suffer a huge air–fuel explosion. The wreckage of the simulated flight deck lay around at WSPG for several years as mute testimony to what might have been.

3.7.3 Polaris

The Navy concentrated its attention on German experiments carried out by the Peenemünde group in which they had launched relatively small solid propellant missiles from a U-boat (submarine). The Navy had subsequently kept abreast of advances in solid propellant rocket motors and had itself completed a considerable amount of new research into the subject. The service began work on its own solid propellant, ship-launched IRBM before pulling out of the joint Jupiter project on 8 December 1956. The new missile was subsequently named 'Polaris'. It would be launched from a submerged submarine lying just off the enemy coast and was therefore given a range of just 2,413 km.

To test the concept, the USS *Scorpion* was cut in two and a new Polaris missile section was fitted between the two halves. When it was launched on 9 June 1959 the enlarged USS Scorpion was renamed the USS *George Washington*. Meanwhile, the Thiokol Chemical Company's XM-15 solid propellant rocket motor had become the base block from which Polaris was developed. The first test of a Polaris Fleet Ballistic Missile (FBM) was made on 11 January 1958, at Point Mugu, California. Compressed air pushed the missile from its flooded launch silo. Tests continued throughout the year.

The first Polaris missile launch took place from Cape Canaveral on 24 September 1958; however, the autopilot failed and the missile exploded. The next four Polaris launch attempts were also failures. Finally, on 20 April 1960 Polaris number six flew the full range outlined for its flight. The first launch from the submerged USS *George Washington* was made on 15 November 1960.

3.7.4 Titan

In 1955 work began on America's second ICBM, the Glen L. Martin Company's Titan. Titan was capable of supporting its own weight without the need for internal pressurisation and was to be stored vertically, in hardened underground silos. If required for action the missile was elevated out of the silo and fuelled before being launched from ground level. This second-generation ICBM burned liquid oxygen and RP-1 in both of its two stages. At launch a two-chamber Aerojet LR87-1 motor in the first stage was used to push the missile to altitude. When it shut down the first stage was jettisoned. The single-chamber Aerojet LR91-1 on the second stage was ignited and continued to push the thermonuclear warhead to the correct combination of altitude and velocity. The warhead was then separated and used its own guidance system to fall back on to its target. The Titan's second stage burned up as it re-entered the atmosphere. Titan's first launch took place with only the first stage live. The second stage was inert and filled with water. That first launch occurred on 6 February 1959. The next two launches were unsuccessful, but the fourth was the first launch on which Titan was successfully staged in flight. Titan was declared operational in April 1962.

3.7.5 Titan II

In 1958 Martin's engineers suggested that an uprated Titan be developed. In June 1960, the Martin Company was given approval to begin development of the improved Titan, to be called Titan II. The new ICBM's propellants were unsymmetrical dimethyl hydrazine and nitrogen teroxide. The combination could be stored in the missile's tanks indefinitely, which meant that the Titan II could be kept fully fuelled in its underground hardened silo. The propellants were also hypergolic – they ignited on contact with each other – dispensing with the need for a separate ignition system. Titan's 15-minute launch time was reduced to 1 minute for Titan II. The new ICBM used an Aerojet AR87 AJ-5 two-chamber rocket motor on its first stage and an Aerojet AR91 AJ-5 single-chamber motor on its second stage. Titan II employed a 'fire-in-the-hole' staging principle, where the second-stage engine was ignited and brought to mainstage before the first stage was jettisoned. The first Titan II was launched in November 1961 and the ICBM became operational in 1963.

3.7.6 Minuteman

In October 1956 a USAF special committee recommended that the service should develop a solid propellant ICBM, similar in principle to the Navy's Polaris missile. In its original guise Minuteman was an IRBM. On 10 February 1958 the USAF chose to develop the Minuteman concept, extending its range to that of an ICBM. Thiokol Chemical Company produced the first stage, Aerojet the second and Hercules the third. A 1.3-megaton Mark 5 thermonuclear warhead completed the ICBM. Minuteman's first silo launch took place on 15 September 1959. On 1 February 1961 the first attempt to throw a Minuteman warhead over an intercontinental distance was a success. Minuteman was deployed operationally in December 1962.

4

Popularising space flight

4.1 THE BRITISH INTERPLANETARY SOCIETY

In the UK the British Interplanetary Society (BIS) was formed in the Liverpool residence of Philip Cleator. The Society's first official meeting took place in a suite of offices in Dale Street, Liverpool, on 13 October 1933. From the beginning the BIS published a magazine called *Journal of the British Interplanetary Society*, which was known to its recipients by its initials *JBIS*. The magazine's purpose was to record the matters discussed at Society meetings and communicate those ideas to any members of the public who shared the members' beliefs in the future of space flight. In the same year the VfR was shut down in Berlin and von Braun began work for Dornberger at Kummersdorf.

There were numerous small space flight societies around Britain at the time. Cleator contacted a number of Europe's space flight societies and was partially responsible for Willy Ley's escape from Germany to America

In 1936 Dr Ralph Morris attempted to get government permission to carry out rocket research in the UK, and wrote to the Secretary of State seeking permission to develop and launch rockets. He was told in a letter signed by a Civil Servant, *'I am to add that the Secretary of State is advised that approval could not, in any event, be given for the design of rocket, or of liquid-oxygen-petrol filling, described by Dr. Ralph Morris on his visit to the Home Office on February 23rd last ...'*. The British government had banned the BIS from performing rocket experiments. Also in 1936, Cleator wrote a book entitled *Rockets Through Space*. One English reviewer wrote:

> Mr. Cleator thinks it is a pity that the Air Ministry evinced not the slightest interest in his idea; provided that an equal indifference is shown by other ministries elsewhere, we all ought to be proudly thankful.

Like Goddard in America twenty years earlier, the BIS were hindered by the shortsightedness of its country's government and the ignorance of its citizens. In Germany, work was beginning on the Peenemünde rocket research centre.

In the same year the BIS opened a London branch. The first London meeting took place on 27 October in the office of Professor A.M. Low, whom Cleator had recruited in 1934. With the Liverpool branch suffering from financial difficulties the London branch became the principal branch. In 1937 a new Constitution was inaugurated and the home of Ralph Smith, in South Chingford, became the society's new Headquarters. Low was elected BIS Chairman in 1937.

Smith was an artist and committed many of the BIS's hopes and dreams to canvas. In 1936 he painted a large rocket being launched from an artificial island in the middle of the ocean. Today Project Sea Launch does just that. Another picture shows the first stage of a large space launch vehicle deploying parachutes to facilitate its recovery and re-use, as now happens to the Space Shuttle's Solid Rocket Boosters.

Unable to develop and launch rockets, the BIS decided to concentrate its efforts on theoretical studies related to the future of space flight. In 1936 the BIS established a Technical Committee to '*provide scientific consideration and exposition of the subject of astronautics*'. The Committee's first study was of a Lunar Spaceship. The results of the study were published in the January and July 1939 issues of *JBIS*, but by that time the world was slipping into World War II. The January issue of *JBIS* began with the following summary of the period:

> Space travel is not a dream of the far future, you idealists! And none of the practical problems is insoluble, you technicians! A voyage to the Moon is possible at this moment. If the rest of the BIS members had worked as hard as certain members of it have, if but a fraction of the money thrown away on armaments had been devoted to this purpose, the lunar trip would be a historical fact by now. Man would be conquering new worlds instead of destroying his own.

At that time Germany was spending millions of Reichmarks to develop the A-4, the first practical large liquid propellant rocket.

The Lunar Spaceship used pre-World War II black powder rockets to power a technically feasible three-man flight to the lunar surface. Many of the procedures to be employed by the BIS Spaceship in 1939 were actually used during Project Apollo 30 years later. To read the report today shows that in many other ways it was the product of its time, highlighting the inadequacies of the contemporary technology.

As Britain lived up to the terms of her alliance with Poland and entered into World War II against Germany, the BIS challenged:

> That instinct to serve some cause which will outlast us is part of the make-up of normal man, and he seeks to satisfy it in various ways, through religion, art, patriotism, or social reform. The BIS has chosen exploration, to help in the work of pushing the boundaries of territories as far as we can, sheer across the universe if we can.
>
> This present civilisation may collapse, as several have before it, and as many may after it. But sooner or later Man will stand astride worlds and the part, however small, the BIS plays in achieving that end will have justified its existence.

As Europe and ultimately the world was engulfed in war the BIS stopped meeting. When World War II was over the BIS amalgamated with the Combined British

Astronautical Societies (CBAS), which was itself an amalgamation of two smaller British societies. The new Articles of Association were signed on 8 September 1945. The CBAS members brought with them a proposal for a re-usable high-altitude meteorological rocket. The project was worked on by many of the BIS members who had worked on the Lunar Spaceship report and then offered to London University. The University saw no use for the rocket and the project came to nothing. In America, captured A-4 missiles were being used for high-altitude atmospheric and astronomical research.

In the post-war years the Lunar Spaceship study was updated to take into account Germany's advances in rockets employing liquid propellants. In the same period R.A. Smith and H.E. Ross suggested that a stretched A-4 missile, minus its four aerodynamic fins, could be used to launch a one-man capsule along a suborbital trajectory to and latitude of 305 km. Following separation from the launch vehicle the capsule would be spun-up as it climbed to apogee. At the target altitude the pilot would stop the spin and experience a brief period of microgravity. The capsule would then be pulled back to Earth by gravitational attraction and descend under a parachute to a mid-ocean landing. Landing loads would be softened by a 'crumple skirt'. Ross and Smith called the project 'Megaroc' and offered it to the Ministry of Supply in 1946, where it was turned down. The same principle would be used in the earliest flights of Project Mercury, but in that case they would trace their line of descent back to the US Army's Man In Space Soonest proposal and not the British Megaroc proposal. (See Figs 6 and 7.) Smith and Ross also studied proposals for 'Orbital Bases' (space stations). Their orbital bases would be assembled in Earth orbit from prefabricated parts launched on numerous individual rockets. This is just how Mir was assembled and how the International Space Station is being constructed today.

One early BIS member was Arthur C. Clarke, who published the principles of using communication satellites in geosynchronous orbit in 1947. His 'long-distance relays' were large manned spacecraft. The year ended with the BIS Lunar Spaceship study being updated to a nuclear powered vehicle. This work was completed 20 years before the American government began work on Project NERVA.

Two years later Ross published a paper detailing the principles of orbital rendezvous and refuelling. The study included the proposal that a lunar spacecraft might leave the propellant and rocket motor required for return to Earth in lunar orbit while the lunar landing took place. The landing vehicle would then lift off from the surface and rendezvous with its propellant supply before returning to Earth. The paper was a plan for Lunar Orbit Rendezvous a decade before NASA engineer John Houbolt began his quest to have the principle adopted for Project Apollo.

Arthur C. Clarke wrote *Interplanetary Flight* in 1951 and followed it in 1952 with *The Exploration of Space*. Both books tried to explain how space flight would occur. In the latter title Clarke stated that space flight, '*must now be regarded as a matter beyond all serious doubts*'. In 1951 Arthur C. Clark, Alan Dixon and Anthony Kunesch wrote a paper entitled, 'Minimum Satellite Vehicle' for the BIS. The paper, which was presented to the Second International Congress in London in 1951, described the launch of a 'metalised' balloon that would be inflated following

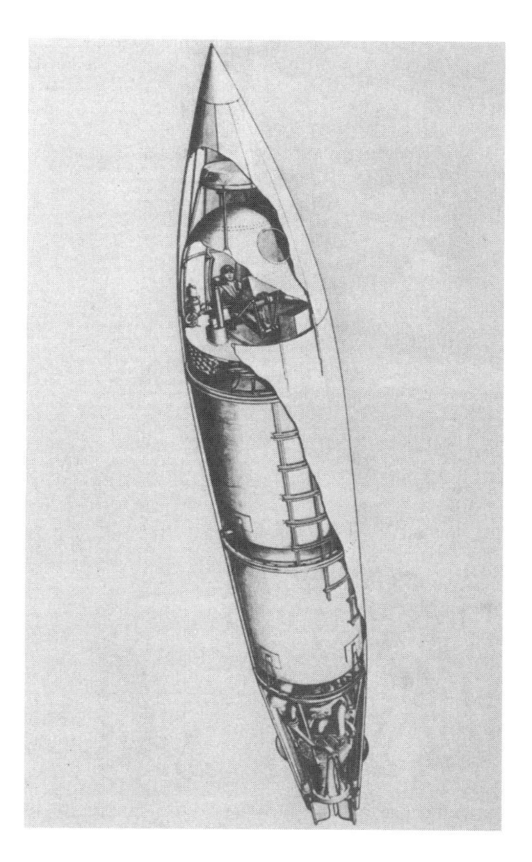

Fig. 6. British Interplanetary Society design for the 'Megaroc' manned rocket based on the German A-4. Note the lack of stabilisation fins at the base of the rocket. (BIS)

separation from the final stage of its launch vehicle. By concentrating on the smallest and simplest imaginable satellite the paper brought the prospect of orbital flight within the realms of technological and financial possibility. This paper would be used in America in the following years to promote the call to launch a small satellite. It also foresaw the Echo 1 communication satellite, which employed the techniques stated in the BIS paper.

Over the following decades the membership of the BIS have watched their dreams come true. Man entered space as they had predicted. Unmanned satellites and unmanned probes to the Moon, the planets and comets were all predicted in Smith's many paintings. He also foresaw re-usable manned spacecraft with wings that lifted off vertically and landed like a standard aircraft. Unlike many dreamers Smith even predicted the control room from where his imaginary space flights were controlled. His painting shows a disciplined room with banks of mainframe computers. He was almost right, only in Project Mercury the control room and the mainframe computers were hundreds of miles apart.

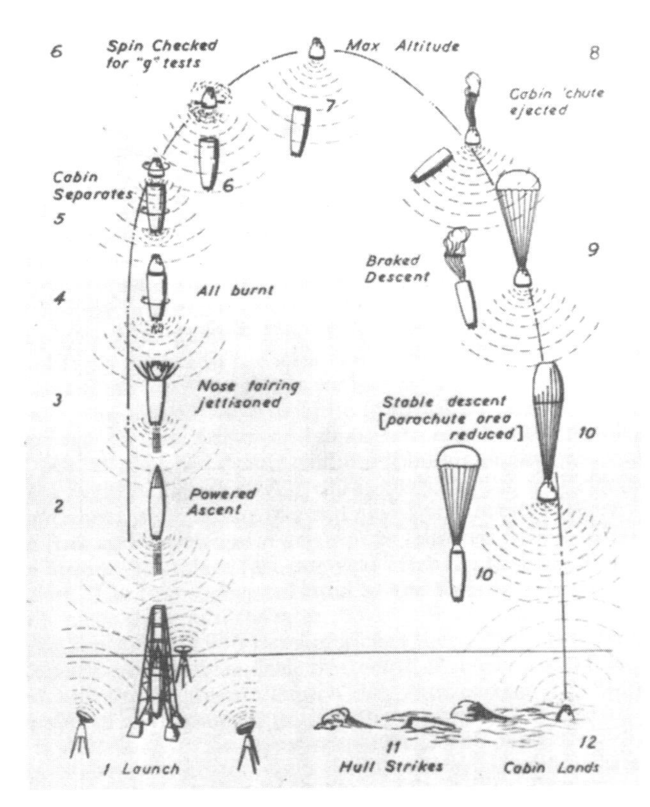

Fig. 7. Flight plan for the British Interplanetary Society's 'Megaroc' project. This project pre-dated the Project Adam and Mercury Redstone proposals in America by at least a decade. (BIS)

In the 1970s the BIS initiated a study into an interstellar space probe. 'Project Daedalus' was a technical study in the manner of the original Lunar Spaceship study in the 1930s.

Many of the BIS's members worked in the British missile programme, until Harold Wilson's Labour government cancelled it in the 1970s. Even so, the BIS continued to do all that it could to educate the public as to how mankind was exploring space and, more importantly, why they were doing so. In 1938 Ralph Smith wrote of the BIS:

> We have decided to concentrate our attention on the task of meriting a reputation for sound scientific work, by undertaking a survey of the whole problem, such as will attract the interest of the scientific world and commend the respect of the layman.

At the beginning of a new century the BIS is the oldest survivor of the European rocket and space flight societies of the 1920s and 1930s. With so many of their members' predictions now a reality and a part of history, the British Interplanetary Society is still proud to live up to Smith's words.

4.2 GERMAN PIONEERS

Hitler's National Socialist Party came to power in Germany in 1933. The German army closed down the country's amateur rocket societies, employing the young Wernher von Braun and beginning the research programme that would result in the development of the A-4 rocket. Willy Ley was a German amateur rocket engineer who immigrated to America in 1935. This was achieved with the assistance of the members of the American Rocket Society. Ley took up residence in New York and made his living writing books on Zoology. On his arrival he found that no one was interested in the possibility of space flight except the writers of science fiction comics, such as the members of the American Rocket Society.

In 1944, as von Braun's A-4 became operational, Ley began a writing career, producing numerous books championing the cause of space exploration, the first of which was *Rockets: The Future of Travel Beyond the Stratosphere*. In 1949 he wrote *Conquest of Space*, which Chesley Bonestell illustrated. The pictures were based on accurate astronomical observations of the Moon.

When World War II ended the German A-4 rocket engineers were brought to America and housed at Fort Bliss, New Mexico. While at Fort Bliss von Braun wrote *The Mars Project*, an attempt to prove that a manned flight to Mars was technically possible at that time. The book explained how a manned flight to Mars might be made, but the current understanding of the planet's atmosphere was incorrect and we now know that von Braun's proposed method of landing on the surface, in large gliders, would not be possible. The book was published in Germany and then America. Following the publication von Braun began a deliberate programme of lectures and writing to convince the American public that their nation should begin the exploration of space as soon as possible.

Hayden Planetarium, New York, organised three symposia to promote space flight as a serious subject. Ley organised the speakers for the first symposium, held on 12 October 1951. Von Braun, who was among the speakers at the second symposium held on 13 October 1952, called for a national space exploration programme, independent of the military missile programmes then underway. Arthur C. Clarke organised the third symposium, held on 4 May 1954, in which he detailed the history of the idea of space flight. R.C. Truax of the Navy Bureau of Aeronautics highlighted the fact that, until the public aligned itself behind the desire to explore space, there was no imperative for the government to commence that exploration.

On 22 March 1952 the magazine *Collier's* published the first in a series of eight articles written by von Braun, Ley and other supporters of space exploration, giving their ideas on how mankind might explore space. Chesley Bonestell illustrated the articles. Von Braun's plan called for the development of a large three-stage rocket with a piloted space plane at its apex. The winged space plane would be launched vertically, as the third stage of its launch vehicle, and land like a conventional aircraft at the end of its orbital mission.

Next in the line of development was a round space station in Earth orbit. The station would be constructed in orbit by large numbers of astronauts working in pressure suits in the vacuum of space, and once the space station was complete the engineers would begin the construction of three huge manned lunar spacecraft. These would be used for a single lunar landing. More details of this scheme can be found in Chapter 25, 'Alternative Paths', in this volume. The *Collier's* series was followed by a number of articles in America's numerous weekly magazines, many of which concentrated on the charismatic von Braun.

In 1952, von Braun, Joseph Kalan, Willy Ley, Heinz Huber and Fred Whipple cooperated on the writing of the book *Across the Space Frontier*. The book expanded on some of the ideas that von Braun had expressed in the *Collier's* magazine series. It proved popular and stimulated early interest in a manned space programme among the American public. A second book, *Man on the Moon* written by von Braun, Whipple and Ley and published in 1953, described a manned flight to the Moon.

On 25 June 1954 von Braun and other space flight pioneers in America backed a proposal to launch a 2.2-kg satellite using a Redstone missile with solid propellant Loki motors as an upper stage. This was the beginning of Project Orbiter and was largely based on Clarke, Dixon and Kunesch's 1951 BIS paper 'Minimum Satellite Vehicle'. The Army would provide the satellite and launch vehicle, while the Navy would provide tracking throughout the flight. In early 1955 the Navy would advance its own plans for an Earth satellite using a launch vehicle based on the Navy's Viking rocket.

In 1954 Walt Disney asked Ley to assist him in the preparation of a series of one-hour television programmes on the future of space flight and, in his turn, Ley recruited von Braun. Using models and cartoons, the three programmes explained how rockets would be used to launch von Braun's winged space plane into orbit, how the crew would be protected from the hostile environment and how the space plane would re-enter Earth's atmosphere and fly to a landing like a conventional aircraft. The first programme aired on 9 March 1955. Programme two concentrated on the construction of an Earth orbiting space station and a manned flight to the Moon and back. It was shown on 28 December 1955. A manned flight to Mars and the development of life in the Solar System were the subjects of the third programme, aired in late 1957. By that time von Braun had been challenged by the American government to place America's first artificial satellite into orbit.

Throughout the 1950s Ley, von Braun, Clarke, and numerous others continued to write books on the way that mankind would begin the exploration of space and how they would travel to the Moon, the planets and even the stars. Most of these books concentrated on von Braun's Grand Scheme – satellites, manned spacecraft, and a doughnut-shaped space station, a manned lunar landing, and then Mars. As the exploration of space became a reality, the books on general sale reflected the technology and advances associated with that exploration. Many of these early books contained explanations of the principles involved in an attempt to educate the reader and thereby pass on the dream of space flight to a new generation of believers.

4.3 THE SATELLITE RACE

Some private American companies as well as government research agencies and the military services had begun research into the possibility of 'a world-circling spaceship' – an artificial satellite – as early as 1945, but those studies came to nothing. It was only the mid-1950s that most people began to accept that the concept of an artificial satellite had passed from science fiction to technical feasibility.

In October 1945 the US Navy's Bureau of Aeronautics had suggested that their engineers develop new upper stages for a captured German A-4 missile and use the resulting vehicle to launch an artificial Earth satellite. As the study advanced it was decided that the Navy should develop a new high-altitude test vehicle (HATV) with a liquid-propelled first stage and solid propellant upper stages. In 1946 the USAAF refused to join the project when asked and, having learned that the Navy was considering such a project the USAAF began studies of its own satellite project. Both services continued their own satellite studies until 1948, when the USAAF satellite was refused funding. The Navy's project died shortly afterwards, also due to lack of funds.

With little or no military information on the interior of the Soviet Union, the possibility of a reconnaissance satellite caught the imagination of some in the USAAF. Cameras had been floated across the Soviet Union beneath small balloons in the late 1940s. Few were recovered and even then the cameras' timing equipment frequently malfunctioned. Many of the photographs that were obtained proved unusable as their subjects could not be identified, or located accurately on a map.

The Lockheed Company developed the U-2 reconnaissance aircraft to fly over the Soviet Union, beginning in the 1950s. The aircraft were controlled by the Central Intelligence Agency (CIA) and flown to predetermined targets, or the pilot could photograph targets of opportunity as they presented themselves. In 1960 a U-2 spy plane was shot down over the Soviet Union and President Eisenhower banned Soviet over-flights.

To many in the military services and the universities and laboratories that supported them, a reconnaissance satellite was the answer to the surveillance problem. Work began on defining such a satellite and this led to Project Corona, a secret USAF/CIA programme to build a photographic reconnaissance satellite. Corona was designed to record images of the Soviet Union from orbit and return them to Earth in a detachable re-entry capsule at the end of its mission. To prevent the capsule falling into the wrong hands the USAF would recover it in mid-air as it made its final descent towards the ground by parachute. Early Corona satellites flew under the cover-name Discoverer and some non-sensitive Earth photographs were released for public consumption.

Reconnaissance from orbit presented the intelligence services with one difficulty – the right to over-fly another country without violating its national airspace. If the Soviet Union argued that its airspace extended beyond Earth's atmosphere and into space, and won recognition for that argument in the United Nations, then they would have the right to destroy any American satellite that violated that airspace.

With the announcement of the International Geophysical Year (IGY), which actually ran for 18 months in 1957/8, Earth became the centre of an unprecedented international scientific investigation programme. On 29 July 1955 the Eisenhower Administration announced that America would launch a scientific satellite into orbit around Earth as part of their nation's IGY programme. On 2 August the Soviet Union announced that they too would launch a satellite during IGY. Few people in the West thought the Soviets were capable of launching a satellite at all, let alone before the Americans.

President Eisenhower did not want America's first satellite to be launched by a military missile and set up a committee to choose which, if any, of the military proposals should become the country's first civilian satellite programme. On 3 August 1955 the committee chose the Navy proposal based on the Viking Rocket. The Naval Research Laboratory, which had been in charge of the scientific payloads carried on the captured A-4 missiles launched out of WSPG, was placed in command of the satellite project. Any development difficulties with the satellite launch vehicle would not be allowed to interfere with the vital ICBM and IRBM development programmes that would have priority at all times. To that effect a DoD committee decreed that the Navy's Viking rocket would be up-rated with an Aerobee-Hi second stage and a new solid propellant third stage.

Named 'Vanguard' on 16 September 1956, the project was scheduled to launch a 1.8-kg satellite. Four or five flight-tests of the new launch vehicle would be made, some with 'test satellites', before the official satellite was launched in March 1958. The scientific satellite effort would be an open programme, with media reporting positively encouraged. When Vanguard was launched it would over-fly the Soviet Union. Depending how the Soviets reacted to that fact would either place the question of over-flight before the UN, or put it to rest.

In 1954 the von Braun team at Huntsville had been invited by the Navy to provide a Redstone missile with a combination of small solid propellant upper stages for Project Orbiter, which was intended to launch an Earth satellite weighing between 2.3 and 3.2 kg. Like earlier satellite programmes, Project Orbiter died for lack of funds. Two years later the Army argued that they could have a satellite in orbit up to a year earlier than Project Vanguard if they were allowed to proceed with Project Orbiter. Their protests were acknowledged but not acted upon, except in entirely negative ways. The Jupiter C missiles used in the warhead re-entry test programme were ballasted with sand to prevent the Army 'accidentally' placing a warhead test vehicle into orbit. Following the completion of the third re-entry test the Army was instructed to ensure that there were no live upper stages stored at Cape Canaveral, thereby preventing them being used to launch an unofficial satellite.

The first Vanguard associated launch took place from Cape Canaveral on 8 December 1956. Only the Viking first stage of the Test Vehicle 0 (TV-0) launch vehicle was live. A second launch, TV-1, including a prototype of the new Vanguard third stage, was launched on 1 May 1957.

Two weeks later the Soviet Union launched the first R-7 ICBM from a supposedly secret new launch site that would soon become known to the world as Baikonur. In fact, U-2 reconnaissance aircraft had photographed the construction of Baikonur from the beginning. The huge R-7 ICBM was in five parts. A central core housed a single four-chamber rocket motor with four vernier steering motors. Four strap-on boosters were mounted around the central core, each with a four-chamber rocket motor and two vernier steering rockets. All five motors were ignited at launch and burned in unison for 2 minutes, at which time the four boosters were shut down and jettisoned. The central core continued to fire for a further 3 minutes before it too shut down. At that time the warhead separated and followed a ballistic trajectory through the atmosphere to its target. The first flight-test failed after just 50 seconds and the missile was destroyed. Seven more flight-tests followed with the last missile throwing its dummy warhead 6,437 km on 3 August 1957. The Soviet Union followed that flight by announcing that they had an operational ICBM capable of hitting a target 'in any part of the world'.

4.4 SPUTNIK

Like the Americans, the Soviets had begun their quest for a ballistic missile in the years directly after the end of what they call The Great Patriotic War (1939–45). When Soviet engineers failed to develop miniaturised electronics the design of the R-7 ICBM was frozen at an early stage, while the Soviet nuclear warhead was still very large and heavy. As a result, when the R-7 was employed as a space launch vehicle, it was able to carry far more massive payloads than its American counterparts, which had been designed to carry smaller and lighter warheads. It was this imbalance in ICBM lift-capability that would allow the Soviets to place much larger satellites and spacecraft into orbit than the Americans in the first decade of space exploration.

The Soviet space programme was controlled by the military. Everything was designed and manufactured in government-run bureaux and delivered to Baikonur Cosmodrome, Kazakhstan, for final preparation and launch. Sergei Korolev's design bureau (OKB-1) developed the R-7 ICBM and then petitioned Khrushchev's government for permission to launch a satellite. When the development of the proposed scientific satellite fell behind schedule, OKB-1 developed the object that the Soviets called *Sputnik*. OKB-1 was charged with designing and launching a series of interplanetary probes and a vehicle capable of carrying a human passenger into orbit.

On 4 October 1957 an R-7 ICBM carried Sputnik into orbit. As it separated from the launch vehicle's central core, a plug disconnected and the satellite began broadcasting its bleeping message to receivers around the planet. The Soviet Union had beaten America in the race to launch the first artificial satellite. The Soviet engineers who had participated in the launch became national heroes, but they were not identified to the public. Khrushchev instructed the engineers of OKB-1 to commence the development of an unmanned reconnaissance satellite with a recoverable film capsule. The Zenit satellite was so large that in time it would be converted into the Soviet Union's first manned spacecraft, Vostok, but in October 1957 Vostok still lay in the future.

With the launch of Sputnik, the Soviets had given Eisenhower the one thing he wanted more that any other from the satellite programme. Sputnik flew directly over

America on each orbit and the Americans did not complain. A Soviet satellite had set the precedent of free over-flight. The way was now clear for Project Corona.

4.5 AMERICA'S REACTION TO SPUTNIK

Following the launch of Sputnik the American military quoted the satellite's 83.6-kg weight as a gauge of the R-7's lift capability, which was obviously far beyond that of any missile then under development in their country. The myth of the large difference between Soviet and American missile capability was thus born. The American military used the perceived 'missile gap' as an excuse to call for more funds for missile development. The media used the missile gap to call for a manned space programme. Finally, the Democratic Party would use the missile gap to prise the Republicans out of the White House. In time, photographs from the Corona satellites would prove that the Soviets had far fewer long-range bombers and even less ICBMs than American intelligence had previously suggested, but in 1957 Corona was still under development. President Eisenhower called for a 'Manhattan District' style approach to be set up to control the nation's military missile programme and ensure that the ICBMs and IRBMs then under development reached operational status as soon as possible

At Redstone Arsenal on 4 October 1957 the Army was asked how long it would take them to prepare a satellite for launch. They quoted a figure of 90 days. On 8 October President Eisenhower agreed to the Army preparing a satellite launch as a back-up, in case Project Vanguard failed. The President passed an additional $4 million funding for the project on 30 October, stating that he had always supported Project Orbiter, but the DoD had advised him to keep the development of the ICBM and IRBM separate from the development of a satellite launch vehicle. The Army's launch vehicle was a Jupiter C with solid propellant upper stages, a combination that the engineers at Redstone Arsenal called Juno 1. A small satellite, Explorer, was rapidly put together at JPL, Pasadena. America was already reacting to Sputnik by speeding up its own satellite programme.

Also on 8 October an internal USAF committee suggested that the service should commence an upgraded programme of launch vehicle and satellite development. The satellites would record reconnaissance, meteorology and communications. Another USAF committee, under the leadership of nuclear physicist Edward Teller, was charged with defining the USAF's role in space. Teller's committee recommended a national space programme under the command of the USAF.

President Eisenhower attended a press conference on 9 October to voice the Administration's official reaction to the Soviet satellite. The President stated that America would continue with Project Vanguard as a civil scientific satellite programme, divorced from military missile development. The programme would not be speeded up in reaction to Sputnik. He did not mention Project Corona, or free over-flight. Nor did he say that the Army had his approval to launch a satellite within 90 days.

Ten days after the Sputnik launch, members of the American Rocket Society suggested that a civilian Space Agency be established to lead America's efforts in space exploration. On 21 November 1957 NACA would also make an appeal for a consolidated National Space Establishment.

As November ended, the Army's engineers at the ABMA were commencing studies into the development of a new heavy-lift missile/space launch vehicle with a lift-off thrust of 6,672,000 N. The original plans for the Juno V called for a cluster of four Rocketdyne E-1 rocket motors. This vehicle would ultimately be developed as the Saturn I.

With the launch of Sputnik the President established a Presidential Committee and gave it two tasks. First, it was to write an easy to read document for the American public explaining how and why America should explore space. Second, it was to look into the feasibility of establishing a national space science programme. This in its turn led the Committee to suggest that a new civil Space Agency be established using NACA as its basis.

In December Project Vanguard's third launch was the first with a live upper stage. It would carry a 1.8-kg 'test satellite'. The test satellite was promptly billed as America's first satellite and an answer to Sputnik. The launch, from the Atlantic Missile Range (AMR), Florida, was shown live on American television channels. On 6 December 1957 the Vanguard burst in to life at the end of its countdown and climbed ponderously off the launch pad. When the first-stage rocket motor lost thrust the whole nation watched in horror as the vehicle fell back on the launch pad, toppled to one side and exploded. The satellite was thrown clear and was later put on public display in Washington. (See Fig. 8.)

President Eisenhower was determined not to let Sputnik interfere with his established economic plans for the country, but he also refused to spend blindly on military/space projects. Rather, he took advice and concentrated his spending on science and education.

On 3 November 1957 the Soviets launched Sputnik II. The launch phase went well, but the world's second satellite failed to separate from the second stage of its launch vehicle. The failure to separate meant that cooling radiators on Sputnik II could not be deployed and, therefore, could not radiate the heat produced within the satellite to the vacuum of space. The problem would have been a minor technical difficulty had Sputnik II not been carrying a smooth-hair terrier bitch named Laika, the first living creature to orbit Earth. Laika was housed in a pressurised container, which maintained her in an artificial environment. There were no plans to return her to Earth, but the failure of the radiators to deploy meant that she suffered terribly as she died slowly from heat exhaustion after a few days. Sputnik II re-entered the atmosphere and burned up on 14 April 1957. It was now clear to Western observers that Soviet space research was heading towards the goal of manned space flight.

With some animal groups in the West protesting against Laika's use in suicidal space experiments the Soviets lied, blatantly. They made up a story that told how Laika had performed well for eight days, throughout which telemetry had transmitted her conditions to the ground. They claimed that she had been put to sleep, painlessly, when she consumed her final, drugged, food pellet. For the next 40 years the Soviets simply forgot to tell the world how Laika had really died.

Fig. 8. The third launch in America's Vanguard satellite programme ends in public humiliation as the launch vehicle loses thrust, falls back on the launch pad and explodes. (NASA)

Inside NACA the staff at the Ames field centre were insisting that the organisation retain its present role by supporting the military as they extended their role in to the exploration of space. On the other hand, the engineers at Langley and its satellite station, PARD, on Wallops Island, wanted to become directly involved in the first-hand development of space vehicles, including a manned satellite. Between 18 and 20 November 1957 NACA's Main Committee met on board the aircraft carrier USS *Forestall* and decided their organisation should '... *Act now to avoid being ruled out of the field of space flight research*'. Despite this decision, the Main Committee encouraged the military services to continue to seek NACA's assistance in developing their own space vehicles.

On 5 December 1957 the DoD brought the Advanced Research Projects Agency (ARPA) into being. The new organisation would oversee and manage all existing military space and missile programmes until the government decided how America might best explore space. ARPA was formally established on 7 February 1958.

Following the launch of Sputnik II the USAF Air Research Development Command (ARDC) prepared a five-year plan and presented it to USAF

Headquarters, Washington, on 13 December. The plan was passed to the Pentagon.

On 31 January 1958 the television cameras were once again present at Cape Canaveral when von Braun's Juno 1 launch vehicle placed Explorer, America's first satellite, into orbit (see Fig. 9). The 14-kg satellite carried Geiger counters in an experiment put together by James Van Allen. When the experiment discovered two vast radiation belts surrounding Earth, the belts were named after the instrument's principal investigator.

On 31 January 1958, the day Explorer 1 was launched, the USAF invited NACA to participate in the development of the X-20 Dyna-Soar winged spacecraft, and on a simple ballistic spacecraft. On accepting both invitations NACA Director Hugh L. Dryden explained to the USAF representatives that NACA Langley field centre satellite, the PARD, was carrying out its own research into the ballistic spacecraft approach to manned space flight. The conical spacecraft would be launched on a 'bare' Atlas ICBM in place of its nuclear warhead. Its shape meant that the spacecraft was streamlined for exit through the atmosphere, while its wide back end allowed the application of Allen's low-lift high-drag re-entry principle. Over the following 12 months NACA spacecraft would undergo three configuration changes.

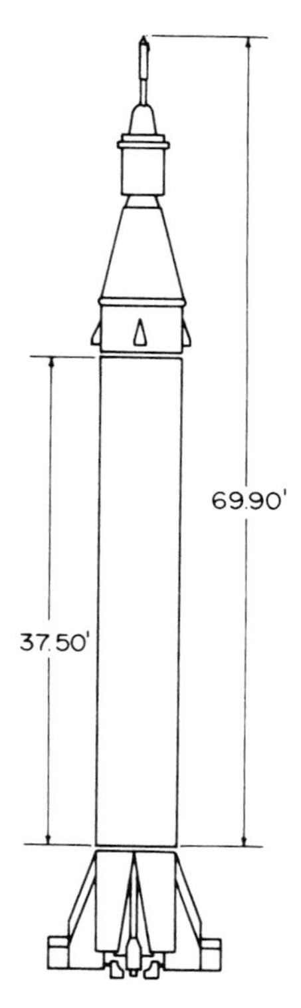

Fig. 9. Redstone Arsenal's Juno 1 launch vehicle as used to launch Explorer 1, America's first satellite. (NASA)

In the same month the USAF hosted a conference during which all manner of possible manned space vehicles were discussed. The range went from the complicated X-20 Dyna-Soar proposed by the Air Force to a number of individual aerospace companies that had already spent many man hours defining relatively simple ballistic spacecraft.

Inside NACA, Project High-Ride was approved for development. The small rocket, proposed by two PARD engineers, grouped seven Sergeant solid rocket motors in an aerodynamic casing. The vehicle would be used to launch the PARD's manned ballistic spacecraft in suborbital trajectories. The name High Ride would later be dropped, as would the plans to use it to launch manned suborbital flights. Ultimately using two different combinations of solid rocket motors and operating under the name Little Joe, the new rocket would prove vital to Project Mercury.

A Soviet satellite launch attempt failed on 3 February 1958, before Sputnik III was launched on 15 May 1958. The orbiting science station was to have been the first Soviet satellite, but its construction fell behind schedule and the first Sputnik was launched. One can only imagine how the American population would have reacted had the first Soviet satellite weighed 1,327 kg as Sputnik III did.

5

Defining the frontier

In 1915 President Woodrow Wilson established NACA with a single field station at Langley, Virginia. A 12-man Main Committee oversaw the work of the new organisation, which would carry out cutting-edge research into aeronautical engineering at the request of the US military. Throughout the 1920s and 1930s, NACA studied aerodynamics, developed innovative aeronautical technology and helped to flight-test numerous new military aircraft. The organisation developed a reputation second to none in the aeronautical industry.

As Europe commenced World War II in 1939, the British Royal Air Force (RAF) was quick to realise the vulnerability of their bombers in daylight raids against German occupied Europe. The RAF turned its attention to perfecting the more difficult role of night bombing. The American military used European advances in military aeronautics to draw improved budgets from Washington. In 1940 NACA opened the Ames Aeronautical Laboratory at Moffett Field, California. The Lewis Flight Propulsion Laboratory, Cleveland, Ohio, was the third NACA field station to commence work in 1942. Engineers at Lewis began work on jet propulsion when news reached America that both the Germans and British had already done so.

When America entered the war, following the Japanese raid on Pearl Harbour in 1941, President Roosevelt agreed to a 'Europe first' policy. At the same time NACA established three flight regimes – subsonic, transonic and hypersonic: the first was any flight regime up to that approaching Mach 1; the second progressed through Mach 1 to Mach 5; and the third was considered to be anything above Mach 5. The last two both represented flight regimes where aerodynamic heating on the aircraft or missile was of major importance.

In 1944 the Allies went on the offensive in Europe. At this time, NACA began to react to the British intelligence that the German Army had developed a practical ballistic missile. Between 1944 and 1946 the organisation turned some of its attention to hypersonic flight and the problems presented by the development of ballistic missiles. The Special Flying Weapons Team was set up to study the development and use of missiles. The Supersonics Branch was established to study the airflow problems associated with the flight of an object travelling through the atmosphere at supersonic velocity. A new Auxiliary Flight Research Station, a satellite of the

Langley Aeronautical Laboratory, was established on Wallops Island. The new site became the home of the Pilotless Aircraft Research Division (PARD). Robert Gilruth, a highly respected Langley engineer, was named as head of the new field centre. Gilruth's engineers launched scale models of new aircraft on solid propellant rockets to obtain the data they required.

NACA was not the only organisation carrying out advanced aeronautical research during World War II. In both Germany and Britain experiments were taking place with jet propulsion. Only Germany perfected the technology and produced an operational jet fighter before the war ended. Germany also applied rocket technology to fighter aircraft and even bought one such fighter into operation.

5.1 MESSERSCHMITT Me-163

The German aircraft manufacturer Messerschmitt developed a rocket-propelled, single-seat fighter aircraft, which the Luftwaffe brought to bear against the formations of Allied bombers flying against the German homeland. The Messerschmitt 163 (Me-163) 'Komet' used a single-chamber, liquid propellant, Walter rocket motor and flew for the first time in August 1941. At that time it achieved a speed of 965.4 km/h, approaching twice the average speed of the propeller-driven fighter aircraft of the time. Over the following three years the aircraft was perfected and entered service as an operational fighter in 1944. Like the A-4 missile it was produced too late and in too few numbers to affect the outcome of the war.

The Komet ignited its rocket motor on the ground, using a contemporary grass military airfield for take-off, and as the aircraft left the ground it dropped its undercarriage. The Komet was only able to remain in the combat zone for approximately 25 minutes and had then to return to the ground before it lost speed, and with it its airworthiness. With no undercarriage, the pilot had to perform a dangerous 'belly flop' landing, bringing the bottom of the aircraft's fuselage into contact with the grass of the airfield each time it landed. At the end of World War II examples of the Me-163 were returned to America.

5.2 NATTER

The Natter (Viper) was a measure of the desperation that the German nation felt in the wake of the round the clock bombing of their nation. World War II ended before Natter became operational. The Natter was a wooden aircraft designed to take off vertically under the thrust of its four solid propellant rocket motors, which burned for 12 seconds. The pilot lay in his seat, in the prone position. At launch the Natter was under the control of an autopilot because the human pilot was expected to experience 2.2*g* and blackout while the solid rocket motors were burning. When the take-off stage fell away the liquid propellant Sustainer motor began an 80-second burn. While it was burning the pilot was expected to experience 0.7*g*. Having regained consciousness, the pilot would take manual control of the Natter, which would continue to climb to an altitude of approximately 12,000 m, at which point the Sustainer motor would shut down. The pilot would then put the Natter into a controlled dive, firing his 20 missiles as he passed through the formation of Allied bombers beneath him. At an altitude of 304 m the pilot would leave the Natter and return to Earth under his personal parachute. The aircraft would crash to the ground, where anything that could be salvaged would be re-used on future Natter aircraft.

The project came under the control of the SS and unmanned vertical flight-tests began in October 1944. Fifteen Mark 1 and four Mark 2 models were flown before the SS demanded a manned flight-test. In February 1945, Oberleutnant Lothar Siebert made the only manned Natter flight. He died when the cockpit cover fell off, rendering him unconscious. With no pilot control at the apex of its trajectory the unpowered aircraft dived into the ground and was destroyed. The project returned to unmanned testing and had not moved beyond that point when World War II ended. American ground forces captured a few examples of the Natter airframe, but the concept was not followed up.

5.3 A-9

In an attempt to extend the operational range of the A-4 missile, the Peenemünde engineers fitted short stubby wings to their missile and designated the new model the A-4B. The intention behind the design was that the A-4B would follow a skip-glide trajectory across the upper atmosphere and hit a target 595 km from its launch site. Flight-tests of the A-4B were carried out at Peenemünde, and met with limited success.

In 1941–42 the Peenemünde engineers took the project one stage further and designed the dart-like A-9, a second winged A-4 model, as a second stage for a huge A-10 missile. The A-10 was to produce 181,500 kg of thrust at launch. When its propellants had been consumed, it would be jettisoned and the A-9's rocket motor would be ignited. The A-10/A-9's target was continental North America, but it could only have been reached if the missile had been launched from occupied France.

Documents captured by the Americans at the end of World War II showed that the German engineers even had plans for a piloted A-4B, which included a tricycle landing gear to allow landing at any medium sized airfield. Wernher von Braun speculated that the replacement of the A-10 with an even more powerful rocket might have allowed a piloted A-9 to be placed into orbit around Earth. Return to the ground would have depended on the vehicle being turned around and its rocket motor restarted, to provide a retrograde thrust to slow it down and allow gravitational attraction to pull it out of orbit. It would also require the vehicle being protected against re-entry through Earth's atmosphere.

5.4 ANTIPODAL BOMBER

Eugen Sanger graduated as an aeronautical engineer in 1929. His work in aerodynamics led him to the conclusion that, where the standard propeller-driven aircraft lost performance as it entered the thin air in the upper atmosphere, a rocket-propelled aircraft would have no such difficulty. In 1933 he published *The Technique of Rocket Flight*, which summarised his research of the subject of rocket propulsion.

With no government assistance Sanger developed his own liquid-propelled rocket motor and tested it at the Vienna Technical School, where he was performing post-graduate work. He continued to develop and test a series of liquid propellant rocket motors until complaints from his neighbours forced him to stop testing in 1936.

By 1932 he had already laid down designs for rocket-propelled aircraft that might be used by the German Air Force to test his theories. These designs were the forerunners of the Me-163 Komet and the American rocket aircraft that would be flight-tested in the decades after World War II. In the same year he defined a rocket aircraft that would be launched vertically and climb rapidly into the ionosphere. With its propulsion unit shut down the aircraft would then glide halfway around the world with centrifugal force supplementing the small amount of lift produced by the wings. Although never developed, this concept would become known as the 'antipodal bomber'.

In 1934 Sanger was invited to establish the 'Aircraft Examination Point' at Trauen, a Luftwaffe rocket development establishment. The new facility was completed in 1937 and its engineers studied how rocket-propelled aircraft would perform at extreme altitudes. Sanger became a German citizen in order to head the new centre, where he employed the outstanding mathematician Irene Brendt. They developed the antipodal bomber concept, on paper. Their plans called for a manned bomber with wedge-shaped wings and a flat underside. The bomber would be launched from a monorail, propelled by a captive booster that would not fly into space. At launch the rocket-propelled captive booster would produce 600 tonnes of thrust to push the antipodal bomber along the 2.9-km monorail track, which was almost vertical at the far end. The antipodal bomber would leave the track at Mach 1.5 and climb to an altitude of 1,700 m. At that altitude the bomber's own rocket motor would be ignited to boost it into the ionosphere. From there it would pass over its target and drop its bomb. It would then skip across the upper atmosphere in order to pass halfway around the world before gliding to a horizontal landing at a standard airfield.

Post-war NACA research on the antipodal bomber concept showed that the proposed low angle of attack as the aircraft passed back through the atmosphere led to the build up of extreme heating levels on the airframe.

5.5 BOMI

When he was freed from prison in Britain, Walter Dornberger, the German Army officer who had led the A-4 rocket programme at Peenemünde, made his way to

America and took employment with the Bell Aircraft Company. At that time Bell was already working on a rocket-propelled aircraft designed for supersonic flight. Dornberger did not want Peenemünde's work with winged missiles – and the possibility of manned space flight – to be abandoned. Enlarging on Sanger's wartime antipodal bomber proposal he developed Bomi, a manned rocket-propelled hypersonic vehicle.

Bomi was a two-stage vehicle with five rocket motors at the base of its winged booster stage and three in its smaller space plane. The space plane was carried mounted on the fuselage of the booster, with its own wings fitting between the two horizontal tail-planes of the booster stage. At launch all eight rocket motors were ignited and burned propellants from tanks in the booster stage to lift the combination vertically off of the launch pad. Two minutes 10 seconds after lift-off, the five booster rockets shut down and that stage was separated. Its crew flew it back to a horizontal landing at a designated airfield. The three rocket motors in the space plane now burned propellants from tanks within that vehicle and climbed to an altitude of 44 km, at which point the rocket motors shut down. A gliding trajectory then allowed the space plane to cross North America in an estimated 75 minutes before landing horizontally at a prepared landing strip. Once again NACA research proved that the space plane would be exposed to an extreme heating regime as it passed back through the atmosphere. As with the antipodal bomber, this was due to the airframe's low angle of attack as it passed through the thickening air.

5.6 BELL X-1

In 1943 the US Army Air Force (USAAF) began development of a rocket-powered aircraft, in an attempt to fly faster than Mach 1. This speed was seen as a 'sound barrier' because the air in front of the aircraft no longer had time to move aside and let the aircraft pass through, as it did at subsonic speeds. The X-1 was designed to pass through the 'barrier' of compressed air in front of it. Like the A-4 rocket, the Experimental Sonic 1 (XS-1, later shortened to X-1) was based on the shape of a bullet, with short straight wings. The aircraft was developed by Bell, with full cooperation of NACA, and was powered by a four-chamber XLR-11 rocket motor developed by Reaction Motor Incorporated. (See Fig. 10.)

Due to the short burn time of its rocket motor; the X-1 was carried into the air beneath the fuselage of a converted B-29 bomber. Flight-testing, without igniting the rocket motor, took place in Florida, before moving to the new NACA High-Speed Research Centre at Muroc Army Air Base, California. The NACA facility at Muroc had been established under the charge of the highly experienced NACA engineer Walter Williams.

On 14 October 1949, following a series of flight-tests, the Bell X-1 was dropped from its B-29 mother ship. Army pilot 'Chuck' Yeager ignited the rocket engine and flew his aircraft to a speed of 1,122 km/h at an altitude of 13,106 m. The speedometer on the instrument panel only went up to Mach 1 and the needle went off the scale. Yeager had broken the 'sound barrier', but in the atmosphere of the Cold War his

Fig. 10. The Bell X-1 rocket-propelled aircraft sits on the apron at Muroc AFB alongside its mother ship. The Bell X-1 was carried aloft beneath the fuselage of the mother ship and air-launched, after which the smaller aircraft's rocket motor was ignited in flight. The Bell X-1 was the first aircraft to pass Mach 1, on 14 October 1949. (NASA)

achievement, and the entire X-1 programme, was classified. Three different X-1 aircraft made 156 flights during a programme that ran from 1946 to 1951.

In January 1952 the designer of the Bell X-1, Robert Woods, wrote to NACA suggesting that they commence research into the problems presented by hypersonic and space flight. On 24 June 1952 the NACA Main Committee resolved that the organisation should,

> ... Devote a modest effort to problems associated with unmanned and manned flight at altitudes from 50 miles [80.45 km] to infinity and speeds from Mach 1 to the velocity of escape from the Earth's gravity.

5.7 DOUGLAS D-558-II

In 1950 the US Navy began developing the Douglas D-558-II 'Skyrocket', its own air-launched rocket-propelled aircraft designed to push the frontiers of piloted flight to Mach 2. The Skyrocket was powered by a Reaction Motors Incorporated LR8RM6 rocket motor. The first Skyrocket carried a combination of jet and rocket engines. The jet and two rocket chambers were lit at take-off from the runway at Muroc. The rockets were then shut down and the aircraft climbed to altitude on the jet engine alone. Finally, the four rocket chambers were fired to push the aircraft through Mach 1.

A second jet/rocket aircraft was designed to be air-launched from beneath a second converted B-29 bomber. Finally, an all-rocket aircraft was constructed and used for an attempt to fly at Mach 2. The skyrocket flight-test programme began in January 1951 and by April 1951 the aircraft had achieved a velocity of Mach 1.88. Despite repeated attempts Douglas test pilot William Bridgeman was unable to achieve Mach 2. When the all-rocket aircraft was handed over to NACA, test pilot Scott Crossfield made further attempts to fly the Skyrocket to Mach 2 and achieved the goal on 20 November 1953.

5.8 BELL X-1A

The Bell X-1A was 1.5 m longer than the X-1, with the additional fuselage space being devoted to carrying more rocket propellant in an attempt to fly the aircraft to Mach 2. Flight-tests began in early 1952. A few days after Crossfield had reached Mach 2 in the Douglas Skyrocket Yeager attempted to exceed his record in the X-1A. He pushed the X-1A past its design limit, to Mach 2.5. At that point it had undergone uncontrollable lateral oscillations and entered an uncontrolled dive, losing 17 km of altitude in 1 minute. Back in the thick, lower atmosphere and flying at subsonic speed, the aircraft entered a flat spin and Yeager was able to regain control and land safely. The supersonic flight regime had presented new difficulties to which NACA addressed itself in order to maintain America's lead in aeronautics. Wind tunnel tests suggested that the aircraft's thin stabilising surfaces lost the ability to produce lift in the thin upper atmosphere.

The X-1A made 21 flights before it was destroyed in August 1955. On that occasion the aircraft's propellant tanks exploded while still attached to the B-29. Despite test pilot Joe Walker's courageous attempts to save the aircraft, it was finally jettisoned and dived, unmanned, to destruction.

5.9 BELL X-1B

In October 1954 the X-1B came on line and proceeded to take aeronautics to altitudes where aerodynamic surfaces no longer functioned in the rarefied air and all manoeuvring had to be achieved by means of a reaction control system – rocket thrusters.

5.10 BELL X-1C, X-1D AND X-1E

The X-1C did not pass the design stage and fire destroyed the X-1D after only one unpowered flight. The X-1E was a successful aircraft but, like its predecessors, was limited to velocities up to Mach 2.

5.11 BELL X-2

The Bell X-2 was late being delivered to Muroc, but in the mid-1950s it was the only aircraft that offered the capability of extending NACA's flight-research programme to velocities up to Mach 3. Following a trouble plagued development the X-2 began flight-tests in October 1952 and commenced the quest for Mach 3 in 1953. The first aircraft exploded while still beneath its B-50 mother ship. The numerous difficulties experienced during the three flights of a second aircraft in 1954 placed the X-2 programme in danger of cancellation. Finally, Captain Melburn Apt, a USAF test pilot, reached Mach 3.5 on 27 September 1956. When Apt shut the rocket motor down the aircraft yawed violently and then pitched down sharply. He was killed when his aircraft struck the desert floor. The X-2 programme came to an abrupt end, but the lessons to be learned from Apt's first flight in the X-2 would make its successor more successful. The crash was caused by the same problems that were experienced during Yeager's December 1953 flight in the Bell X-1A.

5.12 'THE LEAP OF A FISH OUT OF WATER'

In 1952, Langley's David Stone suggested that the Bell X-2 be fitted with two solid rocket motors, one mounted under each wing, and a reaction control system to complement its standard aerodynamic surfaces. Stone argued that such an aircraft could be boosted to orbital velocity following an air launch from beneath the wing of a B-52. The concept was subjected to further study, but was dropped in February 1954.

In 1954 NACA established a Hypersonic Aircraft Study Group at Langley. The group was tasked with studying the possibility of developing a hypersonic aircraft that could boost itself high enough in the atmosphere that it would be subjected to a space-like environment for a short period, at the peak of its trajectory. Hugh L. Dryden would later refer to the flight programme for the hypersonic aircraft as being *'similar to the leap of a fish out of water'*.

The group was originally set up to conceive an aircraft that would allow microgravity to be studied in a space environment. They quickly realised that attitude control outside the atmosphere, and the transition from airless flight to atmospheric flight, were equally important and required more study.

Research showed that re-entering the atmosphere at Mach 7 with a low angle of attack led to the aircraft burning up. Additional research showed that increasing the angle of attack meant that deceleration started higher in the atmosphere and heat formed in a bow wave in front of the aircraft. This meant that the aircraft would already be travelling considerably slower when it encountered the thick lower atmosphere. This was the application of Allen and Egger's re-entry principles to a winged aircraft. Wind tunnel testing showed an 11- to 26-degree nose high attitude to be the best re-entry position for a hypersonic aircraft returning from its 'space leap'.

On 19 July 1954 NACA agreed to develop a hypersonic rocket aircraft capable of approaching Mach 6 at an altitude of 76,200 m. The X-15 aircraft was to be jointly

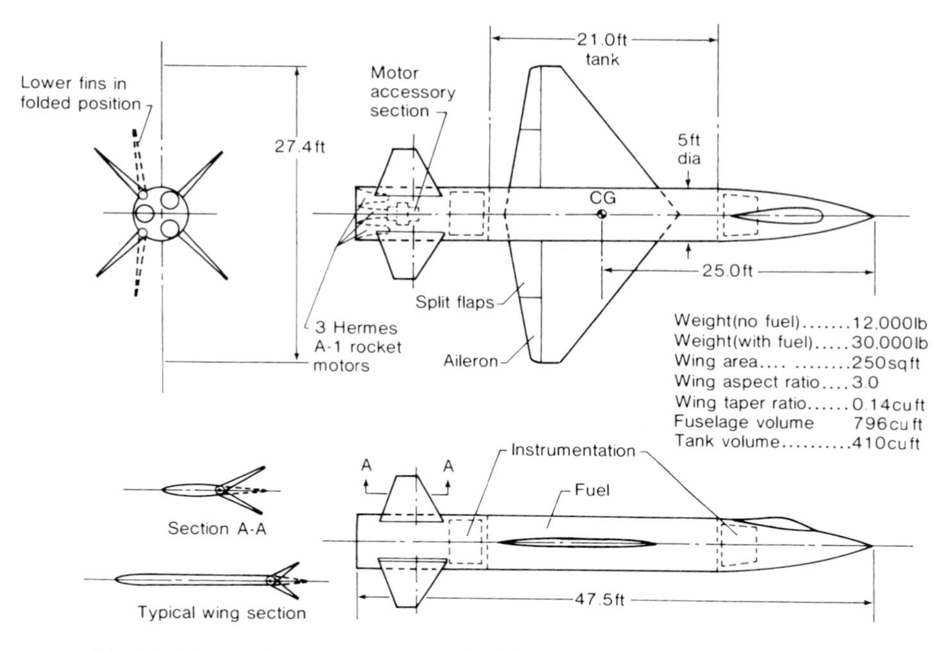

Fig. 11. The preliminary design of the X-15 hypersonic aircraft. (NASA)

financed by the USAF and the US Navy. At the end of the year technical direction of the programme was assigned to NACA. Following initial flight-tests by the contractor, the new aircraft would be handed over to NACA for a research programme. The new airframe and rocket motor were put out to a contractor's competition. In November 1955 North American Aviation (NAA) was selected to develop three X-15 airframes. Reaction Motors Inc., by then part of Thiokol Chemical Corporation, was selected to develop the XLR-99 rocket motor, with a proposed thrust of 25,855 N for periods up to 90 seconds. (See Fig. 11.)

By December the delays in the X-2 development programme had left NACA without an aircraft capable of extending their flight research from Mach 2.5 to the Mach 3 region and beyond. When Mel Apt was killed in the X-2 crash of 1956 the crash investigation and additional NACA research revealed that aircraft flying in the region of Mach 3 and above required a much larger tailplane area to fly in the rarefied upper atmosphere. The X-15's design was corrected to take the new data into account and in September 1957 NAA began work on the first X-15 airframe.

The Soviet Union launched Sputnik on 4 October 1957. The following month Sputnik II carried Laika, a smooth-hair terrier bitch, into orbit. A year later, on 1 October 1958, NACA was redirected to form NASA, which continued to play its part in the X-15 programme. Fifteen days after NASA came into being NAA rolled out X-15-1. With its proposed XLR-99 rocket motor over a year behind schedule, the first X-15 airframe had been fitted with two Reaction Motors Incorporated XLR-11 motors. In time X-15-2 would also be fitted with two XLR-11 motors.

These two machines would be retrofitted with the XLR-99 motor when it became available. X-15-3 was fitted with the XLR-99 rocket motor from the outset.

5.13 HYWARDS

In 1956 the USAF proposed the development of a boost-glide aircraft capable of speeds as high as Mach 18. The new aircraft was referred to as a Hypersonic Weapon and Research and Development System (HYWARDS). NACA's Langley and Ames laboratories disagreed on how the HYWARDS aircraft should be constructed. The initial definition programme settled on a design where the underside of the wing and fuselage gave the aircraft a delta wing with a flat bottom surface. This design offered the best answer to the heating regime experienced during hypersonic flight followed by a 'space leap' and re-entry into the atmosphere.

Research at Ames supported an aircraft high-lift to high-drag ratio offering maximum range for a given initial boost. Langley researched the same theoretical problem and settled on a lower lift to drag ratio, at least for an aircraft of intermediate range. The argument raged throughout the remainder of the year and was a major discussion point at NACA's Round III Steering Committee meeting in February 1957. An impartial report was written offering the good and bad points of both approaches and this was presented to interested parties within the NACA and the USAF. Further research at Langley showed that a reduction in wing size offered increased range. As research into the HYWARDS configuration continued, engineers at the Ames Laboratory came to accept the new Langley data.

Over 16–18 October 1957 NACA held a series of Round III interlaboratory meetings. The Round I meetings had been held during the development of the Bell X-1 rocket aircraft and Round II was part of the X-15 definition process. At the Round-III meetings many NACA engineers expressed the belief that their organisation should turn at least some of its attention to space-related research, including satellite vehicles. The meetings ended with the decision that NACA should commence research into the problem of re-entry through the atmosphere of a ballistic vehicle or a slightly lifting vehicle from a satellite orbit. The new research was to be carried out in parallel with the X-15 and HYWARDS work already in progress.

In November 1957, NACA set up a Special Committee on Space Technology. The Committee included Wernher von Braun and James A. Van Allen and was headed by H. Guyford Stever of MIT. In December Hugh Dryden requested Langley to review aeronautical research *'as it was affected by space-flight problems'*. A new committee was established under PARD director Robert Gilruth and reported to Langley's director on 31 January 1958. Gilruth's committee concluded that aeronautical research was in the process of *'an extensive shift in emphasis towards the fields of hypersonic and space flight'*.

In December 1957 NACA convened its last Conference on High-Speed Aerodynamics. While the hypersonic HYWARDS boost glide aircraft was one concept under discussion, many within NACA repeated the belief that they should be commencing research in other areas, including space flight. Langley presented a series of papers representing a number of different approaches to manned space flight. Maxime Faget, a PARD engineer, presented a paper detailing his work on a zero-lift high-drag ballistic capsule as a method of achieving manned space flight at the earliest opportunity. A converted military ICBM would be used to launch the capsule into orbit; retrograde thrust would be used to slow the vehicle down at the end of the orbital phase; and re-entry would employ Allen's blunt zero-lift, high-drag principle.

Meanwhile, Thomas Wong represented a team of engineers from Ames and presented a paper on a lifting half-cone, high-drag, high-lift vehicle known as a lifting body.

John Becker, one of the engineers behind the Langley HYWARDS study, presented a paper on that project as well as on the concept of winged satellite vehicles. Although a converted ICBM might be used as a suitable launch vehicle, no American missile existed in 1958 with enough throw capability to launch such a vehicle.

5.14 ROCKET SLEDS AND *G* FORCES

Nazi Germany was the first nation to develop and fly a military jet fighter. The Nazis were also the first to test a rocket-propelled ejection seat. German research into acceleration forces had involved sending human subjects down straight tracks on rocket-propelled sleds which were brought to a sudden halt by a combination of mechanical breaking and a large expanse of water across the track. All three of these events occurred during World War II. The Americans captured the German research documents and used them to advance American aviation.

In August 1946 John Paul Stapp, a doctor in the Army Air Corps, was present at Wright Field, Ohio, when the Americans tested their first ejection seat. Stapp realised that jet fighters would fly at greater speeds and carry their pilots to higher altitudes than any previous aircraft. If they malfunctioned, or were shot down at those speeds and altitudes there was no certainty that the pilot could survive ejecting from his aircraft. Over the next decade Stapp devoted himself to proving that such an ejection would be survivable.

Stapp volunteered for a research programme testing an oxygen-breathing system for American bombers. During the test programme he experienced Dysbarism at first hand. In this painful condition the low ambient air pressure caused the various fluids in the body to give off gases such as nitrogen and oxygen. Stapp's research led him to discover that the condition could be avoided if he pre-breathed pure oxygen for 30 minutes before the flight. The pre-breathing of oxygen removed the nitrogen from body tissues and soaked them in oxygen. Almost 20 years later America's first astronauts would pre-breathe oxygen before their flights.

Stapp's research bought him into contact with the details of the German rocket sled tests. In the state of cooperation between the wartime Allies that still existed in 1946, he was able to travel to Moscow and see the original German sled. On returning to Wright Field he began a programme to develop a similar sled, and Northrop Aircraft received the Army contract to build an American sled based on the German design.

Meanwhile, the USAF had been formed and Stapp found himself a member of the new service and transferred to Muroc AFB, California. In 1947 Northrop constructed their rocket sled at Muroc and Stapp began his research programme into high acceleration forces. He started the programme with 35 runs using a test dummy, before he would allow a human test subject to ride the sled. On the first manned run Stapp reached a speed of 145 km/h. On the second run he reached 322 km/h. By May 1948 Stapp had ridden the rocket sled 16 times at increasing rates of positive and negative *g* forces. He had subjected his body to 36*g* and received a number of injuries. He had also written extremely detailed reports after each run. He continued his rocket sled runs using chimpanzees rather than human test subjects and in this he was foreseeing the flight-tests that would precede the first flight of an American astronaut into space. By December 1951 the Northrop sled had made 255 runs – 94 unmanned, 88 carrying chimpanzees and 73 with human test subjects.

Following a transfer to Holloman AFB, Alamogordo, New Mexico, in 1953, Stapp was assigned the leadership of the new Aero Medical Field Laboratory. He constructed a state-of-the-art rocket sled with 1,067 m of track and was able to allocate the professional staff of engineers and doctors that he had been denied at Wright field. He named Northrop's new sled, 'Sonic Wind No. 1'. Its nine rocket motors produced a thrust of 178,000 N and the water-braking pond brought the sled to an abrupt halt when the two met. During the first three flights, in 1954, Stapp suffered numerous serious injuries, but always recovered and wrote his reports. During his final run, on 10 December 1954, Stapp pulled 20*g* during the sled's acceleration phase. As the sled hit the water barrier at the end of the track he was subjected to 46.2*g*.

5.15 BY BALLOON INTO SPACE

Since the 1920s a series of pioneering balloonists had been competing to reach the highest altitude. This became a political race that would be mirrored by the 'Space Race' in the late 1950s. Not only did Russian and American balloonists strive to reach a greater altitude than anyone who had flown before them, but, as in the later Space Race, the American military services also competed among themselves rather than pooling their resources and running a single, well-financed programme.

Most balloon launches took place in the early morning. As the balloon rose through the atmosphere the Sun's heat acted upon the helium within the envelope, causing it to expand as the air around it thinned. When the envelope could expand no further the balloon had reached its maximum altitude. The balloon lost altitude if it stayed aloft through the night as the helium cooled and contracted, thereby reducing buoyancy in the air.

Despite the official definition of space beginning at 100 km, if a pilot could reach an altitude of 30.4 km, then 99 per cent of Earth's atmosphere would be beneath him. Outside his craft at that altitude would be a space-like environment in which his unprotected body not would be able to function.

5.16 STRATO-LAB

Strato-Lab was a US Navy project intended to perform scientific experiments in the stratosphere. It used a two-man pressurised spherical gondola developed by Wizen Research Incorporated, Minneapolis. On 8 November 1956 Strato-Lab I was launched from South Dakota with Lieutenant Commanders Malcolm Ross and Lee Lewis housed in the gondola. The balloon rose to 23.1 km, a new altitude record. After only minutes at that altitude a helium valve malfunctioned and started releasing gas from the balloon's envelope. Strato-Lab I immediately began to lose altitude, and the crew had to eject all of their ballast and much of their equipment, including their radio, in order to control their descent and land safely.

Strato-Lab II flew on 18 October 1957. Ross and Lewis reached 26.2 km during a flight that lasted 10 hours. Strato-Lab III followed on 26 July 1958. Ross and Lewis reached 24.9 km and stayed aloft for 34.5 hours. Strato-Lab IV flew on 28 November 1959 with Ross and Charlie Moore. The balloon reached 24.6 km and the two men spent the night observing Mars. They landed the following afternoon.

Strato-Lab V was launched from the flight deck of the USS *Antietam* on 4 May 1961, the morning before the launch of America's first manned space flight. The gondola was an open framework and Lieutenant Commander Dr Victor Prather, USN, accompanied Ross on the flight. They both wore full pressure suits, plus survival clothing, to protect them from the extreme conditions to which they would be exposed. They handled a number of difficulties during the ascent and came within moments of heating their helmet faceplates to the point of melting them. The balloon rose to an altitude of 34.6 km. Strato-Lab V landed in the ocean and the crew cut the lines connecting the gondola to the balloon. The USS *Antietam* dispatched helicopters to recover the two pilots. Ross was first to be recovered, but Prather slipped off of the hook and fell back into the water, which flooded his pressure suit through his open faceplate. He drowned before he could be recovered.

5.17 PROJECT MANHIGH

While the captured German A-4 missile and the new American sounding rockets were being launched from WSPG the USAF Balloon Research and Development Test Branch, at Holloman AFB, began launching animal test subjects in pressurised containers hung beneath large balloons. The flights were launched from the Tularosa Basin, to test the effects on the test subjects of the cosmic rays found at extreme altitude.

One of the USAF doctors involved with the animal-carrying balloon flights and leading the cosmic ray research was Major David Simons, who was based at Holloman AFB Aero Medical Field Laboratory. Simmons had also been involved in the five unsuccessful *Blossom* monkey launches as part of the A-4 missile programme at WSPG. He reported to Colonel John Paul Stapp and had been responsible for monitoring Stapp's physical condition during the four Sonic Wind No. 1 rocket sled tests at Holloman AFB. Stapp was preparing for a manned balloon flight to carry a pilot to an altitude above 30.4 km, where he would experience space-like conditions at first hand. He invited Simons to be the prime pilot for the Project Manhigh attempt to spend 24 hours in the upper stratosphere. Before that, Captain Joe Kittinger would make a flight-test of shorter duration.

Wizen Research Incorporated constructed the Manhigh balloons and pressurised gondola. The pressurised gondola was an upright cylinder 2.4 m tall and 0.9 m in diameter and pressurised with a mixture of 60 per cent oxygen, 20 per cent helium and 20 per cent nitrogen – a gas mixture that made the Manhigh pilot's voice sound high pitched and squeaky. The used air was passed through chemical scrubbers before being recycled. The gondola carried a 48-hour air supply, primary and back-up batteries, and equipment to monitor and control the gondola systems and the balloon. Radio was the primary communication system, but a Morse code key was included as a back-up. The pilot wore a standard USAF pilot's MC-3 partial pressure suit and sat upright in the gondola. In the event of a malfunction he could jump through a hatch in the floor and descend under a personal parachute. Alternatively, he could jettison the balloon and return to Earth in the gondola under a large military cargo parachute. The Manhigh gondola provided its pilot with every vital resource that a manned orbital spacecraft would have to provide for its astronaut. Much of the Manhigh technology was transferred directly to the USAF's Man In Space Soonest project, which would form the basis of the Project Mercury suborbital flights.

Kittinger made the Manhigh I test flight at dawn on 2 June 1957. The radio transmitter failed, but the pilot made regular reports using the Morse code key throughout the flight. The balloon rose rapidly to a maximum altitude of 29.2 km where it stopped ascending. Outside the sky appeared black and the air was so thin that there was no transmission of sound. Only after reaching altitude did Kittinger discover that the Wizen engineers had fitted the oxygen system aneroid in reverse. The gondola's oxygen supply was being fed into the stratosphere while the incidental bleed off was being fed to the gondola. The oxygen supply was already half consumed. Kittinger turned off the main oxygen supply to the gondola, to prevent the loss of more of the vital gas. From that point on he fed his cabin from the pressure suit oxygen supply. Kittinger descended immediately. The gondola touched down in a muddy river and the balloon was released. Kittinger was in excellent physical health and was heralded as a national hero. He left Project Manhigh shortly after his flight.

Manhigh II, the full duration research flight with Simons as pilot, was launched using the same gondola on 19 August 1957. The balloon rose to an altitude of 31,089 m. The pilot described the view out of his window and exposed a number of photographs. He was able to see the curve of Earth as it stretched away beneath him. Simons described layers in the atmosphere and the effects they had on the colours he was observing.

Fig. 12. Major Clifton McClure, USAF, sits in the Manhigh balloon gondola, thereby illustrating its extremely cramped conditions. (USAF)

Having not slept for nearly 24 hours Simons began making mistakes before eating a chocolate bar to raise his blood sugar level. He then sat and watched in awe as the Sun set over the planet beneath him. After the Sun had passed beneath the horizon he saw stars in the black sky. He described them and the Moon, which he could also see. Next he saw the aurora borealis in the atmosphere below him. As the helium in his balloon cooled, Manhigh II began to descend towards a storm cloud beneath him, cutting the balloon off from the heat being reflected from Earth's surface. Simons tried to sleep. When he awoke he was surrounded by storm clouds and spent 25 minutes struggling to don a thermal suit to combat the cold he was now feeling. As the warmth of the Sun expanded the helium in the balloon, Manhigh II climbed back up into the upper stratosphere. A test of the gondola's atmosphere showed that it contained 4 per cent carbon dioxide, a near poisonous level. Stapp took control of the flight and instructed Simons to begin his descent, close the faceplate on his helmet, and breathe the oxygen in his pressure suit circuit. Manhigh II landed in a ploughed field and was dragged across the dirt before Simons ejected the balloon. He had been strapped in his seat for over 44 hours, 11 of which had been before launch.

Four prospective Manhigh III pilots were sent to Lovelace Clinic, Albuquerque, to undergo a battery of physical and psychological tests. Two of the candidates were disqualified after four days of physical examination. A third was disqualified as a result of the psychological tests. First Lieutenant Clifton McClure was subsequently named as prime pilot for Manhigh III. This approach to pilot selection would be mirrored in the selection of America's first astronauts. (See Fig. 12.)

Manhigh III lifted off at dawn on 7 October. The balloon reached an altitude of 30.3 km but with the internal temperature in the gondola reading over 90 °F and McClure's rectal thermometer giving a reading of 101.4 °F, the panel in charge of the flight on the ground decided to abort the flight. Manhigh III landed in darkness, 1,219 m up in the San Andres mountain range. When the recovery helicopter landed McClure had ejected the balloon and the gondola's top hatch. He was standing alongside the gondola and removing his helmet. He refused to sit down, or be carried to the helicopter on a stretcher, preferring to walk the short distance. The initial physical examination carried out in the helicopter showed that at that time he had a body temperature of 108.5 °F.

5.18 PROJECT EXCELSIOR

Stapp's next project, after Manhigh, was 'Project Excelsior', designed to see what would happen if a pilot were forced to bail out at extreme altitude. The project was set up in 1958, to test a multi-stage parachute system that contained a high-altitude stabilisation parachute to ensure that the pilot did not tumble as he free fell down to the thicker air of the troposphere. Dummies were used in early test drops, but Project Excelsior would use a human test subject – Joe Kittinger, the Manhigh I pilot.

Project Excelsior used an open gondola while Kittinger wore only a standard USAF MC-3 partial pressure suit to protect him from the deadly environment in the upper stratosphere. This was a deliberate decision on Kittinger's part as he wanted his tests of the parachute system to be as relevant to operational USAF jet pilots as possible. To protect him from the extreme cold during the ascent he also wore additional layers of thermal clothing.

On 16 November 1959, Kittinger rode the Excelsior balloon to an altitude of 17.6 km, at which point his gold covered visor had fogged over and he was unable to see his instruments. His helmet then began rising up and threatening to blow off completely. If that happened he would die almost instantly. In his near panic Kittinger missed his planned jump altitude of 18.3 km. The balloon continued to rise until, at 23.1 km, Kittinger began the preparations for his jump. Stepping off the gondola he fell towards Earth, but the air was so thin that he felt no motion and soon succumbed to unconsciousness. Film of the drop proved that Kittinger was lucky to survive. His stabilisation chute and both his main and reserve chutes fouled around his unconscious body as he spun round and round in an uncontrolled flat spin. Finally, the lines of the main parachute snapped, as they had been designed to do in such a situation, and the reserve chute deployed at 1.8 km and lowered the still unconscious Kittinger to the desert below.

Excelsior II was launched on 11 December 1959 and Kittinger left the gondola at 22.7 km. The pilot parachute opened and Kittinger fell through the stratosphere feet first, as planned. The main parachute opened at 5.4 km and Kittinger landed after a successful drop lasting 12 min 32 s.

Excelsior III ascended on 16 August 1960. At 15.2 km Kittinger noticed that his right glove was not correctly pressurised, but he decided to continue the flight, which stopped rising at 39.6 km. Kittinger's right hand had swollen to twice its normal size and was extremely painful when he finally informed the ground of the problem with his glove. Rising from his seat he stood in the doorway of the open gondola. With the words, '*Lord, take care of me now*', he pushed away from the gondola, and fell into the void. In total silence he fell, twisting so that he could watch Excelsior III as he travelled Earthward. Sixteen seconds after leaving the gondola the stabilisation parachute opened and he adopted a feet-first attitude. Fourteen seconds later he was already down to 27.4 km and travelling at 988 km/h, almost Mach 1. After 4 min 37 s of free fall the main parachute opened. He was at 5.4 km. He landed in the New Mexico desert after a fall lasting 13 min 45 s – a man had parachuted back to Earth from the edge of space and survived. Colonel Joseph Kittinger's Excelsior III jump has not been equalled to this day. (See Fig. 13.)

When Project Mercury began in 1959 there were those who believed that Project Excelsior proved that an astronaut could jump out of the Mercury Spacecraft if something went critically wrong during the initial phase of ascent, or during the final descent at the end of the flight. To this end each manned Mercury Spacecraft would carry a personal parchute.

Fig. 13. Colonel Joseph Kittinger, USAF, jumps from the Excelsior III gondola at an altitude of 39.6 km. This record has not been surpassed. (USAF)

5.19 EARLY MANNED SPACE FLIGHT PROPOSALS

In the late 1950s the USAF assumed that space would be classified as an extension of their aerial domain. They assumed that when they were given the go-ahead to militarise space NACA would continue to give its technical support to the programme. Inside NACA there was a different vision of the future. The NACA Main Committee met on the USS *Forestall* over 18–20 November 1957 to discuss the organisation's future. They decided to '*act now to avoid being ruled out of the field of space flight research*'. At the time it was little more than a commitment to extend their research into the region of hypersonic flight.

In January 1958 the DoD established the Advanced Research Project Agency (ARPA) to oversee the large number of manned satellite projects under study by the country's military services and their major contractors. Three weeks later ARPA gave the Air Force permission to develop the proposed 117L liquid propellant upper stage for the Atlas and Thor ICBMs. In time 117L would become famous as the Agena rocket stage. In 1958 it was the top secret 'bus' designed to meet the needs of America's Corona reconnaissance satellite.

In March 1958 eighty members of the Air Force Air Research and Development Command (ARDC), NACA, and the major aeronautical contractors met to discuss what was referred to as the 'quick and dirty' approach to manned space flight. That approach was the ballistic capsule launched in place of the warhead on a military ICBM. Stapp's rocket sled rides and other research had shown that a pilot could survive $12g$ and still perform the flight control tasks required of him. On the other hand, the failure of a single stage launch vehicle at the moment of max-q might result in the pilot being subjected to $20g$. The group suggested that the vehicle be launched on a two-stage launch vehicle, probably a Thor with a new upper stage. The project became known as Man In Space Soonest (MISS). There was little support for the use of Atlas as the proposed launch vehicle, even with an upper stage.

The MISS concept was similar to one under study by NACA's PARD engineers. At NACA-Langley and Wallops Island, Virginia, Maxime Faget had also conceived a ballistic spacecraft that carried the astronaut in the supine position during launch and re-entry. Faget had also developed a prone couch, formed to the pilot's body, to give him maximum support for those periods during the flight when he would be subjected to high acceleration and deceleration forces. Faget's spacecraft was designed for launch on a converted ICBM, and he had also designed a 'tractor rocket' to fit on the top of the spacecraft and pull it away from a malfunctioning launch vehicle. At the end of its orbital flight the spacecraft would be slowed down by firing retrograde rockets against the direction of travel, allowing gravitational attraction to pull it back into the atmosphere. During re-entry the spacecraft would apply Allen and Egger's blunt leading edge theory. The spacecraft would land in the ocean under a large parachute, after which US naval forces would recover the spacecraft and pilot. The second NACA proposal was a lifting body, designed to produce sufficient lift in the shape of its body, without the use of wings, to allow it to fly back to an aircraft-style landing on a runway.

On 14 March NACA accepted the Air Force's offer to cooperate on a manned satellite vehicle. The cooperation ended in May when it became apparent that NACA might form the basis of a new civilian space administration. Unlike the old organisation the new administration would have the power to contract out development and production work and operate the national space programme, rather than just being a research tool for the military.

In April, Convair and AVCO suggested that a manned satellite vehicle might be launched into orbit on a 'bare Atlas', that was to say an Atlas ICBM without an upper stage. The proposal was the subject of disbelief within the USAF and most portions of NACA. It was argued that using a bare Atlas would only save a few months in the preparation of a manned satellite vehicle.

Within one month budget restriction within the Air Force's MISS programme led to that proposal also being designed around a bare Atlas launch vehicle. The revised plan was accepted by ARPA on 16 June, the day after it was submitted. As the USAF and NACA met to refine MISS, both John Paul Stapp and David Simmons were among the USAF officers taking part.

On 25 May 1958 the USAF complied a list of test pilots that might be considered to fly MISS. The list was based on the individual's weight and names were shown against a scale detailing how close they came to the maximum weight allowed for the pilot within the lift capability of a bare Atlas. The names listed were:

Robert Walker	Scott Crossfield	Neil Armstrong
Robert Rushworth	William Bridgeman	Alvin White
Iven Kincheloe	Robert White	Jack McKay

By that time, however, it was obvious that NACA would definitely form the basis of the new space administration and the MISS pilots would probably not get a chance to fly. To that end ARPA began insisting that NACA Director Hugh Dryden also approve any USAF manned satellite proposals. Despite this, MISS went into its sixth draft on 24 July. The proposal stated that the Air Force could have a manned vehicle in orbit by June 1960. Following a two-day presentation ARPA refused funding for the proposal.

In the wake of the Manhigh II balloon flight on 19 August 1957, Wernher von Braun of the ABMA approached Stapp, Simmons and Kittinger with a view of adapting the Manhigh gondola to fit on the top of a Redstone IRBM. The plan was to launch the gondola on a suborbital flight into space and then recover it under a large parachute. The project was referred to as 'Man Very High'. It was refused funding by both the Army and the USAF.

In April 1958 the ABMA proposed a variation of Man Very High called 'Project Adam'. It was an Army plan to launch a man in an especially adapted nose cone along a high-altitude suborbital trajectory on a Redstone IRBM. The flight was seen as a precursor to manned orbital flight but the ARPA refused to fund the project. At the time Hugh L. Dryden, Deputy Director of NACA, described the proposal as being '*as much use as the circus stunt of firing a lady from a cannon*'.

On 29 July 1959 the National Aeronautics and Space Act was signed into law. It called for a new National Aeronautics and Space Administration (NASA) to be formed, using NACA as its basis. NASA would become operational on 1 October

1958, with T. Keith Glennan as Administrator and NACA Director Hugh L. Dryden as Deputy Administrator.

On 3 October 1958, NACA ballistic spacecraft plan was presented to the ARPA. Four days later a similar presentation was made to Glennan. The Administrator listened attentively before closing the meeting with the words '*Let's get on with it*'.

Part II: INFRASTRUCTURE

6

The National Aeronautics and Space Administration

In the wake of the launch of Sputnik, President Eisenhower was sure that American technology and the American military machine were superior to those found in the Soviet Union, and initially refused to enlarge the budget set aside for space exploration. The American media saw things differently and whipped the American public into frenzy of self-doubt. As a result, a public demanded arose for a highly visible reaction to the Soviet satellite. In time it became obvious to Eisenhower that he would have to do something to answer the public's call.

The President was aware of the Corona reconnaissance satellite that was under development by the USAF and the Central Intelligence Agency (CIA) at that time. He considered it to be the most important single piece of space hardware then under development and vital to American national security. Regretfully, military secrecy dictated that he could not tell the public that Project Corona existed. Following the launch of the first Sputnik he increased the budget for Project Corona.

Despite being a retired Army officer Eisenhower called for the development of a civil space programme. The civil space programme would be the public reaction to Sputnik while freeing the military to concentrate on a secret space programme intended to increase their knowledge of the Soviet Union and its war machine.

On 25 November 1957, Democratic Senator Lyndon B. Johnson commenced a series of hearings by a subcommittee of the Senate Armed Services Committee with the aim of reviewing all of America's missile and space programmes. All programmes were found wanting, especially in the area of their allocated budgets. In reply, Eisenhower's Republican Administration allocated more funds to both missile and space programmes in an attempt to speed them up.

On 6 February 1958 the Senate created the Special Committee on Space and Aeronautics with Johnson as its chairman. The Committee was tasked with writing legislation to bring a new civil space agency into being. President Eisenhower favoured an open civil space programme that could be freely reported on by the media and would receive the full support of the nation's scientists. The US military were to be restricted to passive reconnaissance and intelligence gathering from orbit. James R. Killian was Eisenhower's Presidential Science Adviser. On 4 February 1958 he was instructed to convene the President's Scientific Advisory Committee (PASC) and seek their recommendations on a solution to the matter of forming a civil space agency. On 5 March President Eisenhower accepted the PASC recommendation that the new space agency should be founded upon a renamed and reorganised NACA. Such an agency could begin work with the prestige of the original organisation, upon which it could build its own reputation. Eisenhower demanded that a single individual, selected by the President and answerable to him, head the new National Aeronautics and Space Administration (NASA). A National Aeronautics and Space Council, would hold the power to make decisions at the highest levels of all the organisations involved in the national space effort. The President himself would chair the Space Council and thereby be in a position to influence the policies that it recommended.

During a Congressional speech on 2 April Eisenhower instructed NACA's Central Committee to review all space research currently under ARPA's control, to see which programmes should be transferred to NASA when it came into being. The National Aeronautics and Space Bill 1958 underwent its first reading before Congress 12 days later. The Bill's wording was debated and changed over the following months. The Act was passed by Congress on 16 July and signed into law by President Eisenhower on 29 July 1958. It gave the Main Committee of NACA 90 days to prepare for their new role before declaring NASA open for work in the Federal Register.

The first NASA Administrator was T. Keith Glennan, a political appointment by the Eisenhower Administration. Glennan reported to the President through the Space Council. NACA's Director, Hugh L. Dryden, was named as Deputy Administrator. Both posts were confirmed on 15 August. Glennan declared NASA ready for business on 25 September, and the relevant entry was published in the Federal Register dated 30 September. On that evening NACA ceased to exist after a glorious career spanning the history of American aviation back to 1915. In a much-quoted press release Glennan stated:

> ... One way of saying what will happen would be to quote from the legalistic language of the Space Act ... My preference is to state it in a quite different way – that what will happen ... is a sign of metamorphis.

When NACA's employees returned to work on 1 October 1958 they were part of the new NASA. On its first day of operation NASA had nothing more than its headquarters, at Dolley Madison House, Washington DC, and the three NACA research centres. The latter were renamed: Ames Flight Research Centre (AFRC), Langley Flight Research Centre (LaFRC) and Lewis Flight Research Centre (LFRC). There was also a NASA contingent working at Edwards AFB, California. PARD, the LaFRC satellite station at Wallops Island, retained its original name and role.

The Space Act gave NASA the authority to take over a number of organisations, with the Space Council's approval. When NASA came into being ARPA ceased to exist and its staff became part of the new Administration. All military space programmes not directly concerned with national security passed to NASA or were closed down.

Abe Silverstein, the Associate Administrator at LFRC, was appointed to the new post of Director of Space Flight Development at NASA Headquarters (HQ).

Silverstein promptly named three men to fill the positions directly beneath his own: Newell Sanders, Assistant Director of Advanced Technology; Warren M. North, Director, Office of Manned Satellites; and George M. Low, Director, Office of Manned Spaceflight.

6.1 NASA CENTRES

6.1.1 Langley Research Centre

The NACA Langley Memorial Aeronautical Laboratory was NACA's first field station when it became operational in July 1920. The Centre was named after Dr Samuel Langley, an American aeronautical pioneer and the third Secretary of the Smithsonian Institution. Langley remained NACA's only laboratory until 1940 and was the primary location of much of the NACA's aeronautical research until it was transferred to NASA on 1 October 1958. At that time it was renamed the Langley Research Centre. The Centre conducted basic research into a number of aerodynamics-related subjects. Having transferred to NASA it became responsible for the management of a number of unmanned space vehicles. (See Fig. 14.)

Langley Research Centre, Langley Field, Hampton, Virginia 23665.

Fig. 14. Langley Flight Research Centre, the NASA centre responsible for the day-to-day running of Project Mercury. (NASA)

6.1.2 Wallops Flight Centre

NACA's Langley Memorial Aeronautical Laboratory established a satellite station on Wallops Island, Virginia, in May 1945 and named it the Auxiliary Flight Research Station. On 10 June 1946 the station was renamed the Pilotless Aircraft Research Division (PARD). The name changed to the Pilotless Aircraft Research Station in August 1946. The title 'Pilotless Aircraft' was the blanket name given to all of the nation's guided missile research, including cruise and ballistic missiles.

Robert Gilruth, an engineer, led the PARD from LaFRC. It was Gilruth's group who defined the ballistic capsule that would become the Mercury Spacecraft. This group formed the basis of the Space Task Group, which originally led the NACA/NASA manned spaceflight programme.

When the facility transferred to NASA on 1 October 1958, it was shown on organisational charts as Wallops Station and became an independent NASA Centre in May 1959. Wallops Station was named after John Wallops, a seventeenth-century surveyor who owned the island and gave his name to it. Previously the island had carried a number of Native American names. When the facility expanded on to the mainland in 1959, the new portion was named Wallops Main Base. On 26 April 1974 Wallops Station was renamed Wallops Flight Centre.

Wallops Flight Centre, Wallops Island, Virginia 23337.

6.1.3 Ames Research Centre

Congress authorised the construction of the Moffet Field Laboratory on 9 August 1939. It would become the second NACA field station. When the new station opened in early 1941 NACA named it the Ames Aeronautical Laboratory, after Dr Joseph Ames, an aerodynamicist serving in NACA's Main Committee from its inception in 1915. He served as Chairman of NACA from 1927 until his retirement in 1939. Like all NACA laboratories, Ames was dedicated to aeronautical research and, in particular, basic and applied research in the physical and life sciences.

On 1 October 1958 the Ames Aeronautical Laboratory was transferred to NASA and was renamed the Ames Research Centre. Ames continued in its original role, while also taking on new responsibilities as part of the new national civil space programme. The Centre was responsible for pioneering research into re-entry heating and established the high-drag low-lift design used on re-entry vehicles.

Ames Research Centre, Mountain View, Moffett Field, California 94035.

6.1.4 Lewis Research Centre

The NACA Aircraft Engine Research Laboratory was authorised by Congress in June 1940. The establishment was dedicated on 28 September 1948, at which time it was named the Lewis Flight Propulsion Laboratory, after Dr George Lewis. Dr Lewis was a NACA aeronautical engineer and served as NACA's Director of Aeronautical Research from 1919 to 1947. On 1 October 1958 the Centre was transferred to NASA and took the new name of Lewis Research Centre.

Lewis FRC was originally responsible for the development of new aircraft engines and continued to concentrate on propulsion systems following its transfer to NASA.

Lewis Research Centre, 2100 Brookpark Road, Cleveland, Ohio 44135.

6.1.5 Flight Research Centre

The NACA Muroc Flight Test Unit was established at the USAF Muroc Air Force Base, California on 30 September 1946. Walter Williams led a group of 12 other NACA engineers from LaFRC to prepare for the flight-tests of the Bell X-1 and the attempt to fly the first aircraft faster than Mach 1. The staff of the Muroc Flight Test Unit was transferred permanently to their new location in 1947. In 1949 the Unit was renamed the High Speed Flight Research Station. Muroc AFB was itself renamed Edwards Air Force Base in February 1950. The NACA High Speed Flight Research Station moved to new facilities in 1954, on land leased from the USAF at Edwards AFB. It was renamed the High Speed Flight Station at that time.

On 1 October 1958 the facility was transferred to NASA and renamed the Flight Research Centre. From the beginning the facility was responsible for all aspects of NACA/NASA aeronautical flight-testing. It also conducts research in to all aspects of atmospheric and space flight regimes.

In 1976 the Flight Research Centre was renamed the Hugh L. Dryden Flight Research Centre, after Dr Hugh L. Dryden, the man who served as NACA's last Director and NASA's first Deputy Administrator. Dr Dryden was an aeronautical engineer who originally served at LaFRC.

Hugh L. Dryden Flight Research Centre, Edwards Air Force Base, California 93523.

6.1.6 NASA Headquarters

When NASA came into being on 1 October 1958 the above five stations were the only facilities that it owned, along with its Headquarters in Washington. NASA Headquarters was originally based in Dolly Madison House and a number of individual offices around the District of Columbia, while the new Headquarters building was constructed.

NASA Headquarters, 400 Maryland Avenue, Washington DC 20546.

6.2 EXANDING TO MEET NEW GOALS

The National Aeronautics and Space Act allowed the new Administration to take over existing organisations and their facilities as well as to construct new facilities as required. Over the years the following facilities were transferred to NASA or constructed to meet the requirements of NASA remit. The new facilities are listed in the order that they came on line. The list contains only those facilities that were operational in 1963, when Project Mercury closed.

6.2.1 Jet Propulsion Laboratory

Students from the Guggenheim Aeronautical Laboratory of the California Institute of Technology (GALCIT) began developing liquid propellant rocket motors in 1936. During World War II they developed JATO rocket motors for heavily laden aircraft. In 1944 the facility was renamed the Jet Propulsion Laboratory (JPL). After the war

JPL carried out work for the US Army Ordnance and participated in the launch of Explorer 1 and the early American Moon probes. JPL was still independent when NASA was formed on 1 October 1958.

On 3 December 1958 JPL was transferred from the US Army to NASA after which it took responsibility for the unmanned lunar and interplanetary exploration programmes. The facility operates the Deep Space Network to communicate with their probes.

Jet Propulsion Laboratory, California Institute of Technology, 4800 Oak Drive, Pasadena, California 91103.

6.2.2 Robert H. Goddard Space Flight Centre

In August 1958, it was known that NACA would become the basis of the new NASA on 1 October of that year. It was also known that NASA would be responsible for the manned satellite programme, which would become Project Mercury. To meet the requirements of the manned satellite programme Congress authorised the construction of a new Space Projects Centre, in the Department of Agriculture's Beltsville area of Washington DC. The proposed Centre was referred to as the Beltsville Space Centre.

Project Vanguard, the US Naval Research Laboratory satellite programme, was transferred to NASA on 15 January 1958. The Vanguard Division became one of four divisions of NASA Headquarters transferred to the Beltsville Space Centre on 15 January 1959. The other three divisions were the Construction Division, the Space Science Division and the Theoretical Division.

On 1 May 1959 NASA officially named the Beltsville Space Centre the Robert H. Goddard Space Flight Centre, after the American rocket pioneer. Ester Goddard, the pioneer's widow, attended the dedication of the new Centre on 8 February 1961.

The Goddard Space Flight Centre (GSFC) was responsible for unmanned spacecraft, including the operation of the Minitrack satellite-tracking system and the Space Tracking and Data Acquisition Network. The Centre was also responsible for experiments in basic and applied research to be launched on sounding rockets.

GSFC was also responsible for the World-Wide Tracking Network, which was established to support Project Mercury. GSFC was the location to which all WWTN stations relayed their data. That data was then reduced in Goddard's computers before being relayed to the Mercury Control Centre at Cape Canaveral. (See Fig. 15.)

Robert H. Goddard Space Flight Centre, Greenbelt, Maryland 20771.

6.2.3 George C. Marshall Space Flight Centre

In April 1950 the German 'paperclip' rocket engineers were moved from Fort Bliss, New Mexico, to the Ordnance Guided Missile Centre, Redstone Arsenal, Huntsville, Alabama. On 1 February 1956 the Ordnance Guided Missile Centre was formed into the Army Ballistic Missile Agency (ABMA). In this role they developed the Redstone and Jupiter IRBMs.

On 14 March 1960 parts of ABMA were transferred to NASA, including von

Fig. 15. Computers arrive at the new Robert H. Goddard Space Flight Centre, which served as the computing centre for the World-Wide Tracking Network.

Braun's German rocket engineers. The new NASA facility at the Redstone Arsenal was called the George C. Marshall Space Flight Centre (MSFC) after the Army General of that name, who had been Chief of Staff during World War II and Secretary of State in 1948–49. Marshall had been the founder of the Marshall Plan that had supplied money and material to re-establish the European economy after World War II.

MSFC commenced work on 1 July 1960, following the transfer of most of the ABMA staff to NASA. The Centre's role has always been the development of launch vehicles and propulsion systems.

George C. Marshall Space Flight Centre, Huntsville, Alabama 35812.

6.2.4 Michoud Assembly Facility

In 1961 the Government-owned Michoud Ordnance Plant, Michoud, Louisiana, was not in use. On 7 September of that year NASA elected to construct the first stage of the Saturn IB and Saturn V launch vehicles at the site. The work was under the management of MSFC.

The site was renamed the Michoud Assembly Facility on 1 July 1965, and in 1972 it was selected as the site for the manufacture of the Space Shuttle Main Engine.

Michoud Assembly Facility, New Orleans, Los Angeles 70129.

6.2.5 Lyndon B. Johnson Space Centre

When it became obvious that NACA would become the basis of the new NASA, the Space Task Group (STG) was set up within PARD, at Langley Memorial Aeronautical Laboratory, and was charged with developing a ballistic manned satellite vehicle.

On 3 January 1961 STG, which had previously reported to the Goddard Space Flight Centre, became an autonomous NASA field centre. In August 1961 funds were appropriated for the building of a new field station to centralise the nation's recently expanded manned spaceflight effort. After some competition the new Centre was built on a farmland site in Houston, Texas, which was Vice President Johnson's constituency. The establishment was named the Manned Spacecraft Centre (MSC). On 1 November 1961 the STG, which was still based at Langley, was renamed the Manned Spacecraft Centre, in anticipation of its move to Houston.

The new Centre was responsible for the design, development and testing of NASA's manned spacecraft as well as the selection and training of the Administration's astronauts. It was also the home of the Flight Operations Division, which was responsible for the planning and flight-control of each American manned spaceflight. A new Mission Control Centre was built at the MSC and, beginning with the flight of Gemini Titan IV, in June 1965, control of each flight was passed from Cape Canaveral to Houston once the launch vehicle had cleared the umbilical tower. The Centre continues in that role to this day.

Following the death of former President Lyndon Johnson on 17 February 1973, the Manned Spacecraft Centre was renamed the Lyndon B. Johnson Space Centre in his memory.

Lyndon B. Johnson Space Centre, Clear Lake, Houston, Texas 77058.

6.2.6 John C. Stenis Space Centre

On 25 October 1961 NASA announced that it had selected a site in Mississippi to test the various stages of the Saturn launch vehicles when they were developed. The new site went under the unofficial name of the Mississippi Test Facility. On 18 December the site was officially named Mississippi Test Operations, but the title was never accepted. On 1 July 1965 NASA announced the official redesignation of the site as the Mississippi Test Facility.

The site was used to test the first stages of the Saturn IB and Saturn V launch vehicles, before being placed on standby status on 9 November 1970. In March 1971 the site was designated for the testing of the Space Shuttle Main Engine. All work was carried out under the management of MSFC.

With the end of the Saturn launch vehicle test programme NASA offered the facilities of the Mississippi Test Facility for use by other government agencies. At that time NASA established an Earth Resources Laboratory at the facility. On 14 June 1974 the site was renamed the National Space Technology Laboratories and became an independent NASA field centre. In May 1988 the facility was renamed again, this time as the John C. Stenis Space Centre.

John C. Stenis Space Centre, Bay St Louis, Hancock County, Mississippi 39520.

6.2.7 John F. Kennedy Space Centre

The Joint Long-Range Proving Ground (JLRPG) at Cape Canaveral, Florida, was established when the military missiles under test outgrew the White Sands Proving Ground in New Mexico. In 1951 the Experimental Missiles Firing Branch of the Army Ordnance Guided Missile Centre, Huntsville, was established at JLRPG to flight-test the Redstone missile. In January 1953 the Experimental Missile Firing Branch was expanded and renamed the Missile Firing Laboratory.

On 1 June 1960 the ABMA at Redstone Arsenal, Huntsville, was transferred from the US Army to NASA. The Redstone Arsenal was renamed the Robert H. Goddard Space Flight Centre. The Missile Firing Laboratory was merged with NASA's Atlantic Missile Range Operations Office, which was NASA's group responsible for liaising with the military services that ran the Atlantic Missile Range, as the JLRPG had been renamed. The merged Missile Firing Laboratory and Atlantic Missile

Fig. 16. Missile row at Cape Canaveral, Florida. Some of these early missile launch pads were in use long before NASA's Apollo facilities were built on Merritt Island, to form the John F. Kennedy Space Centre.

Range Operations Office was renamed the Launch Operations Directorate and came under the management of MSFC.

On 1 July 1962 the Launch Operations Directorate became an independent NASA field centre and was renamed the Launch Operations Centre. Following the assassination of President John F. Kennedy the area of Florida known as Cape Canaveral, including the land on which the Atlantic Missile Range was built, was renamed Cape Kennedy. On 29 November 1963 President Lyndon Johnson renamed the NASA Launch Operations Centre the John F. Kennedy Space Centre (KSC), in memory of the man who had challenged NASA to land a man on the Moon.

In 1961 NASA selected Merritt Island as the location for its new Project Apollo launch facilities. NASA designated their entire operation in Florida the John F. Kennedy Space Centre in 1965. By that time the Atlantic Missile Range had been renamed the Eastern Test Range. With the close of the Apollo–Soyuz Test Project in 1975 the Apollo launch facilities were converted for use in the Space Shuttle programme. (See Fig. 16.)

John F. Kennedy Space Centre, Florida 32899.

6.2.8 White Sands Test Facility

The US Army's White Sands Proving Ground, New Mexico, was the first long-range missile range to be opened in America. The German A-4 missiles captured at the end of World War II were tested there, as were the first all-American missiles and sounding rockets.

In June 1962 NASA's Manned Spacecraft Centre established their White Sands Operations at the range in order to test propulsion systems for Project Apollo. NASA renamed the White Sands Operations the White Sands Test Facility on 25 June 1965.

White Sands Test Facility, New Mexico.

7

Astronauts

7.1 SELECTING FLIGHT A

When NASA was formed, on 1 October 1958, the STG manned satellite project was already underway. Before the end of the month, NASA Headquarters had appointed Dr W. Randolph Lovelace II as chairman of a Life Sciences Committee. Following selection boards at NASA Headquarters, prospective 'Research Astronaut Candidates' would attend the Lovelace Clinic, Albuquerque, New Mexico, for medical testing. The first military aero-medical staff reported for duty with NASA on 3 November.

NASA assembled a team of experts to oversee the selection of the men who would fly in what was then called Project Astronaut. The group consisted of:

Lieutenant Colonel Stanley White, USAF, physician
Captain William Augerson, US Army, physician
Lieutenant Robert Voas, US Navy, physician
Waren North, NASA test pilot
Charles Dolan, NASA engineer and deputy director of STG
Allen Gamble, industrial psychologist.

A document produced by the Joint Services Aero Medical Selection Committee and released on 22 December 1958 detailed the criteria by which candidates hoping to fly Project Astronaut were to be selected. The criteria, which were expected to attract 'professional risk takers', such as mountaineers, pilots and submariners, included:

Representatives from the military services and industry to nominate 150 men with backgrounds that demonstrated their willingness to accept and withstand great personal risk.
Candidates to meet specified career histories and qualifications.
Applicants to be sponsored by a responsible organisation.
From the 150, 36 to be selected for medical screening at the Lovelace Clinic.
From the 36, 12 to be selected for a 9-month training programme.
From the 12, 6 to qualify for manned space flight in Project Astronaut.

The criteria were submitted to President Eisenhower over the Christmas holiday, but he refused to approve them. Instead, he demanded that the name be changed as it placed too much emphasis on the pilot, and that the Research Astronaut Candidates be selected from the test pilots available within the US military. His reasoning was that the military test pilots would have an engineering background and were trained to record technical data under stressful conditions. They would also have first-hand knowledge of high *g* forces and microgravity. Finally, they were a highly motivated and goal-oriented group of people. The programme's name was subsequently changed to Project Mercury, which had been suggested by Abe Silverstein.

The New Year began with the drawing up of a new set of selection criteria, which were released to the military in early January 1959. Astronauts would now be required to be:

25 years old, or older, but less than 40 years of age at the time of application.
No taller than 1 metre 80 centimetres.
In excellent physical health.
Hold a Bachelor's Degree or equivalent in engineering.
Qualified military test pilot.
Graduate of Test Pilot School.
At least 1,500 hours flight time logged on high-performance jet aircraft.

The criteria prevented any women from becoming Research Astronauts Candidates. In America in 1958 the only way to obtain jet aircraft flying experience and graduate as a test pilot was to do so within the military. At that time there were no female test pilots in the US military. Therefore, no women could meet the criteria.

The selection criteria also cut out many of the individuals who had pioneered stratospheric flight. Chuck Yeager did not have a degree, so he could not apply. He would spend several years leading other test pilots in ridiculing the astronaut's role in Project Mercury. Test pilots Scott Crossfield and Neil Armstrong, both X-15 pilots, were civilians and were therefore unable to apply. Other military test pilots either decided off of their own backs or were advised by their senior officers not to apply, as they did not expect the manned space programme to last much beyond Project Mercury. Many who did meet the criteria did accept NASA's offer of attending a presentation on a new flying project of national importance.

Of those involved with Project Manhigh, John Paul Stapp and David Simmons were military physicians, not test pilots. Simmons tried to get a position at the head of the astronaut biomedical research programme, but failed. Clifton McClure did not meet the criteria and was thus excluded. Kittinger met all of the requirements and discussed his decision whether or not to apply with Stapp. In the end Kittinger decided to remain with Project Excelsior, which he was working on at the time.

As military pilots, successful candidates would retain their military rank and pay and would serve with NASA on a detached duty basis. The initial detachment would be from the astronaut's selection through early 1962. Thereafter, NASA would have to make a year-by-year application to the appropriate military service for each man that they wished to retain.

Beginning on 21 January the records of 508 test pilots were reviewed by the Military Personnel Bureau in Washington. Of those 508, 110 men were found to meet the new selection criteria. The qualified pilots were formed into three groups to assist in the second stage of selection. Each of the 36 men in the first group and the 34 men in the second group were invited to attend Washington in civilian clothes, to be individually briefed on Project Mercury. During their interview the pilots were invited to ask any questions. At the end of the briefing they were asked if they were interested in volunteering for the project. A positive answer meant that they were prepared to undergo the next round of selection procedures. When 24 men from the first group and eight from the second agreed to proceed, it was decided not to offer interviews to the third group of 40 pilots who had met the original selection criteria.

Beginning on 7 February the first five volunteers reported to the Lovelace Clinic, where they underwent a one-week long round of tests designed to highlight their strengths and weaknesses. Four subsequent groups of five and one group of two men attended on each following Saturday until all 32 men had been tested. The volunteers were told at the outset that the results of the tests would not be placed on their military records and therefore could not have any bearing on their future military careers if they were not selected as a Research Astronaut Candidate.

The tests at Lovelace were of a physical nature to determine which of the men were the fittest. The volunteers were tested from 0700 to 1800 every day for 7.5 days and even then some tests continued on into the evening. Testing covered all areas, including, haematology (blood), pathology (tissue), roentgenology (X-rays), ophthalmology (eyes), otolaryngology (ear, nose and throat), cardiology (heart and circulation) neurology (nerves), mycology (muscles) and general medicine.

There were 17 separate eye tests and 30 tests of blood, urine and tissue. Each man's brainwaves and heart performance were monitored and recorded. The men were submerged in warm water to give the specific gravity of their bodies. 'Load tests' defined each individual's capabilities under laboratory conditions. They pedalled a stationary cycle against increasing breaking loads, to allow their pulmonary (lung) capacity to be defined. Their total body radiation was recorded as they lay in a special trough. Cold water dripping into their ears allowed their susceptibility to motion sickness to be defined. Blood mass, water volume and lean body mass were all measured and recorded.

Despite everything, not one volunteer left the programme of his own accord. Only one subject had to drop out of the selection process at this stage. Naval Aviator James Lovell had a temporary liver complaint, which prevented him from progressing further with the selection process. The 31 successful candidates then moved on to a series of physiological and stress tests and anthropological measurements at the Aero Medical Laboratory, Wrights Air Development Centre, Dayton, Ohio. They attended in groups of five, for six days, beginning on 15 February and each subsequent week thereafter.

In 1959, manned space flight represented a new era of exploration and many physicians were of the opinion that humans would be unable to function in the space environment. Some suggested that astronauts would not be able to swallow in microgravity; others feared that the muscles of the eye would be unable to function.

As a result of this lack of solid knowledge, the tests at this stage were some of the most comprehensive that any military personnel had even been subjected to.

There were 25 psychological tests, including Rorschach 'ink blot' tests whereby subjects had to say what they saw in a series of standard printed patterns. They also had to complete numerous written apperception tests, which required them to write stories on what they saw in pictures that they were shown. Everything they did and said was carefully recorded.

Each man had to complete a 566-item questionnaire about himself and 255 pairs of self-descriptive statements, which combined to give a personal preference schedule. Another 20 pairs of self-descriptive statements went into compiling a personal inventory for each man. Finally, a further 52 statements gave the psychologists a list of each man's personal preferences.

One group of stress tests were more physical, in which subjects were expected to maintain their pace on a treadmill that was elevated at one end by 1 degree per minute until their heartbeat reached 180 times a minute. In another test they had to step up and down from a bench every 2 seconds for 5 minutes.

The 'bake chamber' heated them to the point where it was considered medically dangerous to heat the body core any higher. The test consisted of sitting in a chamber heated to 180 °F for 2 hours. They also spent 3 hours in the isolation chamber, where they were denied any external stimuli, including a light in the chamber itself. On the opposite end of the scale, they rode a centrifuge to 9*g*, facing forward (eyeballs in) and backward (eyeballs out), and their ears were subjected to sounds at various frequencies, to determine the limits of their hearing. Reaction to extreme noise levels was tested along with high-frequency vibration testing that made their entire body shake. A 'simulation chamber' offered 12 signals, which had to be pressed in the correct order while subjected to confusing stimuli. The test subjects called the device the 'idiot box'. The subjects' bare feet were placed in ice water while their blood pressure and pulse rate were recorded both before and after the test. Alternatively, they wore partial pressure suits and were subjected to an hour at simulated extreme high altitude in a vacuum chamber.

Throughout everything, each man had been just a number, his name did not appear on any of the test paperwork. At the end of their two weeks of tests each man returned to active service with his unit, to await the results. The results from both sets of test were then screened by a group of military flight surgeons, who selected those who moved on to the final stage of the selection. The records of all the pilots tested were put in order of 'medical suitability' to face the challenges of manned space flight, with the best suited man at the top of the list and the least suited at the bottom. At the end of this period only 13 men were removed from the selection process.

The results achieved by each of the 18 remaining applicants were screened by a group of senior NASA and STG managers, who selected the six men that would become Mercury astronauts. In the end, it proved impossible to select just six, so the list of Research Astronaut Candidates forwarded to NASA Headquarters by Gilruth had seven names on it. The list was passed to Silverstein, for onward transmission to Administrator Glennan.

Fig. 17. From military test pilot to astronaut) The seven Flight A astronauts pose alongside a spacecraft mock-up. (Left to right) Carpenter, Cooper, Glenn, Grissom, Schirra, Shepard, Slayton. (NASA)

On 1 April Dolan, Gilruth's deputy, took it upon himself to telephone each of the seven men in question, to offer them a position and ask them if they still wished to work with NASA and train for manned space flight within Project Mercury. All seven accepted the offer and Glennan sanctioned the list of names on 2 April. The Flight A astronauts reported for duty in Washington, DC, on 8 April 1959. (See Fig. 17.)

Seven young pilots were introduced to the media at a press conference held at NASA Headquarters on 9 April 1959. They were positioned in alphabetical order behind a table, on a stage in the old ballroom at Dolly Madison House. Glennan opened the press conference with the words:

> Ladies and gentlemen, today we are introducing to you and to the world, these seven men who have been selected to begin training for orbital space flight. These men, the nation's Project Mercury astronauts, are here after a long and perhaps unprecedented series of evaluations, which told our medical

> consultants and scientists of their superb adaptability to their up coming flight. It is my pleasure to introduce to you, and I consider it a very real honour gentlemen ... the nation's Mercury astronauts.

He named the seven men in alphabetical order (see Table 2) as:

Malcolm Scott Carpenter	Lieutenant US Navy
Leroy Gordon Cooper	Captain US Air Force
John Herschel Glenn	Lieutenant Colonel US Marine Corps
Virgil Ivan Grissom	Captain US Air Force
Walter Marty Schirra	Lieutenant Commander US Navy
Alan Bartlett Shepard	Lieutenant Commander US Navy
Donald Kent Slayton	Captain US Air Force

Although Glennan had given the press corps the astronauts' real names, it soon became common knowledge that they liked to be known otherwise. The seven men in Flight A would become known to the world at Scott Carpenter, Gordon Cooper, John Glenn, Gus Grissom, Wally Schirra, Al Shepard and Deke Slayton.

It quickly became clear that Glenn was far more expressive in his answers to questions than his six colleagues. While the other six would give short answers, Glenn said things that the journalists wanted to hear and could quote in their articles. Among Glenn's most quoted answers to journalists' questions were:

> I think we are very fortunate that we have, should we say, been blessed with the talents that we have been picked for something like this. I think we would be almost remiss in our duty if we didn't make full use of our talents. Every one of us should feel guilty, I think, if we didn't make the fullest use of our talents in volunteering for something that is as important as this is to our country and the world in general right now.

> I am a Presbyterian, a Protestant Presbyterian, and I take my religion very seriously as a matter of fact ... I was bought up believing that you are placed on earth here more or less with sort of a 50–50 proposition, and that is what I still believe. We are placed here with certain talents and capabilities. It is up to each of us to use those talents and capabilities as best you can. If you do that, I think there is a power greater than any of us that will place the opportunities in our way.

> I don't think any of us could really go on with something like this if we didn't have a pretty good backing at home, really. My wife's attitude towards this has been the same as it has been all along through my flying. If it is what I want to do, then she is behind it, and the kids are too, 100 percent.

The journalists applied Glenn's replies to all seven members of Flight A. The public image of the clean-cut, patriotic, God-fearing astronaut was born at this first press conference. In fact, Glenn was probably the only member of the group who lived up to the public image. The others were more typical of the young military pilots that they were.

Glennan had introduced the astronauts to the press but he was not close to them. Like President Eisenhower, he believed that once Project Mercury had placed the first man into space, the Soviet Union would concede the 'Space Race' to the Americans. At that time Project Mercury would be dismantled and the astronauts returned to more standard military service. Meanwhile, the seven astronauts were detached from their military units to NASA.

On 27 April Flight A began work with STG at Langley, where they reported to Gilruth. The 'Astronaut Office' was in a two-storey building on Langley Field, a USAF base close to the NASA LaFRC. The single office with its seven heavy wooden desks served as their base, but as they became busy with their role in Project Mercury they spent less time in it. The group shared a secretary named Nancy Lowe.

Carpenter and Cooper moved their families into military housing on Langley AFB. Glenn lived in the base's Bachelor Officer's Quarters during the week and commuted to his family home in Washington, DC, at weekends. Grissom, Slayton and Schirra bought houses on a new development close to Langley AFB. Slayton and Grissom lived next door to each other, with Schirra just up the road. Shepard moved his family to military housing on the Naval Air Station at Oceana, where the Navy had posted him.

Table 2: Project Mercury Crew Assignments

Flight	Prime pilot	Back-up pilot	Announced
Mercury Redstone 3	Alan Shepard	John Glenn	21.02.61
Mercury Redstone 4	Virgil Grissom	John Glenn	15.07.61
Mercury Atlas 6	John Glenn	Malcolm Carpenter	29.11.61
Mercury Atlas 7	Donald Slayton	Walter Schirra	29.11.61
	Malcolm Carpenter	Walter Schirra	15.03.61
Mercury Atlas 8	Walter Schirra	Leroy Cooper	27.06.62
Mercury Atlas 9	Leroy Cooper	Alan Shepard	13.11.62

7.2 THE ROLE OF THE ASTRONAUT

Prior to commencing the search for Research Astronaut Candidates, STG had defined what an astronaut would be required do when he joined Project Mercury. The astronaut's role was broken down into ten areas.

7.2.1 Design of Mercury Spacecraft

Although the Mercury Spacecraft was already under development by the time Flight A reported to Langley AFB, the seven astronauts were closely involved in its final design. They took part in spacecraft mock-up reviews (see Fig. 18) and attended the contractor's factory to assist in fit and function tests. Engineering and flight-test were two areas where the astronauts' personal experiences were able influence the spacecraft that they were to fly.

Fig. 18. A wooden Mercury Spacecraft, Launch Escape System and Launch Vehicle Adapter shown during a mock-up review at McDonnell's St Louis factory. (NASA)

During one spacecraft mock-up review, on 10–11 September 1958, the astronauts made several demands for changes to the vehicle that they would ride into space. First was an optional manual control system, throughout the orbital phase of the flight. This would allow them to control the spacecraft if the automatic system failed.

They also wanted a manual option to fly their launch vehicle in the case of a malfunction in its automatic systems. Second, they wanted the two small portholes, one on either side of the astronaut's head, replaced by a large 'heads-up' window. This would allow them to take star sightings and gauge the spacecraft's attitude visually. Third, they asked for the side hatch to be replaced by one that employed shaped charges, to cut through the torque bolts that secured it in position, thereby allowing for a quick escape in an emergency. Such explosive hatches were used in military aircraft. They also requested changes to the main instrument panel that would give them a greater amount of information on how their flight was progressing. Lastly, and least importantly, they demanded that people stop referring to the 'Mercury capsule' and refer to the 'Mercury Spacecraft' instead. Despite this demand, most of the astronauts still referred to 'the capsule' in radio communications during their space flights.

The engineers building the spacecraft and launch vehicles thought it unlikely that the astronauts would be able to control their vehicles manually at the velocities at which they would be travelling. As a result, they argued against the changes. The astronauts counter-argued and, after several months, finally won all of their demands.

7.2.2 Development of operational procedures

No one had ever flown in space before; therefore there were no procedures that could be copied for use in Project Mercury. The astronauts worked with hardware contractors and STG's Operations Division to define the procedures they would use in flight. Many of the procedures would be based on those used in military flight-tests.

7.2.3 Development of in-flight test equipment

In flight the astronaut was expected to carry out flight-test procedures on equipment carried in and on the Mercury Spacecraft. This included performing scientific experiments carried on orbital flights. The astronauts were involved in the development of the equipment concerned.

7.2.4 Public relations

The members of Flight A were involved in public relations events relating to the space programme, including press conferences and public-speaking engagements. Sometimes they appeared as a group, on other occasions they appeared as individuals.

In an attempt to control the astronaut's media exposure and allow them to concentrate on their training, NASA employed lawyer Leo DeOrsey, who refused to be paid for his services, to represent them. DeOrsey proposed that they sell their personal stories to one magazine, the highest bidder in an auction. Bidding began at $500,000 for exclusive access to the astronauts and their families. All articles would have to be cleared through NASA and the astronauts themselves would have final editing rights.

Only one magazine, *Life*, made a firm offer of $500,000, and thereby won the contract, which was signed on 5 August 1959. Under the deal *Life's* $500,000 would be split equally between the seven astronauts, giving them approximately $72,000 each over the three years that Project Mercury was expected to last. Payments would be made to all seven men, regardless of whether or not an individual astronaut flew in space, or if one of them became the first man in space. In return for the money only *Life* could place a reporter and a photographer in an astronaut's home when he was in space. The *Life* reporters would have exclusive rights to the life story of the astronaut and his family. The controversial contract led to a series of articles that perpetuated the myth that all seven members of Flight A, as well as their wives and children, were formed in John Glenn's image.

Life portrayed all seven men and their families as the personification of the American dream. The articles showed the astronauts as God-fearing, hardworking career officers in the US military, who had answered their country's call. They were seen as loving family men, with dedicated wives and happy children. For Glenn at least, the image was true. The patriotism of the other six was never in doubt; they were all serving military officers. Their commitment to their wives, children and religion were less well defined. The other six astronauts were fast-living, hard-working military pilots. They enjoyed flying fast aircraft and driving fast cars. If the stories are to be believed, they also enjoyed the young women who made themselves available. With long hours, often days and weeks away from home, the astronauts saw little of their wives and children. Even so, divorces were rare in the early years, because they threatened an astronaut's career.

Despite the *Life* contract, the astronauts maintained a good relationship with journalists and reporters throughout the world press and media. Only the astronauts' personal stories were restricted by the *Life* contract. In all other areas they remained available to the world media.

In the earliest days of the Project NASA's public relations representative, Air Force Lieutenant Colonel Shorty Powers, made the astronauts available to the press and media throughout the working week. This eased off as the astronauts spent less time at Langley field.

7.2.5 Sequence monitoring

Throughout the flight the astronaut monitored the automatic in-flight procedures. If an automatic system failed he used the manual override to ensure that flight procedures were completed in the correct sequence.

7.2.6 Systems management

The astronaut monitored the onboard systems, reporting their status to Mercury Control regularly while in flight. In the event of a malfunction he attempted to make corrections and, if they failed, he would make an emergency return to Earth.

7.2.7 Attitude control

The astronaut had the capability to manoeuvre the spacecraft's attitude to meet the requirements of the flight. Major attitude manoeuvres, such as turnaround and those for retrofire and re-entry, were controlled by an onboard programmer. In the event of a malfunction the astronaut used the spacecraft's manual control systems to adopt these attitudes and maintain them.

7.2.8 Navigation

The spacecraft's position at retrofire was critical. The astronaut was trained to determine his spacecraft's position in orbit at any time throughout the flight, and to calculate the correct retrofire time.

7.2.9 Communications

The astronaut maintained two-way radio communication with the World-Wide Tracking Network, keeping Mercury Control Centre (MCC) updated on his personal condition and that of his spacecraft.

7.2.10 Research observations

The astronaut observed the spacecraft instruments and reported the condition of his medical and the spacecraft's mechanical condition. He reported those conditions to MCC and acted upon the instructions that he received in reply. During orbital flights the astronaut made observations of a number of scientific experiments.

To meet their expected role the members of Flight A brought a number of the test pilot's professional skills to Project Mercury, including:

- High intelligence.
- A good engineering knowledge and education.
- Knowledge of operational and flight-test procedures.
- Scientific knowledge and research skills.
- Psychomotor skills similar to those required for flying high-performance aircraft.
- Good stress tolerance.
- Good decision-making abilities.
- The ability to work both as an individual and as a team member.
- Emotional maturity.
- Freedom from disease or disabilities.
- Resistance to the physical stresses of space flight.
- Medium size, allowing a spacecraft to be designed to be lifted into orbit on an unaltered Atlas ICBM.

7.3 GROUP TRAINING

The astronauts' training programme was split into two sections. Two years of group training was designed to integrate the seven men into Project Mercury. Group training took place in five specific areas.

7.3.1 Basic astronautical science instruction

Flight A's first week with STG was spent attending a series of general introduction lectures, to give them a basic knowledge of Project Mercury. Table 3 shows the

Table 3: Lectures given to astronauts during first week of training

Subject	Hours
Elementary mechanics and aerodynamics	10
Principles of guidance and control	4
Navigation in space	6
Elements of communication	2
Space physics	12
Basic physiology	8

number of hours the group spent on each of the subjects covered during this period. After the first week each man spent approximately half of his time working at Langley. Mornings were generally given over to lectures, while afternoons were passed in training towards space flight.

7.3.2 Systems briefings

STG personnel provided an initial series of lectures on the Mercury Spacecraft systems. These were followed by lectures on operational areas. More in-depth lectures took place at contractors' plants. Lectures were repeated throughout the two-year group-training period as required.

7.3.3 Contractor visits

Visits to contractor sites were made for two reasons. The astronauts were directly involved with the design and development of their spacecraft and their launch vehicles. Secondly, they gave public relations addresses to contractor employees. This was done with the intention that the employees would perform their work to a higher standard if they had met and spoken to the men whose lives would depend on the machines they were building.

During one solo visit to Convair, where the Atlas ICBM was produced, Gus Grissom, a man of very few words and uncomfortable making public addresses, made a speech consisting of just three words, '*Do good work*'. They became the motto of the Convair workers building the Mercury Atlas launch vehicles. Table 4 shows the time that each man spent in contractor visits during his first year as an astronaut.

7.3.4 Manuals

Each major contractor produced a manual on its hardware to help the astronauts become acquainted with it. The manuals were used extensively during the astronauts' training period.

7.3.5 Specialisation assignments

STG management believed that Project Mercury was too large for one man to take in everything. In July 1959, each member of Flight A was given an area of specialisation. Each astronaut concentrated on his own area of specialisation and

Table 4: Astronauts' contractor visits during first year of training

Contractor	Days
McDonnell, St Louis (Spacecraft)	10
Convair Astronautics, San Diego (Atlas)	5
Cape Canaveral (Launch facilities)	2
Redstone Arsenal, Huntsville (Redstone)	2
Edwards Air Force Base	2
Space Technology Laboratories, El Segundo	2
Air Force Ballistic Missile Division	2
Rocketdyne, North American Aviation (Atlas rocket motors)	1
AiResearch (life support system)	1
Protection Incorporated, Los Angeles (helmet)	1

reported to his six colleagues once a week, or more often if the situation required it. Areas of speciality were matched as closely as possible to each individual astronaut's qualifications and previous work experience and were allocated as follows:

Carpenter: Communication and navigation.
Cooper: Redstone launch vehicle.
Glenn: Mercury Spacecraft Crew Compartment layout.
Grissom: Spacecraft automatic and manual control systems.
Schirra: Spacecraft's life support system and the astronauts' pressure suit.
Shepard: Tracking and recovery.
Slayton: Atlas launch vehicle.

7.4 TRAINING DEVICES

The majority of training for space flight was carried out using a series of training devices that were the forebears of today's computer-driven simulators. In 1959 these training devices were limited by the fact that no one had flown in space and, therefore, no one knew exactly what to expect. They were further hindered by the fact that each device could only reproduce one, or at most only a few, of the conditions that would be experienced by an astronaut in flight. The following list describes the various training devices and methods in the order in which they became available to the astronauts.

7.4.1 Analog Trainer 1 and Trainer 2

Analog Trainer 1 was the first simulator made available to the astronauts as part of Project Mercury, in April 1959. Langley engineers used this trainer to define the Mercury Spacecraft flight control tasks. This was also the trainer on which the seven astronauts were first introduced to those tasks.

Analog Trainer 2 was online by late 1959. It employed a computer from an old F100 gunnery trainer. The second trainer included an old Styrofoam prone couch and a Mercury Spacecraft three-axis hand controller. The trainer was used to offer the astronauts their first practice at flight control tasks wearing a pressure suit, as well as for a number of engineering feasibility studies.

Each astronaut spent approximately 18 hours on these two trainers.

7.4.2 Jet aircraft proficiency training

Despite being heralded in the American press as the country's best pilots, the astronauts originally had no high-performance aircraft on which to maintain their flying skills. This meant that they wasted many hours waiting for scheduled airline flights and using rental cars to meet their many widespread commitments. It also meant that they lost money. Being military pilots, four hours jet flying per month could earn each astronaut an additional $190 'flying allowance'.

Cooper made the group's discontent over this situation known to a reporter from the Washington Post, who made their opinions public knowledge. This was the first of several occasions when Cooper's actions annoyed senior NASA and STG management. Shortly afterwards STG took delivery of two T-33 jet training aircraft. These were replaced in August 1959 by two F-102 jet fighters, which were replaced in their turn by two F-106s. The aircraft became the astronauts' personal transports and were used by them to fly to the many locations that they were required to attend.

Each astronaut spent approximately 460 hours in these aircraft.

7.4.3 Centrifuge

The astronauts completed four programmes of centrifuge training, using the facility at the Aviation Medical Acceleration Laboratory's centrifuge at the Naval Air Development Centre, Johnsville, Philadelphia, commencing in August 1959. The first two programmes combined engineering feasibility and initial astronaut familiarisation runs. The last two programmes offered accurate representations of the *g* force profiles to be expected during lift-off and re-entry during Mercury Redstone and Mercury Atlas flights.

During the tests the interior of the centrifuge gondola was fitted out to represent the Crew Compartment of the Mercury Spacecraft and the test subjects sat in prototype prone couches and operated a spacecraft three-axis hand controller. They wore full pressure suits and some runs were made at simulated extreme altitude.

High acceleration forces required the astronauts to learn a special breathing technique, to stop the loss of peripheral vision due to the lack of oxygenated blood. In the early stages of centrifuge training, concentrating on this breathing technique meant that the test subjects fell behind on their simulated flight tasks. As the breathing technique became 'routine', the subject was able to breathe correctly and concentrate on the simulated flight tasks.

The astronauts were subjected to both 'eyeballs in' and 'eyeballs out' centrifuge rides as well as rides where the gondola moved around an axis so that the astronaut passed rapidly from one to the other. The seven astronauts came to the decision that 16*g* in either direction was the maximum they could take and still perform their tasks. Each man spent approximately 48 hours riding the centrifuge.

7.4.4 Air-Lubricated Free-Attitude (ALFA) Trainer

NASA introduced the ALFA Trainer in October 1959. The trainer moved on an air bearing and offered 360 degrees of freedom in roll and 35 degrees in both pitch and yaw. The astronaut lay in a mock-up prone couch facing a spacecraft-style instrument panel. He operated air-jets using a spacecraft three-axis hand controller to simulate use of the Manual Proportional Control and Fly-by-Wire attitude control systems in the Mercury Spacecraft. Both the view through the spacecraft periscope and the overhead window were simulated, while an actual Mercury Spacecraft gyroscope package and attitude display instruments were mounted on the trainer. Each astronaut spent approximately 12 hours training on the ALFA Trainer.

7.4.5 Slowly revolving room

This device was located at the Naval Air Station, Pensacola, and was used by the Navy to associate their aviators with disorientation. While a closed room revolved slowly around its central axis the test subject had to walk along a straight line from the outside of the room towards the centre. The astronauts used the facility for 1 hour each, in October 1959.

7.4.6 Microgravity aircraft training

Commencing in December 1959, the astronauts undertook three phases of microgravity training, flying in aircraft that followed a parabolic trajectory, resulting in microgravity conditions at the apex of each parabola. At no time was it possible to simulate microgravity for longer than 1 minute in any one parabola.

- Phase 1 took place in late 1959, with the astronauts flying in the rear seat of an F-100 jet fighter.
- Phase 2 was in March 1960, during which the astronauts were able to float around freely in the padded rear compartment of a C-131 transport aircraft.
- Phase 3 took place in September 1960, using a converted C-135 cargo aircraft.

Each astronaut experienced a total of approximately 0.7 hour of microgravity.

7.4.7 Chapel Hill Planetarium

In February 1960 the astronauts took the first in a series of lectures at Chapel Hill Planetarium, which belonged to the University of North Carolina. The course was designed to help the astronaut to recognise the star field that he would see through his spacecraft window while in orbit. It was hoped that the astronaut would be able to recognise when his spacecraft was not in the correct attitude and override the automatic control systems, to bring it back to the correct attitude. Each man spent approximately 28 hours studying star fields.

7.4.8 Multi-Axis Spin-Test Inertia Facility (MASTIF) Trainer

The MASTIF was developed by NASA's LFRC, so that the prototype capsule to be flown on the Big Joe flight could be rotated in three axes, at up to 60 revolutions per minute, to test its attitude control system. When that task was complete consideration was given to using the MASTIF to help train the astronauts. It was introduced into the astronauts' training programme in February 1960. By that time MASTIF consisted of a prone couch-mounted within three gimbals, a three-axis hand controller and Mercury Spacecraft style rate indicators.

The training device spun at 30 degrees per minute in pitch, yaw and roll, all at one time. The astronaut let the machine build up to its full tumbling rate and then used a gaseous nitrogen fly-by-wire control system in conjunction with the rate indicators to bring the MASTIF under control. The exercise represented the actions necessary to bring about a recovery from a loss of control in all three spacecraft manoeuvring axes at the same time. The astronauts considered the situation to be extremely unlikely to arise. Each man spent approximately 4 hours on the MASTIF.

7.4.9 Egress trainer

McDonnell supplied a full-scale boilerplate Mercury Spacecraft (SC-5) for use in egress training, which was completed in three phases. Prior to commencing egress training all seven men underwent a scuba diving training course, which Glenn's son also attended.

- Phase 1 began in February 1960. SC-5 was placed in Centre Hydrodynamic Basin 1 at LaFRC and each astronaut made several egresses through the neck of the spacecraft dressed in military flight suits under static conditions and simulated waves up to 0.6 m in height.
- Phase 2 began in April 1960 and was conducted in open deep water off Pensacola Naval Air Station. During one day at sea each astronaut completed, through the neck and side hatch, egress procedures in pressure suits. A second day was spent rehearsing water survival techniques in the Pensacola Naval Air Station's training tank.
- Phase 3 commenced in August 1960 and took place in the Centre Hydrodynamics Tank 1 at LaFRC. Each astronaut made six egresses from SC-5, which had been submerged in the tank. Three egresses were made in flight suits and three in pressure suits. Each astronaut spent approximately 25 hours in egress training.

7.4.10 Water survival training

Water survival training was generally carried out in conjunction with egress training. In March 1960 the astronauts received a series of survival lectures and a training film. They then practised survival techniques in the training tank and in open water as part of their egress training programme the following month. (See Fig. 19.)

7.4.11 Procedures Trainer 1 and Trainer 2

Two Procedures Trainers were developed and used for the majority of Mercury Spacecraft systems training. These devices in particular were the forebears of today's simulators. The astronauts considered them the most useful of all the training devices that they used.

The Procedures Trainers simulated the spacecraft interior accurately. There was

Fig. 19. The Flight A astronauts are shown during Water Survival Training using a boilerplate spacecraft. Note the astronaut leaving the spacecraft through the Recovery Section tunnel. (NASA)

provision for the astronaut's pressure suit to be pressurised, but no Crew Compartment atmosphere was provided. Rather, normal life support system operation was simulated on the 'spacecraft's' instruments. The trainers were used to train the astronaut in flight procedures. When the astronaut had mastered normal operations, a simulation supervisor seated at a desk nearby could introduce the effects of 275 separate spacecraft systems malfunctions, which the astronaut was expected to overcome.

The spacecraft prime contractor provided the two trainers in June 1960. Trainer 1 was housed in the Full Scale Tunnel at Langley AFB, where it could be used in conjunction with the simulated World-Wide Tracking Network remote site at that location. This trainer could simulate all 22 combinations of manual and automatic control systems available in the Mercury Spacecraft. An active periscope display consisted of a dot on a cathode-ray tube, which was activated by the hand controller and the analog computer. Prior to Mercury Atlas 9, the final manned flight, a Virtual Image Celestial Display was added to this unit.

Trainer 2 was installed at Cape Canaveral, where it was used in conjunction with the Mercury Control Centre. Three months after delivery a small capacity general-purpose computer was fitted to the unit, which limited its use to simulating only the spacecraft's manual control systems. In mid-1963 a second computer made it possible to simulate rate damping during re-entry.

In using these trainers, the astronauts were first introduced to each spacecraft system and its possible malfunctions. Thereafter, they began training for both suborbital and orbital flights. As they became more proficient, so more malfunctions were introduced into each training session. Each astronaut spent approximately 101 hours on the Procedures Trainers. (See Fig. 20.)

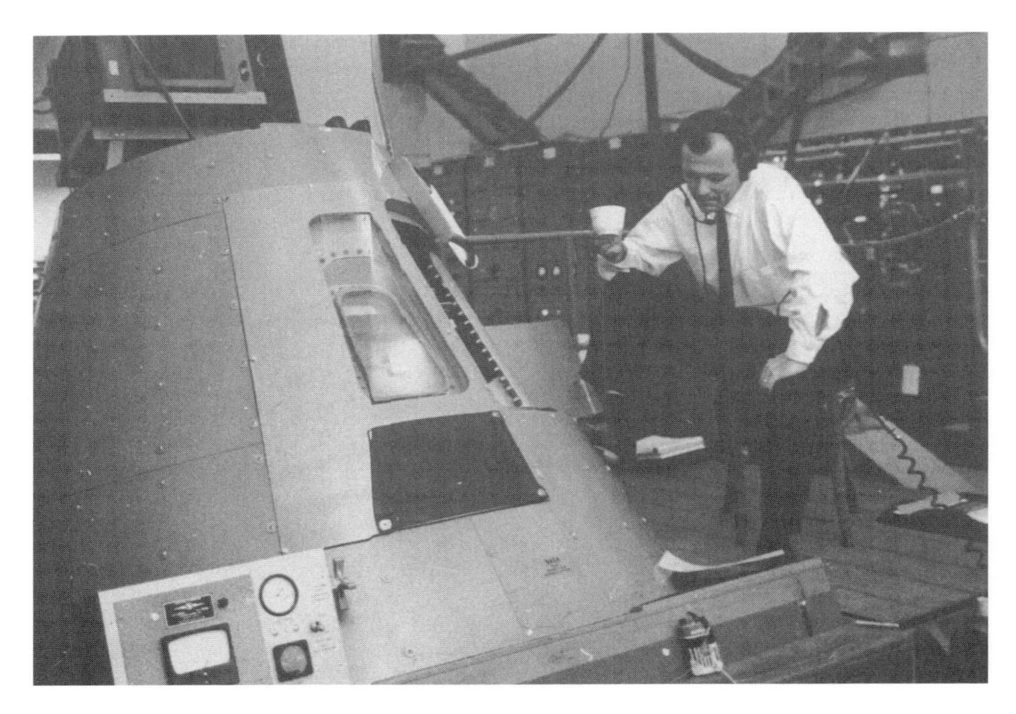

Fig. 20. An external view of the Mercury Procedures Trainer. This was the forerunner of the computer-driven simulators used by today's astronauts. (NASA)

7.4.12 Desert survival training

In the unlikely event that an emergency re-entry resulted in a landing in one of the world's deserts the astronauts were trained to survive until they could be located and recovered. To that end they attended a 5.5-day desert survival course at the Air Force Survival School, Stead AFB, Nevada, in July 1960. The course was in three phases.

- Phase 1 consisted of 1.5 days of lectures related to survival techniques.
- Phase 2 was a 1-day course on the utilisation and care of available clothing, spacecraft, and survival equipment.
- Phase 3 was a 3-day remote site survival exercise, during which the astronauts were left in the Nevada desert and had to apply what they had learned in the previous two phases of the course.

7.4.13 Environmental Control System Trainer

The Environmental Control System Trainer was delivered by the spacecraft prime contractor in November 1960 and installed in a man-rated vacuum chamber at US Naval Air Crew Equipment Laboratory, Philadelphia.

All seven astronauts participated in familiarisation training during December 1960 and January 1961. In that time each man was familiarised with the spacecraft's environmental control system and how it reacted with the pressure suit. Each astronaut was subjected to a simulated re-entry and a 2-hour simulated post-landing phase.

Each man spent approximately 3 hours in this trainer. A capability also existed to rehearse pressure suit functions in the Procedures Trainers and each astronaut participated in the pre-flight checkout of the environmental control system in his own spacecraft.

7.4.14 Attitude instrument display mock-up

NASA provided this device in January 1961. It consisted of a half-scale model of the Mercury Spacecraft mounted on a four-gimbal all-attitude display. The mock-up contained actual Mercury rate and attitude indicators. The cover was left off the attitude gyroscopes, so the astronauts could see how they moved in reaction to the spacecraft's motions.

The device illustrated how the spacecraft attitude display indicators could give incorrect readings when the gyroscope axes and the spacecraft's axes were not parallel. It was used to teach the astronauts how to recover from the situation just described. Each astronaut spent approximately 10 hours working with the device.

7.4.15 Ground recognition trainer

NASA brought this trainer online in April 1961. The astronauts lay in a prototype prone couch and observed a projection of Earth's continents through a Mercury Spacecraft periscope. The view accurately represented that from a spacecraft in orbit.

The trainer was intended to make the astronauts aware of the compression that occurred in the periscope image at the wide-angle view setting. The astronauts only used this trainer for approximately 1 hour each. The periscope image was also available on the ALFA Trainer.

7.4.16 Yaw recognition trainer

NASA developed the Yaw Recognition Trainer prior to the Mercury Atlas 8 flight, bringing it into the astronauts' training regime in September 1962. The trainer was designed to help the astronauts become familiar with the visual clues available out of the spacecraft's overhead window to identify the spacecraft's yaw attitude.

An 11-m diameter convex screen was used to hold a projected display of the Earth's surface, or a constant altitude cloud base. Images of real clouds were projected on the screen so that they were moving at the correct speed. The astronaut viewed the screen from a platform that placed him 0.6 m away from its centre. To make the image more realistic he wore a box over his head with a cutout representing the spacecraft window in the correct size and position relative to his head.

The Mercury Atlas 8 and 9 prime and back-up astronauts used the trainer for approximately 2 hours each. Other astronauts who had orbited Earth on earlier Mercury flights said that it accurately represented the visual yaw motion cues visible out of the spacecraft window when in orbit.

7.4.17 Visual Image Celestial Display

State-of-the-art electronics in 1959 and the requirement to meet an early delivery date had stopped Mercury Procedures Trainer 1 having any simulated external view, other than the simulated periscope view. The Ferrand Optical Company developed the visual Image Celestial display in time to have it fitted to Procedures Trainer 1 in May 1963, for use in the pre-flight training for Mercury Atlas 9, the final manned flight. Setting ball bearings of various sizes inside a 12-inch diameter sphere and illuminating them produced a simulated star field. This was displayed to the astronaut in the Procedures Trainer. The Mercury Atlas 9 prime and back-up astronauts each spent approximately 2 hours training with this device.

7.4.18 Personal physical training

Physical training was left up to the individual after Flight A turned down the offer of a formal programme of callisthenics. In October 1959, Slayton, Shepard, Carpenter and Schirra all gave up smoking. The other three astronauts did not smoke.

7.5 PRE-FLIGHT PREPARATION

Approximately 3 months prior to a flight a dedicated primary astronaut and back-up astronaut were named and began a training programme specifically dedicated to that flight. As most back-up astronauts served as the primary astronaut for the following flight this gave most of the Flight A astronauts 6 months of flight-specific training. Those astronauts flying the last few Mercury flights found that up to 2 years had elapsed since the end of the group training programme and that many significant changes had occurred in the spacecraft configuration in that time. These astronauts found the intensive pre-flight training period particularly useful. Each astronaut was required to work at least six 10- to 12-hour days per week in the period leading up to his flight. During this time frequent reviews were made of his training programme, to ensure that nothing critical was overlooked.

During this time the astronauts also received refresher training on the centrifuge and other items covered in the Group Training period. Refresher courses held at Chapel Hill Planetarium were usually able to simulate the star field that an astronaut would see on the day of his launch. Pre-flight preparation was broken down into five areas.

7.5.1 Integration

Following delivery of the spacecraft to Hangar S at Cape Canaveral the astronaut had an opportunity to operate the actual spacecraft that he would fly in space, rather than spacecraft simulators. He participated in the final preparations of his spacecraft for launch, thus becoming familiar with it.

The astronaut's flight pressure suit was tested inside the spacecraft with which it would operate. The astronaut also had the opportunity to find the most comfortable life support system settings prior to his flight. He also became acquainted with the specific layout of instruments and controls within his spacecraft. Following the mating of the spacecraft to its launch vehicle the astronaut participated in tests and simulations carried out on the launch pad. He became familiar with his role during the countdown to launch and with the particular instrument lights and indications that accompanied those actions.

7.5.2 Systems

Astronauts received a series of lectures on individual spacecraft systems, from engineers involved in the checkout and countdown of his particular vehicle. These lectures were followed by emergency procedure exercises on the Mercury Procedures Trainer 2 at Cape Canaveral. The Procedures Trainer was kept up to date in order to reflect changes to individual spacecraft.

7.5.3 Flight plan

Each pair of astronauts assisted in the development of a flight plan for their flight. This was done to ensure that the astronaut was aware that flight procedures interlocked in such a way that none of them compromised the flight. Procedures were tested, updated and rehearsed in the Mercury Procedures Trainer 2.

7.5.4 Combined training

Using Mercury Procedures Trainer 2 astronauts were able to simulate their flights while communicating with the Mercury Control Centre at Cape Canaveral. At the same time, vehicle data displayed to the astronaut in the Procedures Trainer was displayed in the Mercury Control Centre, as it would be displayed during a real flight. In this way astronauts and controllers learned to work together. Post-exercise debriefing sessions were used to discuss any matters arising from the exercises.

These exercises were found to be of use to the flight controllers, and in particular the medical personnel, who used an astronaut's voice patterns as part of their range of ways of telling his medical condition. The astronaut found the combined exercises less useful as he was required to remain with his spacecraft in orbital configuration for too long at a time. Therefore, the astronauts frequently carried out a launch simulation followed by one orbit, retrofire and re-entry as a combined exercise. He was then free to spend more time practising emergency procedures in solo exercises with the Procedures Trainer simulation supervisor.

7.5.5 Medical

A physician called Charles Berry, who became commonly known as Chuck Berry, ran the NASA medical section. Berry had a number of USAF medical specialists working under him. Among them was Lieutenant Colonel Bill Douglas, who became the nearest thing the astronauts had to a personal physician. USAF nurse Dee O'Hara worked with Douglas and served as the astronauts' nurse. Military pilots usually have an open animosity towards the Flight Surgeons who have the power to stop them flying. Despite this most of the seven astronauts have stated that they considered both Douglas and O'Hara among their close friends.

7.6 GOTYAS AND TURTLES

Training for the early manned space flights was hard work. When they were working at LaFRC the astronauts returned home at the end of the day and spent the evening with their families. Glenn alone stayed in the bachelor officer quarters at Langley AFB. When they were working at Cape Canaveral things were different.

When the chimpanzees that would fly on the early Mercury Redstone and Mercury Atlas flights were installed at Hangar S, the astronauts moved into the Holiday Inn, Cocoa Beach. Henri Landwirth was the manager of the Holiday Inn and the astronauts befriended him. Landwrith turned a blind eye to many of their activities in his hotel. As a result, there were women in the town of Cocoa Beach claiming to have slept with six out of the seven astronauts.

The astronauts also became friends with Jim Rathmann, the owner of a local General Motors dealership. Rathmann gave each of them a deal on the sports car of their choice. Shepard, Grissom, Cooper chose Corvettes and spent much of their free time racing them. At one time Grissom and Cooper had Rathmann adjust Shepard's car so that he could not win. It was a practical joke; a tension breaker among friends who were preparing to do something that no one had ever done before. Similar practical jokes became part of the group's way of life. They went by the name of 'Gotyas'.

The astronauts also befriended the comedian Bill Szathmary, better known by his stage name of Bill Dana. Dana did a routine with a cowardly Mexican astronaut called Jose Jiminez, who had been chosen to become the first man in space but was scared and did not want to fly. The astronauts learned his routines off by heart and often quoted them in moments of stress. Shepard was Dana's biggest fan in Flight A and even had the chance to play Jose Jiminez live on stage with the comedian.

As military pilots the astronauts continued the military tradition of the Turtle Club. The joke lay in catching someone in a public position and then asking him or her, '*Are you a Turtle?*' The correct reply was '*You bet your sweet ass I am*'. The challenge was to ask the question when it was impossible for the person to answer. Incorrect answers, or a failure to answer, meant that the person giving the wrong answer had to buy a drink for everyone present. Several of the astronauts were asked if they were Turtles while they were in flight and broadcasting on an open radio link. They overcame the problem by recording their answer on their personal tape recorder. In their turn the astronauts frequently asked the question of those who worked with them.

8

Operations Division

In July 1959 STG was reorganised into three divisions, one of which was the Operations Division, under Charles Mathews. Walt Williams, the engineer who had headed the NACA Flight Research Centre at Edwards AFB was his deputy (see Fig. 21). Williams established the Operations Coordination Group at Cape Canaveral on 12 February 1960 with Christopher Kraft at its head. Like Williams, Kraft had served as an aeronautical engineer at the NACA's Langley Memorial Aeronautical Laboratory before being named as one of the 35 original members of the STG. Kraft oversaw the development and testing of the Word-Wide Tracking Network (WWTN) and the Mercury Control Centre, as well as the definition of the procedures required to launch, fly and recover the Mercury flights. In his turn, Kraft collected around him a group of engineers who would define exactly how Project Mercury's flight-test programme would be carried out. While many of the Division's personnel were new graduates, others had spent their careers within NACA, or in the military, where they had been involved in flight-test.

Because many of the Operations Division personnel, including Williams, had experience in flight-test, that was where they drew their example of flight operations. In flight-test the Operations Division planned the test and defined what the pilot

Fig. 21. Walter Williams, Head of the Space Task Group Operations Division. (NASA)

should do in order to complete the flight. Once the aircraft was in the air, however, there was little that the Operations personnel could do to assist the pilot if things did not go according to plan.

For the Manhigh III balloon flight a new form of Flight Control had been developed. A 'committee of experts' controlled the flight from the ground. The pilot reported any in-flight incidents to the ground by radio and the committee made a decision on how the incident should be dealt with and radioed their instructions to the pilot, who would act on them. The pilot would only be given control of the balloon in the event of an extremely serious incident over which the committee could exercise no control.

Flight Control during Project Mercury was a combination of these two approaches. The Operations Division defined the flight programme, including flight documentation, the astronauts' training programme, in-flight activities, and emergency procedures. When a manned flight was in space the Operation Division operated in a manner very similar to that pioneered by the Manhigh III panel of experts.

Hardware was delivered from the respective contractors to NASA facilities at Hangar S Cape Canaveral. There the hardware was prepared for launch by a combination of contractor and NASA personnel. The astronaut became acquainted with the hardware that he would fly into space in Hangar S. He rehearsed for his flight using both his spacecraft and the Mercury Procedures Trainer 2. These simulations were tied in to the simulations carried out by the men in the Mercury Control Centre (MCC) also located at Cape Canaveral. (See Figs 22 and 23.)

The Redstone and Atlas launch vehicles were prepared for launch by a combination of contractor, military and NASA Launch Operations Directorate personnel. The countdown to launch was controlled from a concrete blockhouse close to the launch pad, and minutes before launch the control was passed to MCC.

Command of each flight was exercised from the MCC, where Kraft was the Flight Director, but he remained answerable to Williams. Kraft had the people beneath him write many of the procedures that were used in Project Mercury. Those procedures were taken from aviation flight-testing where possible, but in many areas completely new procedures had to be defined simply because no one had attempted to launch a manned spacecraft before.

Eugene Kranz, an ex-USAF combat pilot and flight-test engineer, was responsible for developing the countdown procedures used to launch both the Mercury Redstone and Mercury Atlas vehicles. He also developed the Mission Rules – procedures that had been prepared in advance and would be followed in the event of an emergency. The Mission Rules defined the conditions that had to be met before a launch could take place, or before an ongoing flight would be allowed to pass on to the next phase. They also defined the conditions under which a flight would be aborted.

As the programme progressed and the number of in-flight failures increased, so did the Operation Division's knowledge base. Kranz, became responsible for training the second generation of flight controllers and ensuring that the first generation remained up to the role expected of them. He made each person in the

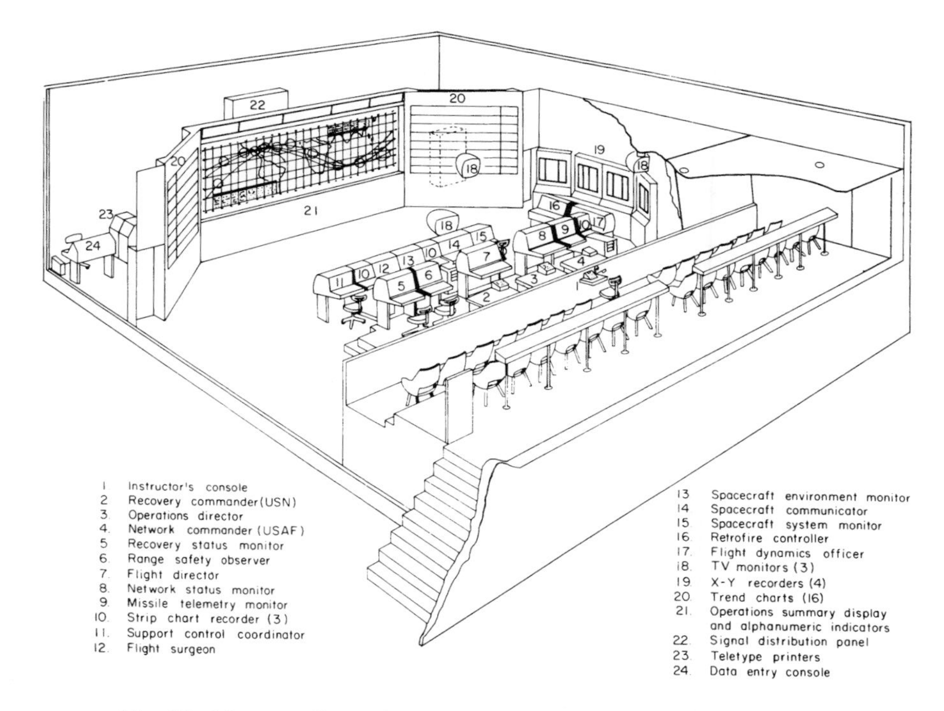

Fig. 22. Mercury Control Centre, Cape Canaveral, Florida. (NASA)

Operations Division responsible for learning one spacecraft system in-depth. That person then briefed all the others on his system. In this way the Operations Division personnel built up a knowledge base on the spacecraft and its two launch vehicles that was far greater than that held by the individual astronauts and at least as great as that held by the contractors that built the hardware. This in its turn led to the Operations Division adopting an approach to each flight that was driven by the desire that the spacecraft should always survive any in-flight failure. Whenever the Division's own knowledge base let them down, telephone calls were placed directly to the contractors that had built the equipment. Those contractors had their own experts on call to answer MCC's questions throughout a flight.

This method led to personnel being assigned to positions within MCC depending on their specialisation. Kranz, who served as Kraft's assistant, was responsible for ensuring that all relevant information was passed to the communications Centre at GSFC for distribution to the remote tracking stations. In their turn each remote site sent their own teletypes to GSFC at the end of each spacecraft pass over their station. All of these messages were received at GSFC Communication Centre where the information contained in them was entered into three huge (by today's standards) mainframe computers.

Other positions in MCC included the Capsule Communicator (CapCom). On manned flights the six astronauts that were not flying filled CapCom positions at

Fig. 23. Mercury Control Centre.

MCC, Bermuda and those tracking stations that were located beneath the spacecraft at points where major flight decisions were made. At the more remote sites a member of the Operations Division filled the CapCom position. Throughout Project Mercury the position of Flight Surgeon was occupied by a serving USAF physician.

A second Control Centre at the Bermuda tracking station was similarly set up, with John Hodge in the Flight Director's role. In the event that MCC went off-line due to mechanical failure during a launch, the Bermuda Control Centre would take over control of the flight and make the Go/No Go decision regarding whether or not to continue the flight following orbital insertion. (See Fig. 24.)

In 1962 Gilruth reassigned Mathews as head of the new Spacecraft Research Division at MSC, Houston, where he would lead the quest to define the Apollo and Gemini Spacecraft. Kraft replaced Mathews at the head of the Flight Operations Division. At the same time John Hodge was tasked with establishing the Flight Control Operations Branch, with Eugene Kranz as his deputy. The new branch was responsible for the WWTN personnel as well as defining Mission Rules for each up-coming flight. Two months later Kraft selected Hodge as his own deputy and Kranz became head of the Flight Control Operations Branch. He remained in that position throughout the remainder of Project Mercury.

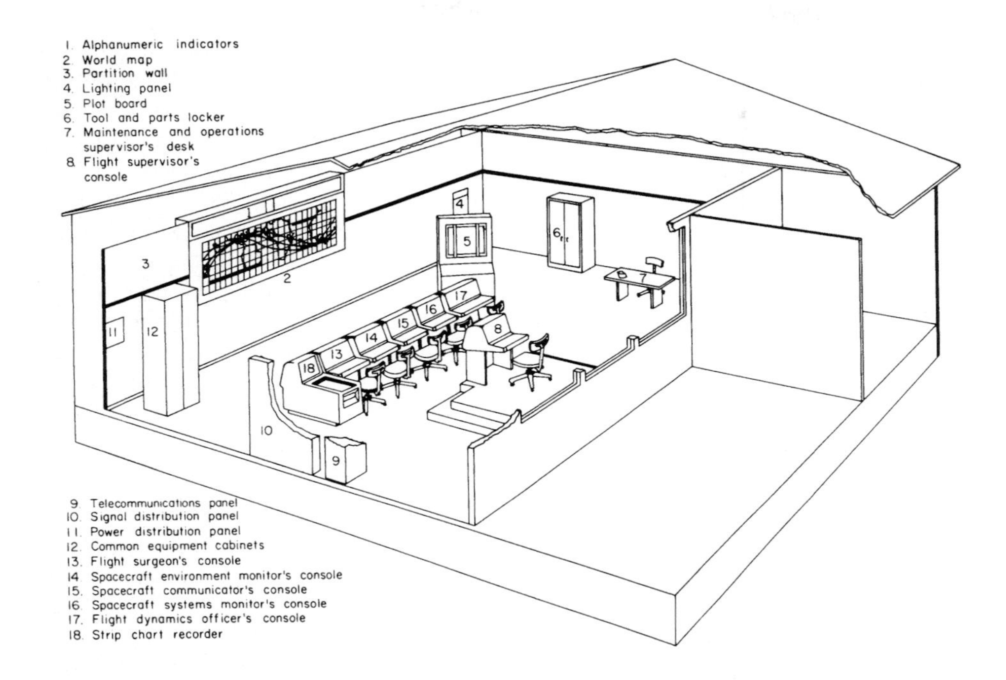

Fig. 24. Bermuda Control Centre. (NASA)

8.1 WORLD-WIDE TRACKING NETWORK

In the earliest months of Project Mercury the decision was taken to develop a new tracking network to handle the manned space flight programme. A series of tracking stations were constructed on land under the Mercury Spacecraft's ground track, to provide tracking, two-way voice communication, telemetry and command. The stations of the World-Wide Tracking Network (WWTN) were constructed under guidelines that included:

- Provision of adequate tracking and computing to determine launch and orbital parameters and spacecraft location of both normal and aborted missions.
- Voice and telemetry communications with the spacecraft with periods of interruption not greater than 10 minutes during the early orbits, contact at least once per hour thereafter, communications to be available for at least 4 minutes over each tracking station.
- Command capability to allow ground initiated re-entry for landing in preferred recovery areas and to initiate abort during critical phases of launch and insertion into orbit.
- Ground communications between tracking stations and Mercury Control Centre at Cape Canaveral.

Tracking station locations were selected to provide maximum coverage during a standard three-orbit flight. Where possible, existing tracking stations were employed. In constructing new stations negotiations were undertaken with governments sympathetic to America. Even so, it was impossible to meet the maximum of 10-minute communication interruption on some orbits, particularly the third.

Two control centres coordinated the tracking network. The Mercury Control Centre (MCC) was located at Cape Canaveral, Florida. It was from there that Christopher Kraft and the staff of the Operations Division of the STG coordinated all activities related to the Mercury flights. A secondary Control Centre was located at the tracking station on Bermuda. Bermuda had its own Operations Room similar to that at the MCC and its own computing facilities. The Control Centre on Bermuda covered the vital period of orbital insertion and was designed to operate independently if communications were lost with the rest of the network. Bermuda's independent computing facility was shut down in 1962, when a submarine cable was laid to carry two-way communication. (See Fig. 25.)

The network's Computing and Communication Centre was based at the new Robert H. Goddard Space Flight Centre, Greenbelt, Maryland. Goddard handled all communications controls, switching and distribution, as well as all computations relating to the vehicle in flight.

Fig. 25. The Project Mercury computer room at Goddard Space Flight Centre. (NASA)

The US Department of Defence (DoD) allowed NASA to use the facilities of the Atlantic Missile Range (AMR), the Pacific Missile Range (PMR), the White Sands Missile Range (WSMR) and the Elgin Gulf Coast Test Range. The British government allowed the construction of tracking stations on Bermuda and Canton Islands. Australia provided existing facilities at ranges such as Woomera and allowed the construction of new facilities at Mucha. Nigeria in Africa allowed the construction of sites at Tunga and Chawaca. Spain provided land for the construction of a site at the Canary Islands. Ground stations covered the majority of the orbital trajectory but there were still large gaps represented by the oceans. Four tracking ships, *Rose Knot Victor*, *Coastal Sentry Quebec*, *Twin Falls Victory* and *Range Tracker*, covered them.

Existing equipment was used throughout the network to save time in its construction. Redundancy was built in to all major systems and many vital functions were then reproduced through other ways to provide additional diversity. All systems were built and tested for maximum reliability, with the overriding factor at all times being the safety of the astronaut in space.

The WWTN was designed to meet the following functions:

- Ground radar tracking of the spacecraft and transmission of the tracking data to the computers at Goddard.
- Computation of launch, orbital and re-entry trajectories during the flight and transmission of the data to the MCC.
- Real-time telemetry data display at tracking stations.
- Command capability at designated tracking stations to allow control of specific spacecraft functions.
- Voice communication between spacecraft and the ground and the maintenance of a network for voice,
- Teletype and radar data communication.

The 19 tracking stations in the WWTN provided the services detailed below.

8.1.1 Radar

Between them the sites at Cape Canaveral and Bermuda provided continuous radar tracking of the launch trajectory and orbital insertion. The Go–No Go decision for the remainder of the flight depended on the data received from these two stations. The remaining stations provided radar coverage within the original requirements of the network's design. All sites employed FPS-16 radar, extended to 926-km range to track the spacecraft's C-band beacon. Verlot radar was used to track the spacecraft's S-band beacon at all stations except Elgin Air Force Base (AFB), which used a MPQ-31 radar with its range extended to meet the requirements of Project Mercury.

8.1.2 Active acquisition aid

The fear that the fast-moving spacecraft would pass through a radar beam before it could lock on to, or 'acquire' it, led to the installation of active acquisition aid at all WWTN sites. The aid acquired the spacecraft in orbit and tracked it in angle with sufficient accuracy to provide initial pointing data for the tracking radars.

8.1.3 Computing systems

The main computing centre for Project Mercury was established at the GSFC. Two IBM-7094 mainframe computers worked independently but in parallel to process data received from all WWTN sites, sort it, prioritise it, perform any necessary calculations and relay the results to MCC. In the event of a computer failure one computer was capable of handling the entire requirements of a manned orbital flight. At MCC the data was displayed on 18 display screens, four plot boards and one large wall map. The electronic wall map displayed a world map showing the positions and range of coverage of the WWTN sites. The spacecraft's location was shown and moved along its planned trajectory in time with the spacecraft in orbit. The screen also displayed where the spacecraft would land if retrofire occurred in the next 30 seconds.

An independent computing facility was established at Bermuda to cover the vital orbital insertion period if contact was lost with MCC and or GSFC. As well as sending data to GSFC, Bermuda's radars sent their data to the IBM-709 mainframe computer at the Bermuda Control Centre. The data was processed in a similar manner to that at GSFC and displayed in the Bermuda Control Centre Operations Room. The Bermuda computing facility was not used following the laying of a submarine cable in 1962.

8.1.4 Telemetry

Ground-based telemetry systems were designed to meet the requirements of the systems installed in the Mercury Spacecraft. Two redundant PAM/FM/FM systems were used with independent receivers and related equipment for both systems. Data from either system was available to MCC at any time. Those tracking stations with command capability had independent decoding and display equipment for both systems, while those with no command capability had only one decoding and display system for both systems.

Both telemetry links, which carried data on the performance of many spacecraft systems and the astronaut's medical condition, were recorded on magnetic tape to provide a playback capability and a permanent record of telemetry received during a flight. Telemetry data was received at all tracking stations from horizon to horizon during a spacecraft's pass overhead. Medical data was recorded in real time on direct writing pen charts. Other data was displayed on monitor displays, meters, lamp indicators and direct writing pens.

When some Mercury flights employed drifting flight to conserve attitude control propellant the result was a misalignment of the spacecraft antennae with the tracking stations below. Also, during re-entry a sheath of ionised air surrounded the spacecraft. At the times telemetry could not be received by the WWTN. (See Fig. 26.)

8.1.5 Air to ground communication

Each tracking station in the network had the capability of two-way voice communication with the astronaut in space. The Mercury Spacecraft contained a primary UHF system and a reserve FM system, both of which were handled by the WWTN. The UHF system was backed up at all sites while the FM system was only

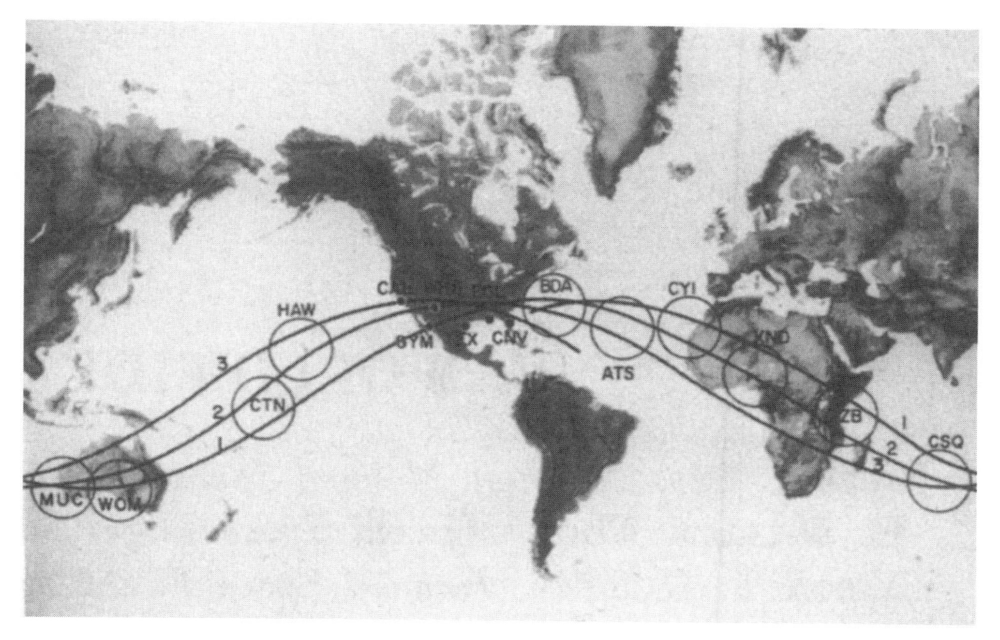

Fig. 26. Location of the World-Wide Tracking Network Earth stations showing a typical three-orbit Mercury flight. The circles show the range of each station's communications systems. (NASA)

backed up at critical locations. All stations were fitted with both local and remote control of their voice communication transmitters as well as two complete sets of antennae and associated equipment for each system. Voice communications received on Earth could be transmitted through the tracking station intercom system and control room loud speakers in real time. All voice communications were recorded on magnetic tape in real time.

8.1.6 Command

Eight ground stations in the WWTN had the capability to send commands to the Mercury Spacecraft to instruct its systems to perform certain operations as a back-up to manual control, or internally programmed events during early flights. When later flights were extended beyond three orbits the tracking ships *Rose Knot Victor* (RKV) and *Coastal Sentry Quebec* (CSQ) were fitted with command capabilities. The command system was limited to line of sight and a range of 1,296.5 km. All command sites were fitted with dual FRW-2, 5000-W transmitters. Bermuda was also given a 10-kW RF power amplifier to ensure that it was capable of sending commands to a Mercury Spacecraft in orbit regardless of its antennae pattern. (See Fig. 27.)

8.1.7 Ground communications

The tracking stations of the WWTN were all linked to MCC and the GSFC computing centre in the following ways:

Fig. 27. Tracking vessels *Rose Knot Victor* and *Coastal Sentry Quebec* tied alongside each other in harbour. (NASA)

- Transmission of acquisition information from GSFC to tracking stations.
- Transmission of commands and instructions from MCC to tracking stations.
- Transmission of digital tracking data from tracking stations to GSFC.
- Transmission of telemetry summary messages from tracking stations to MCC.
- Transmission of mission teletype traffic throughout the WWTN.
- Provide high-speed data transmission between GSFC and MCC for display purposes.
- Provide voice communications capability between certain tracking stations and MCC.

8.2 DEVELOPING THE WORLD-WIDE TRACKING NETWORK

In the earliest days of Project Mercury it was hoped that the project would be able to use existing tracking networks that had been established for unmanned satellites. When it became obvious that sufficient facilities did not exist around Earth's equator it became obvious that a new, dedicated network was required for the manned satellite project. NASA and DoD met in January 1959 to discuss the requirements of the new network, concentrating on fiscal years 1959–60. The meeting also saw the establishment of the NASA–DoD Spaceflight Tracking Resources Committee.

By 16 January Charles Mathews had convinced Abe Silverstein that the tracking network required its own team of specialists, as did the Mercury Spacecraft. Silverstein agreed and relieved STG and Mathews of the responsibility for the new tracking network and on 20 February established the Tracking and Ground Instrumentation Unit at Langley. The new unit had four managers, Hartley Soule, Sherwood Butler, Ray Hooker and G. Barry Graves.

Also in January STG let contracts to study the requirements of tracking the Mercury Spacecraft in orbit. The contracts went to the Massachusetts Institute of Technology (MIT), Ford, Space Electronics and Radio Corporation of America (RCA).

In April representatives from 20 companies attended a briefing for prospective contractors and received a general description of what the tracking network would be expected to achieve. They were told that 18 sites would handle voice communication, biomedical data, telemetry and radar tracking. Each site would be linked to MCC in Florida and the new computing centre at GSFC. A back-up control room and computing facility would be built at Bermuda. The remaining sites would handle radar tracking landing point prediction, monitoring the status of the astronaut and his spacecraft for MCC.

A Bidder's conference was held on 21 May and seven companies placed their proposals before the Review board before the closing date of 22 June. The seven companies concerned were: Brown & Root, Chrysler Corporation, Philco Corporation, Radio Corporation of America, Pan American Airways, Western Electric Company and Astronautics. A Source Selection Panel and Technical Evaluation Board were set up at LaFRC and all seven proposals were closely scrutinised. When the results of the technical evaluation were presented to the Source Selection Panel on 6 July, Western Electric Company was named as the prime contractor for the WWTN. NASA forwarded a letter of intent to Western on 30 July and contract negotiations began after that.

Meanwhile, Soule had established six Site Survey Teams within the tracking unit at LaFRC and sent them out to select prospective locations for new tracking stations in Africa, Australia, the Pacific Islands and North America. Only eight of the proposed 18 sites already existed. Of the remaining ten, three would be constructed on American territory and seven would require international negotiations with governments friendly to America, to allow new sites to be constructed and US military/US civil service personnel to run them. Soule personally undertook a great deal of the negotiations with the governments concerned. On 22 July the US Navy provided NASA with a list of reserve ships that might be turned into tracking vessels. Specific details on the listed ships followed six days later.

Air Force physicians originally demanded real-time voice communication and medical telemetry as well as television coverage of the astronaut throughout the entire flight. They let their argument drop when they were force to admit that, following diagnosis, they could do nothing for a seriously ill astronaut until he had landed and been recovered. Re-entry would require a minimum of 20 minutes and communications would be lost as the spacecraft passed back through Earth's atmosphere anyway. It was agreed that there should be no more than 10 minutes between communication sessions but in the end even that requirement was not met.

The WWTN was designed to give optimum coverage to a three-orbit manned flight inclined at 32.5 degrees to the Equator. On 9 October Kraft, the Project Mercury Flight Director, told the Society of Experimental Test Pilots the four reasons for selecting that inclination. They were:

- Maximum use of existing communications and tracking systems.
- Use of the AMR for both launch and the primary recovery site.
- Ground track should pass over the continental USA so as to provide the maximum period of unbroken communications and tracking, especially during re-entry.
- Trajectory was planned to fly over 'friendly' territory and temperate climate regions.

By the end of November the preliminary design of the WWTN was complete and five companies were working together to build it. Western was prime contractor, with overall responsibility for constructing the network. Bendix Corporation supplied search radars, telemetry equipment and displays. Burns & Roe Incorporated constructed roads and buildings. IBM supplied computers to GSFC, MCC and Bermuda. Bell Telephones Laboratory Incorporated designed and developed MCC, provided overall systems analysis and constructed a Procedures Trainer for the flight controllers. The year ended with NASA Administrator Hugh Dryden offering to cooperate with the Soviets and use the WWTN to track their manned space flights as well as NASA's. The Kremlin refused the offer.

January 1960 began with the contractors being told that the new tracking network would have to be ready in time to support the first manned suborbital flight by 1 June 1960 and the first manned orbital flight by 1 January 1961. Meanwhile, Kraft had established the training routines that would be used to prepare the members of the STG Operations Division to use the WWTN during Project Mercury. Having volunteered to become Flight Director, Kraft was in charge of MCC although Mathews and other STG and NASA Managers would also have input in major decisions. For the final flight of Project Mercury Kraft picked John Hodge from among his flight controllers to become NASA's second Flight Director.

The training for Operations Division staff began with lectures and training sessions using equipment similar to that being installed throughout the new network. Training then progressed to two existing tracking stations within North America, where operators could gain first-hand experience by tracking unmanned satellites and repairing equipment of the same type that the new network used. Personnel who would man the remote sites were named on 18 February. At the same time the DoD aero medical staff began their training. Negotiations for the last of the foreign tracking station sites were not completed until April 1960, but the network was still declared ready for operation on 1 November of the same year.

This was made possible partially by the establishment of a demonstration site at Wallops Island. A complete tracking station was constructed on the island and linked to an existing FPS-16 radar. Equipment from all of the major contractors was installed at the site and tested under simulated Mercury flight conditions. The four tests developed at the demonstration site were then applied to each of the WWTN stations as their equipment was installed, or adjusted to meet the requirements of Project Mercury. NASA had aircraft fitted with the same communications equipment that would be carried in the spacecraft and had them fly over each site and test its capabilities.

Senior engineers within the Operations Division were made responsible for installing equipment at the demonstration site before moving on to oversee the installation and testing of similar equipment at the site to which they had been assigned. Thereafter, they received lectures on space-flight principles, Project Mercury, and theoretical studies in the equipment they would be using. Operation drills, local site simulations and full network simulations completed their training, although training exercises were continued throughout Project Mercury, to keep their skills sharp.

A training centre was established at the demonstration site to continually upgrade training and to provide training for new personnel before they were deployed to the remote sites. The training centre also prepared training packages for use at individual sites throughout the WWTN.

8.2.1 Pre-flight preparation

Remote site control teams consisted of three people. The Capsule Communicator (CapCom) was the leader of the team and spoke directly to the astronaut in space. The Systems Monitor used his 21 instruments to review the status of the spacecraft based on the telemetry that it was transmitting to the ground. The Flight Surgeon had 13 instruments to review the astronaut's health, again based on the telemetry transmitted from the spacecraft. Teams were flown to their stations 14 days prior to the expected lift-off of a Mercury flight. The intervening time was used to perform maintenance, calibrate and test the equipment at the site. Simulations were then run in real time covering launch, three or more orbits and re-entry. At remotes stations the data for the simulation was played through the station's instruments from magnetic tape. An operator at the site read the words spoken by the astronaut from a printed script. The tapes for the simulations were prepared by the Operations Division and might contain a normal flight, or any number of in-flight failures and emergencies. A network countdown to launch began 5 hours 50 minutes prior to the planned launch of a Mercury flight. In that time a shortened version of the full systems checks previously carried out were completed along with voice and data communications checks between all stations in the network.

8.2.2 Flight activities

During an orbital pass, operators at each station used the 75 minutes prior to signal acquisition to calibrate their equipment and annotate their recorders to provide known standards for post-flight data reduction. Twenty-five minutes prior to the pass the station would receive a teletype message from MCC informing them of when and where they might expect to acquire the spacecraft's signal on its next pass over that station. That data was based on computations made at GSFC using tracking data from all sites in the WWTN. The information received was then used to train the acquisition and communication antennae on the correct location in the sky. An updated acquisition message was received 5 minutes prior to the expected acquisition of signal.

The spacecraft's signal was usually acquired within seconds of the spacecraft crossing the site's radio horizon. The wide-beam width of the Active Acquisition Aid normally meant that it acquired the signal first, but that was not always so. Regardless of which instrument was first to acquire the spacecraft's signal, all other antennae were then slaved to that instrument.

As the radar locked on to the signal it was set to automatic tracking. At dual radar sites the data from the more accurate C-band radar was transmitted to GSFC. If the C-band radar malfunctioned the less accurate S-band radar's data would be sent. Following loss of signal (LOS) on the final pass over the station the Post-flight Instrumentation Message was transmitted to MCC on the teletype machine. This gave details of acquisition and LOS times for each instrument at the site along with details of each instrument operating mode and performance.

After a given launch Bermuda, the second station in the chain, would acquire the spacecraft's signal at T + 5 min, while Elgin, the final station in the network, would not acquire the spacecraft for 90 minutes. The average pass over a site lasted approximately 7 minutes, with approximately 85 minutes between successive passes over the same station. During the 22-orbit flight of Mercury Atlas 9 each station in the network had a period of at least three orbits when the spacecraft did not pass overhead and they were therefore unable to communicate with it.

8.3 ADDITIONAL NETWORK SUPPORT FOR EXTENDED FLIGHTS

Mercury Atlas 6 met project Mercury's principal goal of a three-orbit manned flight in February 1962. After a second three-orbit flight Mercury Atlas 8 was extended to six orbits and Mercury Atlas 9 was planned as a 22-orbit, 34-hour mission to end the project. To meet the requirements of these extended flights the following changes were made to the standard WWTN:

Mercury Atlas 7

- No Atlantic Ocean tracking ship.
- Indian Ocean tracking ship relocated to the Mozambique Channel, off the East coast of Africa.

Mercury Atlas 8

- Primary recovery zone in the Pacific Ocean.
- *Rose Knot Victor* fitted with command capability, relocated from Atlantic to Pacific Ocean, South of Japan. Redesignated Pacific Command Ship (PCS).
- Three additional ships, *Huntsville*, *Watertown* and *American Mariner*, added to network and located off of Midway Island to record re-entry data.

Mercury Atlas 9

To meet the requirements of the 22-hour final flight of Project Mercury the WWTN was extensively altered:

Equipment

- All command sites fitted with additional controls over spacecraft systems. RKV's command system upgraded from 600 W to 10 kW. CSQ fitted with command capability.
- Tracking site clocks extended to cover the extended flight.
- Bermuda and Point Agguello upgraded to allow medical data to be transmitted to MCC by landline.
- An AN/UYK-1 computer added to Bermuda to provide automatic telemetry processing.
- Slow scan television receivers installed at MCC, the Canary Islands and CSQ. The Canary Islands system was record only while the other two had both record and display facilities.
- Goddard's two IBM-7090 computers were upgraded to IBM-7094 computers. A third IBM-7094 computer was installed at Goddard at the same time.

Communications

- A submarine cable made the radio links to Bermuda redundant.
- CSQ transmitted through Honolulu or Bassenden.
- RKV transmitted through Honolulu and New York.
- Range Tracker transmitted through Honolulu.
- The method in which data was handled at Goddard was changed.

Relocation of tracking ships

- CSQ relocated to 28° 30′ N latitude, 130° E longitude, to provide retro-sequence command back-up on the 6th, 7th, 21st and 22nd orbits.
- RKV relocated to 25° S latitude, 120° W longitude, to provide command coverage not provided by other sites on 8th and 13th orbits.

Additional support

- *Range Tracker* equipped with C-band radar to provide coverage during re-entry on 4th, 7th and 22nd orbits. Located at 31° 30′ N latitude, 173° E longitude.
- *Twin Falls Victory* equipped with C-band radar to provide re-entry coverage on 2nd and 17th orbits. Located at 31° 03′ N latitude, 75° W longitude.
- Ascension Island site provided with FPS-16 radar tracking during orbit 4.
- East Island Station, part of the AMR, provided with FPS-16 radar.
- Antigua Island station, part of AMR, provided telemetry recording, air-to-ground communication and ECG relays.
- Wake Island, Kwajalein Island and San Nicholas Island provided with voice communications system. The first two provided extended coverage for the Hawaii station while the third provided extended coverage for Point Arguello, California.

8.3.1 World-Wide Tracking Network locations

Cape Canaveral (CNV and MCC)
Grand Bahama Island (GBI)
Grand Turk Island (GTI)
Bermuda (BDA)
Atlantic Tracking Ship (ATS)
Grand Canary Island (CYI)
Kano, Nigeria (KNO)
Zanzibar (ZZB)
Indian Ocean Ship (IOS)
Muchea, Australia (MUC)
Woomera, Australia (WOM)
Canton Island (CTN)
Kauai Island, Hawaii (HAW)
Port Arguello, California (CAL)
Guaymas, Mexico (GYM)
White sands, New Mexico (WHS)
Corpus Christi, Texas (TEX)
Elgin, Florida (EGL)

Goddard Space Flight Centre (computing and communications centre)

Part III: HARDWARE

9

Mercury Spacecraft (Type D)

The McDonnell Aircraft Corporation Mercury Spacecraft was 2.3 m tall from the apex of its wide, curved heat shield, to the point of the protective cone on the Antenna Canister at its narrow end. The Retrograde Package, which was strapped to the heat shield of most spacecraft, added a further 41 cm to the length. A Launch Escape System (LES), which was built by the spacecraft's prime contractor, bought the total length to 7.9 m. The circular heat shield that protected the wide end of the spacecraft during re-entry through Earth's atmosphere at the end of its flight had a diameter of 1.8 m. (See Fig. 28.)

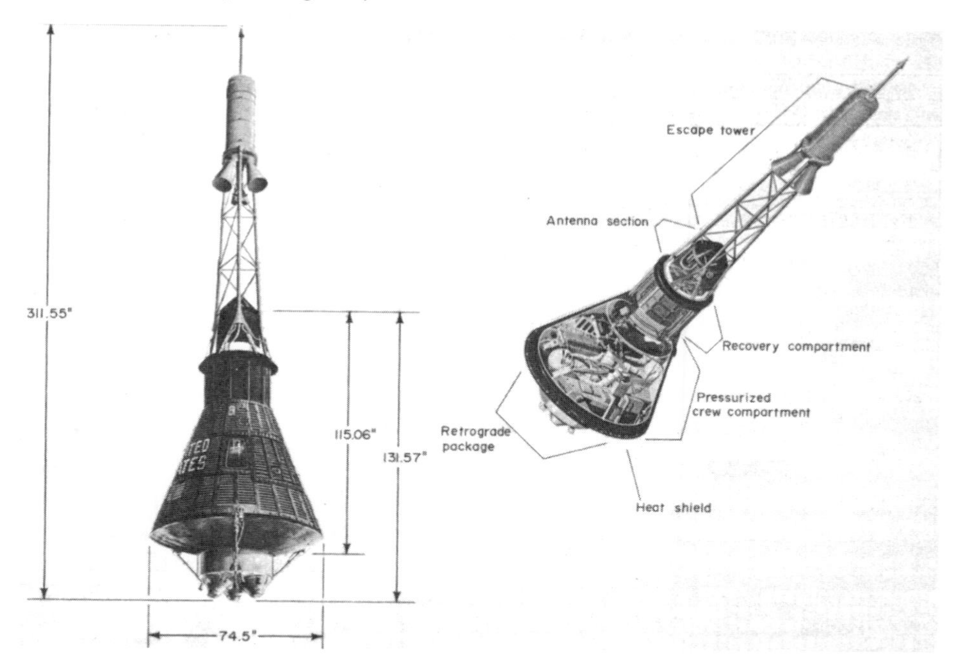

Fig. 28. Two views of the Type D Mercury Spacecraft and Launch Escape System, showing the major proportions and the different sections of the vehicle. (NASA)

The exterior surfaces of all production spacecraft were painted with a black, semi-gloss, heat-resistant paint. They carried national markings, namely the words 'UNITED STATES', in white upper case letters as well as the American flag. The six spacecraft that carried human astronauts also bore the astronaut's personal choice of name and heraldry. This was in keeping with the tradition whereby military pilots name their aircraft, particularly in times of combat.

The spacecraft was specifically designed to support a single astronaut on a flight consisting of three orbits of Earth. It was a sophisticated vehicle, with many of its system pushing the limits of the technology available in the final years of the 1950s. Even so, it was designed to meet a number of relatively simple design principles.

- Capable of launch into orbit by an unaided Atlas Intercontinental Ballistic Missile.
- Streamlined, for exit through Earth's atmosphere.
- Blunt/wide leading face for re-entry through Earth's atmosphere.
- 24-hour orbital life, with natural orbital decay at the end of that period.
- Nominal orbital flight termination to be by means of retrograde rockets.
- Final atmospheric descent to be by parachute.
- Mid-ocean landing, with recovery by US military forces.
- Where possible, existing technology to be used.

Mounted on its launch vehicle the Mercury Spacecraft consisted of five major sections. From the ground up, these were:

- Retrograde Package.
- Heat shield and landing skirt.
- Crew Compartment.
- Recovery Section.
- Antenna Canister.

McDonnell Aircraft Corporation, the spacecraft's manufacturer, was also responsible for the construction, testing and delivery of the launch vehicle adapter, Launch Escape System and the two small, pressurised couches in which two chimpanzees rode in two Mercury Spacecraft.

9.1 RETROGRADE PACKAGE

The Retrograde Package held two separate sets of three solid propellant rocket motors. The three smallest rockets provided thrust along the line of the spacecraft's orbital trajectory and were therefore called posigrade rockets. Three larger rocket motors applied their thrust against the line of the spacecraft's orbital trajectory, in order to slow it down. These retrograde rockets gave their name to the Retrograde Package. (See Fig. 29.)

On a normal flight the LES was not required and was jettisoned during the powered phase of the launch. At the end of the powered phase the launch vehicle's

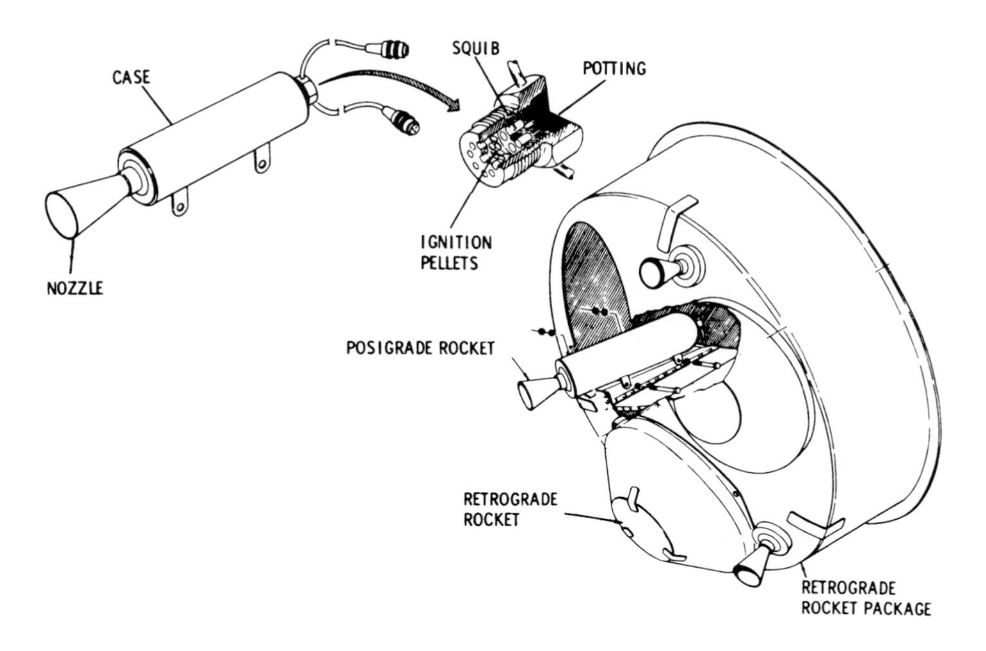

Fig. 29. Details of the Mercury Spacecraft Retrograde Section. (NASA)

propulsion system shut down and the Marman Clamp connecting the base of the Mercury Spacecraft to the neck of the launch vehicle was released.

The three posigrade rocket motors mounted in the Retrograde Package were then ignited. The thrust from these motors provided a positive separation rate between the spacecraft and the inert launch vehicle. The requirement for a positive separation rate was identified during the Big Joe flight, the first launch of the Mercury Atlas combination. Each posigrade rocket produced 1,779.2 newtons (N) of thrust for one second. This was sufficient to propel the Mercury Spacecraft clear of the launch vehicle adapter without causing the explosive destruction of the pressurised Atlas launch vehicle. Having performed this vital role, the spent posigrade rocket motor casings remained in the Retrograde Package until the latter was jettisoned at the end of the flight.

Following positive separation from the launch vehicle, an onboard timer had to complete its 10-second run down before the Mercury Spacecraft's attitude manoeuvring thrusters were fired to turn the spacecraft through 180 degrees, so that it was travelling with its wide end facing the direction of travel. The spacecraft was then manoeuvred to a negative 34-degree pitch, narrow end down, orbital attitude.

At the end of the orbital phase the onboard programmer realigned the spacecraft to a negative 14.5-degree pitch attitude. This position was called the retrofire attitude. With the spacecraft in the correct attitude the programmer fired the three retrograde rockets mounted in the Retrograde Package, one at a time. Each retrograde rocket produced a thrust of 4,448 N for 10 seconds. This reduced the spacecraft's forward velocity to a point where Earth's gravitational attraction bent the spacecraft's trajectory sufficiently that it would now intersect the planet's surface.

The Retrograde Package was positioned in the centre of the heat shield, on the wide end of the spacecraft. It was held in place by three high-tension titanium straps spaced 120 degrees apart. Each strap was connected to the wide end of the spacecraft body by a pyrotechnic guillotine. Each strap also served as a mounting for an external electrical umbilical from the spacecraft batteries to the ignition head on each of the six rocket motors in the Retrograde Package.

At retrofire plus 1 minute 30 seconds, after the third retrograde rocket had stopped firing, the pyrotechnic guillotines were fired to sever the titanium straps and electrical umbilicals. This jettisoned the Retrograde Package and exposed the heat shield for re-entry. If the guillotines failed to fire the astronaut had a manual back-up firing system available, which he could activate from his position in the Crew Compartment.

Under normal conditions the Retrograde Package burned up when it entered Earth's atmosphere. On one occasion it proved necessary to re-enter without jettisoning the Retrograde Package, but it finally fell away when the titanium straps holding it in place burned through. The astronaut lived to give a vivid description of a spectacular re-entry.

9.2 HEAT SHIELD AND LANDING SKIRT

In order to survive the intense heat produced by the layer of ionised air that formed during re-entry through Earth's atmosphere, the Mercury Spacecraft employed the blunt, wide leading face concept developed by Allen and Eggers at LaFRC. The leading face of the spacecraft was protected by a heat shield consisting of ceramic materials with ablative properties. Having fired the retrograde rockets and jettisoned the Retrograde Package, the onboard programmer manoeuvred the Mercury Spacecraft to the correct 1.5-degree pitch up, heat shield forward, re-entry attitude.

On entering the atmosphere, the spacecraft pushed the air molecules before it, causing them to compress and ionise, producing many thousands of degrees of heat in the process. This formed a bow wave in front of the returning spacecraft and a sheath of ionised air around it, which temporarily blacked out radio and telemetry communications. The bow wave ceased to form once the spacecraft's velocity fell below Mach 1 and communications were resumed once more at that time.

The bow wave carried the majority of the heat away from the descending spacecraft. However, where the bow wave and the spacecraft's heat shield met, heat was transferred to the shield, which then ablated. As the ablative material in the heat shield charred and broke away, the heat that had built up within that material was carried away with it. In the earliest days of heat shield research ablative materials were national security secrets because they could be used on the warheads of nuclear missiles, making them lighter and smaller than those employing heat-sink materials.

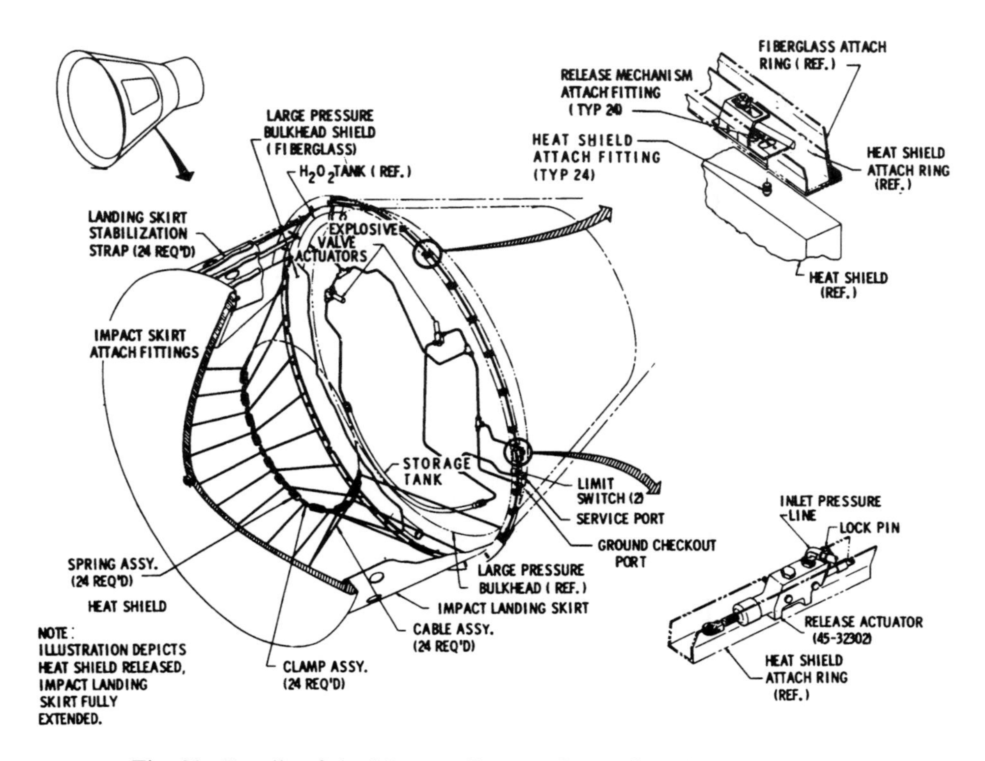

Fig. 30. Details of the Mercury Spacecraft Landing Skirt. (NASA)

For the majority of the flight a Mercury Spacecraft's heat shield was secured to the base of the spacecraft after body. When the main recovery parachute deployed, at an altitude of 3,048 m, the heat shield was released and dropped 1.2 m to deploy the landing skirt. (See Figs 30 and 31.)

The landing skirt was made of rubberised canvas and was attached to the rear of the heat shield and the base of the spacecraft. It was supported internally by 24 spring and cable assemblies and externally by 24-load-bearing straps. On impact with the ocean the impact skirt cushioned the landing forces by allowing the air within it to escape through a series of perforations. During water landing (the media named it 'splashdown') the landing skirt was estimated to reduce the landing loads from 45*g* to 15*g*. This was to protect the astronaut from severe landing loads as well as protecting the rear wall of the Crew Compartment from landing forces that might cause it to rupture, thereby allowing seawater to enter the spacecraft.

Following landing, seawater would enter the landing skirt through the same perforations that had let the air out. This allowed the heat shield to sink and the landing skirt to fill with water and act as a sea anchor, maintaining the spacecraft in an upright position throughout recovery operations. On the last few flights US Navy recovery personnel also placed a Stullken collar around the base of the spacecraft, to assist in maintaining the spacecraft in an upright position.

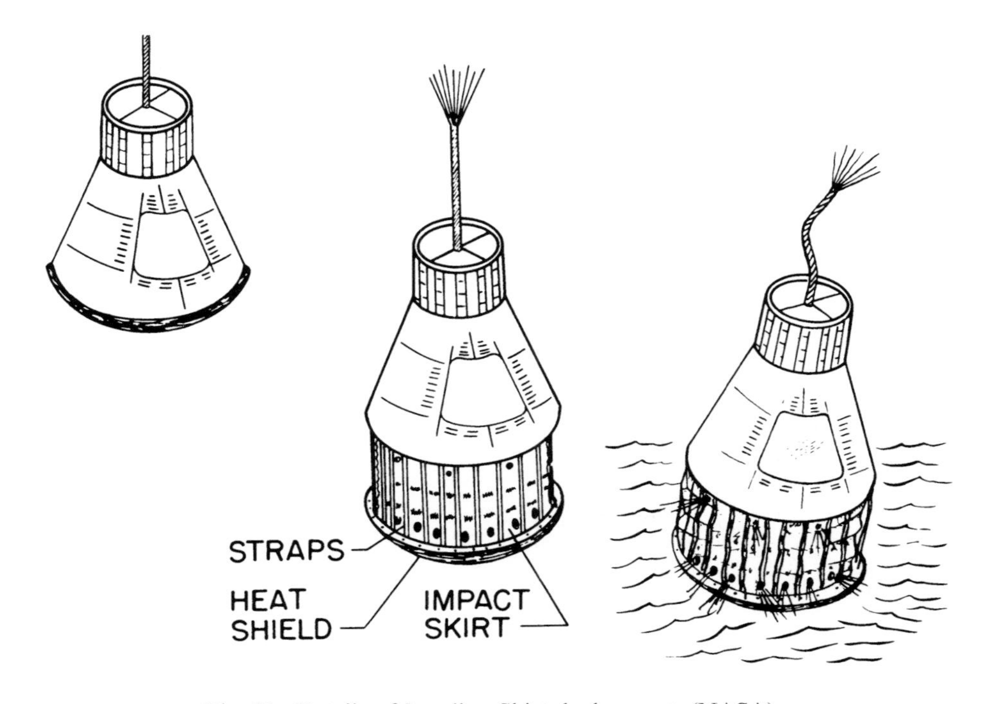

Fig. 31. Details of Landing Skirt deployment. (NASA)

Following recovery, the Mercury Spacecraft was placed on a wheeled dolly, on the deck of the recovery vessel. The landing skirt was folded and stored between the base of the afterbody and the heat shield. Wooden blocks were used as spacers, to prevent unnecessary post-recovery damage to the landing skirt.

9.3 CREW COMPARTMENT

The Crew Compartment consisted of the conical afterbody of the spacecraft, the sloping outer walls of which had a half-cone angle of 25 degrees. The body of the Crew Compartment was constructed with two skins. The inner skin and the forward and aft bulkheads were constructed from titanium 0.25 cm thick. The outer skin was stood off from the inner skin, supported by aircraft-style stringers. The gap between the two was insulated with ceramic fibre insulating material. The outer, heat protective skin was constructed from shingles of 0.0406-cm thick Rene-41 nickel alloy. The shingles were beaded for additional strength.

As the name suggests, the Crew Compartment housed the single astronaut throughout his flight. He entered the spacecraft through a rectangular hatch, positioned to the right of the head support on the prone couch. The hatch lifted away from the spacecraft completely and had to be fitted in place by the launch crew tightening 70 torque bolts. Once in position, all 70 bolts had to be undone from the

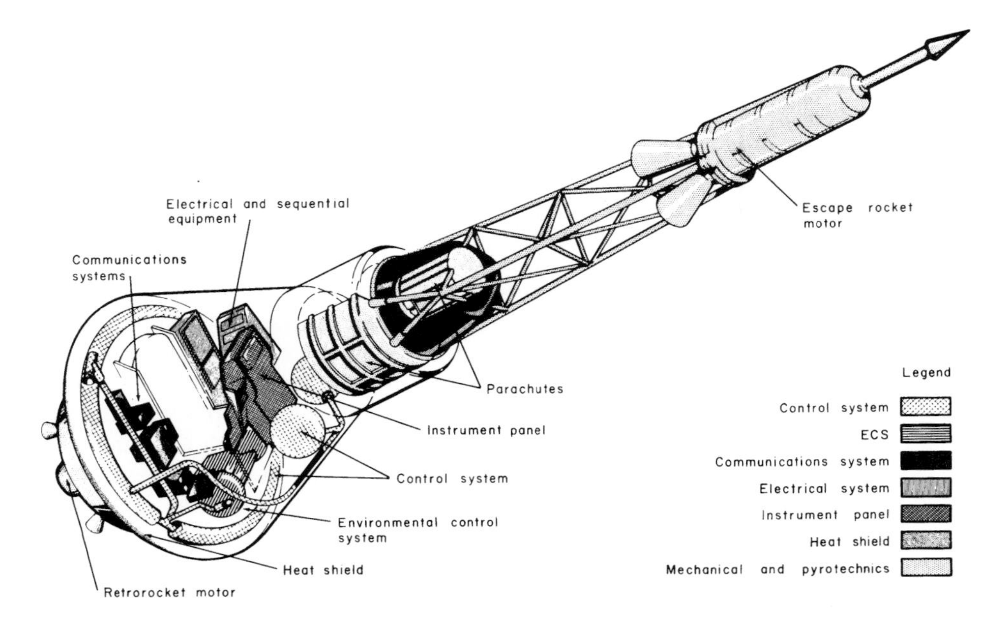

Fig. 32. Major Mercury Spacecraft systems. (NASA)

outside before the hatch could be removed. In a launch pad emergency a single cord of explosive could cut through all 70 torque bolts to remove the hatch and allow the astronaut a rapid escape from the spacecraft. (See Fig. 32.)

The original hatch design was replaced after the first manned flight. The new hatch included a double-shaped explosive charge and the torque bolts were drilled through to provide a weak point. The charge was detonated by the astronaut striking a plunger inside the Crew Compartment, or by the recovery crew pulling an external lanyard. Detonating the charge severed all 70 torque bolts simultaneously and threw the hatch clear of the spacecraft. The new style hatch was only carried on the last five Mercury spacecraft to fly.

Similarly, early models of the spacecraft had two small circular portholes, one on either side of the astronaut's head. These offered only an extremely limited view out from the Crew Compartment. The last five Mercury Spacecraft to fly had a large 'picture window' in front of the astronaut's face. The new window offered a 30-degree field of view with the astronaut's head back against his prone couch.

Dressed in his Mercury pressure suit, the astronaut lay in his prone couch, which consisted of a standard fitting into which was fitted a fibreglass couch former that was form-fitted to the individual astronaut. The first five astronauts to fly the Mercury Spacecraft used a couch former with leg and feet supports included. On the final Mercury flight, stirrups were used to support the astronaut's feet, replacing the original leg and feet supports. The prone couch supported the astronaut in a sitting position, with his back to the heat shield. His legs were bent 90 degrees at the hips and again at the knees. Supported in this position he was best able to endure the *g* forces that were experienced during lift-off and re-entry.

The Environmental Control System (ECS) was located primarily under the astronaut's couch and was developed by the AiResearch Manufacturing Division of Garrett Corporation under a McDonnell subcontract. It was designed to pressurise the Crew Compartment and the astronaut's pressure suit with 351.5 g/cm^2 of 100 per cent oxygen. It was designed to control the pressure and temperature of the environment automatically. The astronauts had the option of over-riding the automatic systems and setting the levels manually. The ECS contained a storage tank for the astronaut's urine as well as providing cooling water to remove the heat produced by the spacecraft's electrical systems. (See Fig. 33.)

The ECS was effectively two subsystems, one for the astronaut's pressure suit and the other for the Crew Compartment. The two circuits were independent and redundant, but operated simultaneously, using the same coolant water and electrical supplies. Oxygen was supplied from two high-pressure tanks, each with sufficient oxygen for 28 hours of operation. The tanks were connected in such a manner that when the oxygen in one ran out the ECS automatically began drawing oxygen from the other. Water was held in a tank with an internal bladder, to ensure a regular flow of the liquid under microgravity conditions. Electrical power came from the spacecraft's batteries, which were charged to capacity before launch.

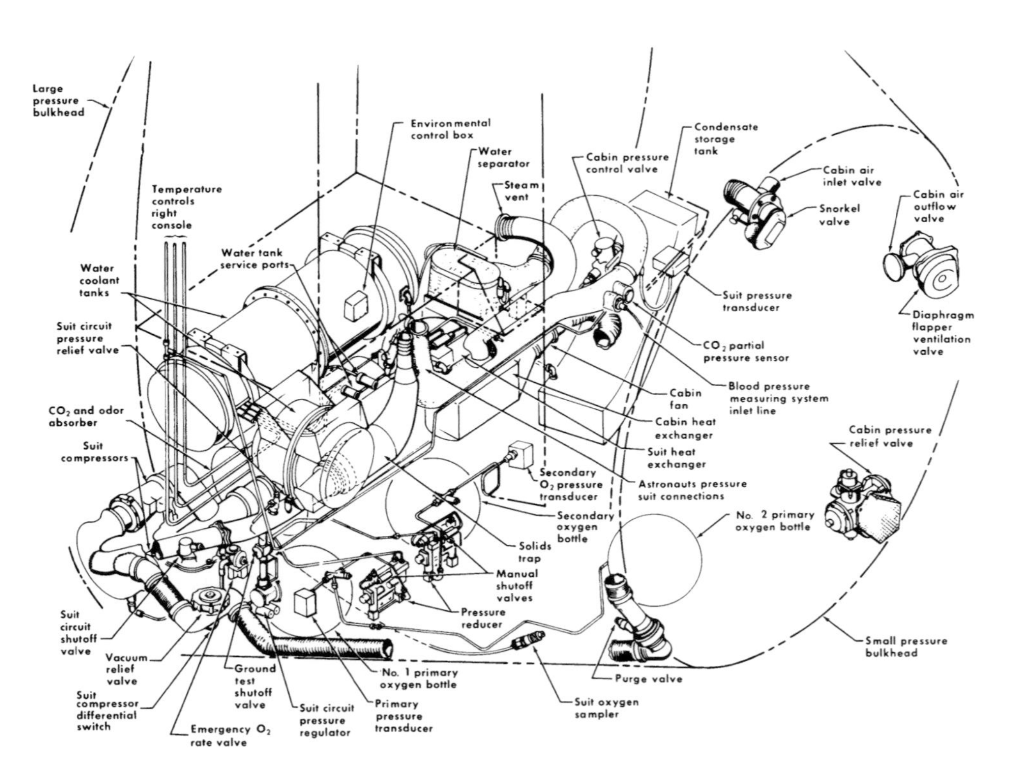

Fig. 33. Mercury Spacecraft Environmental Control System. (NASA)

The pressure suit circuit provided the astronaut with a constant supply of breathable oxygen, maintained pressures vital to the correct functioning of his body, removed metabolic waste products and maintained a comfortable body temperature. Oxygen entered the pressure suit through an umbilical connected between the ECS in the spacecraft and the torso of the pressure suit. It exited through a second umbilical connected to the gas exhaust on the astronaut's helmet. Oxygen was forced into the pressure suit ducts and carried to the body's extremities. It was then allowed to flow freely across the body, in order to remove heat, until it entered the helmet. There, the astronaut's breathing produced metabolic waste products, which were exhaled into the oxygen in the helmet.

On leaving the suit through the gas exhaust the oxygen passed through a debris trap, to remove any particulate matter. It then passed through a canister of activated charcoal and lithium hydroxide, to remove odours and exhaled carbon dioxide. The gas was then cooled by a water evaporative heat exchanger, which employed the vacuum of space to make the coolant water boil at a low temperature. The heat exchanger exit gas temperature was regulated by manual control of the coolant water flow valve. The resulting steam was exhausted overboard. A thermal switch on the steam exit valve regulated the steam exit temperature and lit an indicator in the Crew Compartment if the exit duct's temperature fell below a set level.

On the gas side of the heat exchanger, moisture picked up in the pressure suit was condensed into water droplets and carried by the gas flow to a mechanical water separator. This 'sponge' device was mechanically 'squeezed' periodically to remove the water from the system and direct it to a collection tank. A compressor maintained the constant flow rate of gas through the system. With fresh oxygen added, to make up the lost volume, the gas was recycled back through the ECS.

The Crew Compartment ECS circuit controlled the pressure and temperature. A pressure regulator metered oxygen flow into the Crew Compartment in order to maintain a minimum pressure level. The internal temperature was maintained by means of a fan and a heat exchanger, similar to those used in the pressure suit circuit. A pressure relief valve allowed the atmosphere to be dumped overboard to the vacuum if the Crew Compartment internal pressure reached its upper limit, or in the case of a fire, thereby starving the flames of oxygen. With the dump valve closed the Crew Compartment repressurised automatically.

Under normal orbital conditions the astronaut remained in his pressure suit, but opened the faceplate on his helmet. This allowed him to breathe the Crew Compartment atmosphere. In the event of a failure of the Crew Compartment ECS circuit he could close his faceplate, isolate the malfunctioning circuit and continue the flight relying on the pressure suit ECS circuit, while he made an emergency return to Earth.

Lying in his prone couch, the astronaut faced the main instrument panel, which was less than an arm's length from his face. The main panel housed the flight control instruments, including an artificial horizon, and attitude indicators in all three axes. On the panel to the right of the main panel, the environmental controls were at the top, with communication controls and voltage displays at the bottom. There were also controls for the astronaut's medical experiments and monitoring equipment.

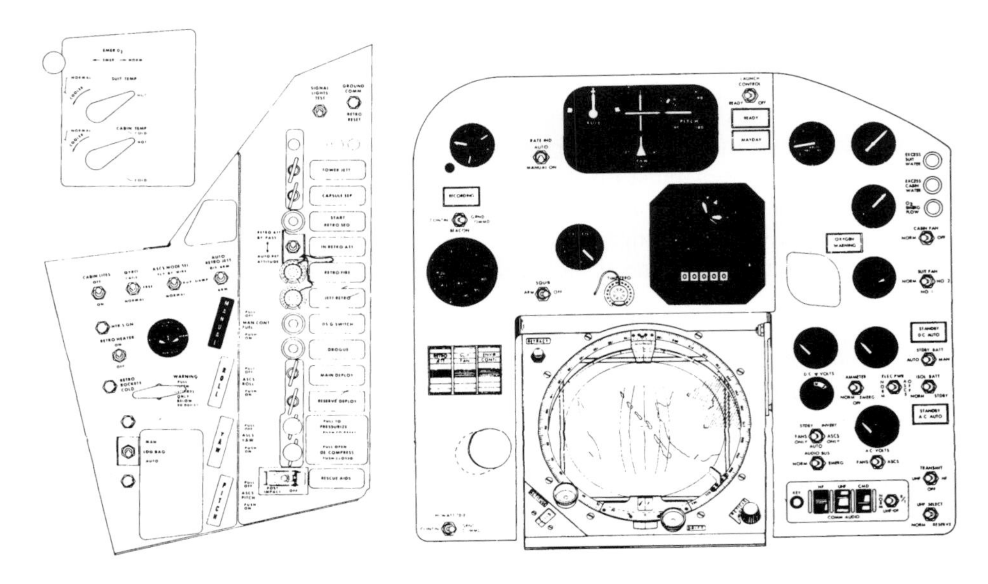

Fig. 34. Mercury Redstone 3 control console. (NASA)

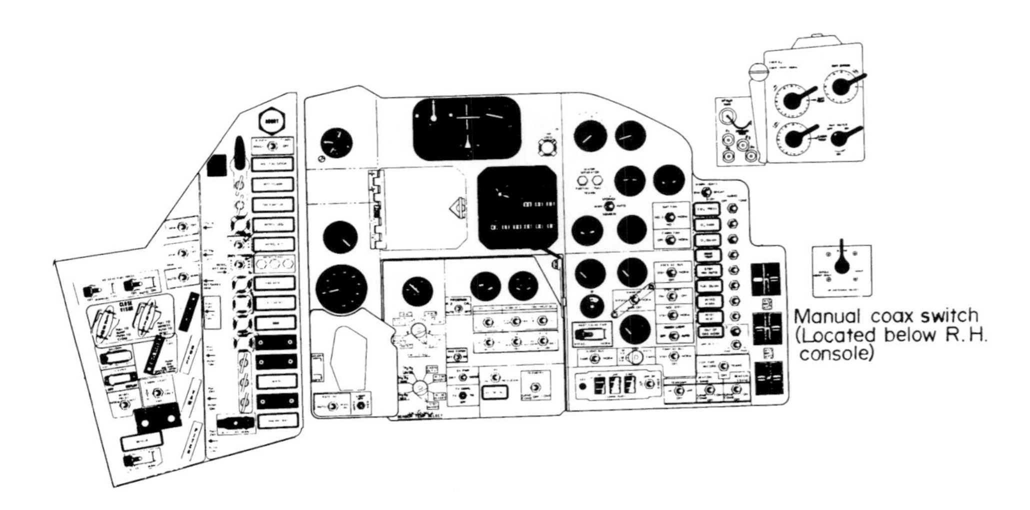

Fig. 35. Mercury Redstone 9 control console. (NASA)

The auxiliary control panel, on the far right, housed controls for the Retrograde Package rocket motors and guillotines, as well as controls for the LES, the periscope, the recovery parachutes and the recovery aids. (See Figs 34 and 35.)

The left-hand panel housed the primary event controls on its right side. These included the abort button and controls for both of the manual spacecraft manoeuvring systems. The left-hand side of the panel offered lighting controls, the controls to depressurise and repressurise the Crew Compartment through the overboard dump valve, and the manual control switch. The fuse circuit breaker panel was to the astronaut's far left. A panel next to the astronaut's right thigh offered manual overrides for the pressure suit and Crew Compartment temperature controls.

The main instrument panel was supported by the periscope, which was carried on all manned spacecraft except the last one. The periscope was available to astronauts while on the launch pad, but was retracted and covered with a small door during the powered launch phase. It was then deployed again once the spacecraft was in orbit. The view of Earth's surface that it offered was shown electronically on a display between the astronaut's legs and beneath the main instrument panel. If the display was too bright a grey filter could be put in place to reduce the glare.

The periscope passed through both walls of the Crew Compartment and its external end was exposed to the vacuum of space. The astronaut could turn the periscope through 360 degrees and at any one time the High, or telephoto, setting offered a view some 128.7 km in diameter, while the low or wide-angle setting offered a view some 3,057 km in diameter. The periscope was retracted prior to re-entry, but could be deployed again once the main parachute had deployed for landing. A storage bag for maps, charts and flight documents was attached to the body of the periscope, below the display screen, where they were readily available to the astronaut.

To the astronaut's right were boxes of electronics designed to record voice and telemetry data. Telemetered data included the physical and physiological condition of the astronaut recorded through his biomedical harness. The data also included that on the condition of the spacecraft's systems. All of this data was recorded on board and then 'dumped' to certain tracking stations in the WWTN. To his left were communication equipment and many layers of black boxes containing avionics.

The spacecraft communications system consisted of voice, radar, command, recovery and telemetry links. The voice system used high-frequency (HF) and ultra-high-frequency (UHF) systems, with the UHF system offering shorter range, but slightly better quality than the HF system. The system was used to carry two-way voice communications between the astronaut in space and the stations of the WWTN. A back-up UFH system was also carried and back-up voice communication links existed through the command receiver channel and the low-frequency telemetry carrier.

The radar system consisted of C- and S-band radar beacons that were active throughout the flight. Tracking stations could interrogate either or both beacons when they were within line of site communication.

Commands could be sent from the ground to make the spacecraft perform certain operations. These commands were received in the spacecraft through the two identical receivers and decoders of the command system. This option was available to ensure safe recovery of the astronaut in the case of the major automatic systems failing. The command system offered flight controllers the option to command a launch abort, retrofire, a spacecraft clock update, or instrumentation calibration.

Two almost identical FM–FM telemetry subsystems offered redundancy in the telemetry system. The system relayed medical data from the astronaut's medical harness and technical data from spacecraft systems to tracking stations on the ground. The most important data was displayed in real time in the MCC and other tracking stations. Other data was recorded for post-flight analysis.

The recovery system consisted of a HF transceiver, and a recovery package consisting of a CW SEASAVE beacon, a pulse-modulated SARAH beacon, and a pulse-modulated SUPERSARAH beacon.

Immediately to the left of the astronaut's head was a motion picture film camera that was used to record the main instrument panel throughout the flight. A second motion picture film camera was mounted above the main instrument panel and filmed the astronaut's head and shoulders. A round, convex mirror worn on the astronaut's chest allowed this camera to also film the main instrument panel, as a back-up to the primary camera.

Seated in his prone couch, the astronaut's hands fell naturally on two hand controllers. The left-hand controller was the abort handle, or 'chicken switch'. If the automatic systems failed this handle allowed the astronaut to activate the LES, to escape from a malfunctioning launch vehicle. His right hand rested on the three-axis hand controller (Fig. 36) that allowed him to manually control his spacecraft's

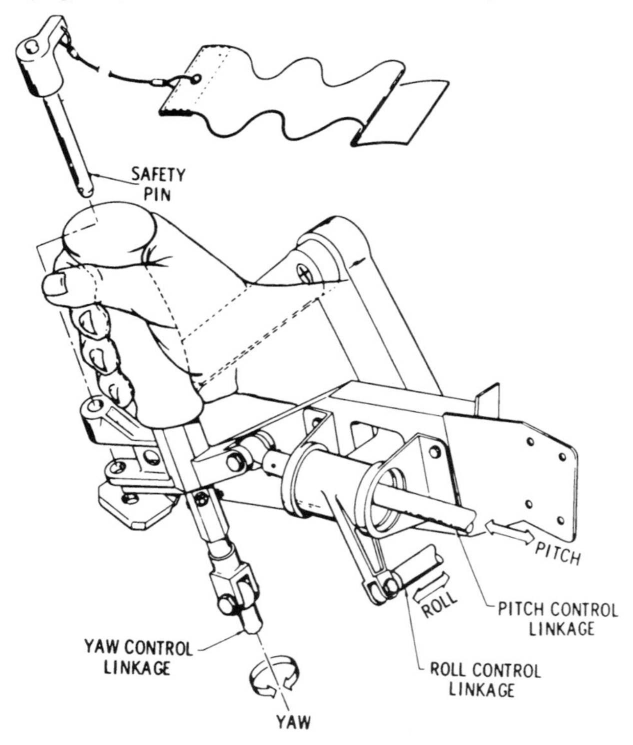

Fig. 36. Mercury Spacecraft three-axis hand controller. (NASA)

attitude in space. The controller took its name from the fact that it offered control around three axes, each spaced at 90 degrees to the other two. Pushing the controller forward pitched the narrow end of the spacecraft down. Pulling back pitched the narrow end up. Moving the whole controller to the left rolled the spacecraft about its long axis in that direction. Moving it to the right rolled the spacecraft in that direction. Twisting the controller handgrip to the left yawed the narrow end of the spacecraft to the left. Twisting the handgrip to the right yawed the narrow end to the right. The spacecraft could be manoeuvred through 360 degrees in any of these six directions. By combining two or more controller movements the spacecraft could be manoeuvred into any attitude between these positions.

The spacecraft offered the astronaut the option of any one, or any combination of four separate control systems. There were two separate propellant supplies and thruster systems. Each system used the decomposition of hydrogen peroxide over a catalyst, to produce steam and oxygen, which was then released through a series of mechanical valves to a solenoid and thus through the thrusters. Individual control systems could be combined to overcome malfunctions.

System A offered the astronaut the choice of the Automatic Stabilisation and Control System (ASCS), and the Fly-by-wire (FBW) option. System A contained four 26.6-N thrusters mounted around the exterior of the Crew Compartment and Recovery Section that were used for spacecraft positioning during turnaround and at retrofire. Four pitch and yaw thrusters mounted in pairs on the Recovery Section, and four roll thrusters mounted in pairs on the wide end of the Crew Compartment offered a thrust of 4.4 N. These low thrust units were used for attitude control.

System B offered the Manual Proportional Control System (MPCS) and the Rate Stabilisation System (RCS). System B contained a further six thrusters, in pairs, to control pitch, yaw and roll. These thrusters were in two thrust ratings and offered variable thrust. The pitch and yaw thrusters offered a thrust of between 17.7 and 106.7 N, while the roll thrusters offered a thrust of between 4.44 and 26.6 N.

Originally, the Mercury Spacecraft was designed to fly automatically, with the astronaut being carried as a medical test subject and observer. The ASCS employed three spinning gyroscopes; one aligned in each of the spacecraft's three axes, as a fixed attitude reference. By comparing data received by horizon sensors mounted in the Antenna Canister with this fixed reference, the ASCS made adjustments to the spacecraft's attitude, to correct any discrepancies between the two sets of data, by firing the System A thrusters. This system required no input from the astronaut's hand controller. The original spacecraft design also offered the astronaut manual control through the MPCS. This used the astronaut's hand controller to fire the six System B thrusters.

Fly-by-wire offered a semi-automatic manual control system that used electrical links from the hand controller to operate the solenoids on the 12 System A thrusters. The RSCS used mechanical, rather than electrical links, from the hand controller to operate the System B thrusters. These two systems were added to the spacecraft design in reply to the Flight A astronauts' demands for manual control of the spacecraft in orbit.

Three pairs of 1 × 3,000 watt-hour (W-h) and 1,500 W-h rechargeable silver-zinc batteries supplied the Mercury Spacecraft with a total of 13,500 W-h of electrical power. In the event of an individual battery failing the astronaut could switch it off. Likewise, six batteries were connected in such a way that a defective or low-voltage battery could not drain power from a higher charged battery. Inverters provided alternating current to the ASCS and ECS.

9.4 RECOVERY SECTION

The Recovery Section formed the cylindrical neck of the spacecraft. It was constructed from similar materials to the Crew Compartment, except that the outer heat protection layer was constructed from panels of Beryllium, just 0.55 cm thick and beaded for added strength. The Recovery Section housed the pitch and yaw thrusters for both the System A and System B thruster sets, as well as their self-contained propellant tanks. Internally, the Recovery Section was separated from the Crew Compartment by the latter's forward bulkhead. It was also split in half by an internal bulkhead that crossed its centre line.

One side of the internal bulkhead provided storage space for the primary and reserve main parachutes. These were 19.2-m ringsail parachutes. The primary parachute was pulled from its container by a 9-m long lanyard attached to the Antenna Canister when the latter was jettisoned at an altitude of 3,048 m. If the primary parachute failed to deploy then the astronaut could manually deploy the back-up main parachute.

The main parachute was initially opened in a reefed condition, for 4 seconds, to lower the shock of opening. The reefing lines were then released and the canopy inflated to its full diameter. At the same time, the heat shield dropped away from the wide end of the Crew Compartment to deploy the landing skirt. The primary and reserve main parachutes were both jettisoned following landing.

At main parachute deployment a series of recovery aids were deployed from their storage position within the Recovery Section. A Sofar bomb was dropped into the ocean to explode underwater and alert ships searching for the capsule with radar and sonar. A cylindrical container of fluorescent green dye was dropped over the side of the spacecraft, but remained attached to the Recovery Section by a line. On impact the dye stained the surrounding seawater, making the small black spacecraft easier to see from aircraft searching for it. A UHF electronic homing signal and a flashing beacon were also activated on the spacecraft to aid aircrew searching for it. The recovery whip antenna deployed following landing. The Recovery Section also housed the UHF descent and recovery antenna, which was used for communications following the loss of the Antenna Canister at main parachute deployment. (See Fig. 37.)

Following landing, the astronaut had a choice of two recovery methods. He could choose to remain in his spacecraft, or exit through the Recovery Section and wait for recovery in his personal inflatable life raft.

In the first method the astronaut remained inside his spacecraft until the recovery helicopter made its approach. One winch line from the helicopter was then being

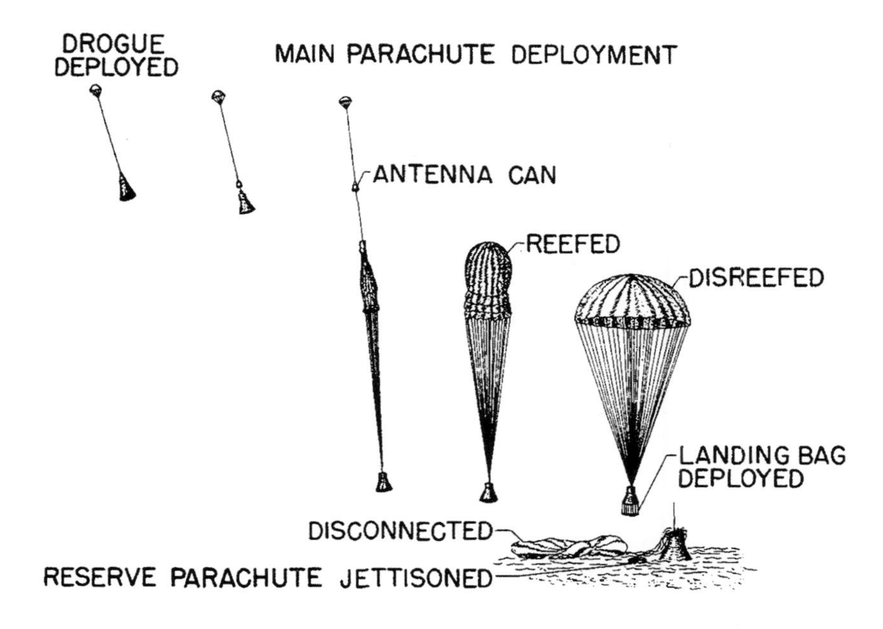

Fig. 37. Earth Landing System deployment sequence. (NASA)

connected to the recovery loop at the top of the Recovery Section. Only when the slack had been taken up on that line, did the astronaut explosively jettison the side hatch and exit using the 'horse collar' which had been lowered on a second winch line from the helicopter. When the astronaut was safely on board the helicopter the spacecraft was lifted out of the water and carried to the recovery vessel. (See Fig. 38.)

Prior to the definition of the explosive side hatch the mid-ocean egress was the preferred method of exiting from the spacecraft. This method required the astronaut to remove the left-hand wing of the main instrument panel and one half of the Crew Compartment forward bulkhead. He then had to push the main parachute container out of the top of the neck of the Recovery Section. Finally, taking his personal life raft and survival kit with him, he exited through the egress tunnel that he had just cleared (see Fig. 39), consisting of one half of the Recovery Section. Having inflated his life raft, he secured it to the spacecraft before climbing on board to await recovery.

As the project evolved, some astronauts elected for recovery inside their spacecraft, only leaving the spacecraft after it had been lifted from the ocean, secured on the deck of the recovery ship.

9.5 ANTENNA CANISTER

The Antenna Canister was a cylindrical container on the narrow end of the Spacecraft. It was constructed from Vycor, covered in a layer of 0.78 cm thick,

Fig. 38. Training exercise to familiarise the Flight A astronauts with the helicopter recovery method favoured at the beginning of Project Mercury. (NASA)

Fig. 39. Scott Carpenter illustrates the small size of the Mercury Spacecraft's access tunnel in the Recovery Section. Carpenter would be the only Flight A astronaut to use this method of egress at the end of his flight. (NASA)

vertically ribbed, Rene-41 nickel alloy. The canister's role was to protect the vital equipment in the Recovery Section.

The Antenna Canister took its name from the fact that it provided a mount for the biconical antenna mounted on its front face. During launch a fibreglass cone covered the biconical antenna. The cone prevented exhaust from the LES rocket from polluting or even destroying the antenna and was jettisoned when the spacecraft entered space.

With the protective cone jettisoned, the semi-circular Destabilisation Flap hinged through 180 degrees, from its stored position, in front of the biconical antenna. Had the spacecraft re-entered Earth's atmosphere in a stable, narrow end forward, attitude, air resistance would have acted against the stabilisation flap and turned the spacecraft through 180 degrees to a heat shield forward attitude.

Two horizon scanners were mounted in the Antenna Canister. One scanner faced the narrow end of the spacecraft and the other faced to the astronaut's right, as he sat in his prone couch. The scanners provided spacecraft attitude information for the ASCS.

The final piece of equipment housed in the Antenna Canister was the drogue parachute. Following re-entry two barostats, or atmospheric pressure switches, fired the drogue parachute mortar at an altitude of 6,400 m, deploying the 2-m diameter drogue parachute to slow, and stabilise the spacecraft. At the same time, radar chaff, packed with the drogue parachute was deployed to help recovery ships and aircraft to locate the returning spacecraft.

At 3,048 m a second barostat jettisoned the Antenna Canister, which remained attached to the bottom of drogue parachute harness. A 9-m long lanyard attached to the Antenna Canister then pulled the Main Parachute from its canister in the now exposed neck of the Recovery Section. Following Main Parachute deployment the lanyard separated and the Antenna Canister was lowered to the ocean under the drogue parachute. It was not usually recovered.

10

Developing the Mercury Spacecraft

10.1 DEFINING THE TYPE A CAPSULE

The ballistic capsule design studied by Maxime Faget (Fig. 40) and his colleagues in the early 1950s was based on Allen's re-entry concept developed for nuclear warheads. It was a zero-lift, high-drag vehicle that was aerodynamically shaped for exit through the atmosphere, while carrying a broad, flat leading face for re-entry. Early models for a manned satellite vehicle were designated Type A and were commonly referred to as capsules.

The Type A capsule was 2.3 m in diameter across its base and 3.3 m tall. Overall, the capsule's shape reflected that of a missile nose cone, with a half-cone angle of 15 degrees. Research in the LaFRC wind tunnels had proved that a curved face heat shield on the wide end gave greater heat protection than a flat-faced heat shield. Type A capsules reflected research that proved that a heat shield was heated more evenly if its curved face formed part of a circle, the radius of which was 1.5 times the diameter of the face to be heated.

This was the design that Faget presented to the NACA conference on the subject at NACA–Ames on 18–20 March 1958. His paper was entitled 'Preliminary Studies of Manned Satellites, Wingless Configuration, Non-Lifting', and called for the cone-shaped Type A capsule to exit the atmosphere thin end first on top of a converted Atlas intercontinental ballistic missile, in place of the nose cone. The astronaut would lie on his back, facing the thin end of the capsule, with his knees drawn up to help him withstand the acceleration forces during the launch. At the end of its orbital flight retrograde rockets would fire against the direction of travel to decrease its forward velocity and allow Earth's gravitational attraction to pull it out of orbit. The broad leading face would result in the majority of re-entry heating being directed away from the spacecraft in a bow wave. Even so, the heat shield would reach a temperature of many thousands of degrees. During re-entry through the atmosphere the broad-end-first approach would produce a high drag rate allowing the spacecraft to be slowed to the point where a parachute landing system could be safely employed in the thick lower atmosphere. Landing in the ocean would absorb final landing loads.

Fig. 40. Maxime Faget, principal designer of the Mercury Spacecraft, the astronaut's prone couch and the Launch Escape System. (NASA)

To help the astronaut withstand the high *g* loads of launch and re-entry he lay in one of Faget's form-fitting prone couches, with his back to the heat shield. Research had shown that when a vertically seated astronaut underwent high *g* forces his blood pooled in his abdomen. Lying horizontally on a prone couch, with the legs positioned higher than the head, gave a more evenly distributed blood supply along the entire body. The close-fitting couch allowed the astronaut to withstand the 7 to 8*g* of the standard Atlas launch and proposed re-entry trajectories.

During the last run in the Sonic Wind-1 rocket sled series, on 10 December 1954, Stapp had pulled 20*g* as the rocket sled was propelled down the track. When it hit the water brake at the end of the track he was thrown forward in his harness and subjected to 46.2*g*. That was twice the 20*g* that Faget believed the astronaut would have to survive if his Type A capsule was rocketed away from a failing Atlas ballistic missile at the moment of maximum aerodynamic pressure on the ascending vehicle. On 29 July 1958 Carter Collins, an Air Force test subject, successfully pulled 20*g* on the centrifuge at the Johnsville Naval Base, thereby adding credence to Faget's belief that the astronaut could withstand such an emergency.

During the same conference at NACA–Ames, John Bird of Langley presented a paper on the re-entry characteristics of the Type A capsule. Wind tunnel testing on scale models had shown that the design became aerodynamically unstable at the beginning of re-entry through Earth's atmosphere. Beginning at an altitude of 121,920 m, rates in yaw and pitch slowly decreased throughout the fall to 60,960 m. After that point they built up once again until, when the capsule's velocity was reduced to Mach 1, the yaw and pitch rates were back-up to where they had been at 121,920 m. This reduction in instability rates during the period of highest re-entry heating suggested that the Type A design might stabilise sufficiently to survive re-entry if it was suitably protected against the extreme heating experienced during that period.

Further wind tunnel testing revealed that a re-entering Type A capsule's yaw and pitch rates peaked at 45 degrees in any single axis, or combination of axes. The tests revealed that these rates could be reduced to 25 degrees by deploying a drogue parachute in the upper troposphere, prior to deploying the main recovery parachute.

A number of full-scale wooden Type A capsules were constructed at NACA–Langley. A single sheet steel capsule, known as a boilerplate (BP) model, was also constructed. Engineers from Gilruth's PARD used the boilerplate capsule in a series of impact tests in the water surrounding the island.

As the impact tests were being carried out another series of wind tunnel tests showed a major deficiency in the Type A capsule. Although the high-drag shape led to the build up of the bow wave in front of the vehicle during re-entry, as predicted by Allen, the long afterbody received an undesirable amount of heating. This deficiency led to the design of the Type B capsule, in May 1958.

10.2 THE SHORT-LIVED TYPE B CAPSULE

At launch the Type B capsule had a similar shape to the Type A model. After leaving the thickest portion of Earth's atmosphere an aerodynamic shroud would be jettisoned to reveal the rounded top of the new capsule. Two cylindrical parachute containers were mounted, one on top of the other, in the centre of the rounded upper surface. The Type B capsule had a wide end diameter of 2.13 m; its conical sides had a half cone angle of 15 degrees.

In mid-1958 engineers planned to fit the manned capsules with a heat-sink heat shield on their wide end. The heat shield would be thick enough retain the heat produced during re-entry and prevent it from reaching the main body of the capsule. At that time the engineers defining the capsule began to question what would happen when the extremely hot heat shield hit the cold water of the Atlantic Ocean. In an attempt to alleviate the problem engineers began studying the possibility of dropping the heat shield after re-entry, but before the capsule hit the water.

During the summer a number of scale model Type B capsules were lifted into the air underneath balloons. When they reached an altitude of 2,000 m, they were dropped and filmed as they fell through the atmosphere. The test series suggested that the design tumbled continuously during descent.

One 8 June PARD launched a scale model Type B capsule on a Nike-Recruit solid propellant rocket, out of Wallops Island. The model displayed the rate-damping properties that Bird had suggested for the Type A capsule and experienced none of the continuous tumbling that the balloon launched models had. A second Nike-Recruit launch, on 4 August, produced similar results. Despite these successes the Type B capsule was already on its way out by August 1958.

Meanwhile, the National Aeronautics and Space Act 1958 had been signed into law on 29 July 1958. President Eisenhower assigned the nation's manned satellite programme to the new civilian administration on 18 August. Of the numerous military proposals for a manned satellite vehicle, only the USAF's X-20 Dyna-Soar programme survived the new law that made NACA the basis of the new NASA. The new administration began work under Administrator T. Keith Glennan on 1 October.

Seven days later, on 8 October, an unofficial group calling themselves the Manned Ballistic Satellite Task Group was established at LaFRC. The group's title was quickly shortened to the Space Task Group (STG). The STG became a separate body within the LaFRC organisation on 5 November 1958.

10.3 BRING ON THE TYPE C CAPSULE

October was also the month when capsule studies moved from the Type B to the Type C design. The new capsule had a heat shield radius of 2.03 m and a wide end diameter of 2.03 m. The half cone angle of the afterbody had increased to 23 degrees. A cylindrical neck was added to the top of the afterbody to hold the parachute recovery system and a tunnel that would allow the astronaut egress from the capsule following landing. A small conical cover protected the flat top of the neck during launch. (See Fig. 41.)

All STG manned satellite studies concentrated on the Type C capsule throughout the end of 1959. Boilerplate versions of the design were produced and employed on drop tests of the proposed parachute recovery system, as well as in full-scale recovery exercises with the armed forces. Type C capsules were also used in full-scale tests of the Launch Escape System, a tractor rocket that Faget had suggested as a means of emergency escape from a launch vehicle failure.

While these tests were underway, STG engineers were preparing for the first Bidder's Conference, at which details of the proposed manned satellite vehicle would be explained to aerospace companies that might bid for the contract to develop and construct it. Details of the Type C capsule were mailed to 40 companies on 23 October. All but two of those companies sent representatives to the Bidder's Conference on 7 November.

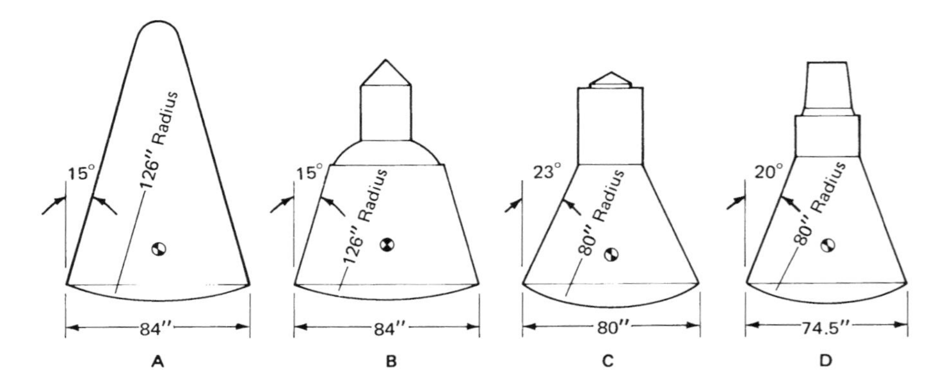

Fig. 41. Comparative views of the four designs considered for the Mercury Spacecraft. (NASA)

At the Bidder's Conference STG staff gave briefings on the Type C capsule and defined its role as a manned satellite vehicle. When the presentations were complete, 19 companies asked to take part in the competition to become the prime contractor for what it was hoped would be the vehicle to make an American citizen the first man in space. Those 19 companies were mailed STG's 'Specifications for Manned Space Capsule', on 14 November. The STG laid down only minimal rules for the definition of the capsule, including:

High-drag, zero-lift shape.
Heat-sink, non-ablation heat protection.
Astronaut escape system.
Orbit termination to be by 'retro thrust'.
Retrorockets not to be used as part of the astronaut escape system.

In the meantime, despite the American public's reaction to Sputnik, President Eisenhower's Administration refused to grant the manned satellite programme a DX rating, the nation's highest national priority. Despite this set back, on 26 November Glennan officially announced that the programme had been officially named Project Mercury, after the winged messenger to the Roman gods.

Eleven companies had submitted their proposals to STG by the closing date of 11 December. Complete proposals were received from AVCO, Chance Vought, Convair, Douglas, Grumman, Lockheed, Martin, North American, McDonnell, Northrop and Republic. STG personnel were surprised to learn that eight of these companies had been studying manned satellite configurations for over a year using their own money to fund their individual studies. A twelfth, incomplete, proposal was received from Wizern Research Laboratories, the company that had designed the pressurised gondola for the Manhigh balloon project. Boeing and Bell did not bid, but they were both involved in the X-20 Dyna-Soar programme.

Each bid was assessed for technical content in nine areas. Each area was assessed by teams of four to six STG personnel and was awarded a score from one to five. The scores in each area were then added together to give an overall assessment rating. The nine areas of assessment were:

Power supply
Instrumentation
Systems integration
Environmental control systems
Structure and heat shield
Escape systems
Retrograde and landing systems
Astronaut support and restraint
Navigation and display instrumentation.

The assessments were complete by 9 January 1959 and the evaluation board passed their recommendation to Glennan at NASA Headquarters. The prime contract for the Mercury Spacecraft went to McDonnell Aircraft Corporation, St Louis, Missouri, on 12 January. The negotiation of contractual details began two days later and were completed on 26 January. McDonnell was to deliver the first three capsules within 10 months, but technical difficulties would extend that period by a further two months. The contract originally called for McDonnell to deliver twelve identical capsules, with a separate contract for the development and delivery of ground support equipment. Capsules would be delivered with LES and adapters to mount them on the Little Joe, Redstone, or Atlas launch vehicle, as required. McDonnell would ultimately deliver 20 individual capsules, two Procedures Trainers, an ECS Trainer and seven further training devices.

McDonnell's proposal had been based on STG's Type C capsule. The astronaut sat in a moulded prone couch, with his back to the heat shield covering the wide end of the capsule. His pressure suit was connected to the ECS, the majority of which was located behind his legs. The Crew Compartment was a self-contained pressurised area within the capsule's external skin. A periscope protruded through the capsule's two walls between the astronaut's feet. It could be rotated through 360 degrees and presented an external view on a screen positioned below the centre of the main instrument panel. The only direct external view was through two small round portholes, located on either side of the astronaut's head.

A drogue parachute and a primary and back-up main parachute were housed in the neck of the capsule. Following re-entry this Earth Landing System (ELS) supported the capsule during its final descent through the thick lower atmosphere. When the main parachute deployed the heat shield was released to drop 2 m and deploy a rubberised canvas landing skirt. The skirt was intended to help lower final landing loads, before filling with seawater and acting as a sea anchor. Following landing a number of standard location and recovery aids were deployed. The astronaut exited the spacecraft by removing part of the instrument panel and a portion of the forward pressure bulkhead. After pushing the empty main parachute canister out of the neck of the capsule he would egress through the one half of the capsule's neck that he had just cleared.

A tractor rocket, known as the LES, was mounted on the neck of the spacecraft at launch. In the event of a catastrophic launch vehicle failure, either on the launch pad, or in the first few minutes of flight, the LES could be activated to pull the capsule to safety. The LES was then jettisoned and the capsule descended to the ocean under its parachutes. If no emergency occurred the LES was jettisoned once it was no longer required. McDonnell manufactured and supplied the LES.

Despite McDonnell's receipt of the prime contract to develop the Mercury capsule STG's Robert Gilruth was sceptical of McDonnell's plan to deploy a landing skirt to reduce landing loads and act as a sea anchor after landing. Gilruth preferred the installation of a crushable honeycomb structure between a fixed heat shield and the aft bulkhead of the capsule. The final vehicle would carry a combination of both systems.

On 16 January STG opted to negotiate for a capsule that could take either a heat sink, or ablative heat shield. Both types of heat shield would be flown on early flight-tests until one proved obviously better than the other. When that occurred, all future capsules would be delivered with that type of heat shield. In order to meet the heat shield requirement McDonnell let two subcontracts. Brush Beryllium Company,

Cleveland, Ohio, would develop and deliver six heat-sink heat shields to the prime contractor. The General Electric Company and the Cincinnati Testing and Research Laboratory (CTL) would cooperate on the development of 12 ceramic ablative heat shields. By the time the contracts were let some engineers at STG were demanding that the ablative heat shield should be fitted to the prototype capsule to be flown on the Big Joe flight, a suborbital Mercury Atlas flight intended to prove that the combination was sound.

Just 13 days after receiving the prime contract to build the Mercury capsule, McDonnell delivered an egress trainer to STG at Langley. The sheet metal 'boilerplate' capsule was a full-scale mock-up of a Type C capsule. The trainee entered the trainer through a hinged side hatch that could then be latched closed. Once inside, he could rehearse leaving the capsule through the tunnel in the neck and deploying his personal life raft, as he would be expected to do at the end of a flight in to space.

McDonnell let subcontracts to companies with the relevant experience to develop and construct the individual systems to be incorporated into the capsule. The ASCS contract went to Minneapolis Honeywell, while the contract for the RCS thrusters went to Bell Aircraft Rocket Division. Barnes Instruments developed the horizon scanners.

Having defined the biological requirements of the single astronaut, McDonnell let the contract for the ECS go to the AiResearch Division of Garrett Corporation. Despite the experience with a mixed gas atmosphere in Project Manhigh, it was decided to provide the Mercury capsule with a pure oxygen atmosphere at 351.5 g/cm^2. This simplified the ECS compared with that used in the Manhigh gondola, but greatly increased the risk of fire. Oxygen was carried in three high-pressure bottles and supplied to the Crew Compartment at the required heat and pressure. Used oxygen was then removed from the Crew Compartment, cleansed of moisture and exhaled carbon dioxide before being supplemented by additional oxygen and recycled.

At this time STG initiated a series of development programmes in support of McDonnell's capsule. A multi-establishment wind tunnel programme began using the Type C capsule design. At the same time a programme of tests to develop Faget's LES began at Wallops Island. The tests showed that Faget's original design required considerable alteration before a reliable system could be constructed. A programme of high-altitude parachute tests was also begun, in order to prove the reliability of the ELS.

A shortage of industrial beryllium delayed the development of the heat-sink heat shield for the early flight-tests. Throughout the first year of capsule development a growing number of STG engineers placed their support behind the new ceramic ablative heat shield technology. Their confidence came from the result of wind tunnel testing that showed that the capsule's shallow re-entry path through the atmosphere would result in extremely high heating rates over a prolonged period. It was feared that the use of a heat-sink heat shield might result in sufficient heat being transferred to the Crew Compartment to physically and mentally incapacitate the astronaut.

LaFRC and LFRC began subjecting ablation materials to conditions predicted for the Mercury capsule's re-entry. In a parallel programme commencing on 9 March LaFRC began trying to define how much noise the astronaut might be expected to withstand during an orbital space flight.

With no DX rating for the manned satellite programme, McDonnell found difficulty in competing with other military programmes for materials. With its DO rating the Mercury capsule had to give way to military programmes that carried the higher rating. On 10 March McDonnell informed STG of their difficulties and warned that continued difficulties in obtaining vital materials might lead to a slippage in the planned delivery dates for the first Mercury capsules.

McDonnell held a Capsule Mock-up Review at their St Louis plant over 17–18 March. A full-scale wooden and cardboard mock-up of the proposed Mercury capsule was displayed along with a full-scale Mercury Atlas Launch Vehicle Adapter and a mock-up of the LES.

The capsule mock-up displayed an Antenna Canister mounted on a shortened neck. A small conical cover protected the horizon scanners housed in the Antenna Canister. The cover would be jettisoned once the capsule achieved the vacuum of space. This modified Type C capsule became known as the Type D capsule and was very close to the design that would ultimately fly in Project Mercury. The mock-up also contained a wooden copy of Faget's prone couch and a main instrument panel that was an inverted U shape. A wooden copy of the periscope was included, as were the two round portholes and a side hatch that would require a large number of individual torque bolts to secure it in place.

The LES mock-up showed the design as modified as a result of the flight-tests at Wallops Island. It had a braced three-point support tower, three exhaust outlets for the Escape Motor and a single exhaust nozzle for the Jettison Motor. Further testing, as part of the Mercury Little Joe flight-test programme, would lead to the Jettison Motor also being fitted with three exhaust nozzles, to prevent impingement of the exhaust plume on the support structure with a resulting loss of effective thrust.

Following a thorough review of all systems and the mock-ups over the two-day period, STG issued McDonnell with instructions to redesign the astronaut egress system, instrumentation layout, sequencing lights, clocks and other displays. A second review was arranged to inspect the changes when they had been made.

While the question of the landing skirt verses a crushable honeycomb impact system remained unresolved, the capsule's recovery aids began acceptance testing in March. During the same month the capsule's 19.2-m extended skirt main recovery parachute was replaced with a ringsail parachute of the same diameter.

On 23 March, McDonnell compiled a list of 23 items that required the DX rating if the Mercury capsule was to stay on schedule. The list was forwarded to NASA Headquarters in Washington. In the same month McDonnell quoted the requirement to supply spare parts and test equipment for the Mercury capsules as their justification for attempting to raise the value of their contract from $18 million to $41 million. Abe Silverstein, NASA's Director of Spacecraft Development, refused the increase and made the Administration's discontent with the application clear.

Meanwhile, on 26 March, LaFRC was allocated funds to commence a series of hypersonic flight-tests. PARD would conduct heat transfer rate tests at velocities up to Mach 17 and dynamic behaviour tests at speeds between subsonic velocities and Mach 10. The following day Glennan announced that all NASA satellites and launch vehicles would be identified with the words UNITED STATES in upper-case letters.

The month ended with an instruction to deliver all Mercury capsules intended for Mercury Redstone flights directly to Hangar S, Atlantic Missile Range. By deleting an original requirement to deliver the capsules to Huntsville, where they were to have undergone fit tests with their launch vehicles before delivery to Florida, STG cut two months off of each capsule's preparation time. In April it was decided that the capsule intended for Mercury Redstone 1 would spent two weeks at Huntsville before delivery to Florida.

The seven Flight A astronauts, the men who would fly the Mercury capsule into space, were introduced to the public in Washington, DC, on 2 April. From the earliest days of their tenure at STG the astronauts were encouraged to participate in the design and building of the hardware that they would fly into space. This gave them a certain amount of power, which they used to have the capsule's design changed to meet their own requirements. One week later, on 9 April, a 16-month long series of wind tunnel tests related to the development of the Mercury capsule began. The series involved 25 different wind tunnels and 103 individual test runs. As a result, of continued delays in the development of a heat-sink heat shield, the ceramic ablation shields began to find favour within the engineering community studying the problem of heat protection during re-entry through Earth's atmosphere.

In April McDonnell engineers strapped pigs to their bags in a mock-up of the prone couch that the astronauts would sit in inside the Mercury capsule. The pigs were then used in drop tests and survived up to 58*g* for short periods without permanent injury. At LaFRC NASA engineers worked on developing a crushable honeycomb to fill the space between the main heat shield and the aft bulkhead of the Crew Compartment. NASA wanted the honeycomb to replace the landing bag included in McDonnell's original design. As the month drew to a close McDonnell named John Yardley as the manager responsible for their Project Mercury activities. (See Fig. 42.)

President Eisenhower assigned DX rating to Project Mercury on 28 April and Congress approved the move on 4 May. Achieving DX rating placed Project Mercury on a par with the Atlas ballistic missile and other military projects for material procurement.

The second mock-up review took place at St Louis over 12–14 May. NASA engineers originally reviewed McDonnell's changes made as a result of the earlier mock-up review. When engineers in pressure suits entered the updated wooden mock-up of the capsule they discovered that they were not able to reach many of the switches on the main instrument panel. McDonnell were again instructed to make changes. When the Flight A astronauts visited McDonnell in May, they too were encouraged to enter the wooden capsule mock-up and pass comment on what they found.

In March 1959 a Project Mercury Flight Schedule was published, which included

Fig. 42. Mercury Spacecraft under manufacture at McDonnell's St Louis factory.

plans to build upon the US Air Force and US Navy's experience in stratospheric ballooning. Balloons were to lift two Mercury capsules to an altitude of 24.3 km, where they would 'soak' in the space-like environment for up to 24 hours. Following the completion of its period of soaking the capsule was to have been released and allowed to fall back to Earth, employing its ELS to land in the ocean. Contracts were given to the US Weather Bureau, the Office of Naval Research and the USAF Cambridge Research Centre to support this programme, which was to be under the control of the USAF. The Mercury balloon programme was cancelled in May 1959 when it became apparent that the altitude wind tunnel at LaFRC could simulate the conditions at 24.3 km without the spacecraft having to leave the ground.

James Chamberlin was named as head of the Mercury Capsule Coordination Office on 19 June. The new office had been set up to oversee both McDonnell and NASA's efforts in developing the Mercury Spacecraft. Chamberlin would replace Faget as NASA's representative in meetings with McDonnell's John Yardley. In its turn, a Capsule Review Board was established to oversee the work of the Coordination Office.

The first ablation heat shield was delivered to McDonnell on 21 July. Employing top-secret military technology, the heat shield was heavily guarded during its journey from the Cincinnati Testing Laboratory, but this first ablation heat shield never flew in space. During the same month McDonnell subcontracted the capsule's ELS to Northrop Corporation, Radioplan Division, Van Nuys, California. The ELS consisted of a high-altitude drogue parachute that stabilised the spacecraft prior to the deployment of a 19-m diameter ringsail main parachute. Northrop developed and tested the ELS before delivery to McDonnell for installation on the production capsules.

In July McDonnell opened the 'clean room' in which the Mercury capsule would be constructed at their St Louis plant. The air filtration system in the clean room (white room) ensured that the air within was cleaner than that in a hospital operating theatre. Everything entering the clean room was carefully inventoried and controlled. Engineers working in the room wore white overalls, overshoes and hats. In the same month McDonnell took delivery of the first ASCS, and the horizon scanners associated with the ASCS entered qualification testing immediately.

10.4 TYPE D: THE FINAL CONFIGURATION

A 0.25 scale model of the Type C capsule was launched on a rocket from Wallops Island in July. Following separation from its launch vehicle it tumbled wildly and failed to establish a stable glide attitude before it hit the water. As a result of that flight, the Type C capsule was replaced by a similar, but new design. The Type D spacecraft had a half-cone angle of 25 degrees and a diameter of 189.2 cm. The heat shield radius was 203.2 centimetres.

During the first week in September McDonnell moved part of their Project Mercury effort to the USAF's AMR in Florida. Along with the rest of Project Mercury, they were based in Hangar S, the building where the US Navy had prepared Project Vanguard. When Project Mercury first arrived at Hangar S, Project Vanguard had not been completed. The manned satellite programme was assigned one fourth of the Hangar S floor space. As Project Vanguard was completed so Project Mercury was assigned a greater portion of Hangar S until it occupied the whole building. Yardley, and the engineers who moved to Florida with him, lived in caravans on the open ground behind Hangar S. Personnel from other contractors added to the McDonnell vans when they too moved to Florida.

STG provided a copy of the Preparation Ground Rules for a Mercury Atlas launch on 3 September. The document allowed McDonnell to define exactly what ground support equipment (GSE) they would have to develop in order to meet the rules and prepare a Mercury Spacecraft for launch.

Six days later Big Joe lifted off from Launch Complex 14 (LC-14) AMR. The Atlas launch vehicle carried a heavily instrumented prototype Mercury capsule with an ablation heat shield. As a result of the Big Joe flight, ablation heat shields would be fitted to all production Mercury Spacecraft.

On the same day, 9 September, NASA provided McDonnell with details of what was required to be in the Astronaut's Handbook that they were required to produce to introduce the Flight A astronauts to their spacecraft. The final publication was in three sections, Normal Operating Procedures, Emergency Operating Procedures and Failure Analysis Procedures.

A third mock-up review was held at St Louis over 10–11 September. The Flight A astronauts attended and rebelled against their passive role on the proposed flights of Project Mercury. They did not like the word 'capsule' and its suggestion that they were to be helpless passengers. They introduced the word 'spacecraft' and insisted that everyone use it. Even so, all of the men that flew the Mercury Spacecraft would refer to it as a 'capsule' at some point during their flight, or in their official flight documentation.

Most of all, the astronauts resented the fact that they were intended to ride the Mercury Spacecraft as a passenger and medical experiment. The sole extent of their engineering input during a flight would be to read back the instruments on the main control console. As military test pilots, the astronauts were the butt of many jokes from the test pilot fraternity. Chuck Yeager, the pilot who had ridden the Bell X-1 rocket plane through Mach 1 for the first time in 1949 had come up with the phrase 'Man in a can' to describe the astronauts' role in Project Mercury. He had also made it clear to the press that it would be the Mercury capsule's monkey test subjects that would be first to fly and would therefore be the real test pilots in the project, not the human astronauts that would follow them. Yeager had told one reporter that he would not want to fly any vehicle that required him to '*brush the monkey shit off the seat before I climbed in*'.

The astronauts demanded manual control of all portions of their flight. They wanted to control the activation of an abort, as well as their attitude in flight. They wanted to be able to jettison the LES and fire the retrograde rockets if the automatic systems failed. They also demanded manual overrides on the automatic systems responsible for deploying each of the spacecraft's parachutes. The manual overrides would be installed along with two new manual attitude control systems to complement the two automatic systems that were included in the original McDonnell design. Work began on the RSCS in July while work began on the FBW in August.

They also voiced concern over the small size and location of the two portholes, one on either side of the astronaut's head. In order to be able to control their attitude in space, especially if they had to perform a manual retrofire, they demanded that the portholes be replaced with a large heads up, or 'picture window', directly in front of the astronaut's face. Another subject of concern was the requirement to exit through the neck of the spacecraft at the end of a space flight. The astronauts felt that an explosive hatch, similar to those employed on military aircraft, should replace the spacecraft's original side hatch, which was bolted into place with 40 torque bolts.

In the event of an emergency on the launch pad, a single explosive charge would be detonated in order to blow the hatch off and allow the astronaut to make an emergency escape. This model was replaced with a new side hatch that contained 70 torque bolts and two shaped explosive charges around its exterior. The hatch could be jettisoned from outside by pulling a lanyard. Alternatively, the astronaut could strike a plunger with his hand to set off the explosive charges.

Meanwhile, LaFRC spent the month testing nine different ablation materials for use in the heat shield for the Mercury Spacecraft. Following the Big Joe flight, STG had made the decision that the few beryllium heat shields that had been produced would be used on the early unmanned Mercury Redstone flights. All Mercury Atlas flights and the manned Mercury Redstone flights would use ablation heat shields.

When the proposed retrograde rockets proved unstable in testing, consideration was given to methods of getting the spacecraft back to Earth before the astronaut's oxygen ran out if the rockets failed in orbit. At the same time a hinged 'destabilisation flap' was added to the Antenna Canister (see Fig. 43). In the case

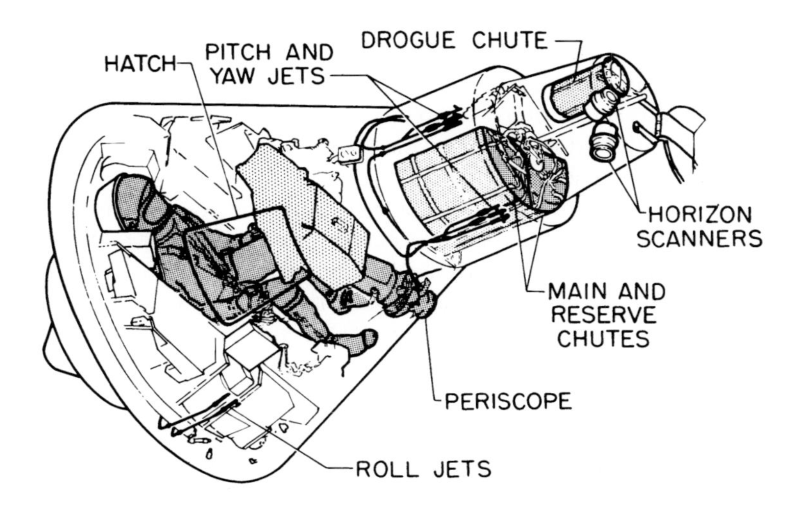

Fig. 43. Mercury Spacecraft systems, note the deployed Destabilisation Flap on the Antenna Canister. (NASA)

of a spacecraft re-entering the atmosphere in a stable Antenna Canister forward attitude the flap would cause sufficient drag to turn that spacecraft through 180 degrees to a heat shield forward position. Elsewhere the development of the drogue parachute was proving difficult and consideration was given to deleting it completely from the ELS. The drogue parachute was only retained because Gilruth insisted that no alternative system could be developed in time to retain Project Mercury's already delayed flight schedule.

One alternative form of ELS that was considered for use on the Mercury Spacecraft during this time was the flexible wing, developed by Francis Rogello at LaFRC. PARD carried out four launches of small-scale Rogello Wings on Nike-Cajun solid propellant launch vehicles out of Wallops Island on 6 October, 25 November, 7 December 1959 and 2 February 1960. Only the first model was recovered, but all four were tracked by radar and performed well. Although the Rogello Wing never flew on a NASA manned spacecraft its development was continued for several more years. It would ultimately enter popular use as the hang-glider and micro-light aircraft popular with some sportsmen and women. (See Fig. 44.)

September saw the completion of the first ECS test bed system. By the end of the year neither the pressure suit circuit nor the Crew Compartment system had completed a full 28 hours of continuous operation, as demanded by the design specification. McDonnell's Warren North, who had had a Mercury pressure suit produced to fit him, participated in the ECS tests. A mechanical man simulator was also produced to simulate the demands placed on the ECS by a human astronaut. A combination of tests employing either North, or his mechanical partner, continued through early 1961 before the ECS was declared ready for flight.

Between 22 September and 10 October a series of centrifuge runs were performed to finalise which design of hand controller should be fitted in the spacecraft to operate the manual control systems demanded by the astronauts. The final design was a single three-axis controller that offered full manoeuvring in pitch, roll and yaw.

As October began the spacecraft was 72 kg over its maximum design allowance, as dictated by the payload lifting capabilities of its two launch vehicles. During the month NASA Headquarters allocated the money to develop and install the overhead window, an explosive hatch, a RSCS, a reefed ringsail main parachute and reconfiguration of the main instrument panel. This allocation of funds did away with the requirement to deliver identical spacecraft, as dictated by McDonnell's contract with NASA. During the same month McDonnell received an amendment to their contract for an additional six Mercury Spacecraft.

Meanwhile, LaFRC and LFRC cooperated on a series of test programmes in support of the Mercury Spacecraft. These included:

- Full-scale spacecraft/launch vehicle separation tests, including the effect of the posigrade rocket motor exhaust on the launch vehicle and spacecraft.
- Manned wind tunnel tests of astronaut control techniques.
- Retrograde rocket calibration tests.
- RCS testing.
- Determining the effects of the LES exhaust plume on the spacecraft.

Fig. 44. A Mercury Spacecraft mock-up is shown in a hanger attached to a deployed Rogello flexible wing. The Rogello Wing was considered as an alternative to the standard parachute Earth Landing System. (NASA)

In November 1959 the STG was re-named the Manned Spacecraft Centre.

Between November 1959 and January 1960 the seven Flight A astronauts and several NASA and McDonnell personnel had Plaster of Paris moulds made of their bodies dressed in Mercury pressure suits. The moulds were used in the manufacture of the fibreglass inserts that were fitted into their prone couches when their individual spacecraft were manufactured. Additional fibreglass inserts were also used in mock-up prone couches installed in the gondola of the centrifuge at Johnsville and in McDonnell's two procedures trainers at LaFRC and at Hangar S.

The year 1959 ended with McDonnell engineers in pressure suits obtaining excellent results during tests of the spacecraft's ECS at AiResearch. At the Arnold Engineering Development Centre testing was also underway to determine the ignition characteristics of the retrograde rockets. Bell Aircraft Corporation began a series of tests on the Attitude Stabilisation and control system in December 1959. The Bell tests continued until February 1960.

10.5 PREPARING FOR FLIGHT

On 15 January 1960 the spacecraft onboard programmer began testing at Wheaton Engineering, where it was being developed. The programmer contained references for the ASCS as well as attitude manoeuvres required at various times throughout a given flight. Testing was not completed until 26 March.

January saw the astronaut and instrument recording camera complete its test programme. The camera was mounted above the main instrument panel, from where it would film the astronaut's head and torso. A large convex circular mirror attached to the astronauts parachute harness allowed the camera to film the instrument readings on the main panel at the same time.

McDonnell delivered Production Spacecraft 1 to MSC on 25 January 1960. It was used in the Beach Abort 1 flight. Following delivery, MSC installed the flight instruments before delivering the spacecraft to Wallops Island. MSC also took delivery of Spacecraft 4, for use on Mercury Atlas 1 in January 1960. It was delivered to Hangar S, AMR, on 23 May.

The spacecraft periscope was qualified for flight on 1 February. On the same day MSC completed a series of studies into the internal and external noise levels that were to be expected during a Mercury flight. Their data was inconclusive and they suggested that real time noise data be recorded during the project's early flight-tests. Four days later beryllium was selected as the material to be used to protect the spacecraft's Recovery Section during re-entry. That same day the spacecraft's telemetry system passed its final design approval. The telemetry system completed reliability testing on 27 February.

The spacecraft battery and the spacecraft's post-landing aid completed qualification for flight on 15 February. On 20 February the ASCS completed testing and was qualified for flight. The spacecraft Command Receivers followed suit on 27 February. The ELS completed a series of 56 drop tests on 11 August.

McDonnell informed MSC on 5 April, that they were commencing research into the possibility of designing the Mercury Spacecraft for a controlled, rather than purely ballistic, re-entry. By designing a lift capability into the spacecraft's shape the astronaut would be able to manoeuvre the spacecraft to increase or reduce drag as it passed through the atmosphere. Increased drag would decrease lift and shorten the length of a spacecraft's re-entry trajectory, while decreased drag would increase lift and lengthen the re-entry trajectory. This work ultimately led to the total redesign of the Mercury Spacecraft by a group of engineers led by James Chamberlin. Chamberlin's masterpiece was originally called Mercury Mark II. History records it as the Gemini Spacecraft.

On 9 April the construction of a vacuum chamber was completed at Hangar S. It was used during pre-flight-testing of spacecraft at that location. The Flight A astronauts participated in many of these tests, thereby becoming familiar with their spacecraft prior to flight. Also on 9 April, Spacecraft 1, the first production model, was delivered to Wallops Island on Beach Abort 1. When the flight took place, on 9 May, it would lead to a major redesign of the LES.

During April John Yardley moved to Hangar S, where he joined McDonnell's contingent at Cape Canaveral preparing the Mercury Spacecraft for the early flight-tests. When numerous major problems were discovered in the Spacecraft delivered to Florida, Yardley instructed managers and engineers at St Louis to pay particular attention to upgrading systems reliability. Meanwhile, McDonnell's commitment to Project Mercury peaked at 880 engineers during April 1960. As the terms of the prime contract were met, this number would steadily decrease over the next three years. The month also saw the retrograde rocket pass its qualification testing. McDonnell engineers participated in ECS tests and LaFRC completed a series of tests on a number of different ablation materials.

McDonnell delivered Procedures Trainer 1 to LaFRC on 4 May. Procedures Trainer 2 was delivered to Hangar S on 5 July. Procedures Trainer 1 was moved to the Manned Spacecraft Centre, Houston, in 1962. The spacecraft's new explosive side hatch was qualified for flight on May 15. Because Spacecraft 7, the first manned spacecraft, was already well advanced in the manufacturing process, the first new hatch and overhead window were fitted to Spacecraft 11, the second manned spacecraft. Thereafter, they were fitted to all remaining Mercury Spacecraft.

NASA's Nicholas Golvin was named as head of a group contracted to oversee attention to reliability at McDonnell's St Louis plant. Beginning in June, when Golvin's group started work, reliability of components fitted to Mercury Spacecraft increased.

June ended with technical difficulties in the ECS of Spacecraft 2, intended for Mercury Redstone 1. As a result, Project Orbit was established at McDonnell and one spacecraft was removed from the production line and assigned to a simulation programme designed to qualify the Mercury Spacecraft for a manned three-orbit flight. To meet the objectives of Project Orbit, a vacuum chamber was constructed at the McDonnell plant. That chamber was large enough to hold a Mercury Spacecraft and was capable of simulating orbital conditions, in order to test the integrity of the spacecraft.

MSC had delivered Spacecraft 4 to Hangar S in advance of Mercury Atlas 1 on 23 May. McDonnell shipped Spacecraft 2 to Marshall in June, for use in vibration testing along with the Redstone 1 launch vehicle. On 29 July, Spacecraft 3 was delivered to Langley for noise and vibration testing. MSFC's spacecraft checkout facilities were transferred to Hangar S on 18 October.

Spacecraft 7 had been assigned to the first manned suborbital flight, on a Redstone launch vehicle. The Spacecraft was the subject of an engineering review over 16–18 August. At the end of the review Spacecraft 7 was accepted for manned flight. At this late stage MSC's proposal to install flotation bags in the neck of the spacecraft were dropped from the design in favour of McDonnell's original idea of dropping the heat shield at main parachute deployment, in order to deploy a combined impact skirt/sea anchor. The new design required the spacecraft's centre of gravity to be moved to ensure a stable, upright flotation attitude following landing. These changes caused a delay in Spacecraft 7's delivery date.

At the same time engineers at Langley began work on a Stullken Collar, to be fitted around the wide end of the spacecraft as it rested in the ocean. Recovery swimmers would then inflate the collar with gas held in two high-pressure bottles. In its inflated condition the Stullken Collar would maintain the spacecraft in a near vertical position with its side hatch clear of the water. The collar, designed by Donald Stullken of MSC, would not complete qualification testing until 3 January 1962 and would be used during the recovery of each of the last four manned flights in Project Mercury.

McDonnell completed their testing of Spacecraft 7 in the last week of November. It was certified for manned suborbital flight on December 1, at which time it was 20 days behind schedule. The spacecraft was delivered to Hangar S on 9 December.

10.6 CONTINUED DEVELOPMENT

Project Orbit was authorised on 26 September 1960, to test a flight-ready Mercury Spacecraft under simulated orbital conditions. The new vacuum chamber at McDonnell's St Louis plant took six months to construct. With its completion Spacecraft 10 was assigned to Project Orbit, on 16 March 1961. The project's original goal was to simulate in real time a three-orbit, 4.5-hour flight, which represented Project Mercury's primary goal.

As Project Orbit began, Spacecraft 7 was at Cape Canaveral being prepared for Mercury Redstone 3, a manned suborbital flight. Both NASA and McDonnell recognised the fact that the Mercury Spacecraft was still a long way from being qualified for manned orbital flight. Five Mercury Little Joe flights at Wallops Island had failed to qualify the LES for the worst-case abort during a Mercury Atlas launch. The first Mercury Atlas flight vehicle had exploded in flight, while the second had flown successfully, but the launch vehicle had been of a unique configuration that would not be repeated.

The first simulated flight of Project Orbit occurred on 2 April 1961. Over 1,000 hours of testing followed and revealed numerous faults within the spacecraft's RCS and ECS. Each problem was diagnosed, corrected and tested before the new design could be qualified for orbital flight.

Alongside Project Orbit a series of other test programmes continued. The ELS, which had been qualified for manned suborbital flights, was subjected to a new series of drop tests, to qualify the parachutes to support the greater weight of a manned orbital spacecraft. US Navy recovery teams practised their role in Project Mercury to ensure their familiarity with the procedures that they would use to recover astronauts at the end of their flights.

On 12 April 1961, a Soviet Red Air Force Major became the first man in space. He was launched from the Baikonur Cosmodrome in a Vostok Spacecraft launched on a converted R-7 ICBM.

Mercury Atlas 3, an unmanned qualification flight, was launched on 23 April 1961. The Range Safety Officer destroyed it. The LES was finally qualified for manned Mercury Atlas orbital flights on 28 April 1961. The seventh and last Little Joe airframe had subjected the LES and Mercury Spacecraft that it carried to greater than planned acceleration forces at the moment of 'abort'.

On 5 May 1961, Spacecraft 7 carried the first Flight A astronaut on a suborbital flight out of Cape Canaveral. With the exception of one indicator light, all systems performed well. Meanwhile, McDonnell hosted a Development Engineering Inspection for the Mercury Spacecraft over 23–24 May. The review resulted in 45 requests for changes.

Spacecraft 11 carried a second astronaut on the manned Mercury Redstone 4 suborbital flight, on 21 July. Following premature loss of the side hatch at the end of the flight, Spacecraft 11 sank in the Atlantic Ocean. Between 5 August and 12 October the spacecraft side hatch was tested extensively under conditions far more severe then those experienced during the flight. At no time did it jettison prematurely.

As a result of the Spacecraft 11 side hatch event, definition began of a side hatch with a mechanical latch. The finished article was too heavy for inclusion on the remaining spacecraft, so they flew with explosive hatches similar to those used on Mercury Redstone 4. Instead of using the mechanically latching hatch, post-landing astronaut procedures were changed so that the astronaut did not remove the safety pin from the hatch detonation plunger until after the recovery helicopter already had its winch line secured to the recovery loop on the neck of the spacecraft.

During a 27–28 July meeting between NASA and McDonnell it was decided that Project Mercury would end with a single 18-orbit manned flight. The plan was approved by NASA management at Headquarters in Washington on 25 September. The two organisations also agreed to commence studies into an advanced two-man spacecraft capable of prolonged orbital flight. The two-man spacecraft would be based on Chamberlin's Mercury Mark II studies.

Spacecraft 5 was used in sea-worthiness testing off Wallops Island, on 28 June. A second series of tests, in larger waves, were completed successfully over 1–3 August. Three days later, on 6 August, the Soviet Union launched Vostok II. The flight lasted 25 hours, during which the cosmonaut orbited Earth 17 times. Eleven days later, Spacecraft 13 was shipped to Hangar S. It would become the first Mercury Spacecraft to carry an astronaut into orbit.

In preparation for the manned Mercury Atlas flights three rocket sled tests were made at the Naval Ordnance Test Station, China Lake, California. As the sled passed down its track the explosive bolts fitted to the spacecraft/launch vehicle Marman Clamp were fired under high acceleration forces. The three tests were made on 5, 9 and 14 September. Minor modifications were made after each of the first two runs. By the completion of the third run the separation sequence had been qualified for manned orbital flight.

On 26–27 October ships of the US Navy's Project Mercury recovery fleet practised methods of recovering a Mercury Spacecraft directly from the sea, without the use of a recovery helicopter. To meet that requirement, specialist equipment had been developed to fit onto the standard lifeboat davit fitted to US Navy ships. Several Destroyers were fitted with the equipment and located at secondary recovery sites, ready to recover any Mercury Spacecraft that did not land in its primary recovery area.

Mercury Atlas 4 was launched on 13 September. Only three spacecraft anomalies were recorded during the one-orbit flight. Mercury Atlas 5 followed on 29 November. The Mercury Spacecraft completed only two of the three planned orbits. After both flights MSC representatives told the press that, had a human astronaut been on board, the mission would have run the full course. The year ended with two drop tests, on 14 and 18 December, using two Mercury Spacecraft without impact skirts. No structural damage was sustained.

On 3 January 1962, further recovery tests were carried out, concentrating on the deployment of the Stullken Collar by US Navy recovery swimmers. In January a series of 20 parachute drop tests were begun to qualify the main parachute for the heavier spacecraft that would be used on the final 1-day flight. This series was called Project Reef and was completed on 26 June.

Mercury Atlas 6, the first manned orbital flight of the project, was launched on 20 February 1962. The flight completed Project Mercury's principal goal of three orbits. With the project's principal goal achieved, a memorandum was issued on 4 May requesting proposals for experiments to be flown on later manned orbital flights. Mercury Atlas 7 was the first flight to carry experiments when it was launched in October. They would be responsible for the astronaut falling behind his flight plan. Mercury Atlas 7 had the highest ASCS propellant usage of all the manned Mercury flights. As a result, the astronaut's suggestion that the spacecraft be fitted with a switch to allow the high-rate ASCS thrusters to be turned off during orbital flight was acted upon. If the astronaut on future flights forgot to re-enable the high-rate ASCS thrusters prior to re-entry an automatic switch would do so for him. The Mercury Atlas 7 astronaut was also the first to suggest that the spacecraft periscope was not necessary and could be removed.

Following a series of tests carried out between 29 April and 5 May, it was decided to remove the leg supports from the astronaut's prone couch. The test programme had shown that just the toe and heel supports were sufficient to support the astronaut's legs during lift-off and re-entry.

On 27 May it was announced to the public that Mercury Atlas 8 would make six orbits of Earth, rather than the three originally planned for that flight.

The Design Engineering Inspection for a mock-up of the spacecraft to be used on the 1-day mission was held at St Louis on 7 June. Negotiations for spacecraft changes continued throughout August and September. On 8 August, Spacecraft 9 was redesignated Spacecraft 9A and transferred to Project Orbit, where it served as the test bed for the 1-day flight that would end Project Mercury.

Mercury Atlas 8 was launched on 3 October. Despite the Mercury Atlas 7 astronaut's recommendation, Mercury Atlas 8 carried a periscope. The decision to carry the periscope resulted in an experiment that had been designed to fit into the periscope well being deleted from the flight. In order to complete a six-orbit flight the spacecraft's oxygen leakage rate was reduced from 1,000 to 600 cm^2/min. The spacecraft's lithium hydroxide bottle was filled with 2.54 kg of the oxygen scrubbing chemicals for the first time during the longer flight. On previous flights the bottle had only carried 2 kg of lithium hydroxide.

The astronaut conserved manoeuvring propellant by caging the spacecraft's gyroscopes and turning off the ASCS. This resulted in periods of drifting flight when the spacecraft was free to assume any attitude. This also conserved electrical power and allowed a six-orbit flight to be made without installing a second battery in the spacecraft. A second power-saving method included switching the spacecraft's telemetry and radar beacons over to ground command at various phases during the flight. The flight of Mercury Atlas 8 was the only manned orbital flight in the project on which the astronaut was able to rely on the ASCS throughout retrofire and re-entry.

With one of the Flight A astronauts grounded for medical reasons, Mercury Atlas 9 became the last flight of the project and was therefore assigned the 1-day flight. To meet the requirements of the extended flight the last Mercury Spacecraft was fitted with additional batteries, another 1.8-kg high-pressure oxygen bottle, 2 kg of drinking water, 6.8 kg of thruster propellant. A plan to replace the astronaut's prone couch with a hammock-like structure was not implemented due to fears that the material might stretch and therefore not support the astronaut correctly during periods of high *g* forces. The spacecraft only carried one UHF and one telemetry transmitter and the RSCS thruster system was not installed along with the spacecraft periscope. With so much additional weight McDonnell established Project Reef to requalify the main parachute for the heavier spacecraft. By this time Project Orbit had simulated over 1,000 hours of manned orbital flight. It suggested that the spacecraft thrusters might freeze during the long periods that they were inactive. The 34-hour flight was completed in May 1963.

In the 55 months of Project Mercury the spacecraft had changed from a passenger-carrying nose cone to a fully controlled manned spacecraft. Time and again the four men who flew orbital flights were given the opportunity to prove the importance of having an astronaut onboard. If the astronauts had not demanded manual control over all aspects of their flight then only Mercury Atlas 8 would have completed its entire flight plan. The others would have been returned to Earth early after the ASCS had malfunctioned and consumed high rates of propellant that was vital for re-entry.

Faget's basic spacecraft design had proved sound and, with numerous modifications, had met all of the demands of Project Mercury. Its basic shape would be incorporated in both the Apollo Command Module and the Mercury Mark II spacecraft. In both of those spacecraft as much equipment as possible was placed outside of the Crew Compartment. It was a lesson that had been hard learned during the long days and weeks preparing the Mercury Spacecraft for launch. Faget's designs for the astronaut's prone couch and the LES were also carried over to Apollo and Mercury Mark II, although the LES was ultimately replaced by aircraft-style ejection seats on the latter.

11

Mercury Spacecraft Earth Landing System

On 22 August 1958, STG officially requested permission to commence a series of parachute drop tests in order to qualify the ELS planned for use on their piloted satellite vehicle. Five days later representatives from NACA Langley's Model Engineering Section visited the Pioneer Parachute Company to negotiate the development and production of 14.1 m extended skirt parachutes for use within their proposed test programme. The smaller diameter extended skirt parachutes were employed during the drop test programme because the test vehicle was lighter than the production satellite vehicle, which would use larger, ringsail parachutes. The NACA Committee approved the drop test programme on 11 September.

The programme was completed in three phases:

Phase 1 tested the parachute pyrotechnics by making several drops of an active parachute canister attached to a 208-litre oil drum filled with concrete. The test vehicle was dropped from a helicopter from an altitude of 914 m above West Point, Virginia. During each drop the pyrotechnics were fired to open the parachute canister and deploy the parachute.

Phase 2 dropped a single boilerplate mock-up of the satellite vehicle from the back of a C-130 transport aircraft over USAF Pope Field, North California, on 29 September. This drop was a test of the system used to extract the test vehicle from the aircraft. The boilerplate vehicle was mounted on a cargo pallet in the rear of the aircraft. When the rear cargo door was opened, an extended skirt parachute attached to the cargo pallet was deployed into the aircraft's slipstream. This pulled the cargo pallet out of the aircraft, after which the test vehicle separated from the pallet and deployed its own extended skirt parachute. The test vehicle made a successful descent. Two days later, on 1 October, NACA became the nucleus of the new NASA.

Phase 3 consisted of five operational tests using the same deployment sequence as the Phase 2 drop. These drops were designed to test the stability of the satellite vehicle in free flight and with its deployed parachute. Shock loads were recorded within the spacecraft as the parachutes deployed. Finally, the drop tests were used to develop recovery techniques for the production vehicles. All five drops took place over the water off Langley's PARD facility at Wallops Island, Virginia.

The first drop took place on 25 November 1958, using a Type C boilerplate spacecraft with a simulated heat shield. The spacecraft left the C-130 at an altitude of 3,048 m and was filmed by the crew of a T-38 chase plane. During the 24-second free fall the spacecraft oscillated up to 90 degrees in both positive and negative pitch before the onboard timer fired the pyrotechnics to open the parachute canister, jettison the heat shield and deploy the single extended skirt parachute. Following landing a saltwater-activated switch jettisoned the parachute. A Marine Corps helicopter on station for the test recovered the spacecraft, while a hired fishing boat recovered the parachute. The spacecraft was flown to Chincoteague Air Station, before being returned to Langley Field in the rear of a C-130 cargo aircraft. The test was successful.

The second drop occurred on 1 December, at an altitude of 1,524 m. Having exited the aircraft and separated from the cargo pallet, the second boilerplate spacecraft was even less stable than the first. It was travelling through the air, apex down, when the parachute canister opened and the single parachute deployed. The pyrotechnics intended to jettison the simulated heat shield failed to fire. The saltwater switch also malfunctioned and the parachute lines failed to separate from the neck of the spacecraft after landing. The parachute was cut free by the crew of the fishing vessel hired to recover it.

A Marine Corps helicopter got a line on the spacecraft's recovery loop but the aircraft's hoist safety release failed and the spacecraft fell back into the water. The fishing boat secured a line to the spacecraft and began towing it towards Wallops Island. The spacecraft sank in shallow water and the vessel's crew was forced to cut their line. The following day the US Navy attempted to recover the spacecraft, only to lose it again, this time in deep water. Finally, on 4 December, US Navy divers recovered the spacecraft and returned it to Wallops Island, where the flight data recorders told the story of the in-flight instability

As a result of the first two Phase 3 drops, a high-altitude drogue parachute was added to the spacecraft's ELS. The drogue parachute would be deployed first, to steady the spacecraft in an apex-up attitude and to reduce loads on the main parachute when it opened. (See Fig. 45.)

Drop three took place on 10 December, from an altitude of 7,010 m. Once again the spacecraft experienced severe oscillations right up to the moment of drogue parachute deployment. As the main parachute deployed the simulated heat shield was jettisoned. It re-contacted the spacecraft, striking the parachute canister where it dislodged a camera. On landing, the spacecraft floated on its side and could not be righted for helicopter recovery. The fishing vessel recovered the parachute and towed the spacecraft back to the Coast Guard Station on Wallops Island. The flight data recorder showed that all parachute deployment loads were within acceptable limits.

The least successful test in the series was No. 4, made on 17 December, from an altitude of 6,096 m. The drogue parachute deployed as planned and the main parachute began its deployment at the correct time. When the simulated heat shield was jettisoned it impacted the parachute canister and closed the parachute release circuit. The main parachute was jettisoned partway through its deployment sequence. The spacecraft tumbled continuously as it fell towards the water below.

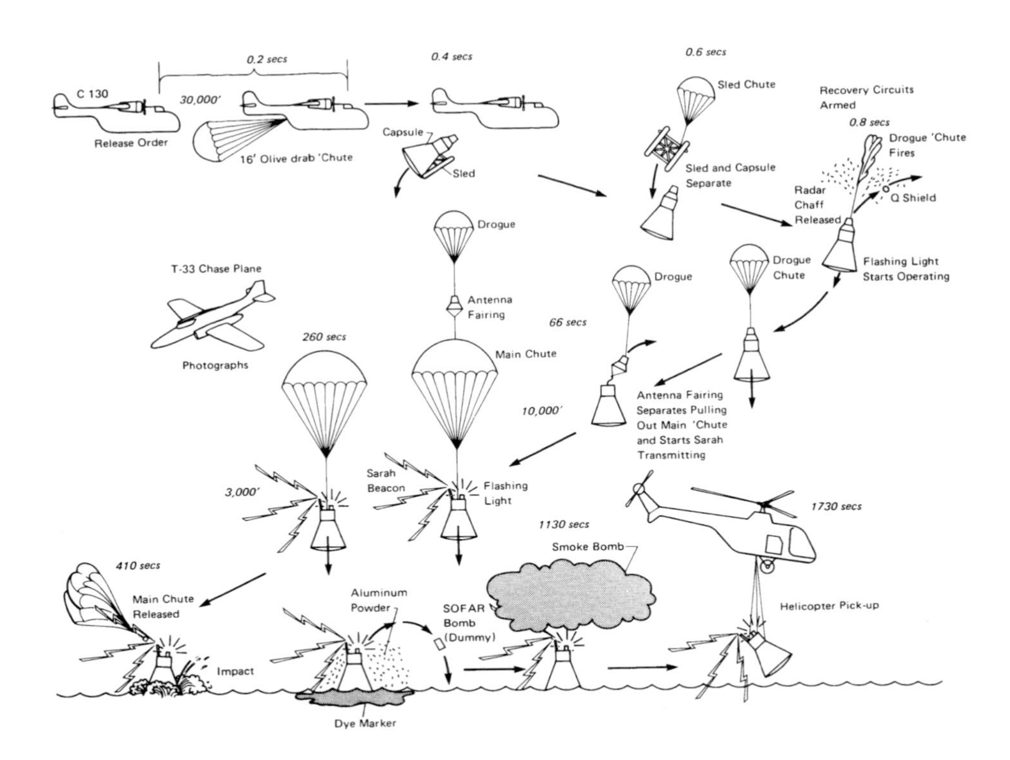

Fig. 45. Mercury Spacecraft Earth Landing System qualification drop test. (NASA)

It was destroyed on impact, although it was recovered. The flight data recorder revealed what had happened.

As a result of the two consecutive flights where the heat shield recontacted the spacecraft, plans to jettison the heat shield were abandoned. Instead STG accepted McDonnell's original idea of jettisoning the heat shield from the base of the spacecraft and allowing it to drop 1.2 m in order to deploy an impact skirt. Following landing the skirt would fill with water and the weight of the heat shield would act as a sea anchor, holding the spacecraft upright in the ocean.

The final drop test in the initial series occurred on 24 March 1959. The drogue parachute deployed as planned, but the main parachute continually opened and closed in a manoeuvre known as 'squidding'. The spacecraft was destroyed when it hit the water. Further testing was delayed while STG took advice from the USAF on how best to overcome the problems that the Phase 3 drop tests had encountered. The USAF experts suggested the change to a 19.2-m diameter ringsail main parachute and a 2-m diameter fist ribbon drogue parachute. Testing of the new ELS took place over Edwards Air Force Base, California, throughout 1959 (Table 5).

Alongside the mainstream ELS development, STG also ran a series of 10 drops aimed specifically at qualifying the ELS to be used on the first Mercury Atlas launch, Big Joe, planned for 9 September, 1959. In all 10 drops the boilerplate spacecraft

Table 5: Big Joe parachute qualification test series

Test	Date	Result
1	22 April 1959	Successful drop. Helicopter recovery.
2	23 April 1959	Successful drop. Helicopter recovery.
3	29 April 1959	Successful drop. Helicopter recovery.
4	30 April 1959	Successful drop. Fog forced a recovery by towing the spacecraft behind a boat.
5	6 May 1959	Successful drop. Helicopter remained on the ground at Wallops Island until spacecraft was in the water. Spacecraft was lifted from the water within five minutes.
6	5 June 1959	Successful drop. Helicopter recovery.
7	11 June 1959	Successful drop. Spacecraft fell away from helicopter during recovery and was destroyed upon impact with the water.
8	22 June 1959	Successful drop. Helicopter recovery. (Full rehearsal using US Navy personnel who would recover the Big Joe spacecraft.)
9	2 July 1959	Successful drop. Recovered by boat.
10	28 July 1959	Successful drop. Spacecraft used in beacon ranging tests with five Neptune tracking aircraft. Helicopter recovery.

was pulled from the rear of a C-130 aircraft in a similar manner to the original Phase 2/3 tests.

Mainstream ELS qualification resumed in early 1959 using the new style parachutes. The bomb-shaped test vehicle was dropped from an F-104 aircraft at Edwards AFB and the Naval Parachute Test Centre, both in California. The tests were aimed at qualifying the fist ribbon drogue parachute and were initiated between 3,048 and 21,336 m and at both subsonic and transonic velocities. The fist ribbon parachute quickly proved unsafe at high altitude and was replaced by a 2-m diameter conical drogue parachute in June.

A new series of F-104 drops, using the new drogue parachute, began on 4 August. Fifteen drops later the drogue parachute was qualified for deployments at altitudes up to 21,336 m and velocities up to 1.5 Mach. Under normal conditions the drogue parachute would be deployed at 12,192 m.

Qualification tests of the 19.2-m main parachute began in May and were completed in August. Thereafter, a test series began to qualify the complete ELS under flight conditions. Fifty-six drops of a boilerplate spacecraft were made from the rear of a C-130 aircraft over the Salton Sea, California. Drops were made up to an altitude of 9,144 m. Low-level drops were also made from beneath a helicopter, to simulate an off-the-launch pad abort. The series was successfully completed on 11 August 1960.

A three-phase series of tests over water and land began in August 1960. A boilerplate spacecraft complete with landing skirt was dropped in order to define the landing dynamics of the new design. When the spacecraft continually floated on its side, STG and McDonnell commenced an investigation that resulted in the spacecraft's centre of gravity being moved. The spacecraft completed a number of successful land landings despite the remote chance of a Project Mercury flight ending in a land landing.

As difficulties continued to haunt the ELS, STG began looking for alternative recovery methods for the Mercury Spacecraft. One possibility under development at LaFRC was Francis Rogello's flexible wing. Rogello made four rocket-launched tests of scale models of his flexible wing from Wallops Island between October 1959 and February 1960. Due to time constraints STG decided to stick with the original ELS design and testing proceeded.

Phase 2 testing took place between 21 and 30 November 1960. A boilerplate capsule was dropped in wind conditions up to 10 knots. All test results were within design limits.

Phase 3 was flown between 20 March and 13 April 1960 and consisted of six-helicopter drops into open water. The tests were designed to define the danger of heat shield making contact with the base of the spacecraft on water impact. Two additional drops onto land were made on 24 and 26 April. Finally, two solid-bottomed spacecraft (no impact skirts) were dropped onto water, in order to simulate the failure of the landing bag to deploy. The tests took place on 14 and 18 December 1960 and neither spacecraft suffered structural damage. These tests completed the qualification testing of the ELS for manned suborbital and orbital flights.

The final phase of ELS testing was Project Reef. Twenty drop tests qualified the 19.2-m ringsail main parachute to support the heavier spacecraft that would fly the project's final 1-day manned orbital flight.

11.1 TRAINING FOR RECOVERY

The engineers of STG designed their piloted satellite vehicle to be recovered in mid-ocean for a number of reasons, not least of which was the fact that oceans were huge and relatively easy to hit. Nor did they contain large, high-population areas. Also, water would absorb the final landing loads easily, while rapidly cooling the spacecraft, which would be subjected to severe heating during re-entry.

NASA invoked the terms of the National Aeronautics and Space Act 1958 to provide US Naval and Marine Corps support for Project Mercury recovery operations. Negotiations between the two organisations began on 12 February 1959. As a result of the first meeting the NASA–Navy Committee was formed, with the task of establishing recovery techniques for Project Mercury. The Committee met for the first time on 17 February, and agreed that the Navy was capable of supporting initial recovery activities off Wallops Island. The Navy also agreed to support recovery operations down the length of the Atlantic Missile Range and in the Atlantic Ocean, between Cape Canaveral and Bermuda, where emergency landings would be made in the event of a failed launch attempt.

In March 1959 STG prepared a document titled *Recovery Operations for Project Mercury*. It covered plans for both Mercury Redstone and Mercury Atlas launches. The document was sent to the DoD.

In the first week of April the Chief of Naval Operations directed Destroyer Flotilla Four (DESFLOT-Four), part of the US Naval forces on station in the Atlantic Ocean, to support the requirements of Project Mercury in the following areas.

- Landing and recovery tests at NASA's LaFRC facility in Norfolk Virginia (Wallops Island), in order to develop and rehearse spacecraft pick-up and handling techniques.
- Recovery of Beach Abort and Mercury Little Joe prototype spacecraft launched from Wallops Island.
- Atlas launch vehicle development flights from the Atlantic Missile Range, Cape Canaveral, Florida.

A DoD working group met at Patrick AFB over 27–28 April to establish the areas of responsibility assigned to each organisation during the first two unmanned Mercury Atlas launches. On 5 May, STG and DoD personnel met to discuss the Project Mercury recovery procedures training programme. Both parties agreed that joint NASA, Navy, Marine Corps training exercises should begin as soon as possible. In June, STG supplied the Navy with a number of boilerplate spacecraft and the specialist equipment required to support recovery operations, thus allowing the men and machines of DESFLOT-Four to engage in a series of recovery training operations. Training also included DESFLOT-Four personnel taking part in the recovery of boilerplate spacecraft used in the ELS development and qualification drop test series and the recovery of the prototype spacecraft used on the unmanned Mercury Little Joe, Redstone and Atlas launches. Also in May, STG let a study contract to Grumman Aircraft Corporation to study the requirements for recovery operations at the end of a three-orbit manned flight. Grumman's report was released in June.

On 25 June members of DESFLOT-Four participated in the drop test of a boilerplate spacecraft, off the coast of Jacksonville, Florida. The spacecraft was deliberately dropped 74 km from the 'predicted landing point' and 83.3 km from the nearest ship. The spacecraft was located and recovered in less than 2.5 hours. The year ended with the release of the DoD's instructions on how to handle the various foreseen recovery situations, on 15 December.

The Navy's School of Aviation Medicine spent the second quarter of 1960 attempting to convert a standard 20-man life raft into a Stullken Collar. When their attempts failed a Mercury unique Stullken Collar was developed. The Stullken Collar was qualified for use at the end of manned orbital flights on 24 September. The following month, ship retrieval tests confirmed the validity of the procedures developed for use at the end of manned orbital flights.

On 27 July the launch site recovery forces exercised off of Cape Canaveral using a boilerplate spacecraft that had been preplaced in the water and had not been dropped from an aircraft. The exercises were successful. Two ships of DESFLOT-Four also exercised recovery operations 222 km off Florida in October.

The first successful recovery of a Project Mercury astronaut returning from space occurred during the suborbital flight of Mercury Redstone 3, in May 1961. The first manned orbital flight was recovered in February 1962, with the final Mercury flight being recovered after 22 orbits, in May 1963.

11.2 PROJECT MERCURY RECOVERY PHILOSOPHY

The idea behind the Project Mercury recovery philosophy was to provide for the safe recovery of the astronaut and his spacecraft under any conceivable landing conditions. The action to be taken would depend on the real-time situation and the location of the landing zone. To allow for planning to take place in advance of real launches, prospective landing zones were placed in one of five types:

1. *Launch Site Abort Landing Area.* Landings that might occur from the time of arming of the LES to a point in the flight trajectory where an abort would result in a landing within 22.2 km from the launch pad.
2. *Pre-orbital Abort Landing Areas.* Landings as a result of in-flight aborts taking place beyond the limits of an abort resulting in a Launch Site Abort Landing Area and orbital insertion.
3. *Planned Landing Areas.* Landing areas along the orbital ground track at approximately 100-minute intervals, with pre-deployed recovery forces.
4. *Primary Planned Landing Area.* The planned landing area to be used at the end of a successful flight or, if at all possible, at the end of a shortened flight. The primary recovery forces were located in this area.
5. *Preferred Contingency Landing Areas.* Designated landing areas along the ground track, but between the Planned Landing Areas, to be used in an emergency that would not allow the landing to be delayed long enough to let the spacecraft reach the Primary Planned Landing Area. Reduced recovery forces were maintained in these areas, the size of which was dictated by the probability of the spacecraft landing in each area.

The recovery operation itself was broken down into two parts:

1. *Location* began with the notification to the recovery forces that a landing was about to occur and the general area in which it should take place. This information was backed up by data from the World-Wide Tracking Network. Search aircraft then swept the ground track in that area in an attempt to locate the spacecraft's recovery beacon, or the astronaut's personal survival beacon, using UHF Direction Finding (UHF/DF) equipment. Having located one of the beacons the aircraft flight crew began a visual search in an attempt to definitely locate the spacecraft or astronaut in the water. The spacecraft's green marker dye and flashing light beacon were visual location aids intended for day and night use, respectively. If mechanical failure meant that the location beacons were not available then standard military open water search techniques were employed to locate the spacecraft and/or its astronaut.
2. *Retrieval* required that all ships in the recovery fleet had the capability to lift the spacecraft and its astronaut from the water. NASA developed a lifting cradle

that fitted on the standard US Navy cranes and lifeboat davits with only the minimum of modification. Helicopters with the capability to lift both the spacecraft and the astronaut were fitted with specialist 'man-rated' lifting slings and hooks for the task.

The Project Mercury retrieval operations evolved in four distinct stages.

Parachute drop tests employing boilerplate spacecraft and the unmanned prototype spacecraft used on Mercury Little Joe flights out of Wallops Island were, on the whole, recovered by Marine Corps Helicopters from DESFLOT-Four, based temporarily on the island. A fishing boat was hired locally to recover the parachute. On the few occasions that the helicopter could not recover the spacecraft as planned the fishing boat towed it ashore, landing it at the Coast Guard station on Wallops Island.

In the earliest days of Project Mercury the astronaut was to be sealed in his spacecraft with a side hatch that required as many as 70 torque bolts to hold it in place. A single round of explosive chord could be used to remove the hatch in the case of an emergency situation arising while the launch vehicle was still on the launch pad. At the end of his flight the astronaut was expected to use a pair of pliers to remove part of the main instrument panel and the forward Crew Compartment bulkhead. He then had to push the primary and emergency main parachute canisters out of the egress tunnel in the neck of the spacecraft. Only then could he push his life raft and emergency survival in front of him as he left the spacecraft through the neck. Once in the water he had to inflate his raft, climb into it and secure it to the spacecraft to ensure that he did not drift away. During training the astronauts practised using the large concave mirror strapped to their personal parachute harness to signal their location to approaching aircraft. Recovery was then to be completed in an astronaut first, spacecraft second order by a single helicopter.

When training revealed the difficulties inherent in the 'through the neck' method of astronaut egress a new recovery method was developed. A helicopter crew cut away the spacecraft's recovery whip antenna and secured one winch line to the recovery loop mounted on the neck of the spacecraft. That line was pulled tight before the astronaut detonated the explosive chord around the side hatch of his spacecraft. The helicopter crew then lowered a second line until the 'horse collar' was directly outside the now open side hatch. The astronaut pulled the horse collar inside and placed it under his armpits. He then exited the spacecraft through the side hatch and was winched up to the helicopter. Only when the astronaut was safely in the helicopter was the spacecraft lifted out of the water and carried to the prime recovery vessel.

Following the development of the Stullken Collar, procedures changed again. The spacecraft was located in the water and swimmers were dropped close by to deploy the collar and inflate it to support the spacecraft in an upright position with its side hatch clear of the water. With the collar in position the astronaut had the choice of jettisoning the side hatch and being recovered by helicopter, or remaining in his spacecraft until it was hoisted directly aboard the recovery vessel.

Post-recovery activities included medical assistance if required, physical examinations, a medical debriefing and scheduled rest periods for the astronaut. On early manned flights the astronaut made a personal debriefing into a tape recorder before being flown to Hawaii, where he participated in a 3-day technical debriefing. On later flights the astronaut remained on the prime recovery vessel and was debriefed while the vessel steamed to Hawaii. Following the recovery of the spacecraft trained personnel secured its systems, carried out an initial post-flight inspection and removed relevant flight documents and data recorders for transport to Cape Canaveral. The spacecraft was then transported back to Cape Canaveral by aircraft.

If a Preferred Contingency Landing Area had been required, normal search and location procedures were used to locate the spacecraft. The aircraft then circled the spacecraft until a second aircraft, with a team of swimmers, arrived on the scene and dropped them to place a Stullken Collar around the base of the spacecraft. If the original aircraft had to leave the spacecraft before the second aircraft arrived, due to fuel shortage, or technical difficulties, it would drop electronic marker buoys into the ocean around the spacecraft. The buoys transmitted on the same frequency as the search aircraft's UHF/DF equipment.

The basic recovery methods developed for Project Mercury (see Table 6) were sufficiently sound that they served Projects Mercury, Gemini, Apollo, Skylab and the American half of the Apollo–Soyuz Test Project in 1975. They were not replaced until the Space Shuttle Orbiter 'Columbia' made its first landing at Edwards AFB, California, in April 1981.

Table 6: Project Mercury Spacecraft recovery fleets

Flight	Ships	Aircraft	Helicopters	Recovery vehicle
LJ (X9)	3–4	2	3	Ship or helicopter
BJ	13	7	3	Ship
MA-1	8	5	3	Salvage ship
MR-1	8	6	3	Salvage ship
MR-1A	8	4	4	Helicopter
MR-2	8	6	5	Helicopter
MA-2	8	14	5	Helicopter
MA-3	15	12	7	Helicopter
MR-3	8	7	7	Helicopter
MR-4	8	7	7	Helicopter (pilot) spacecraft not recovered
MA-4	9	34	6	Ship
MA-5	18	49	9	Ship
MA-6	24	49	14	Ship
MA-7	20	49	14	Helicopter (pilot) ship (spacecraft)
MA-8	26	69	14	Ship
MA-9	28	110	14	Ship

12

Launch Escape System

In its final configuration, the Mercury Spacecraft LES consisted of a Grand Central 1KS52000 solid propellant Escape Motor, mounted on a tripod tower structure. Three exhaust nozzles, spaced 120 degrees apart and angled at 19 degrees from the vertical, replaced the Grand Central's more usual single exhaust nozzle. Once ignited, the rocket motor consumed its propellant in just 1 second, producing a thrust of 231,296 N. The high thrust rate and short burn time were sufficient to pull a Mercury Spacecraft and its astronaut clear of the fireball expected from the catastrophic break up of a Redstone or Atlas launch vehicle. (See Fig 46.)

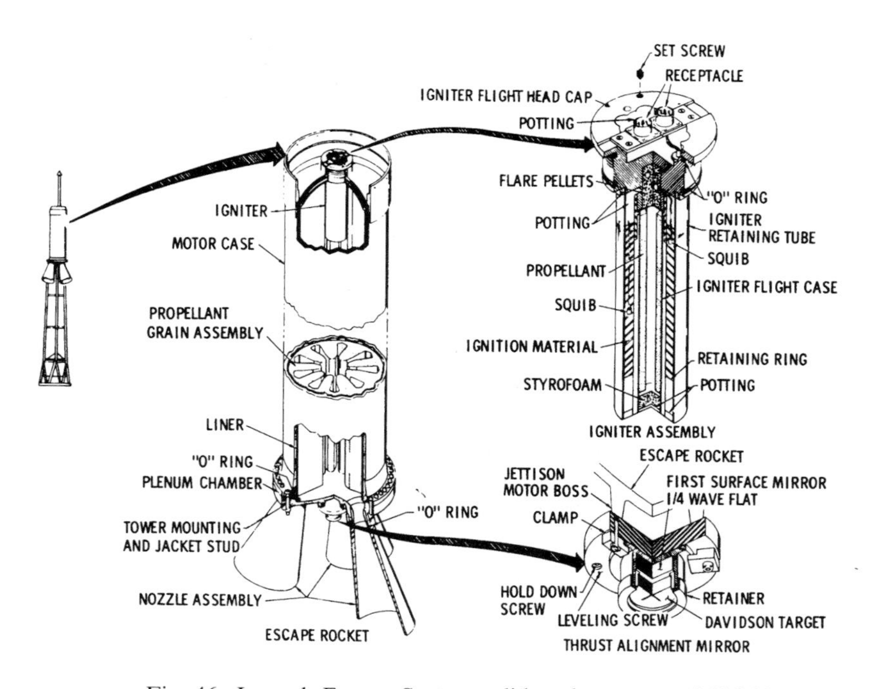

Fig. 46. Launch Escape System solid rocket motor. (NASA)

With its propellant consumed the Escape Motor rocket stopped thrusting, although the spacecraft/LES combination would continue to climb with the momentum already built up. Following shut-down of the Grand Central motor the Marman Clamp holding the LES to the neck of the spacecraft fired, releasing the base of the LES tower structure. Thereafter, a single Atlantic Research Corporation 1.4KS785 tower jettison solid propellant rocket motor fired, producing 3,558 N of thrust to push the LES away from the spacecraft. Like the Escape Motor, the Jettison Motor originally only had one exhaust nozzle, but flight-testing showed the requirement for three exhaust nozzles, spaced 120 degrees apart. When the Jettison Motor had consumed its propellant and shut down, the LES followed a ballistic trajectory until it impacted the ocean below. The LES tower was a tripod structure braced by cross-members. From the base of its tubular steel tower, to the tip of its aerodynamic spike, the LES was 5.1 m in length.

Both the Mercury Redstone and Mercury Atlas launch vehicles were fitted with systems that measured a number of launch vehicle performance and trajectory parameters against preset limits. In each case a maximum allowable deviance was defined. When the performance reached that maximum allowable deviance an abort signal was issued and the Escape Motor was ignited to pull the spacecraft clear of its launch vehicle. Alternatively, the astronaut could activate the LES by striking the abort button in the Crew Compartment. The astronauts referred to the abort button as the 'chicken switch', but there is no doubt that any one of them would have used it if the situation had called for it.

Abort situations were identified in two main groups. The first was an 'off the pad abort', which was initiated by a situation arising within the launch vehicle that required the LES to be activated while the launch vehicle still stood on the launch pad at Cape Canaveral. In this situation the LES would have pulled the spacecraft along a ballistic trajectory to an altitude from which it could make a parachute descent, using its ELS. Recovery would then be carried out by helicopter from the Atlantic Ocean off Cape Canaveral.

In-flight aborts activated the LES if they occurred within the first 2 minutes 23 seconds of the powered ascent, before the LES was jettisoned because the spacecraft was capable of making a parachute descent without its help. If the abort occurred within the first 30 seconds of flight the launch vehicle's engines were not shut down, in order to prevent it falling back on the launch pad and destroying it. As a result, the launch vehicle would continue to follow its trajectory over the ocean to an altitude where the Range Safety Officer could safely command its destruction. Despite this precaution, the Range Safety Officer had the ability to destroy the launch vehicle at any point during those first 30 seconds if the situation demanded it.

All in-flight abort signals resulted in the same action being taken by the vehicle's automatic systems. If the flight had passed the 30-second point the launch vehicle's engines were shut down and the Marman Clamp mating the spacecraft to the launch vehicle was released. The Escape Motor then fired and pulled the spacecraft clear of the failing launch vehicle. When the Escape Motor had consumed its propellant the Marman Clamp holding the LES to the neck of the spacecraft was released and the tower Jettison Motor fired, to pull the LES clear of the spacecraft. The spacecraft then made an ocean landing under the parachutes of its ELS, after which the astronaut would await recovery by the men and machines of DESFLOT-Four. The LES followed a ballistic trajectory and sank on impact with the ocean.

Under normal flight conditions (no abort situation) the LES was jettisoned at T+2 min 23 s. On a Mercury Redstone flight the LES was jettisoned following the Redstone's main propulsion shut down. On a Mercury Atlas flight the LES was jettisoned following booster engine cut off and booster skirt jettison. The Atlas Sustainer was pitched down slightly prior to LES jettison and then pitched back up to continue on its planned ascent into orbit. At LES jettison the Marman Clamp holding the LES to the neck of the spacecraft was released and both rocket motors fired to pull the LES clear of the spacecraft. When the rockets had consumed their propellant the inert LES fell into the ocean below. During operational flights no attempt was made to recover the LES.

12.1 DEVELOPING THE LAUNCH ESCAPE SYSTEM

In the early 1950s ballistic missiles exploded frequently, either on the launch pad or within the first few minutes of flight. If a human astronaut was to be expected to ride such a missile into space a means had to be developed to pull him clear of the expected fireball produced by an exploding, fully fuelled, missile and then allow him to make a safe landing using the spacecraft's ELS.

For their MISS programme the USAF had suggested the use of solid propellant rockets at the base of the spacecraft to push it clear of a failing launch vehicle. McDonnell's original design for the Mercury Spacecraft contained a similar system with rockets mounted in fins around the wide end of the spacecraft.

For the NACA's piloted Satellite Project, Maxime Faget suggested the use of a tractor rocket mounted on the narrow end of the spacecraft, to pull it clear of a failing launch vehicle. In the event that it was not required, the tractor rocket could be easily jettisoned during the ascent into space. Faget made his suggestion in July 1958.

On 8 March 1959 a boilerplate Type C spacecraft stood on a wheeled dolly alongside the water at Wallops Island. The tractor rocket consisted of a Recruit solid propellant rocket motor, with a three-point exhaust outlet. Each outlet was spaced at 120 degrees to the other two and angled at 15 degrees to the vertical, to prevent the exhaust gasses impinging on the three tubular steel support struts that held the rocket motor away from the spacecraft. An electronic signal ignited the rocket motor and it pulled the spacecraft along an arcing trajectory over the water. This type of flight-test was called a Beach Abort (BA) and represented a simulation of an off-the-pad abort. The flight was stable for the first 60 seconds, after which the spacecraft tumbled end over end three times before making a perfect landing in the water some 304 m offshore. (See Fig. 47.) Post-flight analysis revealed that the graphite throat in one of the three exhaust outlets had burned through, resulting in an incorrectly aligned thrust vector, which caused the tumbling. The graphite throat was redesigned.

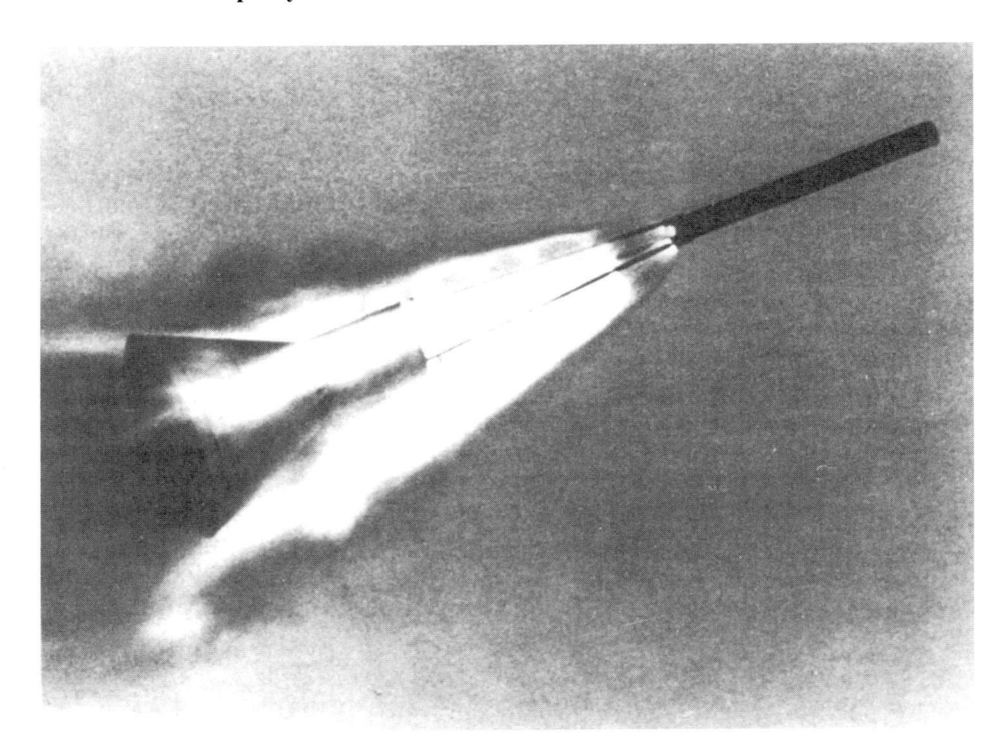

Fig. 47. A Type C capsule in flight during an early Launch Escape System. (LES) flight-test. Note the original design of the LES support structure. (NASA)

At LaFRC an alternative theory blamed the tumbling on the angle of the three exhaust outlets. To test the theory, five 0.33 scale models were prepared in the LaFRC workshops. Each model spacecraft was fitted with an 8.25-cm rocket motor to simulate the Recruit rocket motor in the full-scale tractor rocket. Each model had its three exhaust outlets angled 5 degrees further from the vertical than the model before it. All five models were launched over 13–15 April 1959. As a result of these five launches the three exhaust outlets on the Grand Central Escape Rocket Motor that would be used on the production tractor rocket were set at 19 degrees from the vertical (see Table 7).

Table 7: Tractor rocket exhaust nozzle definition launch programme

Designation	Outlet angle	Launch date	Result
U-1	10 degrees	13 April 1959	Very unstable
U-2	15 degrees	13 April 1959	Very unstable
U-3	20 degrees	14 April 1959	Unstable
U-4	25 degrees	15 April 1959	Stable
U-5	30 degrees	15 April 1959	Stable

On 14 April the second full-scale Type C boilerplate spacecraft was launched in a Beach Abort, from the waterside at Wallops Island. The Recruit rocket motor's thrust was deliberately offset, to simulate a simultaneous pitch and translation away from a launch vehicle standing on the launch pad. The Recruit rocket motor was rated at 154,078 N, with a burn time of 1.5 seconds, to simulate the Grand Central rocket motor. Following launch the combination tumbled once prior to the tractor rocket being jettisoned. The spacecraft was recovered by helicopter, and a boat sailing out of Wallops Island picked up the parachute. The flight was recorded as successful.

Meanwhile STG engineers undertook a comparative review of Faget's tractor rocket and McDonnell's original posigrade rocket and fin design. The Arnold Engineering Development Centre, Tullahoma, Tennessee, was contracted to run a series of comparative wind tunnel tests of both systems. On 22 April 1959 STG came out in favour of the tractor rocket, which was slowly taking on the name 'Launch Escape System'. The STG management agreed to continue studies of McDonnell's idea until the LES was proved reliable.

The first Beach Abort flight-test of a production LES with a Grand Central Escape Motor and a production spacecraft took place from Wallops Island on 22 July 1959. The spacecraft reached an altitude of 609 m and tumbled once prior to LES jettison. It then continued to tumble until the drogue parachute was deployed and only stabilised completely after main parachute deployment. A second Beach Abort using a production LES occurred at Wallops Island on 28 July 1959. The flight was completely successful.

Between March and July 1959 Arnold Engineering subjected the Grand Central rocket motor to a series of wind tunnel tests in order to confirm its ignition and burning characteristics. NASA–Lewis also subjected the Grand Central rocket motor to wind tunnel tests to qualify its ignition at a simulated altitude of 3,048 m. It passed all of the tests.

The LES and the Atlas abort-sensing equipment were qualified for manned flight and a Mercury Atlas abort at the moment of maximum aerodynamic pressure on the vehicle during the series of seven Mercury Little Joe launches. Two further Beach Abort launches were also made to qualify the LES for manned off-the-pad aborts. During the first of these two Beach Abort flights, the LES failed to separate sufficiently from the spacecraft. As a result, the single exhaust outlet on the Jettison Motor was replaced by a three-point outlet, with each outlet spaced at 120 degrees from the other two and angled at 30 degrees to the vertical.

Despite its elaborate development and flight-test programme, the LES was never called upon to save the astronaut's life during Project Mercury.

13

Handling the Mercury Spacecraft at Cape Canaveral

Following its delivery to Cape Canaveral each Mercury Spacecraft was subject to a number of tests to ensure its readiness to fly and complete its planned mission. Any discrepancies found were the subject of review before any corrective action was taken. Any changes to the original configuration were installed as rapidly as was safely possible so as not to delay the launch unnecessarily. Both the prime and back-up astronauts participated in the many tests completed at 'The Cape'. This was the time when they became familiar with the spacecraft they would fly, rather than the generic interior of the two procedures trainers, which often did not accurately represent their real spacecraft.

On delivery from McDonnell the spacecraft was moved into Hangar S, the centre of NASA's Project Mercury operations at Cape Canaveral. The spacecraft was initially given a thorough inspection and any open work items were completed. Also at this time any design changes dictated by previous flights were carried out, if they had not been incorporated at McDonnell's factory. The spacecraft was then subjected to a series of individual and combined systems tests. Any discrepancies were reviewed and repairs, or changes, were made before that test was repeated. Each test had to be successfully completed before the next one could begin. Systems testing fell into nine groups:

1. *Electrical power*. The first test applied electrical power to the spacecraft systems through an external cable, to ensure that the electrical systems were functioning as designed and were capable of carrying the loads required for their normal functions.
2. *Instrumentation*. The spacecraft instrumentation system was tested and calibrated prior to its installation. It was then retested following installation.
3. *Sequential*. The automatic and manual sequencing systems were subject to test by inputting a signal and ensuring that the correct sequence of events was activated. The proposed flight into orbit was roughly broken down into four sequences of events that had to be completed correctly. They were: launch, abort, orbit and recovery.
4. *Environmental control system* (ECS). The full test of each of the individual systems making up the ECS was performed in order to prove the correct

function and reliability of the complete ECS prior to subjecting the primary and back-up astronauts to altitude chamber testing in the spacecraft.

5. *Communications system.* Testing was completed to determine the performance of individual components comprising the communications system in the spacecraft.
6. *Reaction control system* (RCS). The RCS system was tested in a test cell designed specifically for that purpose. A number of gas checks were performed first, to test the integrity of the individual RCS systems. The system was then filled with 35 per cent hydrogen peroxide and left for 24 hours to precondition it for use with 100 per cent hydrogen peroxide. This period was followed by 24 hours with the system filled with 90 per cent hydrogen peroxide. Finally the system was filled with 100 per cent hydrogen peroxide and each individual thruster was static fired to prove overall system performance.
7. *Communications radiation tower.* The spacecraft was installed on the outdoor communications tower so that the HF characteristics of the biconical antenna could be determined. Tests of the other communication systems were completed at the same time using AMR communications antennae at Cape Canaveral. All tests were completed with the spacecraft on its own internal power source, so as to simulate flight conditions as closely as possible.
8. *Automatic stabilisation and control* (ASCS). The ASCS was first subjected to a static test before the spacecraft was mounted in a test structure for a dynamic test of the ASCS. The spacecraft was rotated in the structure and the ASCS had to detect the amount of movement and fire the correct thrusters to correct it and return the spacecraft to the correct attitude for the portion of flight being simulated.
9. *Altitude chamber test.* The final test was performed with the spacecraft in the Hangar S altitude chamber. During this test the prime or back-up astronaut was dressed in his pressure suit and seated in the spacecraft. The altitude chamber test was the first time that the astronaut was connected to the ECS of the spacecraft that he would fly in space. During the test the chamber was pumped down to simulate orbital conditions as closely as possible.

The altitude chamber test was run with the spacecraft sitting on the launch vehicle adapter and electrically connected to it. The LES was also installed, but without its Escape and Jettison rocket motors. All separation pyrotechnics were replaced with 0.20-amp fuses, which were monitored to ensure that they 'fired' at the correct time. Only the minimum of ground support equipment was connected to the spacecraft in order to simulate flight conditions as closely as possible.

The altitude chamber test had four primary functions:

- To verify the correct operation of all individual spacecraft systems.
- To ensure the correct operation of the sequencing system through all flight modes.
- To demonstrate the correct operation of all systems in conjunction with each other.
- To show the correct operation of all spacecraft systems when flight conditions were simulated as closely was practicable.

Following the successful completion of the altitude chamber test the spacecraft was declared ready for transport to the launch complex and mating with its launch vehicle. Launch complex activities involved a second series of nine checks and a further three stages of preparation for launch. The 12 activities involved in the preparation of a Mercury Atlas launch vehicle were:

1. *Launch complex checkout*. Prior to the spacecraft being moved from Hangar S to the launch complex, the launch vehicle was erected on the launch pad and subjected to its own series of tests, to bring it to the point where it was ready to receive the spacecraft. The launch complex checkout was a test of the launch complex's electrical systems to ensure that they were ready to support the installation and testing of the spacecraft on the launch vehicle.
2. *Interface inspection*. The launch vehicle adapter was the first portion of McDonnell's hardware to be mated to the launch vehicle. This was then inspected to ensure its condition and its readiness to receive the spacecraft. (See Fig. 48.)

Fig. 48. The final Mercury Spacecraft is hoisted to the top of the launch complex prior to mating with its launch vehicle. (NASA)

3. *Mechanical mating*. Prior to hoisting the spacecraft above the launch vehicle the Retrograde Package pyrotechnics were tested and then shorted-circuited with shorting plugs. The spacecraft was then raised and positioned above the launch vehicle adapter, before being lowered on to it. Installation of the Marman Clamp and all electrical connections completed the mechanical mating process. The access panels in the launch vehicle adapter were then sealed. They would only be opened again to support emergency work prior to launch.
4. *Spacecraft systems tests*. Tests were run on all spacecraft systems to check that they had not been damaged in the move to the launch pad, or during mating with the launch vehicle. The tests also proved the capability of the blockhouse at the launch complex to communicate with and command the spacecraft systems.
5. *Electrical and interface aborts*. Ground support electrical cables were attached to the spacecraft for this test. The test validated of the spacecraft/launch vehicle interface compatibility and redundant electrical and radio frequency paths. The use of external cables also allowed for the test redundant abort signal paths to be used during the countdown and launch phases of the flight.
6. *Flight acceptance composite tests*. This test proved the compatibility and non-interference of the spacecraft radio frequencies with the communications equipment installed in the launch vehicle and along the length of the AMR.
7. *Flight configuration sequence and aborts*. This provided for testing of the compatibility of the spacecraft systems and the launch vehicle autopilot. It also allowed a further test of the abort modes with the spacecraft in a simulated flight configuration.
8. *Launch simulation*. The launch simulation was designed to allow everyone involved in the launch to rehearse their role in real time. It also served to validate the spacecraft and launch vehicle systems in the launch configuration. The launch simulation ended in the astronaut rehearsing the emergency egress procedures, without detonating the explosive chord holding the spacecraft side hatch in position.
9. *Flight safety review*. Approximately five days prior to launch a flight safety review was convened in order to confirm the spacecraft's readiness for launch. Board members reviewed the documentation produced throughout the spacecraft-testing phase before coming to their decision and committing the spacecraft to launch.
10. *Simulated flight*. The simulated flight allowed the entire spacecraft/launch vehicle to be checked in a real-time simulation of the flight from pre-launch activities through launch, orbit, re-entry and landing. The simulation also allowed the automatic launch vehicle abort system to be checked, at lift-off + 1 min 40 s.
11. *Spacecraft servicing*. Following all testing the spacecraft was prepared for flight. It was cleaned and checked to ensure that nothing had been left inside that should not be there. The spacecraft oxygen bottles were filled and the flight data recorders and the astronaut's voice tape recorder were installed. Numerous other tasks were completed at this stage to make the spacecraft ready for launch day.

12. *Pre-count*. The pre-launch countdown was a series of timed activities that had to be performed in the correct order, and at the correct time, to prepare the spacecraft and launch vehicle for ignition and lift-off when the countdown reached zero. The countdown was performed in two parts, the first of which was called the pre-count. During this period all spacecraft systems were checked to ensure their readiness for flight. The pre-count was followed by an approximately 15-hour hold in the countdown, during which the all pyrotechnics were installed and checked and their electrical connections were made. The peroxide system was also serviced.

13.1 COUNTDOWN

The countdown to launch followed a set procedure. The following list is taken from the Mercury Atlas 6 countdown:

T − 390 min:	LES escape rocket igniter installed. Launch pad evacuated and electrical power applied to the spacecraft to ensure no inadvertent ignition of the LES. Launch team return to launch pad.
T − 250 min:	Static firing of Reaction Control System, followed by preparation of the spacecraft for the astronaut to board it.
T − 120 min:	Astronaut boarded spacecraft.
T − 90 min:	Side hatch bolted in place.
T − 90 min to	
T − 55 min:	Final mechanical work completed to prepare spacecraft for launch.
T − 55 min:	Launch Complex 14 evacuated for launch.
T − 35 min:	Commence filling Atlas liquid oxygen tank. Final spacecraft systems check begun.
T − 10 min:	Spacecraft transferred to internal power.
T − 3 min:	Launch vehicle transferred to internal power.
T − 35 s:	Spacecraft umbilical ejected.
T − 4 s:	Atlas Booster and Sustainer engine start.
T = 0:	Mainstage. Clamps released. First movement.
T + 2 s:	Roll programme initiated.
T + 15 s:	Roll programme ends, pitch programme initiated.
T + 130 s:	Booster Engine Cut-Off (BECO).
T + 134 s:	Staging.
T + 156 s:	Sustainer guidance initiated.
T + 300 s:	Sustainer Engine Cut-Off (SECO). Ten-second timer initiated.
T + 310 s:	Spacecraft separation.

14

The Mercury pressure suit

14.1 DYSBARISM AND RELATED PROBLEMS

The human body has evolved to live in the thick, lower atmosphere of Earth. During the breathing process the lungs produce carbon dioxide and water vapour under a constant pressure, while inhaled air is at external pressure. As an unprotected body passes up through the atmosphere the external air pressure decreases. This results in carbon dioxide and water vapour occupying more of the lungs' internal volume, leaving less room for inhaled air. Around 9,130 m the body tissues and fluids begin to out-gas nitrogen and other gases as a result of low external pressure. The condition is called dysbarism, or air embolism, and can be overcome by breathing pure oxygen. Above 12,000 m oxygen at increased pressure is required, but beyond 13,000 m even this is not sufficient to overcome dysbarism.

The alveoli of the human lungs are at 37 °C. At 15,000 m the air-pressure is 87 mm of mercury, equal to the combined pressures of carbon dioxide and water vapour at 37 °C. At that altitude the unprotected lungs are full of carbon dioxide and water vapour and there is no internal space left for inhaled breath. This is the condition called anoxia, or oxygen starvation. By 19,000 m the surrounding air pressure is down to 47 mm of mercury, the pressure at which fluids in a healthy human body will boil, even if that body is at rest.

To overcome these conditions the Crew Compartment of the Mercury Spacecraft was pressurised with 100 per cent oxygen at 351.5 g/cm^2 in order to ensure the correct function of the astronaut's body. The astronaut also wore a pressure suit to protect him in the event of a loss of pressure in the Crew Compartment. The pressure suit consisted of four major parts: torso, helmet, gloves and boots. Beneath it the astronaut wore a number of biomedical sensors and a one-piece undergarment.

The International Latex Company, B.F. Goodrich and the David Clark Company competed to develop the Mercury astronauts' pressure suit. The contract ultimately went to B.F. Goodrich in July 1959 and called for the construction and testing of 21 pressure suits, three for each astronaut, for $98,000.

The B.F. Goodrich Company developed the Mercury pressure suit from the Mark IV pressure suit worn at that time by the US Navy's aviators. Originally, four Mark

IV suits, designated XN-1 to XN-4 were manufactured and underwent testing at the US Air Force Aeromedical Laboratory. The suits suffered a number of difficulties, not least of which was the fact that the fabric stretched. They were also uncomfortable to wear and oxygen circulation was poor. After a number of changes were made the pressure suit was accepted for Project Mercury in 1959. The astronauts and a number of McDonnell engineers made visits to Goodrich to be fitted for their pressure suits as well as visits to the subcontractor that produced the helmet. Three pressure suits were produced for each astronaut – a training suit and a primary and back-up flight suit.

In order to manufacture an accurately fitting suit the potential wearer was dressed in the long underwear that he would wear under the suit and then wrapped in strips of wet brown paper tape. When it was dry the tape became hard. The tape and underwear were then cut off and the pieces of the underwear garment were thrown away. The hard paper body mould was fitted to a dummy and used in the manufacture of that individual's pressure suit. (See Fig. 49.)

Fig. 49. The Flight A astronauts in early pressure suits. (Left to right) Schirra, Shepard, Slayton, Grissom, Cooper, Glenn, Carpenter. (NASA)

The final pressure garment consisted of the six main items detailed below, which worked together to protect the astronaut from the space environment and to record and telemeter his physical condition to the tracking stations of the WWTN.

14.1.1 Biomedical sensors

Following his pre-launch physical examination the astronaut was fitted with the biomedical sensors that would telemeter his physical condition to Earth during his flight. Some, but not all, of the Mercury astronauts had their bodies tattooed so that the sensors were always fitted in the same location. The body sensors were taped into place and covered with a square of moleskin. The sensor leads were plugged into one side of a connector built into the torso section of the pressure suit. A second lead was plugged into the same connector, to complete a permanent connection to the spacecraft telemetry section. A rectal thermometer recorded body temperature, while body sensors provided information on the astronaut's heartbeat.

14.1.2 Undergarment

The one-piece undergarment had long legs and sleeves. Thumb loops were fitted at the ends of the sleeves to prevent them pushing up the astronaut's arms when he donned his pressure suit. The garment contained patches of waffle-weave material where the wearer's body would be pressed too tightly against the pressure suit to allow proper ventilation circulation. The waffle-weave patches lifted the suit material sufficiently to allow the oxygen to circulate and cool those areas of the wearer's body. These patches were located at the upper and lower arms, thighs, back and chest

14.1.3 Torso

The torso represented the main portion of the pressure suit and covered the wearer's entire body except his head and hands. The inner gas-retaining layer was constructed from a ply of neoprene and neoprene-coated nylon fabric in the basic shape of the individual wearer. The outer layer was manufactured from heat-reflective aluminised nylon fabric. The helmet was attached to the torso by a rigid neck ring and locking mechanism. A tie-down strap was provided to prevent the helmet moving up when the suit was pressurised. Straps on the torso section allowed for a certain amount of individual sizing adjustment of the length and circumference of the arms and legs. The wearer's feet were encased as part of the torso

The wearer entered the torso through a diagonal pressure zip running across the front of the garment from the left shoulder to the waist. Two zips at the neck and a circumferential waist zipper were also used in the donning and doffing procedure.

An oxygen inlet port was mounted above the left waist to give access to an internal manifold, from which perforated tubes carried the oxygen from the spacecraft ECS to the body's extremities. The oxygen was used for both breathing and cooling.

The torso also included a connector for the equipment that would allow the astronaut's blood pressure to be telemetered to Earth during the flight. The blood pressure cuff was wrapped around the astronaut's upper arm during suit donning.

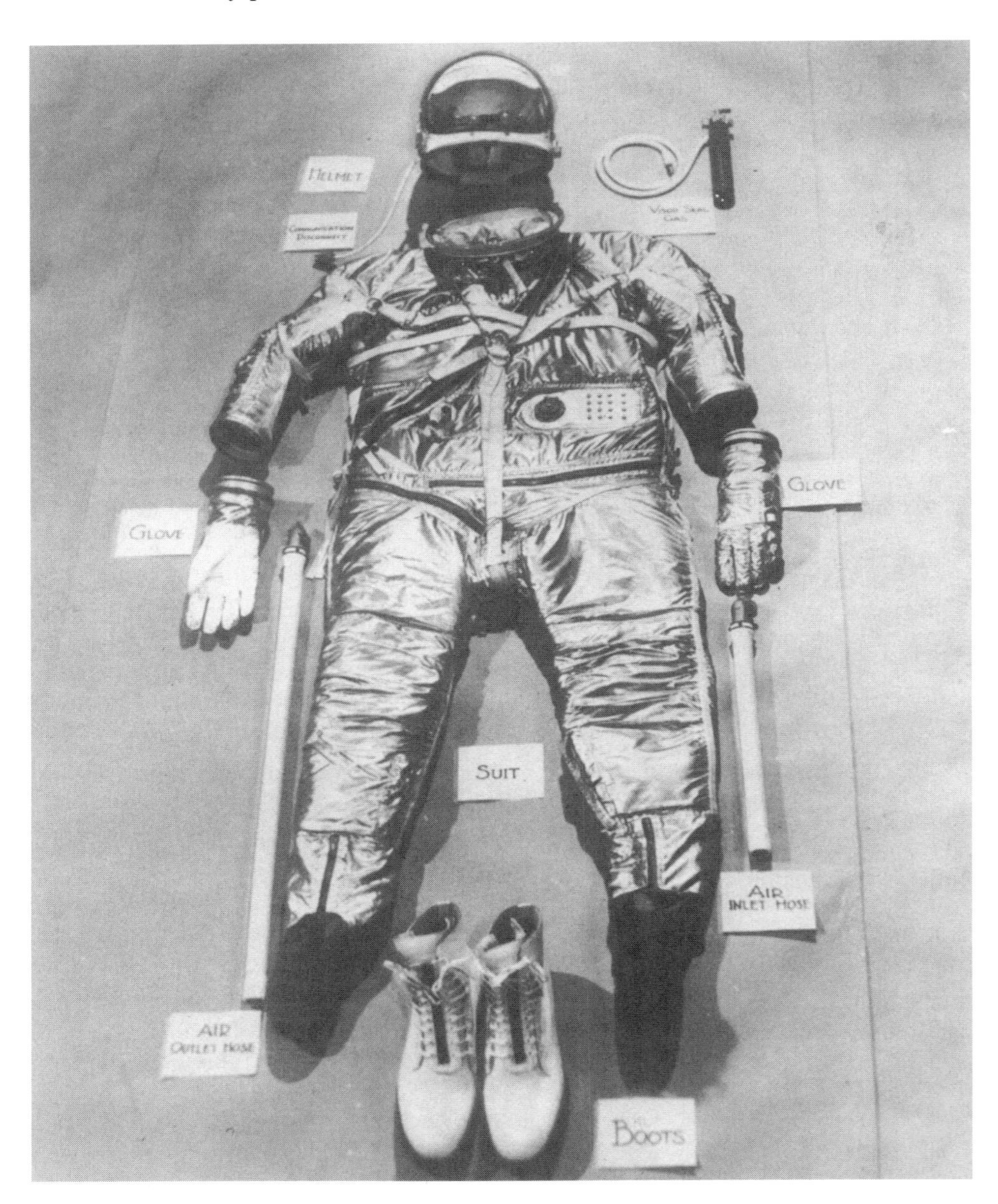

Fig. 50. The principal parts of the Mercury pressure suit laid out for the camera. (NASA)

The pressurisation hose was then connected to the interior of the connector on the chest of the torso. When the astronaut entered his spacecraft the external portion of the pressurisation equipment was connected to the exterior of the connector. A pressure indicator giving a visual representation of the suit's internal pressure was mounted on the left wrist. Finally, a rubber neck dam was attached to the neck ring and could be rolled against the wearer's neck to prevent water entering the pressure suit if he exited the spacecraft after landing without wearing his helmet. (See Fig. 50.)

14.1.4 Helmet

The helmet consisted of a hard fibreglass shell with an individually moulded crushable impact liner, and a pivoted clear plexiglass visor. Closing the visor activated a switch that caused the automatic inflation of the visor seal. This was achieved using gas from a pressurised bottle that connected to the helmet by as short hose. A manual switch was activated to deflate the visor seal, to allow the visor to be opened. The helmet impact liner was designed around a mould of each individual's head. The liner was covered in an outer skin of leather. The helmet communication system consisted of two independently wired earphones and two independently wired microphones, one of which provided the mounting for a respiration sensor. Having been carried to the neck ring from the inlet manifold, oxygen was allowed to circulate around the interior of the face area of the helmet for cooling and breathing. Exhaled gasses were removed through an outlet valve in the right side of the helmet, from where an umbilical returned them to the spacecraft ECS.

14.1.5 Boots

Many early photographs show the Mercury astronauts wearing their training pressure suits with white- or silver-coloured military boots. These were worn over the sock-like feet of the pressure bladder and offered strength in everyday use while also preventing the wearer's feet from swelling when the suit was pressurised. These boots were coloured white or silver for increased heat resistance.

During a flight into space the astronaut wore aluminised nylon fabric boots, with tennis-shoe-like soles, with the torso. These fitted over the gas bladder that surrounded the wearer's feet and were laced to keep them in place. During the walk out to the spacecraft plastic overshoes protected the boots. The overshoes were removed before the astronaut entered the spacecraft. The final Mercury pressure suit included boots that were integral to the outer heat-resistant layer of the suit.

14.1.6 Gloves

Gloves attached to the lower forearm of the torso by a pressure tight ball-bearing lock ring. The gloves were formed in two layers. The inner layer was made from estane, while the outer layer was constructed from stretch nylon on the back of the hands and fingers and neoprene injected nylon fabric on the inside of the hand to provide good grip when operating the spacecraft instruments.

All fingers were formed with a permanent curve to them. This placed the hand in the correct position to operate the spacecraft three-axis hand controller, even when the glove was fully pressurised. The middle left finger was the one exception to this rule and was left un-curved to allow the wearer to operate the spacecraft instruments. On orbital flights the index and middle fingers on both hands were fitted with small red lights, powered by a battery pack mounted on the back of the glove. The lights let the astronaut read the spacecraft instruments in the Earth's shadow before his eyes became fully dark-adapted.

14.2 PRESSURE SUIT UPGRADES

Following Mercury Redstone 3, the first manned flight in Project Mercury, the astronaut's pressure suit underwent a series of upgrades after each subsequent flight.

14.2.1 Mercury Redstone 4

The MR-4 astronaut was the first to wear a convex mirror attached to his parachute harness. The mirror was positioned such that, when the astronaut was seated on his prone couch, the astronaut-observer movie camera positioned above the main instrument panel could also record the instrument readings on that panel throughout the flight. The Flight A astronauts named the mirror the 'Heroes Medal' and it was worn on all subsequent manned flights during Project Mercury.

The pressure suit itself carried additional nylon-sealed ball-bearings in the wrist seals. The zips that had been used to attach the gloves to the sleeves on earlier models of the suit were replaced with twist-lock wrist connectors to increase mobility. A urine collection device was also added to the suit for the first time. All of these and all later updates were carried on all subsequent suits worn on Mercury flights.

14.2.2 Mercury Atlas 6

MA-6 was the first flight on which small electric light bulbs were added to the ends of the middle and index finger of both gloves. The suits worn by the astronauts on Mercury Atlas 7 and Mercury Atlas 8 were the same as that worn on Mercury Atlas 6.

14.2.3 Mercury Atlas 9

The pressure suit worn on the final Mercury flight contained the highest number changes. This was because the suit had been designed to support the astronaut for between 24 and 34 hours in orbit. New soft boots were incorporated into the torso's legs and did away with the requirement to wear separate flight boots over a sock portion of the pressure bladder. The shoulders of the Mercury Atlas 9 suit were reconstructed and the construction of the gloves was also improved. The helmet contained a new oral thermometer, to replace the rectal thermometer used on all earlier manned flights. It also contained a new design of microphone and a mechanical visor seal that dispensed with the need to inflate the visor seal.

14.3 SURVIVAL EQUIPMENT

Each manned Mercury Spacecraft carried a survival kit consisting off:

4 × Medical injectors
Food container
Life raft
Matches
Pocket waterproof
Sea dye marker
Signal mirror
Soap
Survival flashlight
Water container
First-aid kit
Jack knife
Life vest (MA-6 to MA-9)
Nylon lanyard
SARAH beacon
Shark chaser
Signal whistle
Sun glasses
Survival knife
Zinc oxide

14.4 POST-FLIGHT RECOVERY KIT

Each of the DESFLOT-Four recovery ships was supplied with a post-flight recovery kit. This was for use by the astronaut after his flight, and consisted of:

Comb
Gym shoes
Post-flight coverall (flight suit)
Razor, blades and cream
Soap
Sunglasses
T-shirt
Wrist watch
Flight jacket
Handkerchiefs
Pressure suit helmet carrying case
Shorts
Socks
Toothbrush
Wash cloths

15

Mercury Little Joe

In January 1958, NACA engineers Maxime Faget and Paul Purser were working for Gilruth at PARD, on Wallops Island. At that time they conceived a launch vehicle grouping seven Sergeant solid propellant rocket motors within its body, for use in the developmental stages of the proposed NACA piloted satellite vehicle programme.

The new vehicle was designed to launch a piloted satellite vehicle from Wallops Island, along a suborbital trajectory, to an apogee of 241.3 km. The satellite vehicle would then descend under a 19.8-m diameter parachute to a landing in the water off Wallops Island, from where Navy personnel would recover it. To test the new vehicle before a human astronaut was launched on it, Faget and Purser suggested that a monkey be used first.

The idea went under the name Project High Ride, but the proposal was withdrawn when NACA Director Hugh L. Dryden compared Project Adam, a similar US Army proposal for a suborbital space flight launched on a man-rated Redstone missile, to the circus stunt of *'firing a woman from a cannon'.*

In August 1958, the prospect of launching a prototype spacecraft as an early part of an incremental programme to place a man in space returned to favour within NACA. The intention was to place the NACA piloted satellite vehicle into orbit on the top of a man-rated Atlas Missile. Suborbital flight-tests would be required to qualify the Atlas Abort Sensing and Implementation System (ASIS) and the tractor rocket that Faget had proposed to pull the satellite vehicle away from a failing Atlas launch vehicle. Faget and Purser redesigned their High Ride launch vehicle so that its flight profile mimicked that of an Atlas launch vehicle. The Atlas ASIS and the tractor rocket could now be tested under as near to realistic flight conditions as possible. The new High Ride proposal did not include any flights with human astronauts.

The modifications made to the original High Ride design included the removal of three Sergeant rocket motors from the launch vehicle and their replacement with three Recruit solid propellant rocket motors, which were less powerful. The change was made because the new configuration was only required to reach an apogee of 160.9 km. The launch vehicle was given the PARD designation F-57 and slowly assumed the name Little Joe. Faget assigned W.M. Bland to oversee Little Joe, while he concentrated his efforts on defining the piloted satellite vehicle itself.

On 1 October 1958 NACA became NASA and Project Mercury was officially established seven days later. On 5 November the STG was established at Langley and both Faget and Bland were assigned to the new department. R.D. English took over direct management of Little Joe, although Bland still monitored the project and regularly reported his findings to Faget.

Following the definition of the Little Joe launch vehicle, a Bidders' Conference was held on 21 October. On 26 November representatives from STG presented their proposal to use Mercury Little Joe vehicles to qualify the Atlas ASIS to the management at LaFRC and received approval and backing for the idea. North American Aviation Corporation was awarded the contract to build Little Joe on 29 December. The contract called for seven airframes, one of which would be maintained at the contractor's plant for load testing. The contractor was also called upon to provide one launcher that would be used to support of the Little Joe launches at Wallops Island.

As 1958 drew to a close, Thiokol Chemical Corporation, the company that manufactured the Sergeant and Recruit rocket motors, began a programme to upgrade the propellant and improve the casing used on the Sergeant motor, increasing its performance by 20 per cent. LaFRC supported the programme and ordered the new motors for use on both Little Joe and the Scout launch vehicle that was also under development at that time. When the new motors were delayed STG suggested that the new propellant be put in-to-old style Sergeant casings as a stopgap measure. The new motors would then be integrated into both programmes when they became available. The motors that used the new propellant in old-style casings were named Pollux, after one of the twin stars in the constellation Gemini. The all-new motor was named Castor, after the second star in the pair. Both motors had exhaust outlets angled at 11 degrees to the vertical, one-degree less than the Sergeant motor.

The Little Joe flight-test programme was drafted on 29 January 1960 (see Fig. 51). The rocket was to be used to launch prototype Mercury Spacecraft on suborbital flights in order to qualify the Atlas ASIS for manned flight. The ASIS would 'read' the normal Little Joe engine shutdown signal as if it was a premature shutdown. It would then initiate the abort sequence resulting in the tractor rocket, which was becoming known as the LES, pulling the prototype spacecraft away from the launch vehicle, as it would in a real abort situation. When the Escape Motor shut down, the LES would be jettisoned and the prototype spacecraft would land in the water off Wallops Island by parachute. Animal payloads would be carried on some of the later flights, after the system had proven itself reliable.

The first two operational airframes were delivered to Wallops Island on 28 May 1959. At that time the first Mercury Little Joe launch was planned for July 1959. These two airframes consisted of four Recruit and four Pollux rocket motors inside a cylindrical skin. On later flights the hybrid Pollux motors would be replaced with the all-new Castor model. Four large aerodynamic fins spaced at 90 degrees around the base of Little Joe offered a stabilisation system similar to that offered by the flights on an arrow. With no guidance system, the vehicle's trajectory was dictated by the

Fig. 51. The Little Joe 1 launch vehicle stands on its launch complex while its prototype spacecraft and Launch Escape System are prepared for mating in the foreground. (NASA)

angle of the launcher, the power and ignition sequence of its rocket motors and the weather conditions on the day of launch. The apogee of the trajectory was dictated by the angle of the launcher on which the Little Joe rested prior to flight. (See Table 8.)

Table 8: Effects of launch angle on apogee of Mercury Little Joe trajectory

Launch angle	Expected apogee
70 degrees	16.9 kilometres
80 degrees	111 kilometres
88 degrees	225.2 kilometres

Depending on the trajectory required, either all eight rocket motors, or the four Recruits and just two of the Pollux/Castor rocket motors were ignited at lift-off. The remaining motors were ignited in flight. Following the simulated abort, the Little Joe launch vehicle followed a ballistic trajectory and sank on impact with the water below. No recovery was attempted unless it was required for post-flight data reduction.

15.1 LITTLE JOE PROTOTYPE SPACECRAFT

The prototype spacecraft used on the Little Joe launches were constructed at LaFRC. The prototypes were all constructed to the Type D pattern, but without a Retrograde Package, or Landing Skirt. Their shape, size, mass and centre of gravity all mimicked the production spacecraft as closely as possible. They were sophisticated vehicles constructed in two parts.

The pressurised lower portion contained the majority of instrumentation to record the flight environment and any biological payloads that were carried. Data recorded included acceleration rates, angular motions, internal temperature, external skin temperature, surface air pressure, radiation levels and internal noise levels. Data was transmitted to receivers at Wallops Island in the form of telemetry. Two camera pods protruded beyond the lower bulkhead of the instrument compartment and the heat shield, to film the launch vehicle as it separated from the prototype capsule. The afterbody, Recovery Section and Antenna Canister were formed from sheets of corrugated Inconol. A conical fibreglass shield protected the top of the Antenna Canister from the LES exhaust plume. The LES and Marman Clamps used in the Mercury Little Joe programme were standard production items.

15.2 A MONKEY WILL BE FIRST

Faget and Purser included plans for both biological payloads and human astronauts in their original proposal for the High Ride launch vehicle. When High Ride became Little Joe and was incorporated into Project Mercury the flights with human astronauts were not carried forward to the new programme, but the idea of subjecting biological payloads to a simulated Mercury Atlas max-q abort was too good to miss.

LaFRC and STG personnel discussed Mercury Little Joe biological payloads on 26 March 1959. Final plans for the programme's biological payloads were drawn up at a meeting of interested personnel from STG, LaFRC and the Air Force School of Aviation Medicine, over 2–16 April.

Plans for Mercury Little Joe originally included a number of flights carrying pigs, but these were removed from the programme on 6 May, when it was discovered that a pig would not survive a long period on its back. Even so, McDonnell did use pigs in a number of drop tests of the Mercury Spacecraft at their St Louis plant. The drop tests were carried out in an attempt to better understand a biological subject's ability to withstand the loads associated with a Mercury Spacecraft landing at the end of its flight into space.

Monkeys were chosen as STG's biological payload, because they were similar in many aspects to a human astronaut and could be trained to perform certain basic tasks that would act as an indication of their condition throughout the flight, even if the medical instrumentation attached to their bodies failed. Rhesus monkeys were flown on Mercury Little Joe tests, while Chimpanzees made suborbital flights on the Mercury Redstone and orbital flights on the Mercury Atlas out of Cape Canaveral before human astronauts were risked on similar flights. The monkeys were taken from the African jungle at an early age and sent to the Air Force School of Aviation Medicine, Holloman AFB, New Mexico. There, each animal was assigned an individual trainer who attempted to befriend it, looked after its needs and oversaw its training. Several monkeys had to drop out of the programme and be replaced. In particular, some Chimpanzees could not be trained to perform the necessary tasks, while others grew beyond the 22.6-kg upper weight limit, or passed the four-year upper age limit set on the animals.

Training the monkeys for their flights began in May 1959. They were strapped into a seat and taught to react to the illumination of different coloured lights on the panel in front of them by pushing a particular button, or pulling a lever within a set time limit. If they reacted correctly they were rewarded with a banana flavoured food pellet. If they reacted incorrectly they received a mild electric shock to the soles of their feet. Advanced training consisted of the same tests being performed while under high *g* forces, either in the Johnsville centrifuge, or on a rocket sled at Edwards Air Force Base. Microgravity training took place with the test rig mounted in the back seat of a military jet fighter flying parabolic arcs to produce the condition for a few seconds at a time.

When the time for flight drew near a colony of Rhesus monkeys was established at Wallops Island to support Mercury Little Joe. A colony of Chimpanzees was established at the rear of Hangar S in Cape Canaveral in support of Mercury Redstone and Mercury Atlas launches. Training continued at the new locations right up until the morning of launch, when the healthiest animal, with the best results in training, was selected for flight.

16

Mercury Scout

The Solid Controlled Orbital Utility Test System (SCOUT) launch vehicle was developed as a joint venture between NACA and the USAF (Fig. 52). Its development period included 1 October 1959, when NACA became the nucleus of the new NASA. The four stages of the new vehicle all used solid propellant rocket motors and together they were 21.6 m in length.

Stage 1 was an Aerojet General Corporation Jupiter Senior rocket motor a modified and redesignated Algol motor. It was 9.1 m long and 1.2 m in diameter. 'Base A' was attached to the bottom end of the motor casing and supported four stabilising fins with tip controls and four jet vanes, which acted in the exhaust plume for in-flight steering. It also contained an FM/AM telemeter and hydraulic system for operating the flight control surfaces. The stage produced 444,800 N of thrust for 40 seconds.

A tapering interstage, called Transition B, mated the top of the first stage to the base of the second stage, which was a Castor rocket motor similar to those used on the later models of the Little Joe launch vehicle. The Castor was 6 1 m long and 78.7 cm in diameter. During the second stage burn, flight control was maintained by an RCS using hydrogen peroxide under the control of the Scout's guidance system.

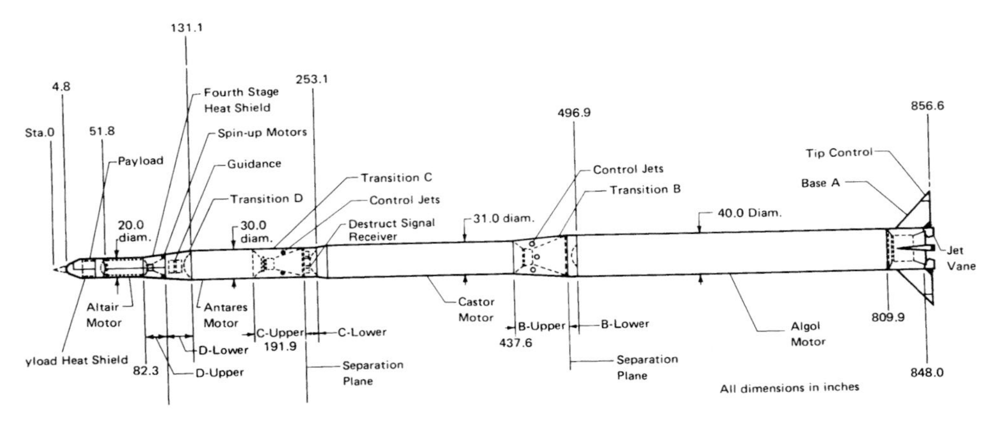

Fig. 52. SCOUT launch vehicle. (NASA)

The third stage was an Antares rocket motor, developed specifically for the Scout launch vehicle by the Allengany Ballistics Laboratory from the Altair rocket motor. The Antares stage was 2.8 m long, 78.7 cm in diameter and used a hydrogen peroxide RCS for attitude control. A cylindrical Transition C interstage linked the second and third stages and was 1.7 m in length and 78.7 cm in diameter.

The 1.2-m long, tapered, Transition D interstage mated Transition C to the Altair fourth stage and housed the Scout's guidance system and an FM/FM telemetry system. The Altair stage was 77 cm long and 50.8 cm in diameter and was spin-stabilised. The payload was contained in a nose-cone/heat shield 1.9 m tall and 50.8 cm in diameter. The payload employed two telemetry systems, an FM/AM and an FM/FM system.

16.1 MERCURY SCOUT 1 PAYLOAD

Ford Aeronautic manufactured the small rectangular payload for Mercury Scout 1 under contract to STG. The payload's mission was to test the WWTN prior to the first manned orbital flight of a Mercury Spacecraft. To complete that mission, it carried:

- 1 × C-band beacon
- 2 × Minitrack beacons
- 1 × 1,500-watt battery
- Relevant antennae
- 1 × S-band beacon
- 2 × command receivers
- Scout fourth-stage instrumentation package

Where equipment installed in the Mercury Spacecraft was carried, Ford Aeronutronic used technical drawings supplied by McDonnell to manufacture it. GSFC supplied the Minitrack equipment, which was similar to that used on unmanned satellites at that time. The accuracy of the WWTN was to be gauged by comparing data received from the new network with that received from the Minitrack system, which was known to be reliable.

The mission plan called for all equipment to be operated for three periods of three orbits, nine orbits in total. Between each block of three orbits the equipment would be turned off for a period, to allow initial data reduction to be undertaken on the ground. In this way three separate manned flights, each of three orbits, were simulated in one test flight.

16.2 DEVELOPING MERCURY SCOUT

In 1957 PARD began the definition and development of a four-stage, solid propellant launch vehicle to increase the range and velocity available for use in their model flight-test programme in support of NACA–Langley. The original intention was that the new launch vehicle should employ existing rocket motors in all four stages, even if those motors were grouped together in a sequence that had not been tried before. Maxime Faget was among the group of PARD engineers that headed the design effort.

The group began their design with the Jupiter Senior rocket motor, which had been designed as part of the US Navy's programme to develop a solid propellant missile that could be launched from a submarine. In 1957 Jupiter Senior was the largest solid rocket motor in existence. The upper stages consisted of the improved Sergeant rocket motor, know as Castor, and two X-248 rocket motors designed for the US Navy's Vanguard launch vehicle. In March 1958 PARD ordered four Castor rocket motors for use in the four-stage rocket project if the NACA Committee approved it later in the year. The design of the new vehicle was finalised in mid-1958, but there were no funds available to develop it.

With rumours rife that NACA was to be replaced by a new Space Administration the design group were told to repackage their proposal as a launch vehicle that could orbit small payloads as part of a programme of research into re-entry through Earth's atmosphere. At that time the new rocket took on the name 'Solid Orbiter' in reference to its solid propellant rocket motors. Claims for the new launch vehicle stated that it would place a 68-kg payload into an 804.5-km orbit. The new proposal was put before the NACA Committee on May 1958 and accepted. A budget was established for five 'small-scale recoverable orbiters', as part of a re-entry research programme related to the future development of a manned satellite vehicle.

On 4 June Gilruth and Faget met with representatives of the USAF, who were considering the development of a solid propellant launch vehicle capable of placing small payloads into orbit. The intention was to encourage the USAF to join PARD in developing the four-stage launch vehicle approved in May, which would serve the requirements of both organisations. The Air Force agreed to participate.

A two-day meeting at NACA Headquarters finalised the design of the new launch vehicle. It would have a Jupiter Senior first stage, a Castor second stage, a new 9,785-N third stage and a Grand Central Meteor fourth stage. The vehicle would fly in both three- and four-stage versions. The four-stage vehicle would place a 45-kg payload into orbit. The USAF placed an initial order for eight vehicles, the first three of which would be test vehicles, flown from the PARD establishment at Wallops Island.

No further action was taken on the new launch vehicle until after 1 October, when NASA came into being. The new Administration decided to develop the new launch vehicle as its own project and placed it under the control of William Stoney, one of the original designers. It was Stoney who gave the vehicle its complicated name, Solid Controlled Orbital Utility Test System, giving the acronym SCOUTS. The second S was dropped from the acronym and the vehicle assumed the name Scout. Once NASA had made it operational, the USAF would be invited to convert Scout to meet their needs.

LaFRC procured Scout's rocket motors from three contractors. Following the signing of a contract on 23 October the Allegany Ballistics Laboratory (ABL) provided 12 motors, four each for the first, third and fourth stages through the US Navy's Bureau of Ordnance. ABL agreed to enlarge the existing X-248 motor to form the new third-stage motor. The new model was designated X-254 and the PARD's engineers called it Antares. The original X-248 motor was assigned to the fourth stage, in place of the proposed Grand Central Meteor motor. PARD assigned the X-248 the name Altair.

On 1 December the contract for the first-stage rocket motor was signed with Aerojet General Corporation. It covered several static firings and the delivery of nine Jupiter Senior rocket engines, which PARD renamed Algol. At that time the Jupiter Senior rocket motor was not fully qualified. As a result temperature limitations were placed on it during transportation. The Algol motors travelled from the contractor's plant to Wallops Island in an environmentally controlled container designed specifically for the job and referred to as the 'Transtainer'.

NASA negotiated with Thiokol Chemical Corporation for the second-stage Castor rocket motors through the Army Ordnance Department. The contract, signed on 5 December, covered the development and delivery of Castor motors for Scout, Little Joe and a number of other launch vehicles used by NASA at the time.

When the programme to update the Sergeant rocket motor into the Castor motor fell behind schedule in December, a number of Jupiter Junior motors were purchased from Aerojet and renamed Aerojet Junior in case they were needed to fly as interim second stages on early Scout launch vehicles. The Aerojet Junior used a Castor casing with different propellants to give a very similar thrust rating to the Castor motor. Three static test motors and four flight motors were added to Aerojet's Algol motor contract. The Aerojet Junior motors later flew on Air Force Scout launches.

A launch programme for the first eight Scout vehicles was published in December. All flights were assigned payloads, developed within NASA, that they would place into orbit. The launch programme was simplified in January. Although all vehicles would still carry a scientific payload, the first two launches, planned for October and December 1959, were reduced to four-stage vertical launches, so that the fourth-stage ignition would take place within range of the tracking stations at Wallops Island. The third and fourth flights would be orbital attempts and were scheduled for January and March 1960.

Scout's guidance system was the subject of an eight-company competition, with Minneapolis-Honeywell being named as prime contractor on 12 January 1959. The first stage was guided by means of vanes acting in the exhaust plume. Stages two and three used hydrogen peroxide thrusters and the fourth stage was spin stabilised.

The fifth contract, for four airframes and one launcher, was competed for by 22 companies, including all of America's major aircraft companies. The contract went to Vought Aircraft Company on 21 April. In May the USAF agreed to procure 10 Scout launch vehicles, for flights using various stage combinations between March and October 1960. A further 65 Air Force launches were predicted over the next two years, and the USAF stated that they intended to make all of their Scout launches from the Atlantic Missile Range, Florida, despite earlier promises to launch at least eight vehicles from Wallops Island. Ford Aeronutronic would handle all aspects of the Air Force Scout launches.

NASA's payloads would continue to be handled by the Armed Forces Special Weapons Centre (AFSWC), which had handled many of PARD's payloads for launch from Wallops Island in the past. AFSWC offered to develop a payload for the first Scout launch, an offer that NASA accepted only after making it clear that

GSFC would supply the payload for the second launch. The first launch also carried an Air Force payload – a flare package.

The first NASA launch slipped to January 1960, but that date was not met because of technical difficulties in developing the Antares motor. The PARD was pressured into launching a test vehicle with a correctly weighted but dummy second stage. The launch was not part of the official flight programme and was designated SX-1. Some PARD engineers referred to the incomplete launch vehicle as the 'Cub Scout'. The 18 April launch was a failure. Scout's first official launch occurred on 4 October and was successful. Thereafter, Scout was added to the growing list of NASA and military space launch vehicles, where it rapidly earned an excellent reputation.

17

Mercury Redstone

On the launch pad Mercury Redstone stood 29 m tall, 1.7 m in diameter and 3.36 m across any to opposing fins. It employed the elongated propellant tanks developed for the Jupiter C, which was itself derived from the Redstone IRBM. The longer tanks provided an extra 20 seconds of powered flight over the Redstone IRBM. The single North American aviation A-7 rocket motor produced 346,944 N of thrust for 2 minutes 23 seconds. That was insufficient to place the Mercury Spacecraft into orbit, but allowed the astronauts the prospect of rehearsing all other aspect of the Mercury mission before attempting an orbital flight. It was expected that all seven astronauts would fly a suborbital Mercury Redstone flight before making an orbital flight on a Mercury Atlas combination. (See Fig. 53.)

NASA procured the Mercury Redstone launch vehicle that it required through the Army Missile Command. After initial manufacture the airframe was shipped to Redstone Arsenal, Huntsville, for final flight systems installation and testing. The first Mercury Redstone was subjected to a fit-test with its Mercury Spacecraft at Huntsville, but this practice was discontinued thereafter to save time on the flight schedule.

Following completion of the work at Huntsville, the launch vehicle was shipped to Cape Canaveral, where it was installed in Hangar S for further testing. It was then transported out to Launch Complex 5, where it was erected on the launch table. When the launch vehicle was ready the spacecraft was mated to it with a Marman Clamp, which contained three bolts that each held a small amount of explosive charge to ensure that they severed at the moment of spacecraft separation. A failure to do so resulted in the spacecraft remaining attached to its launch vehicle until it struck the ocean at the end of its suborbital trajectory. Such an impact would have destroyed the spacecraft and killed the astronaut.

A second Marman Clamp mated the LES to the neck of the spacecraft. The LES rocket motor was not installed until the morning of launch, and only when everything had been erected on the launch table at Launch Complex 5 could joint systems testing begin.

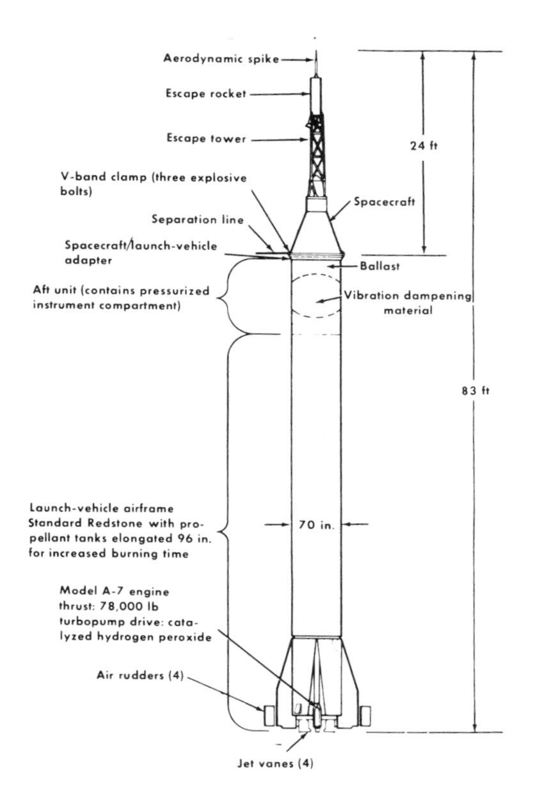

Fig. 53. Mercury Redstone. (NASA)

17.1 MERCURY REDSTONE COUNTDOWN AND LAUNCH EVENTS

T − 10 h 40 min:	Countdown commences. Power checks. Launch vehicle telemetry turned on. WWTN systems/communications checks.
T − 6 h 30 min:	Planned hold prior to fuel loading and pyrotechnic arming. LES igniter installed. Weather/range clearance checks. Countdown resumes after propellant fuel loading is complete.
T − 5 h 50 min:	Astronaut woken up. Performs ablutions, takes breakfast.
T − 4 h 20 min:	Astronaut takes final physical examination followed by donning of pressure suit.
T − 3 h 45 min:	Astronaut leaves crew quarters for launch pad in transfer van.
T − 2 h 30 min:	Commence planned hold for astronaut ingress. Countdown resumes after completion of oxidiser loading.
T − 2 h 15 min:	Astronaut in elevator.
T − 2 h 10 min:	Astronaut in white room.

T – 2 h 05 min:	Astronaut in spacecraft.
T – 2 h:	Commence liquid oxygen loading. Cherry picker in position next to spacecraft side hatch for emergency egress.
T – 55 min:	Cherry picker moved back to 7.6 m from spacecraft.
T – 6 min:	Cherry picker moved to parked position, with platform on the ground next to the launch pad.
T – 20 s:	Commence automated countdown.
T – 3 s:	Spacecraft umbilical disconnected and umbilical support tower dropped to the ground. Astronaut introduced into abort loop.
T = 0:	First movement. Engine cut-off inhibited for first 30 seconds of flight to prevent launch vehicle falling back onto launch pad.
T + 15 s:	Commence pitch manoeuvre to correct exit attitude.
T + 30 s:	Abort plus engine cut-off abort sequence initiated.
T + 2 min 9.5 s:	Inhibition of normal shutdown procedure deactivated.
T + 2 min 11 s:	Velocity cut-off switch armed.
T + 2 min 15 s:	Abort pressure switch deactivated.
T + 2 min 23 s:	Redstone shutdown. LES jettisoned. In-flight abort systems disarmed. Ten-second timer initiated.
T + 2 min 32.5 s:	Redstone/spacecraft separation. Spacecraft turnaround. Flight activities.
T + 8 min 6 s:	Redstone impacts Atlantic Ocean.
T + 15 min 11 s:	Spacecraft lands under multiple parachute system. Recovery by US Naval forces.

17.2 DEVELOPING MERCURY REDSTONE

On 1 October 1958 NASA was formed to lead America's push into space. A few days later Project Mercury was established with the goal of a three-orbit manned flight. NASA's engineers began looking for a launch vehicle on which their astronauts could rehearse for space flight on suborbital trajectories before attempting an orbital flight. The launch vehicle had to be able to lift the Manned Satellite Vehicle to an apogee of 185 km on a trajectory that offered at least 5 minutes of microgravity. Most of all it had to have a proven reliability record, clearly established in time to meet Project Mercury's flight schedule, which meant that the launch vehicle would have to be a military missile. The ABMA's Redstone missile did not meet the criteria. The Redstone derived Jupiter C met most of the criteria, although it did not meet the safety requirements demanded of a manned launch vehicle. On 16 January 1959 NASA instructed ABMA to supply eight Redstone missiles for use in Project Mercury.

A series of hardware changes converted the Redstone missile into the Mercury Redstone launch vehicle. The new launch vehicle employed the larger propellant tanks and attached instrument compartment of the Jupiter C and thereby gained an extra 20 seconds of powered flight over the Redstone missile. The Jupiter C's toxic propellant was replaced with the Redstone's liquid oxygen and alcohol/water, but the change caused rapid deterioration of the carbon steering vanes acting in the exhaust plume. Only vanes of the highest quality carbon were used on the Mercury Redstone. With plans already in place to install the North American Aviation A-7 rocket motor in the Jupiter C, Mercury Redstone was fitted with the A-7 from the outset. The Redstone missile's ST-80 guidance system was replaced by the simpler LEV-3 guidance system.

With the spacecraft in the launch position the astronaut lay on his back, the position that allowed him to best endure the high *g* forces of launch. In the event of a launch pad, or in-flight abort, the Marman Clamp at the wide end of the spacecraft was released and the LES pulled the spacecraft clear of the malfunctioning launch vehicle. Having escaped the potential fireball from the exploding launch vehicle the LES was jettisoned and the spacecraft descended to the ocean under its ELS.

To control the abort sequence Mercury Redstone was fitted with an Automatic Abort Sensing System (AASS). The abort criteria set for the new launch vehicle was the result of reviewing the data from 69 Redstone, Jupiter C and Juno 1 launches. To simplify the AASS a series of criteria that were recognised as precursors to a catastrophic failure were monitored. Each of the criteria monitored represented several different types of failure. The abort sequence was initiated if:

- the pressure in the A-7 rocket motor's combustion chamber fell below 14,765 g/cm^2 at any time between ignition and nominal shutdown;
- the launch vehicle's in-flight attitude deviated from the programmed flight trajectory by $\pm$ 5 degrees in pitch or yaw and $\pm$ 10 degrees in roll;
- the spacecraft electrical systems failed, or power was lost to the launch vehicle/spacecraft electrical interface.

As well as the AASS, the Range Safety Officer in MCC and the astronaut also had the ability to activate the abort sequence and end the flight. In the event that the abort sequence was activated the launch vehicle destruct command was blocked for 3 seconds, to allow the LES to pull the spacecraft to a safe distance. These changes altered the Redstone's flight characteristics. The launch vehicle was predicted to become unstable after 88 seconds of flight, but this was overcome by placing approximately 311.5 kg of ballast in the Instrument Compartment.

Each individual Mercury Redstone system was subjected to vibration, humidity and temperature testing, as part of an exacting reliability programme. Each completed launch vehicle was then subjected to bending loads up to 150 per cent of those that it could expect to experience during handling and flight. All vital systems were subjected to extensive testing. At that time the A-7 rocket motor was found to experience a combustion instability that required the fuel injector to be modified. The static test stand at Redstone Arsenal was also redesigned when it was found to be the cause of a low-frequency oscillation during rocket motor static firings.

Quality assurance during the construction, handling and pre-launch preparation of the Mercury Redstone launch vehicle was much greater than that shown to the Redstone missile, because the Mercury Redstone would ultimately carry a human astronaut. Each component of the Mercury Redstone launch vehicles was required to pass a stringent quality control inspection and receive a Mercury unique stamp on its paperwork to say that it had done so.

Four unmanned flight-tests were made of the Mercury Redstone launch vehicle. On each occasion failures occurred. Those failures were investigated and corrected on all subsequent launch vehicles before the next launch took place. In this way the two launch vehicles that carried Flight A astronauts on suborbital space flights in 1961 were the most reliable that could be produced at that time.

18

Mercury Atlas D

Mercury Atlas D consisted of a single set of monocoque propellant tanks forming the body of the Sustainer stage. The Sustainer was constructed from stainless steel varying in thickness from 0.12 to 0.03 cm. The propellant tanks had to be pressurised at all times, or they would have collapsed as a result of external air pressure.

Standing 15.2 m high, the Mercury Atlas Sustainer had a dome at the top, which housed an Atlas C style liquid oxygen boil-off valve. The base of the dome was fixed to the first skin of the upper conical section, with a diameter of 1.7 m. The base of the upper conical section was 3 m in diameter and mated with the main cylindrical section. Two bulkheads, with insulation between them, separated the liquid oxygen and kerosene tanks. The insulation was not carried on the last four Mercury Atlas launch vehicles. The propellant tanks carried liquid oxygen and kerosene (RP-1). Finally, the lower edge of the main cylindrical section was mated to the top of the lower conical section. A single Rocketdyne LR105-5 rocket motor, capable of producing 253,536 N of thrust for 275 seconds, was gimbal mounted at the base of the Sustainer with plus or minus 3 degrees of movement in both pitch and yaw for in-flight steering, but only when the sustainer was in solo flight.

Two downward-facing vernier steering rockets, each capable of producing 4,452 kg of thrust, were mounted on the Sustainer at 90 degrees to two equipment pods. The verniers had their own start tanks but thereafter took their propellants from the main Sustainer tanks to provide roll attitude control throughout powered flight and pitch and yaw control during staging. They were capable of 140 degrees of movement in pitch and 50 degrees in yaw. The equipment pods carried electronic systems, and the upper equipment pod – the longer of the two – also contained connection points for a number of launch vehicle umbilicals when Mercury Atlas was on the launch pad. The shorter of the equipment pods was called the lower pod, or stub pod. (See Fig. 54).

The Booster section was mechanically mated to the base of the sustainer thrust ring so that its two Rocketydyne LR89-5 Booster motors formed a straight line with the Sustainer engine at launch. The Booster motors burned propellant from the Sustainer tanks to produce 684,992 N of thrust for 120 seconds. The Booster skirt

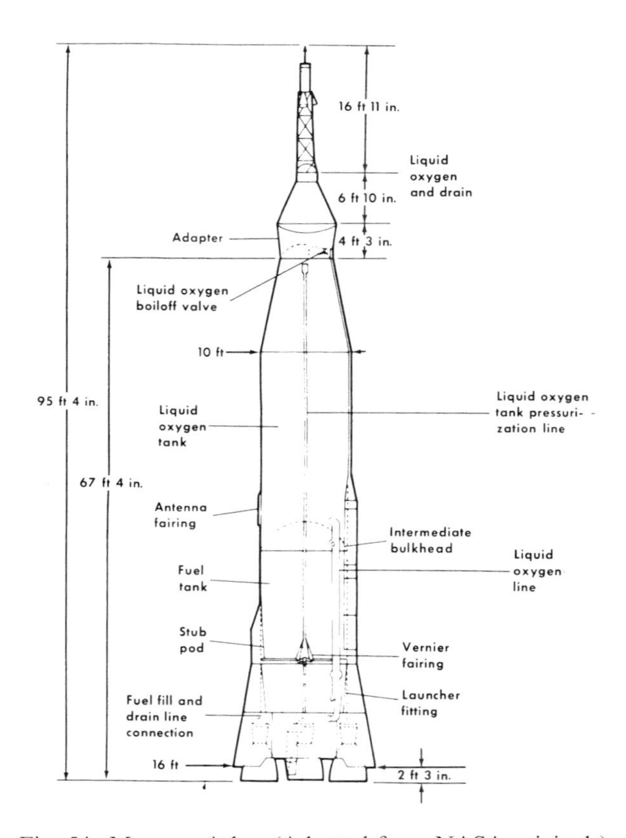

Fig. 54. Mercury Atlas. (Adapted from NASA originals)

gave the Atlas D a maximum diameter of 4.8 m. All five rocket motors were ignited at lift-off to give the launch vehicle a total lift-off thrust of 1,730,603 N.

When the motors ignited on the launch pad the vehicle was held in place by large clamps while the Booster and Sustainer motors built up to full thrust (mainstage), only then were the clamps released and the Mercury Atlas freed to begin its flight. If any one of the three main propulsion units failed to achieve mainstage, the Atlas was shut down and the launch was cancelled.

At launch time (T) + 2 min 20 s the two Booster engines shutdown and the Sustainer engine was locked in its neutral position. At T + 2 min 35 s mechanical clamps were then released and the Booster skirt fell away, guided by two rails mounted on the Sustainer to ensure that it did not interfere with the sustainer motor, which was still firing. After Booster skirt separation the Sustainer pitched down and the LES was jettisoned. The Sustainer then pitched back-up to its planned trajectory and continued the climb into orbit for a further 2 minutes.

18.1 DEVELOPING MERCURY ATLAS D

NASA came into being on 1 October 1958 and one week later PARD's manned satellite programme became Project Mercury. With the spacecraft contract under review, LaFRC commenced negotiations with the USAF for the purchase of Atlas launch vehicles. The USAF would buy the missiles from Convair and oversee their conversion into Mercury Atlas launch vehicles. Prior to their acceptance for use in Project Mercury the USAF inspected each vehicle. Following transportation to Cape Canaveral the Atlas was prepared and launched by a joint USAF–Convair team.

NASA order HS-24, dated 24 November 1958, instructed the USAF to provide them with one Atlas C, nine Atlas Ds and four Thor missiles for Project Mercury. STG was informed that the Atlas C, which would be used for a suborbital test of ablation heat shield material, to be called 'Big Joe', would be ready within six months. The requirement for Thor missiles was later deleted when NASA decided to replace them with the Army's Jupiter missile. An order was placed with ABMA for two Jupiters.

NASA negotiated for its Atlas launch vehicles through the Air Force Ballistic Missile Division (BDM). Meetings occurred between 23 October 1958 and 7 April 1959 to establish a Memorandum of Understanding. BDL established a Programme Office to handle the Project Mercury requirements. They brought in the Air Force Space Technology Laboratory (STL) to provide support and develop the majority of Mercury unique Atlas systems. In 1960 STL became the basis for the Aerospace Corporation. Convair were responsible for the airframe, while the North American Aviation Rocketdyne Division supplied the propulsion units. GEC supplied the Atlas guidance equipment, while Burroughs Corporation provided the guidance computer. Convair launched the missiles from the Atlantic Test Range, Florida, with the assistance of the USAF 6555th Aerospace Test Wing.

NASA Order HS-36 was received by the USAF on 8 December and requested that 10 Atlas Ds be procured for Project Mercury. The order changed the single Atlas C ordered previously for an Atlas D and added a further four Atlas Ds to NASA's original order. It also cancelled the four Thor ICBMs originally ordered for Project Mercury.

On 18 December 1958 the Sustainer of a stripped down Atlas went into orbit around Earth. As it circled the planet it transmitted a recorded Christmas message from President Eisenhower. This was Project Score.

In May 1959 ABMA attempted to add $8 million to the cost of the Redstone and Jupiter missiles required for Project Mercury. This price rise came shortly after similar rises from McDonnell, the Mercury Spacecraft prime contractor, and Convair, the Atlas contractor. The new price for the Jupiter launch vehicle made a Mercury Jupiter suborbital flight more expensive than the launch of a Mercury Atlas orbital flight. Having sought backing from NASA Headquarters in Washington, STG cancelled the two Mercury Jupiter flights. At the same time STG also cancelled two Mercury balloon flights designed to subject the spacecraft to the space-like environment found in the stratosphere prior to manned flights into space.

The Atlas D ICBM was flight-tested out of Vandenburg AFB, California, between April and August 1959. Of the first five launches two failed, two were partially successful and only one was a success. The eighth and final Atlas C was launched in August. Its warhead was protected by ablative material similar to that used in the Mercury Spacecraft's heat shield, and the dummy warhead survived re-entry in good condition.

When NASA moved into Hangar S in Cape Canaveral in 1959, they were informed that Launch Complex 14 was allocated for the first Mercury Atlas launch, Big Joe, but after that they would have to use whichever of the five Atlas launch complexes at AMR was available for their future launches. When they tried to have Launch Complex 14 allocated for all of their launches they were told that the Air Forces Missile Defence Alarm Satellite was already allocated to Launch Complex 14 and had priority over Project Mercury. Further negotiations between NASA and the USAF finally resulted in Launch Complex 14 being allocated for all Mercury Atlas launches.

The major difference between the Mercury Atlas D launch vehicle and the Atlas D ICBM was the Abort Sensing and Implementation System (ASIS). (See Figs 55 and 56.) Like the AASS in the Mercury Redstone, this system monitored the performance of a number of critical systems and initiated the abort sequence if they passed preset parameters. Some USAF Atlas development flights carried the Mercury Atlas ASIS in a 'piggy back' mode, allowing it to be tested in flight before it was employed on the NASA launch vehicles. The systems monitored by the ASIS were:

Liquid oxygen tank pressure.
Differential pressure across the intermediate bulkhead.
Launch vehicle attitude rates in all three axes.
Rocket engine manifold pressures.
Sustainer hydraulic pressure.
Primary electrical power supply.

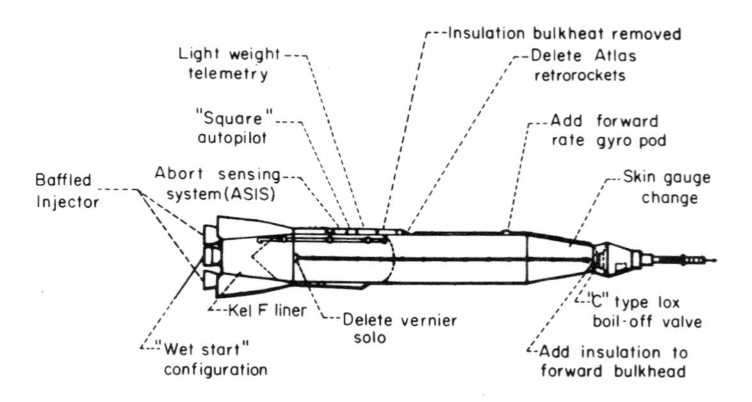

Fig. 55. Project Mercury adaptation to the standard Atlas D ICBM. (NASA)

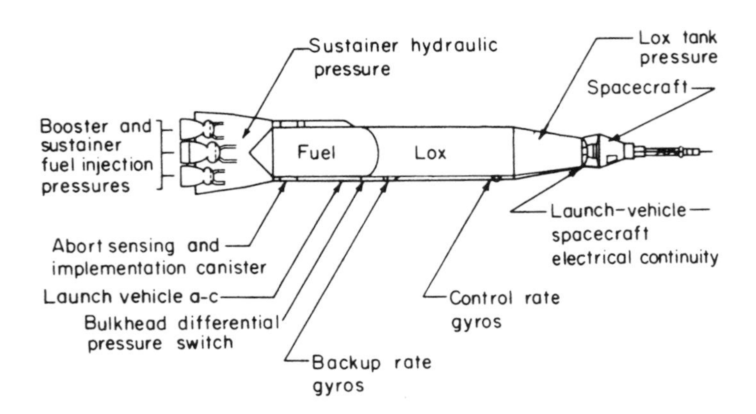

Fig. 56. Mercury Atlas parameters monitored by the Abort Sensing Implementation System. (NASA)

To compensate for the fact that Mercury Atlas was 7 m taller than the Atlas ICBM, the auto rate gyro package was moved that amount higher up the vehicle. The new position allowed it to gauge the bending stresses acting on the vehicle more accurately. New antenna patterns were developed for Mercury Atlas, to prevent interference between individual antennae. Convair had developed a new, lighter, telemetry system for the Centaur launch vehicle. It flew on all Mercury Atlas vehicles after Mercury Atlas 2. The ICBM's electromechanical autopilot was replaced with a modularised electronic system that was easier to programme and to service on the launch pad.

The Mercury Spacecraft carried its own posigrade rockets to provide positive separation from the launch vehicle in the microgravity environment of space. As a result, the forward firing pair of vernier rockets on the Atlas ICBM were deleted from the Mercury Atlas launch vehicle. The Sustainer dome was covered with a fibreglass blast shield, to protect it from the exhaust plume from the spacecraft's posigrade rockets. The ICBM's vernier solo burn, made to place the warhead in the correct attitude prior to separation, was removed from the Mercury Atlas flight profile.

Atlas 12D was launched from the Vandenburg Air Force Base, California, on 9 September by the personnel of the 6555th Aerospace Test Wing. When the missile flew the full intercontinental distance the ICBM was declared operational. At that time the only two operational missiles stood in training galleries at Vandenburg AFB. The USAF ultimately formed 13 Atlas ICBM squadrons.

Originally, the missile was launched from 'soft' sites, like the launch pads at Cape Canaveral. 'Hardened' coffin-silos, capable of withstanding a thermonuclear strike, were developed for the Atlas E. The Atlas was stored in the coffin in a horizontal position. If required for launch, the lid of the coffin opened and the missile was raised to the vertical position before propellant loading and launch. Fifteen minutes were required before the missile was ready for launch. Atlas E and Atlas F models were flight-tested in 1960 and 1961, the latter models being launched from vertical hardened underground silos.

President Kennedy's Secretary of State for Defence cancelled the Atlas ICBM. The operational missiles were phased out in 1966 and replaced by the silo-launched Titan ICBM. By that time, however, the Atlas had a secure role as a space launch vehicle.

Having been constructed and tested by Convair, the Mercury Atlas launch vehicle was delivered to Cape Canaveral, where it underwent an initial inspection. The launch vehicle was then transported by road to Launch Complex 14. The launcher was lowered through 90 degrees, until it lay flat on the ground. The transporter was backed into the launcher and the Atlas was attached to the launcher. A rope and mechanical pulley system was then used to rotate the launcher back through 90 degrees, until the Atlas stood vertical on the launch pad. All systems on the launch complex and the launch vehicle were tested. The launch vehicle systems were tested in the launch vehicle flight acceptance composite test (FACT). This had to be successfully completed before the Mercury Launch Vehicle Adapter, Spacecraft and LES were mated.

On 29 September 'Big Joe' stood ready for launch at Launch Complex 14. It would be the first major launch of Project Mercury.

Part IV: PROJECT MERCURY FLIGHT PROGRAMME

19

The race to be first

Project Mercury cannot be isolated from the events of the Cold War and American domestic policy. To this end I have included in this chapter a 'diary of events' for the period 1959–63, in an attempt to place in their contemporary context the flights carried out as part of Project Mercury. During a period of growing tension between America and the Soviet Union, over Germany and Cuba in particular but also in other regions of the world, America was undergoing its own domestic upheaval with Negroes demanding the Civil Rights promised them by the country's Constitution, but denied them in everyday living. While consecutive American governments addressed the Civil Rights programme as an internal political matter, the Soviets continued to meet internal discontent inside the Eastern Bloc with military power and the Gulag. America also had no choice but to address the expansion of Communism on the world stage. It is to the credit of the political leaders of both America and the Soviet Union that they chose to show their superiority in missiles and advanced technology in the form of a 'Space Race', without applying those missiles to their primary purpose.

19.1 COMPETITION

In the Soviet Union the space programme developed against a background of national paranoia, obsessive secrecy, and the ever-present threat of the Secret Police and the dreaded Gulag prison system. It was controlled by the military. Everything was designed and manufactured in government-run bureaux and delivered to Baikonur Cosmodrome in Kazakhstan for final preparation and launch. No matter how bad the Negroes' situation in America, it was favourable to the prospect of what might happen to the workforce in a Soviet secure military Design Bureau that failed to meet the demands of the country's leaders.

Like the Americans, the Soviets began their quest for a ballistic missile in the years directly after the end of what they called The Great Patriotic War (1939–45). The design of their R-7 ICBM was frozen at an early stage, while the Soviet nuclear warhead was still very large and heavy. Therefore, when the R-7 was employed as a space launch vehicle it was able to carry far more massive payloads than its American counterparts, which had been designed to carry smaller and lighter warheads employing miniaturised electronics. It was this imbalance in ICBM throw weight that allowed the Soviets to place much larger satellites and spacecraft into orbit than the Americans in the first decade of space exploration. Korolev's design bureau (OKB-1) developed the R-7 ICBM before petitioning Khrushchev's government for permission to launch a satellite. At the same time that permission was given for the satellite, OKB-1 was also charged with designing and launching a series of interplanetary probes and a vehicle capable of carrying a human passenger, a cosmonaut, into orbit.

When the development of the proposed scientific satellite fell behind schedule OKB-1 developed and launched the object that the Soviets called *Sputnik*. The following month Sputnik II was launched, with the canine bitch Laika in a pressurised compartment. One year after the first R-7 flight the Soviets launched the scientific satellite that Sputnik had replaced. Today that satellite is identified as Sputnik III.

September 1958: A new third stage for the R-7 was initially used to launch unmanned probes to the Moon. Between September and December 1958 four attempts to launch Luna probes failed to reach Earth orbit.

1 October 1958: NACA ceased to exist, but formed the nucleus of NASA, a civilian organisation that purchased its hardware from the private sector, namely, the nation's aeronautical manufacturers. Prime contractors then subcontracted work they could not complete themselves to companies with the relevant expertise. Hardware was delivered to NASA for pre-launch testing before the contractors themselves prepared their machines for launch. Throughout each flight contractors maintained a group of experts that were available to answer any questions that NASA might have relating to their hardware.

NACA had carried out aeronautical research in reply to demands from the US military and their aeronautical contractors. NASA was charged with both developing and flying the unmanned satellites and manned spacecraft under their charge. The engineers, who had previously concentrated solely on their engineering capabilities, now had to be involved in all aspects of the space programme, including management, dealing with contractors, as well as overseeing the development, testing, launch, flight, and recovery of their vehicles. NASA also had to establish a working relationship with the military services. In the early years the military provided the majority of NASA's launch vehicles. They also provided physicians and medical staff to oversee the health of the Flight A astronauts. Each arm of the military services provided NASA with a Liaison Officer as a single point of contact.

5 November 1958: Some of LaFRC's PARD engineers were formed into the Manned Ballistic Satellite Space Task Group, a semi-autonomous group within the LaFRC organisation. The name was subsequently shortened to the Space Task Group (STG). The 33 men in the original STG were under the control of PARD

Director Robert Gilruth. Silverstein added a further 15 men from LFRC to STG in July 1959.

7 November 1958: The JPL suggested that NASA commence a programme of unmanned probes to Venus and Mars. This was the beginning of a programme of lunar and interplanetary exploration that continues to this day. JPL was transferred to NASA on 5 December 1958.

November 1958: In the Soviet Union, OKB-1 began the design of a manned spacecraft in late 1958. The Council of Chief Designers approved the project in November 1959. The 'Vostok' spacecraft consisted of a spherical Re-entry Module in which the cosmonaut, dressed in a pressure suit, sat on an aircraft style ejection seat. The Re-entry Module was attached to a Retrograde Module that contained the main air supply and the rocket motor required to slow the spacecraft down at the end of its flight and allow gravitational attraction to pull it out of orbit. Following retrofire the metal straps holding the two modules together were released, along with a separate electrical umbilical connector. The Re-entry Module entered the atmosphere and adopted the correct attitude as a result of gravitational attraction acting on its carefully positioned mass. Following re-entry the astronaut was ejected from the spacecraft and landed separately under a personal parachute. The Re-entry Module then deployed its own parachute and landed nearby. All manned Vostok recoveries took place on land, within the Soviet Union.

14 November 1958: NASA requested DX rating, the nation's high procurement priority, for its manned satellite vehicle and the F-1 rocket motor. The question was deferred to the 3 December meeting of the Space Council. At that time the manned satellite vehicle was awarded the DX rating but the F-1 remained on the lower DO rating. When Dryden objected to the latter decision both programmes remained on the lower rating.

26 November 1958: Glennan and Dryden selected the name Project Mercury for the manned satellite programme. The name, which had been suggested by Silverstein, was announced to the public on 17 December. In the same month a Senate Select Committee proposed that Project Mercury be followed by a manned circumlunar flight, with an orbital station as an alternative. Glennan accepted their recommendations. The year ended with President Eisenhower demanding that the astronauts selected to fly the Mercury Spacecraft should be selected from the nation's military test pilots. NASA HQ and STG staff redefined the criteria by which the astronauts were selected. At the same time they were also busy assessing the bids received for the Mercury Spacecraft contract. McDonnell Aircraft Corporation was named as prime contractor for the spacecraft on 12 January 1959.

2 January 1959: Luna I was launched, but passed the Moon at a distance between 5,000 and 6,000 km.

March 1959: A second request for DX rating for the Mercury Spacecraft and the F-1 rocket motor resulted in the approval of the same ratings as in December 1958. This time Dryden accepted the decision. In April President Eisenhower approved the DX rating for the Mercury Spacecraft.

9 April 195l: The seven Flight A astronauts were introduced at a press conference held at NASA HQ. April also saw the assignment of John 'Shorty' Powers, a Colonel in the USAF, as Public Relations Officer for Project Mercury. During the same month STG gained 25 ex-employees of the Canadian company AVRO. Among them were a number of engineers who were deigned to make their names in STG's Operations Division and in the design of future manned spacecraft.

17 April 1959: With continuing delays in the procurement of materials Gilruth informed the World Congress of Flight that, *'The first manned orbital flight will not take place within the next two years'.* Looking beyond Project Mercury, NASA requested an additional $3 million to research orbital rendezvous and docking. Harry Goett, from Ames, was placed in charge of a committee to define the future of American manned space flight.

May 1959: Silverstein announced that NASA's new field centre in Maryland would be named the Robert H. Goddard Space Flight Centre. Gilruth's STG would be based there and it would become the heart of NASA's manned space flight programme. GSFC would be the home of Project Mercury's World-Wide Tracking Network's computer system. In the meantime STG remained based at LaFRC.

July 1959: NASA HQ authorised STG to employ a further 100 people. At the same time STG was reorganised into three divisions. Maxime Faget headed the Flight Systems Division, while the Operations Division came under Charles Mathews, and the Engineering and Contract Division under James Chamberlin.

23 July 1959: NASA gave preference to unmanned lunar flights over unmanned interplanetary flights to Venus and Mars. Flights to the Moon were possible every month, while flights to Venus and Mars were only possible every two years.

27 July 1959: The Titan ICBM was deleted as a possible upper stage for ABMA's Saturn launch vehicle. A new S-IV upper stage would be developed. On 30 July the prime contract for the S-IV was awarded for the World-Wide Tracking Network.

12 August 1959: STG's Project Panel met on 12 August to define long-range goals for America's manned space programme. The result of this and a second meeting six days later was a call for the development of a three-man spacecraft capable of Earth orbital flights and journeys to the Moon. The panel suggested 1970 as a realistic date to attempt the first manned lunar landing if the new spacecraft was developed. During the month Robert Gilruth's title was changed to Director of Project Mercury.

19.2 MERCURY LITTLE JOE FLIGHT PROGRAMME

The solid propellant Little Joe launch vehicle was launched from the NASA facility at Wallops Island. The first series of Mercury Little Joe flights used prototype spacecraft manufactured in two parts, by LFRC and LaFRC. A production LES was used on all Little Joe flights. This combination was used to qualify the in-flight abort procedure under conditions similar to those expected during a Mercury Atlas abort at the moment of maximum aerodynamic pressure on the exterior of the ascending vehicle, known to engineers as max-*q*. When the launch vehicle's propulsion system shut down, the abort sequence was activated as if the Little Joe had suffered a propulsion failure in flight. The Marman Clamp securing the spacecraft to the launch vehicle was released and LES's Grand Central rocket motor ignited to pull the spacecraft clear of the inert Little Joe. Following burnout of the Grand Central rocket motor, the Marman Clamp holding the LES to the neck of the spacecraft was released and the Jettison Motor fired to pull the LES away from the spacecraft. The spacecraft then followed a ballistic trajectory towards the water below, deploying its ELS and recovery aids.

When the primary objective had been met with prototype spacecraft the test series was repeated using production Mercury Spacecraft. None of the spacecraft used in the Mercury Little Joe programme was fitted with a landing bag. As part of the second series of Mercury Little Joe flights a single Beach Abort was flown. This called for a production LES to pull a production Mercury Spacecraft into the air from a standing start on a wheeled dolly. This was done to simulate an abort from a launch vehicle standing on the launch pad, before first movement. Two of the second series Mercury Little Joe flights carried Rhesus monkeys. None of the vehicles carried human astronauts.

19.3 LITTLE JOE 1 (LJ-1)

Five months after the Little Joe contract was awarded, North American Aviation Corporation delivered the first airframe to NASA. The airframe for LJ-1 contained four Pollux and four Recruit rocket motors. On arrival a team of LaFRC and PARD engineers checked the airframe before it was erected on its launcher at Launch Area 1 (LA-1), Wallops Island Flight Test Range. LA-1 was little more than a slab of concrete with suitable consumables available to support the launch. Sitting vertically on its launcher LJ-1 was surrounded by a scaffold work structure from which the engineers could conduct pre-launch testing. Having experienced premature rocket launches in the past, PARD engineers had designed LJ-1 so that all self-contained systems had their own chemical batteries. Individual batteries were charged only for tests requiring that system. The batteries were not all charged at the same time prior to the day of launch. It was hoped that this would prevent Little Joe launches suffering from out-of-sequence rocket firings. (See Fig. 57.)

The LJ-1 flight plan was designed to prove the Mercury LES could operate and pull the spacecraft clear of a catastrophic failure of an Atlas launch vehicle at max-*q*.

Fig. 57. Little Joe 1 on the launch stand at Wallops Island. (NASA)

The flight was to be controlled by the spacecraft's own onboard timer. The same timer also contained a rapid abort and launch vehicle destruct sequence, which would save the spacecraft if the LJ-1 airframe suffered a real in-flight failure that required the Range Safety Officer to destroy it. A pre-launch test of this system revealed that some events had happened out of sequence. A separate test revealed that one wire from the battery powering the rapid abort sequence had been routed to the solenoid coil that armed the destruct system along a path that led to the power terminal for the rapid abort plug. Engineers questioned this when it was discovered but it was found to be correct in accordance with official drawings of the LJ-1 airframe.

On 20 August 1959 the first prototype Mercury Spacecraft was rolled out to LA-1 on a dolly similar to those used for the earlier beach abort tests. Due to difficulties in producing the prototype spacecraft, this first LJ launch was approximately one month behind schedule. As a result NASA was paying overtime and encouraging their personnel at Wallops Island to work long hours in the run up to the LJ-1 launch.

At LA-1 a production LES was mounted on top of the spacecraft. The combination was then hoisted into position and fastened to the neck of LJ-1. With the mechanical and electrical mating of the separate parts, and their joint testing, complete, the scaffold work structure was removed and the launcher was lowered to its 80-degree launch position, facing out to sea.

On 21 August with just 31 minutes to go until the scheduled launch time, personnel were evacuating LA-1 and the vehicle's batteries were in the process of being charged. At that point, what the post-flight investigation report would call, 'a transient current' flowed down the wire attached to the solenoid coil and triggered the rapid abort sequence. Subsequently, everything happened exactly as it should have, had the Range Safety Officer radioed the signal to destroy LJ-1.

There was a loud explosion and LA-1 was covered in a cloud of thick smoke. Everyone in the area ran for cover, thinking that LJ-1 had launched early, or was about to explode. The Marman Clamp holding the two vehicles together separated and the Escape Motor in the LES fired, pulling the spacecraft clear of the launch vehicle. For one second the LES motor produced 231,296 N of thrust, only stopping when its solid rocket propellant had been consumed. At shutdown the Marman Clamp holding the base of the LES tower to the neck of the spacecraft also separated. That was followed by the ignition of the Jettison Motor, which drove the LES away from the top of the spacecraft. Having reached an apogee of 609.6 m the spacecraft began a ballistic fall back towards the water below. The drogue parachute deployed and opened. At that point the partially charged spacecraft batteries ran out of power and the spacecraft's Antenna Canister failed to jettison. The main parachute did not deploy. The spacecraft fell into the sea, where it was severely damaged on impact and sank. The spacecraft's data recorders had not been activated at the time of the premature launch, so no data was recorded during the 20-second flight. When the smoke cleared LJ-1 was still on the launcher, facing out to sea. The spacecraft, which was recovered by the US Navy the following day, was damaged beyond repair.

Post-flight investigation revealed the transient current as the most possible cause of the previous morning's events. The re-routeing of the single wire between the destruct system solenoid coil and the rapid abort plug was all that was considered necessary to prevent such an event happening again. The engineers preparing the vehicle for launch had questioned the original routeing of that wire but it had been found at that time to be in accordance with official drawings. NASA had been encouraging the engineers preparing the launch to work overtime in order to catch up on a launch schedule that was running late. Even so, the official report lay the blame for the LJ-1 débâcle at the feet of the over-tired engineers for not recognising the potential for such an event when they had questioned the original routeing of the wire.

19.4 MERCURY ATLAS FLIGHT PROGRAMME

The Mercury Atlas combination was used to achieve Project Mercury's goal of manned flights into Earth orbit. The first launch of this combination was called Big

Joe, as opposed to Little Joe, and was intended to prove that the Mercury Atlas configuration was a viable one. No LES was used and the flight employed a prototype spacecraft similar to those carried on the first series of Mercury Little Joe launches. Later Mercury Atlas flights used production spacecraft to qualify the combination for manned orbital flights.

Two minutes after lift-off the two booster rockets on the launch vehicle shut down and were jettisoned. The central Sustainer engine continued to fire. If nothing went wrong the LES was not required. The Marman Clamp holding it to the neck of the spacecraft was released and both motors fired, to pull it clear of the ascending vehicle. When, 2 minutes later, the Atlas Sustainer shut down the Marman Clamp holding the spacecraft to the launch vehicle was released and the spacecraft's posigrade rockets were fired to provide positive separation. The spacecraft was then turned through 180 degrees and manoeuvred to the correct orbital attitude. At the end of the orbital portion of the flight the spacecraft's retrograde rockets were fired against the direction of travel to slow it down sufficiently to allow Earth's gravitational attraction to pull it out of orbit. During re-entry through the atmosphere the spacecraft was protected by its heat shield. Once within the lower atmosphere, the spacecraft deployed its ELS and recovery aids. Of ten Mercury Atlas flights, one carried a chimpanzee and four carried Flight A astronauts on orbital flights.

19.5 BIG JOE 1 (BJ-1)

The principal objective of the first Big Joe flight was to test the new ablative heat shield fitted to the wide end of the Mercury Spacecraft. If BJ-1 failed, Big Joe 2 would be launched in a repeat of the test.

NASA's LFRC and LaFRC were given the task of designing and developing the prototype spacecraft that would fly on BJ-1. That spacecraft was a boilerplate model with a half size instrument compartment, rather than the full size Crew Compartment that the production Mercury Spacecraft would carry. The spacecraft was manufactured in two halves. The lower half was made by LFRC and the upper half was the work of LaFRC. The afterbody was manufactured from thin sheets of corrugated Inconel alloy, using a monocoque construction. Five microphones were fitted inside the spacecraft, along with other instruments. Over 100 thermocouples were fitted throughout the interior and over the exterior of the spacecraft, including behind the heat shield, to measure temperatures during re-entry. Engineers from LFRC had designed a telemetry link to handle data from the instruments.

The flight profile called for BJ-1 to climb vertically before pitching over to the horizontal prior to attaining an apogee of 160.9 km. Following Sustainer Engine Cut-off the Sustainer would manoeuvre to an attitude where the spacecraft was angled slightly below the horizontal. The Marman Clamp holding the spacecraft in place would release and the spacecraft would drift away from the spent Atlas and fire its unique cold thruster system that used high-pressure nitrogen to initiate a 180-degree turnaround and place it in a heat shield forward attitude. Thereafter, gravitational attraction would pull the spacecraft back towards Earth along a ballistic curve. The spacecraft would then re-enter the atmosphere wide-end first. The heat shield was designed to be thick enough that, despite ablation, enough material would survive intact to protect the spacecraft. If the ablation method proved successful, as it had in military missile nose cone tests, it would be used on all future Mercury Spacecraft. When the spacecraft entered the thick lower atmosphere the ELS would deploy and landing would take place in the Atlantic Ocean.

Lift-off was originally planned for 4 July 1959 but was put back to mid-August by USAF personnel when Atlas 10D did not pass inspection on the first occasion. A second launch date was cancelled by STG when their engineers had difficulty overcoming problems in the telemetry system. The launch was finally set for 9 September.

The final countdown for BJ-1 began at midnight on 8 September 1959. It proceeded on schedule until a hold was called at 0530 on 9 September, to investigate a rogue reading on the Boroughs computer that was to guide the launch vehicle in flight. A malfunctioning transistor in the Azusa impact prediction beacon – a transponder in the Atlas – had caused the reading. As there were a number of redundant systems for making the landing point prediction the fault was ignored and the countdown resumed after a hold lasting 19 minutes. Everything then proceeded smoothly until launch at 0819 (all American launch times are shown in Eastern Time). The Atlas ignited on schedule and climbed away from LC-14 (see Fig. 58). All appeared well as the Atlas's three exhaust plumes lit up the still dark sky and a thunderous crackling roar filled the night air. For the first 2 minutes Atlas 10D performed by the book.

At T + 2 min the two booster engines should have shut down and the booster fairing should have been pyrotechnically released and jettisoned from the base of the launch vehicle, leaving the central Sustainer engine and the two vernier engines still burning. Booster Engine Cut-Off (BECO) occurred on time, but the booster fairing did not jettison. The Sustainer engine and the two verniers continued to fire, but their performance was degraded by the dead weight of the now inert booster section. The resulting trajectory was both steeper and lower than had been planned for the flight. The launch vehicle's two vernier engines finally shut down, at T + 5 min. The final velocity obtained was 91.44 m/s below the intended velocity. The computer predicted an impact point some 804.5 km short of the original target. As a result, the spacecraft was placed into a steeper re-entry trajectory than originally planned.

The Marman Clamp fired and the spacecraft separated physically from the Atlas Sustainer. Telemetry received on the ground showed that the spacecraft's thrusters were firing, but could not confirm that the spacecraft had turned through 180 degrees and assumed the correct re-entry attitude. Ionised air that built up in front of the spacecraft as it re-entered the atmosphere caused a radio blackout and the spacecraft's attitude remained unconfirmed.

In space, the physical link between the two vehicles had been severed but, with no posigrade rockets to provide a positive separation force in the microgravity environment, the spacecraft did not drift away from the launch vehicle, but remained in place on the top of the adapter skirt. The spacecraft's onboard programmer had

Fig. 58. Big Joe at Launch Complex 14 Eastern Test Range, Florida. (NASA)

fired the cold nitrogen thrusters in an attempt to turn the spacecraft to the correct re-entry attitude. The entire supply of nitrogen was depleted without having any effect whatsoever. The flight peaked at an apogee of 152.8 km.

The spacecraft finally separated from the launch vehicle 2 minutes 18 seconds after the Marman Clamp had fired. At that time the combination was at an altitude of 105.15 km. The spacecraft was now a free-falling object which was forced to rely on Maxime Faget's design principles and, in particular, the location of the centre of gravity and the Destabilisation Flap mounted on the narrow end of the Antenna Canister to turn it around to the correct re-entry attitude before it hit the thick layers of the lower atmosphere. Re-entry heating proved to be shorter and hotter than planned, proving that the ablation heat shield was up to the task.

The spacecraft's ELS performed as planned. The drogue parachute was ejected and deployed. It pulled the antenna canister away from the neck of the spacecraft and that in turn dragged the main parachute from its canister, in a reefed condition. When the reefing lines were cut the main parachute deployed and lowered the spacecraft to the ocean below. Ships and aircraft of the recovery fleet picked up the underwater explosion of the spacecraft's Sofar bomb, which indicated that it had deployed its recovery aids. Landing occurred 2,407.06 km from the launch pad, after a flight lasting 13 minutes.

The crew of a US Navy aircraft tracking the spacecraft's Sarah Beacon reported that they had visually located the spacecraft, floating upright in the Atlantic Ocean. The green marker dye released into the water after landing had helped the airmen to locate the small black object as it rode up and down in the swells. Circling above the spacecraft, the aircraft's crew guided the remainder of the recovery fleet to their location. Seven hours after the spacecraft had first been located, the destroyer USS *Strong* reported that they had recovered it.

Following a 12-hour journey back to Cape Canaveral the STG engineers in Hangar S examined their spacecraft. The ablative heat shield had performed perfectly. Post-flight-tests showed that only 0.33 of the heat shield's thickness had ablated away, which resulted in considerable weight saving on later Mercury Spacecraft when their heat shield were made only 0.5 the thickness of the one flown on BJ-1. The shingles on the spacecraft's afterbody were hardly marked. A strip of masking tape deliberately left on the afterbody was still in place. The engineers separated the two portions of the spacecraft and retrieved a letter of congratulations on a successful flight that had flown aboard the spacecraft. They handed the letter to Robert Gilruth, the head of STG.

The BJ flight led to a number of design changes that were incorporated on future Mercury Spacecraft. Every spacecraft would now fly with an ablative heat shield manufactured using the same methods as the one used on the BJ spacecraft. The beryllium heat shield programme was cancelled. Shingles on the spacecraft's afterbody were thickened as a result of the temperature readings recorded during re-entry on this flight. Beryllium shingles would be added to the spacecraft neck.

The true story of what had happened to BJ -1 only became apparent when the flight data had been reviewed, at the end of the first week following the launch. While STG engineers were pleased with the performance of their spacecraft, USAF and Convair personnel were left with the task of discovering why their launch vehicle had failed to stage.

Big Joe 2 was cancelled three weeks after the BJ-1 flight.

9 September 1959: JPL released details of standard bus that could be used as a lunar or interplanetary spacecraft. The design would form the basis of the Ranger lunar impact spacecraft and the early Mariner Venus and Mars flyby probes.

12 September 1959: The Soviet Luna II probe was launched after another failure earlier in the year. Two days later the small probe became the first man-made object to impact the Moon. The world media were informed of the successful flight.

OKB-1 continued work on three projects. A new generation of robot lunar probes was under development. Their weight would require the development of a second, larger third stage, to enable the R-7 to place the probes into Earth Parking Orbit. A small fourth stage would then provide the thrust to achieve escape velocity. Between 1960 and 1962 the new stages failed on eight out of ten attempts to launch interplanetary probes. OKB-1 also continued the quest to place the first human being into orbit, and then to send humans to the Moon, which would require the development of a new heavy lift launch vehicle.

16–18 September 1959: Dryden supported the development of ABMA's Saturn over the USAF's Titan C as a space launch vehicle. His support was dependent on ABMA being transferred to NASA. In his turn, von Braun agreed to support the transfer to NASA only if the Administration would develop Saturn. The DoD placed all military space flight in the hands of the USAF on 23 September. The USAF elected to develop their Titan C over the Saturn that had been transferred from the Army.

18 September 1959: Khrushchev addressed the United Nations on nuclear disarmament.

22 September 1959: China was refused admission to the United Nations.

25 September 1959: Khrushchev visited Communist China.

October 1959: From October 1959 to January 1960 a number of Red Air Force Jet pilots were subjected to a selection process to identify possible Soviet cosmonauts. At the same time a Cosmonaut Training Centre was established in a secure suburb of Moscow, under the command of Colonel General Kamanin. Twenty pilots were selected to begin cosmonaut training in February 1960. They reported for duty in small groups between March and June of that year. Unlike their American counterparts the Soviet cosmonauts remained anonymous until they flew in space, at which time their name was announced to the world and they became instant international heroes. The eight members of this team who did not fly in space remained unidentified in the West until after the collapse of Soviet Communism in the late 1980s.

October 1959: Chamberlin established the STG-McDonnell Co-operation Panel during the month. The panel served as the direct link and negotiating point between STG and McDonnell and came into its own when the changes to the Mercury Spacecraft demanded by the astronauts caused major delays and resulted in the spacecraft becoming the pacing item in Project Mercury.

4 October 1959: Lunik III was launched by the Soviet Union. Three days after launch, the unmanned probe returned the first views of the far side of the Moon, which is always turned away from Earth.

2 November 1959: Gilruth told STG's engineers to begin the development of a three-man circumlunar spacecraft. The following month Glennan accepted the Goett Committee's recommendations for a manned circumlunar flight as NASA's next manned space project after Project Mercury. On 29 July 1960 the new programme would be called Project Apollo. When Vought Aeronautics, Dallas, Texas suggested a spacecraft design that 'staged' like a missile, jettisoning excess equipment as it was no longer required, Gilruth instructed his engineers to develop the Apollo Spacecraft along similar lines. The Apollo Spacecraft would be sent from the launch pad, around the far side of the Moon and back to Earth, without orbiting either body. This trajectory was referred to as the Direct Method. At the same time a number of people both within and outside NASA, including the ABMA engineers at Huntsville, were expressing an interest in developing orbital rendezvous techniques.

19.6 LITTLE JOE 6 (LJ-6)

Programme managers ultimately decided that it had been unrealistic to expect the LJ-1 airframe to qualify the Mercury Atlas abort sequence at max-q on its first flight. It was agreed that the LJ launch vehicle should undergo its own qualification flight, before they tried to repeat the LJ-1 flight. Having been added to the flight schedule after the first five airframes had already been allocated, the new flight was designated LJ-6, despite being the second launch in the programme. The LJ-6 airframe would be launched with a boilerplate Mercury Spacecraft that contained no pressurised compartment, no flight instruments and no ELS. The LES would consist of an empty Grand Central rocket motor casing filled with ballast. There was no Jettison Motor. The flight's sole objective was to prove that Little Joe could perform the boost phase of the flight profile without malfunctioning.

LJ-6 was erected in the launcher at LA-1 Wallops Island and prepared for launch. The wire to the solenoid coil that armed the rapid abort plug had been re-routed since the LJ-1 flight. On 3 October 1959, the inert LES was mated to the neck of the boilerplate spacecraft and the combination was mounted on the LJ airframe. The launcher was then lowered to its 80-degree launch position.

Lift-off occurred the following day, with four Recruit and two Pollux motors firing in combination. The second pair of Pollux motors was due to be ignited at T + 24.5 s. Hot gases in the base of the airframe melted the polystyrene nozzle pressure seals in the second pair of Pollux motors and ignited their solid rocket propellant prematurely. The first motor ignited at T+ 9 s and the second at T+18 s.

At an altitude of 59.92 km the last Pollux motor shut down and the combination began to tumble as it fell back towards the ocean. At T+2 min 50 s the Range Safety Officer sent the command to safe LJ-6. A charge in the head cap of each rocket motor was exploded, releasing the pressure in the rocket casings. The combination hit the water 127.75 km downrange from the launcher and was destroyed on impact, after a flight lasting 5 minutes 10 seconds. No recovery was attempted.

The joint STG–PARD team returned their attention to qualifying the Mercury abort sequence under Mercury Atlas max-*q* conditions.

19.7 LITTLE JOE 1A (LJ-1A)

Little Joe 1A was a repeat of the LJ-1 flight that had ended in fiasco on 21 August 1958. The LJ-1A airframe, which was delivered to Wallops Island on 28 May 1959, contained four Recruit motors and two Pollux motors. Within a few days the prototype spacecraft and the production LES had also arrived at Wallops Island. Following post-delivery checks LJ-1A was installed in the launcher in a vertical position and months of preparatory work began. The LES was mated to the prototype spacecraft on 3 November, after which the spacecraft was mated to LJ-1A. With all pre-launch checks complete the launcher was lowered to its launch position.

LJ-1A climbed away from the launcher on 4 November and arched out over the water. The six rocket motors burned out and the abort sequence was activated at max-*q*. The Marman Clamp holding the spacecraft in place was released and the Escape Motor in the LES fired at T+30 s, but took a further 12.7 seconds to achieve full thrust. During those 12.7 seconds the combination continued to gain altitude under the momentum it had built up during the boost phase. The drop off in forward velocity after LJ-1A's rocket motors shut down also meant that the air pressure acting on the prototype spacecraft was reduced. As a result, by the time the LES pulled the prototype spacecraft clear of the LJ-1A airframe the combination had passed through max-*q* and was only subject to 0.10 of maximum aerodynamic pressure for this flight.

At burnout the Marman Clamp holding the LES to the neck of the spacecraft fired and the LES Jettison Motor ignited to pull the now useless tower clear of the spacecraft. The now inert LES fell into the water below. The spacecraft reached an apogee of 14.48 km before beginning its own fall towards the water below. Barostats deployed the drogue parachute, which pulled the Antenna Canister clear and deployed the main parachute. Splashdown occurred 18.5 km downrange from the launcher, after a flight lasting 8 minutes 11 seconds. The spacecraft was recovered by the salvage vessel USS *Opportune* and returned to Wallops Island. It would be refurbished and re-used in the final unmanned Mercury Redstone flight – Mercury Redstone Booster Development.

LJ-1A had failed to meet its primary objective and would have to be repeated.

19.8 LITTLE JOE 2 (LJ-2)

While engineers worked on correcting the problems experienced on LJ-1A, it was decided to continue with the other objective of the LJ programme. LJ-2 was intended to qualify the LES under high-altitude, low-pressure conditions. The flight saw the first to use the new Castor solid rocket motors developed by Thiokol Chemical Company.

In May 1959, STG had commenced negotiation with the United States School of Aviation Medicine, for the provision of biological payloads to be carried on LJ flights. Despite the failure of two previous LJ flights to qualify the hardware for a max-*q* abort, STG agreed to allow the School of Aviation Medicine to provide a biological payload for the LJ-2 launch.

That biopackage consisted of an open cylindrical framework containing avionics as well as barley seeds, tissue cultures, rat nerve cells, neurospora, insect eggs and insect larvae. A sealed, pressurised container held a Rhesus monkey in a small version of the Mercury astronaut's prone couch. A sheet of Airtex-like material was attached over the front of the couch, to keep the monkey in place. A camera recorded the monkey's reactions throughout the flight by filming its face, which protruded through a leather bound hole in the Airtex cover. The monkey's arms were also free to move. The monkey was expected to perform simple tasks throughout the flight, in an attempt to prove that a human astronaut could perform monitoring tasks during space flight. When a light illuminated in front of the monkey's face the animal was to be given several seconds to extinguish it by pulling a lever. If the light were still on at the end of the allotted time the monkey received a mild electric shock to the soles of its feet. The animals were trained to perform these tasks in a laboratory before moving on to performing them while riding centrifuges to simulate the LJ's gravity gradient during launch. In the prototype spacecraft the monkey occupied the same 'back to the heat shield', 'knees drawn up' attitude that the human astronauts would in the production spacecraft. An 11-channel oscillograph recorded the monkey's physiological experiences throughout the flight. If all went as planned the little monkey would experience 4 minutes of microgravity at the apogee of the trajectory.

The LJ-2 launch vehicle, its prototype spacecraft and production LES were delivered to Wallops Island and checked out for launch. The spacecraft and LES were mated and then mounted on the launch vehicle on 3 December 1959.

The following morning one of the group of Rhesus monkeys under training was selected for the flight and secured in the bio package. The animal was given the name SAM, after the initials of the School of Aviation Medicine. The bio package was taken out to LA-1 and secured inside the prototype spacecraft. LJ-2 was then lowered to the correct launch angle. This flight, with a monkey onboard, was the first flight of a biological specimen on Project Mercury hardware. It attracted nationwide media attention within America, which concentrated almost solely on the passenger, SAM. Flight A astronauts Al Shepard and Gus Grissom were present at Wallops Island to watch the launch of LJ-2. (See Fig. 59.)

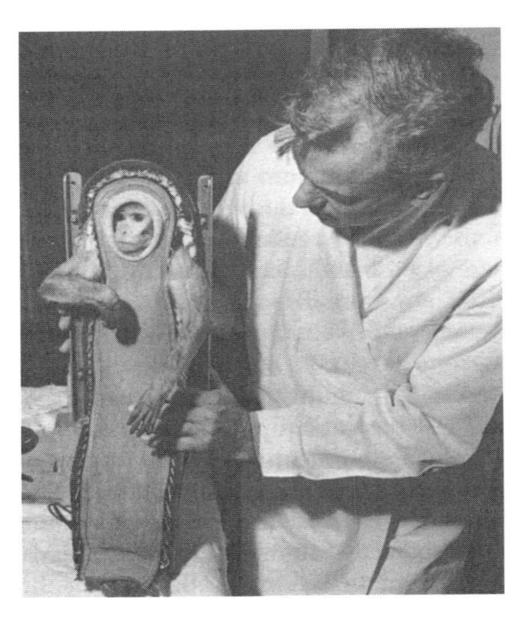

Fig. 59. An aero-medical handler poses SAM for the camera in his prone couch. (NASA)

LJ-2 lifted off at 1115 on 4 December, with four Recruit and two Castor rocket motors burning. The second pair of Castor motors fired 0.5 second before the first pair burned out. The final pair of motors burned out at an altitude of 28.9 km having imparted a forward velocity of Mach 5.7 to the airframe. The Marman Clamp at the base of the prototype spacecraft fired and the Escape Motor in the LES ignited and burned for one second, during which the combination climbed to 85.3 km. SAM was subjected to acceleration forces of 14.8 *g*. At burnout the Marman Clamp at the base of the LES fired and the Jettison Motor performed its task without fault. Due to a pre-launch miscalculation of the wind speed, LJ-2's apogee was 30.4 km lower than planned. As a result, SAM was only subjected to 3 minutes of microgravity.

As the prototype spacecraft fell back towards the water its thrusters fired in an attempt to stop the build up the oscillations that were caused by the spacecraft's aerodynamic instability. At 20.7 km the oscillations were some 25 degrees in each direction. By the time the spacecraft had fallen to 10.2 km the oscillations were as much as 100 degrees in each direction.

The spacecraft's drogue parachute deployed at 6,019 m and reduced the oscillation rate considerably. The main parachute was deployed by a barostat activating its mortar, at 3,048 m. With no landing bag, the prototype spacecraft made a heavy landing in the water off Wallops Island, after a flight lasting 11 minutes 6 seconds. On board recovery aids led the aircrew searching for the spacecraft to locate it easily and the Destroyer USS *Borrie* picked it up 1 hour 45 minutes after launch. To everyone's relief, SAM was found to be alive and well.

8 December 1959: NASA Headquarters reorganised. The Office of Space Flight Development became the Office of Space Flight Programs. It continued to operate with three divisions, Advanced Technology, Space Science and Space Flight Operations. The Office of Launch Vehicle Programs replaced the old Propulsion Division. This new office operated in three divisions of its own, Propulsion, Launch Vehicles and Launch Operations. In January 1960, STG followed suit and reorganised. Faget's and Mathews' divisions both remained unchanged. Chamberlin's division was renamed the Operations Division and acquired the new Launch Operations Division to oversee Project Mercury launch operations at the AMR, Cape Canaveral, Florida. In May 1960 the Office of Technical Information and Educational Programs was established at NASA HQ.

19.9 VOSTOK TEST PROGRAMME

The Soviet Vostok Spacecraft was launched by a three-stage version of the R-7 ICBM. At launch the four boosters (first stage) and central core (second stage) of the R-7 all ignited at once. After 2 minutes the propellant supply in the boosters was depleted and their rocket motors shut down. The boosters were then jettisoned to fall back to the desert below. Thirty seconds later, the vehicle left the sensible portion of the atmosphere and the aerodynamic shroud covering the spacecraft split in two and fell way. The second stage continued to thrust for 3 minutes after booster separation before shutting down and separating. It re-entered the atmosphere over the Pacific Ocean. The third stage then fired for a further 3 minutes 15 seconds to place itself and the spacecraft into orbit.

The spacecraft separated from the rocket stage and oriented itself in orbit. Radiator fins deployed on the Retrograde Module, which held the gas bottles that contained the oxygen and nitrogen that was mixed to provide the air that the cosmonaut breathed. The Retrograde Module also contained the retrograde rocket motor that would be used at the end of the flight. The biological payload, or human cosmonaut, was only a passenger in the Vostok Spacecraft. All flight sequences were automatic, controlled by the spacecraft's onboard timer. Like their American counterparts the Vostok cosmonauts monitored the spacecraft systems and made observations regarding the environment in which they found themselves.

At the end of the orbital flight an infrared sensor aligned with Earth's horizon to place the spacecraft in the correct attitude for retrofire. In the event that the rocket motor failed to fire at the correct time the cosmonaut could enter a three-digit code to allow him to override the automatic system and make the retrofire burn manually. Following the retrofire burn the straps holding the Re-entry Module and Retrograde Module together were released, along with the electrical umbilical and the two modules drifted apart. Gravitational attraction pulled the two modules out of orbit and back into the Earth's atmosphere. Unprotected, the Retrograde Module burned up. Meanwhile, gravity acted on the equipment inside the Re-entry Module to roll it to the correct attitude, so that the cosmonaut was lying with his back to the direction of travel.

The landing loads associated with the Re-entry Module landing on the Kazakhstan Steppes at 10 m/s were thought too much for the cosmonaut to withstand. Therefore, the circular side hatch jettisoned at an altitude of 8,000 m and the cosmonaut was ejected through the opening. Once the cosmonaut was clear, the main parachute deployed and lowered the Re-entry Module to the ground. The cosmonaut separated from his ejection seat and descended to the ground dressed in his pressure suit under a bright orange coverall under his personal parachute. He touched down at a rate of approximately 6 m/s. Landing as they did, inside the Soviet Union, the cosmonaut was frequently met by the local inhabitants before the official military recovery team. When the recovery team did arrive, usually by helicopter, they removed the cosmonaut, the spacecraft and the ejection seat from the landing site.

19.10 VOSTOK SUBORBITAL FLIGHT T-1

In January 1960 an incomplete prototype of the Vostok manned spacecraft was launched from Baikonur Cosmodrome, Kazakhstan. The R-7 launch vehicle with a Luna upper stage launched the prototype on a suborbital flight that reached an apogee of 1,000 km. It travelled 10,000 km downrange. The Re-entry Module landed in the Pacific Ocean approximately 1,700 km south-east of Hawaii, as planned, and was recovered by the Soviet Navy.

19.11 LITTLE JOE 1B (LJ-1B)

The third attempt to qualify the abort sequence under Mercury Atlas max-*q* conditions carried the second and last bio package in the Little Joe programme. The female Rhesus monkey carried on LJ-1B was called Miss SAM (Fig. 60). As with the LJ-2 flight, STG and NASA public relations personnel tried to play down the emphasis on Miss SAM and concentrate media attention of the max-*q* abort. They failed miserably once again and almost all media coverage concentrated on the little monkey.

Following delivery and pre-flight checkout at Wallops Island the prototype spacecraft and production LES were mated to LJ-1B on 20 January 1960. The following morning Miss SAM was mounted in her couch and the bio package was installed in the spacecraft. LJ-1B was lowered to the launch position.

Lift-off occurred at 0928 on 21 January, with LJ-1B's four Recruit and two Pollux motors burning as planned. The remaining two Pollux motors ignited in flight as planned. At final shutdown the combination was 14.48 km high and subject to an aerodynamic pressure of 72,231 g/cm^2. The Marman Clamp at the base of the prototype spacecraft released and the Escape Motor in the LES fired, adding a further 76.2 m to the apogee. One second after ignition the Escape Motor shut down and the Marman Clamp at the base of the LES fired. The Jettison Motor ignited to pull the LES clear of the prototype spacecraft.

Fig. 60. Miss SAM is prepared for flight. (NASA)

Miss SAM had performed tasks inside her container whereby she had to pull a lever within a set amount of time after a light came on in front of her face. At the moment the Escape Motor had fired she stopped performing for 30 seconds. Thereafter, she resumed her tasks with the light and lever and performed well throughout the decent and landing.

The monkey's failure to perform at that vital time led some physicians to predict that human astronauts would also be incapable of performing vital tasks such as manually deploying their spacecraft's parachutes in the event of a malfunction in the automatic system after a Mercury Atlas max-q abort. Noise levels recorded within the prototype spacecraft also led to concerns of the astronaut's ability to withstand it and still communicate clearly with the WWTN.

Both parachutes deployed as planned and the prototype spacecraft splashed down 18.82 km downrange from the launcher, after a flight lasting 8 minutes 35 seconds. A US Marine Corps helicopter plucked it from the water and returned it to Wallops Island 45 minutes after launch. The Mercury Atlas max-q abort sequence had finally been proven, on the third attempt. Following the success of LJ-1B, STG had used four of its six available Little Joe airframes. A seventh airframe remained at NAA's factory, where it was used for testing. NAA were instructed to bring the seventh airframe up to flight standard and prepare it to support an additional flight test.

With the two in-flight abort sequences already proven using prototype spacecraft, STG progressed on to flights carrying production spacecraft. The new flight test programme would consist of three Little Joe flights, to simulate in-flight aborts and one Beach Abort, to simulate an abort from either a Redstone or Atlas launch vehicle standing on the launch complex.

19.12 VOSTOK SUBORBITAL FLIGHT 2

On 30 January 1960 the Soviets launched a second Vostok flight-test. The spacecraft separated from the third stage as it approached apogee. The Re-entry Module separated from the Retrograde Module and gravitational attraction pulled it back towards the ground. Details of this flight remain extremely limited and differing versions of what happened are in the public domain. One version claims that this was a second suborbital flight-test prior to the commencement of unmanned orbital testing. The alternative theory claims that this was an attempted orbital flight-test that failed to achieve the required final cut-off velocity, which resulted in the spacecraft following a suborbital trajectory.

2 February 1960: America's Negroes commenced a series of sit-ins at whites-only lunch counters in shops and diners. They were demonstrating in favour of Civil Rights and non-segregation.

17 February 1960: American Negro Civil Rights leader Dr Martin Luther King was arrested.

12 February 1960: The Operations Co-ordination Group was established at Cape Canaveral with Kraft at its head (see Fig. 61). He oversaw the development and testing of the WWTN as well as defining the procedures required to fly and recover Mercury flights. This group was part of the STG Operations Division and would be responsible for manning the tracking network and MCC.

April 1960: Douglas Aircraft were named as prime contractor for the S-IV upper stage for the Saturn C-1 launch vehicle.

Fig. 61. Christopher Kraft, Flight Director throughout Project Mercury. This photograph was taken during Project Apollo. (NASA)

Fig. 62. A Beach Abort spacecraft waits on its dolly at Wallops Island. (NASA)

19.13 BEACH ABORT (BA)

The first production standard Mercury Spacecraft was delivered to NASA on 1 April 1960, less than one year after the prime contract had been awarded to McDonnell. To mark the change from boilerplate and prototype Spacecraft to production Spacecraft only STG personnel handled Spacecraft 1 (SC-1) during its pre-flight preparations at Wallops Island. Following post-delivery checks SC-1 was prepared for a Beach Abort test. The Beach Abort was similar to the original LES flight-tests, in that SC-1 was mounted on a dolly and wheeled into position at LA-1. There the LES was mounted on the neck of the spacecraft. The dolly was then jacked up, so that the combination angled slightly towards the sea. (See Fig. 62.)

SC-1 did not carry most of its internal subsystems, but it was carefully weighted so as to be representative of a manned Mercury Spacecraft. In that way the data collected from the flight was directly transferable to later manned flights. Inside the SC-1 some electrical cables had been accidentally fitted in reverse and would cause the telemetry transmitters to perform poorly throughout the flight.

On 9 May 1960 the radio command was sent to ignite the Escape Motor on the LES. The motor fired as planned and pulled SC-1 into a cloudy sky, between three tongues of flame and a plume of dark red smoke. At shutdown, after a 1 second burn, the combination was 751.33 m high. The Marman Clamp holding the LES in place fired and was followed by the ignition of the LES Jettison Motor. The LES moved clear of the spacecraft and following the shut down of the Jettison Motor, fell into the sea. The exhaust plume from the Jettison Motor had impinged on the LES tower structure, thereby degrading its performance to 0.42 of the desired thrust. This resulted in considerably shorter than the planned separation distance between the LES and SC-1 when the Jettison Motor shut down. To overcome the problem the single Jettison Motor exhaust nozzle was replaced on all future LES by three nozzles canted at 30 degrees to the vertical and spaced 120 degrees apart. This was a similar configuration to that used on the LES Escape Motor.

The Spacecraft oriented itself correctly for the descent and began the fall back towards the water. The drogue and main parachutes both performed as planned and SC-1 landed 1.6 km downrange, after a flight lasting 1 minutes 16 seconds. SC-1 was recovered by a Marine Corps helicopter 17 minutes later and returned to Wallops Island, where it was found to be in excellent condition.

1 May 1960: President Eisenhower's relations with Premier Khrushchev reached a new low on this date, when the Soviet Union claimed that it had shot down an American U-2 reconnaissance aircraft over the Soviet Union. They backed-up their claims by displaying wreckage with English markings and the CIA pilot, Francis Gary Powers, on Soviet television. Powers was tried for espionage and indicted on 8 July. He was imprisoned in the Soviet Union. President Eisenhower originally instructed NASA to claim that the U-2 was one of their atmospheric research aircraft. The Administration did as their President requested and issued a press statement to that effect. President Eisenhower later admitted that the downed aircraft was indeed on a reconnaissance flight.

19.14 KORABL SPUTNIK 1

The first unmanned orbital flight-test of the Vostok Spacecraft was launched at 0300 (all Soviet launch times are shown in Moscow Time), 15 May 1960. When the launch was announced to the world the Soviet news agency used the name Korabl Sputnik 1 (Spacecraft Satellite 1). In keeping with an American demand that every object that achieved orbit be identified, the Soviets officially identified the spacecraft as Sputnik IV. The Spacecraft was injected into a 312 × 368-km orbit inclined at 65 degrees to the equator and with a period of 91 minutes. The Re-entry Module contained an inert ejection seat with a human mannequin in a Vostok pressure suit. During the flight, tapes of a Soviet choir were played to test the communication system.

On 19 May at the end of 64 orbits, the spacecraft was aligned for retrofire. The infrared sensor that should have aligned with Earth's horizon had failed earlier in the flight. As a result, the spacecraft was misaligned by 180 degrees when retrofire occurred and was boosted into a 290 × 675-km high orbit. The Re-entry Module separated from the Retrograde Module as planned. The two modules continued to orbit Earth for a further five years before their orbit decayed sufficiently to allow the Re-entry Module to re-enter the atmosphere at 0021 MT, 15 October 1965. With no heat shield the Re-entry Module burned up taking its human mannequin with it.

31 May 1960: North American Aviation Rocketdyne Division was named as prime contractor for the J-2 cryogenic rocket motor, to be fitted to the upper stages of a number of different Saturn launch vehicles.

1 July 1960: The ABMA transferred to NASA, taking with it von Braun's group of pioneering rocket engineers from Germany. This transfer took place 18 months after Glennan had originally requested it. The F-1 rocket motor and the heavy lift Saturn CI launch vehicle were transferred to NASA at the same time. Redstone Arsenal, in Huntsville, Alabama was subsequently enlarged and renamed the George C. Marshall Space Flight Centre (MSFC).

15 July 1960: Glennan approved the Mariner interplanetary programme consisting of unmanned fights to Venus and Mars in 1964.

19.15 KORABL SPUTNIK

An unmanned Vostok Spacecraft stood on the launch pad on 28 July 1960. Strapped to the live ejection seat was a pressurised container holding two dogs named Chiaka and Lisichka. At the end of the flight the seat was to be ejected to land under a separate parachute. With all systems apparently performing as planned the countdown proceeded towards launch. When the count reached zero nothing happened. The launch vehicle failed to ignite and did not leave the launch pad. The R-7's propellants were removed and the launch structure was placed back around the launch vehicle. Engineers removed the side hatch from the Re-entry Module and then removed the pressurised container holding the two dogs. The launch attempt was cancelled while the problem was diagnosed.

19.16 MERCURY ATLAS 1 (MA-1)

Following the successful flight of BJ-1 Convair engineers had undertaken a weight reduction programme on the Mercury Atlas D launch vehicle. Atlas 10D (BJ-1) had upper sections constructed from stainless steel approximately 0.50 cm thick, the same areas on Atlas 50D (MA-1) used stainless steel that had been reduced to 0.25 cm. The programme succeeded in reducing the weight of the Mercury Atlas launch vehicle by 45.36 kg. MA-1 would fly a similar suborbital trajectory to BJ-1. The launch vehicle carried no LES and the ASIS was flying in an open loop that would allow it to monitor the Atlas's systems without being connected to the spacecraft.

SC-4 had a production exterior and ablative heat shield with a half size Crew Compartment similar to that used on the BJ-1prototype spacecraft. The instrument unit was built by LFRC and contained two cameras, two tape recorders and a 16-channel telemetry system. More than 50 thermocouples would record both internal and external temperatures throughout the flight. Other instruments would record noise and vibration levels, pitch yaw and roll rates, internal and external pressures and acceleration loads. The spacecraft contained no ECS, no prone couch, and no instrument panel. Its Retrograde Rockets contained an inert paste, but the three Posigrade Rockets were active. The flight plan called for a launch at 0825 with MA-1 climbing to an apogee of 181.4 km, at which altitude SC-4 would be released. The spacecraft would re-enter the atmosphere in a simulated return from orbit. Following a worst possible scenario re-entry, the spacecraft would land in the Atlantic Ocean, 1,107.6 km downrange from the launch pad. The flight would last approximately 16 minutes.

The early morning weather in Florida on 29 July 1960 consisted of heavy cloud and rain. When the rain stopped the decision was taken in MCC to continue with the launch attempt. On LC-14 preparations for launch were picked up following an overnight planned hold in the countdown. Programme managers made the final decision to proceed with the launch at 0725. The gantry was rolled back to its parked position and MA-1 stood ready for launch, but exposed to the rain that had started to fall. (See Fig. 63.)

At T−35 min the first of several minor problems caused the countdown to be held. Difficulties topping off the Atlas liquid oxygen tank, and with receiving some of the launch vehicle's telemetry and also the recurring bad weather ultimately led to a total of 45 minutes during which the countdown remained in the 'hold' condition.

At 0913 the MA-1's three propulsion units and two vernier engines burst into life. Clamps held the vehicle on the launch pad as it made the transition to mainstage. With the three main engines firing at full thrust, the clamps swung back to their parked position and MA-1 climbed away from LC-14. Among the watching crowd the seven Flight A astronauts observed the launch of a vehicle similar to those that would hopefully carry each of them into orbit before Project Mercury was complete. Lift-off was perfect and the ascending launch vehicle passed through the low cloud deck within a few seconds. Observers could hear the Atlas's engines, even though they could no longer see the vehicle.

By T+57.6 s MA-1 was at an altitude of 3,352 m and 9,144 m downrange. The vehicle was passing through max-q when the new thin-skinned liquid oxygen tank conical section collapsed at a point close to where the spacecraft adapter was mated to the launch vehicle. In MCC the reading on the instruments recording the pressure in the launch vehicle's two propellant tanks dropped to zero. The ASIS failed to register the loss of propellant tank pressure, but did register the loss of propulsion thrust and the loss of AC electrical power at the appropriate time. An abort signal was generated but, as the ASIS was not connected to the spacecraft, and as no LES was fitted, it was not acted upon. All Atlas telemetry was lost at T+58.5 s.

The spacecraft separation systems were not due to be activated until T+3 min, so no separation took place when the launch vehicle failed. It is possible that the spacecraft, the adapter section and the crown of the Atlas liquid oxygen tank dome remained attached to each other when the launch vehicle broke up. Observers on the ground, including the astronauts, heard the engines fall silent and then saw chunks of debris pass back through the cloud deck and fall into the ocean below. The spacecraft telemetry system recorded violent motions during the fall back towards the ocean, suggesting that the spacecraft was in one piece. The parachutes did not open and SC-4 was destroyed on impact with the water at T + 3 min 18 s. (See Fig. 64.) Most of the debris from MA-1 fell into water that was only 12 m deep. The official report on the flight concluded that the MA-1 flight-test suffered a catastrophic termination approximately 58.5 seconds after launch by an in-flight failure of an undetermined nature.

James Chamberlin, head of STG's Command Operations Division, became head of a joint NASA–Convair investigation team tasked to discover what had happened to MA-1. With the solid cloud cover at low altitude on 29 July 1960, there was no film of the moment of destruction. The investigation team was forced to base their work on the telemetry recorded during the 3.5-minute flight.

A recovery operation returned pieces of the vehicle to Cape Canaveral. Portions of the spacecraft, the adapter and the upper area of the Atlas liquid oxygen tank were recovered and reconstructed in a hanger. The investigation began by looking at the possibility, held by many engineers, that the Atlas liquid oxygen boil-off valve was badly supported within the liquid oxygen tank dome and may have been fatigued to the point that it failed under the stresses of max-q. Convair's engineers and metallurgists had disproved this theory by early September 1960.

Investigators then concentrated on the fact that Atlas 50D had been constructed after the weight restriction programme and had failed at max-q. On the other hand, Atlas 10D had been constructed before the weight restriction programme and had not failed. The investigators drew the conclusion that the thin skin of Atlas 50D's liquid oxygen tank cone had suffered a catastrophic failure under the stresses of max-q. Chamberlin's group recommended that all future Mercury Atlas launch vehicles be constructed to the same design specifications as Atlas 10D. They also suggested strengthening the spacecraft adapter, by redesigning its stiffeners. The only difficulty with this solution lay with Atlas 67D (MA-2), which was of the thin-skinned design, but was far advanced in its preparations for launch and would cause unacceptable delays to Project Mercury if it was delayed until a thick-skinned Atlas became available.

One solution to the Atlas 67D problem was originally discussed in derogatory terms. It consisted of placing a belly strap, referred to as a 'horse collar', around the liquid oxygen tank dome, where the spacecraft adapter was mated to the launch vehicle. The effect of the 20.3-cm wide horse collar would be to move the stresses of max-q that distance higher up the oxygen dome, where the vehicle was capable of withstanding them without failing. Asbestos insulation was to be packed between the horse collar and the launch vehicle's skin.

On 15 December 1960 a second thin-skinned Atlas failed at max-q, 70 seconds into an attempt to launch a Pioneer Moon probe. NASA Headquarters established another Investigation Committee. The new Committee's report was released on 19

Fig. 63. Mercury Atlas 1 at Launch Complex 14. (NASA)

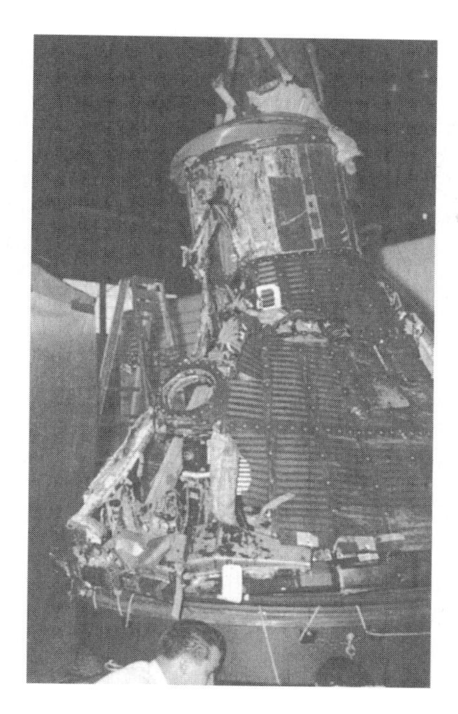

Fig. 64. The reconstructed wreckage of Spacecraft 4 following the loss of Mercury Atlas 1. (NASA)

January 1961 and supported the findings of Chamberlin's MA-1 investigation, including the use of the horse collar on MA-2. The horse collar solution was subjected to a series of wind tunnel tests at NASA–Ames, while the Arnold Engineering Development Centre, Tullahoma, Tennessee, subjected the solution to ground testing, before its final acceptance for MA-2.

7 August 1960: During 1960 President Fidel Castro of Cuba began purchasing his country's oil and military arms from the Soviet Union. This led to America boycotting the purchase of Cuban sugar, the country's main export. In its turn, the boycott resulted in Castro nationalising all American companies in Cuba. Cuba also began to 'export revolution' to other South American countries. The image of Cuba's Che Guevara became an icon to many of the Western nations' disaffected youth.

19.17 KORABL SPUTNIK 2

The Soviet Union's next attempt to launch an unmanned Vostok Spacecraft was made on 19 August 1960. The two pressurised compartments strapped to the ejection seat contained the dogs Strelka and Belka along with two rats, 28 mice and samples of human skin and a number of plants and seeds.

Lift-off occurred at 1138 and Korabl Sputnik 2 entered a 339 × 360-km orbit inclined at 65 degrees. The spacecraft, which the Soviets identified as Sputnik V, separated from the third stage of its launch vehicle. Live television pictures showed that both dogs were distressed during launch and only settled down after the spacecraft entered free-fall. On the fourth orbit Belka vomited, making her the first living creature to suffer from Space Adaptation Syndrome, a form of motion sickness that would affect 50 per cent of future human space travellers.

After 18 orbits the spacecraft was aligned for retrofire. The rocket motor fired as planned against the direction of travel and slowed the Korabl Sputnik 2. The Re-entry Module separated from the Retrograde Module and was pulled out of orbit. At the same time gravitational attraction aligned the spherical module for re-entry. At an altitude of 8,000 m the seat ejected, taking the dogs and the biological samples with it. The Re-entry Module's main parachute deployed shortly afterwards. The ejection seat's parachute deployed and lowered the animals and samples to the ground at a rate of 6 m/s. The Re-entry Module landed at 10 m/s, some 9 km from the intended landing site at 1402 on 20 August.

1 September 1960: Robert Seamans was named as NASA's Associate Administrator, the third most senior post in the Administration. Seamans was a strong supporter of orbital rendezvous and believed that it would be vital to the future of space flight.

Hidden Soviet failures

On 10 October 1960 the Soviet Union launched the first unmanned Mars probe. The R-7 launch vehicle's third stage failed to achieve full thrust. The flight reached an apogee of 120 km and then followed a suborbital trajectory back to the ground. The second Mars probe suffered a similar failure on 14 October.

On 23, October the first R-16 military missile was on the launch pad at Baikonur. When the first-stage electrical system failed engineers were sent out to the launch pad to make the necessary repairs. Further difficulties led to the launch being delayed until the following day. Leaks in a number of propellant lines delayed the new launch attempt. At T − 30 min a fire started which led to a massive explosion. One hundred and sixty-five people died, including Marshal Mistrofan Nedelin, who was in charge of the launch attempt. To this day the events of 24 October 1960 are known in the West as the Nedelin Disaster. None of these failures was announced to the Soviet people, let alone the Western media. Although the R-16 missile was not connected with the manned space programme, the explosion and its horrific after-effects led to the first manned space flight, which was planned for December 1960 at that time, being delayed until some time in 1961.

19 October 1960: America ceased shipments of goods to Cuba.

7 November 1960: For the first time the Soviet Union displayed a number of large military missiles as part of Moscow's Red Square parade.

19.18 LITTLE JOE 5 (LJ-5)

Originally, LJ-5, with SC-3, was scheduled to carry a chimpanzee, in a bio package supplied by the School of Aviation Medicine, through a Mercury Atlas max-q abort profile. Robert Gilruth, the head of the STG, was particularly keen to see the flight completed successfully before manned flights began in Project Mercury. SC-3 was a complete production spacecraft with the exceptions that it would not carry non-essential parts in its communications system and the ASCS was fitted but not active. Water replaced the propellant that the ASCS would carry on later flights.

The flight was originally scheduled for December 1959, but was delayed due to difficulties in the manufacture of the production spacecraft, which resulted in the flight running almost 12 months behind schedule. It was rescheduled for 8 October 1960, but was then pushed back, first to 11 November and then to 16 November. In the intervening period the bio package was deleted from the flight and priority was given to the investigation of the vehicle's in-fight aerodynamics. The bio package would fly on Mercury Redstone 2.

SC-3 was delivered to Wallops Island on 27 September, earlier than had been

Fig. 65. T. Keith Glennan (right), Hugh L. Dryden (left) with R.L. Krieger at Wallops Island. (NASA)

expected, but the launch date was not immediately advanced. Difficulties experienced in preparing a production spacecraft for launch on MR-1 had led the crew at Wallops Island to expect similar problems with SC-3. The preparations at Wallops Island went much more easily than those at Cape Canaveral. (See Fig. 65.)

LJ-5 was delivered to Wallops Island and checked out in the vertical position. With everything going well the launch was brought forward to 7 November. To meet that launch date, the LES was mated to SC-3 on 6 November and the combination was mated to the LJ-5 launch vehicle. The launcher was then lowered to the correct position to support the launch. The 7 November launch attempt was cancelled due to bad weather and a new attempt was scheduled for the following day.

On 8 November preparations began for the launch of LJ-5. The rocket climbed into the sky powered by four Recruit and four Castor rocket motors. For 16 seconds all went as planned, but at T + 16 s the Marman Clamp connecting the spacecraft to the launch vehicle was deflected by the air loads acting upon it until the abort switch was prematurely closed. This resulted in an abort signal being sent and the Escape Motor in the LES firing, despite the fact that the LJ-5's rocket motors were still burning. Despite receiving the abort signal the limit switches on the spacecraft's Marman Clamp failed to operate, possibly due to their having been incorrectly wired. The spacecraft did not separate from the launch vehicle and when the Escape Motor shut down the Jettison Motor fired. The limit switches in the Marman Clamp connecting the LES to the neck of the spacecraft failed to operate, possibly because they too had been wired incorrectly. The LES did not separate from the spacecraft.

Having obtained an apogee of 16.25 km the complete vehicle began to fall back towards the water. It impacted the ocean 21.88 km downrange from the launcher, after a flight lasting 2 minutes 22 seconds. Had a chimpanzee been on board, as originally planned, it would almost certainly have been killed. The salvage vessel USS *Opportune* recovered 60 per cent of the launch vehicle and 40 per cent of the spacecraft. Post-flight investigations were inconclusive, but the reasons given above for the numerous failures were accepted as the most likely course of events. The flight had failed to prove the production spacecraft under max-q conditions and would have to be reflown.

As a result of this flight the spacecraft Marman Clamp was strengthened to ensure that air loads could not deflect it sufficiently to close the abort switch on future flights. The new Marman Clamp underwent extensive rocket sled testing at Holloman USAF Base, San Antonio, Texas.

19.19 AMERICA ELECTS A NEW PRESIDENT

Throughout 1960 Democratic Senators John F. Kennedy and Lyndon B. Johnson fought Republicans Richard M. Nixon and Hubert L. Humphrey in the American Presidential Election. Nixon was Eisenhower's Vice President and was fully supportive of the President's economic policies and the desire to keep America's reaction to Sputnik as low-key as possible. Kennedy, on the other hand, was openly derisive of Eisenhower's policies claiming that he had failed to make a suitable reaction to the threat represented by Sputnik and its military launch vehicle.

The election was fought on many fronts, but time and again Kennedy returned to the perceived 'Missile Gap' and the apparent lead that the Soviet Union had in space exploration as a result of their development of a huge ICBM. The 'Missile Gap' seemed very real to the American public. To the average American civilian the Soviets' large missiles suggested that they had much larger and more powerful nuclear warheads than the Americans did. It was not explained to them that the American military had used new, miniaturised electronics in their vehicles while the Soviets had failed to develop similar technology.

For the first time television was used extensively during the campaign and the two politicians went head to head in a series of televised debates. Kennedy came across better in each of the debates. When, in one of the debates, Nixon commented that America was ahead of the Soviet Union in the development of colour television, Kennedy replied that he would take his television in black and white if it ensured that America led the Soviets in missile technology. Only many years later would the truth become public knowledge – the 'Missile Gap' was a myth. The evidence of this fact came from the one space project to which Eisenhower had never failed to assign top priority – the Project Corona reconnaissance satellite.

With the Project Mercury flight programme already a year behind schedule and launches continually failing to meet their objectives, Senator Kennedy told his followers bluntly, *'We are in a strategic race with the Russians and we have been losing ... If a man orbits the Earth this year his name will be Ivan.'*

Election Day was 8 November. Kennedy won by the slimmest majority in a Presidential Election to that date (Kennedy 334,226,731 votes, Nixon 334,108,157 votes). The new President enjoyed a clear majority in Congress.

President-elect Kennedy had no space policy and even considered dismantling NASA when he came to power. He tasked his scientific adviser Jerome Wiesner to lead a Scientific Advisory Team review of Project Mercury. Wiesner personally favoured the exploration of space by unmanned probes, rather than human astronauts. Predictably, the so-called Wiesner Report was damning in its condemnation of both Project Mercury and the national space launch vehicle programme. Wiesner went so far as to suggest that Kennedy should not *'effectively endorse this program and take the blame for its failures'*. If Project Mercury failed to place the first man in space, or if the astronaut was killed in the attempt, Kennedy could then lay the blame at the feet of the Eisenhower Administration, claiming that Project Mercury had been too far advanced to cancel when he came to office.

11 November 1960: South Vietnamese paratroopers tried but failed to assassinate Ngo Dinh Diem, South Vietnam's President. Diem believed that America was behind the attempt.

19.20 MERCURY REDSTONE FLIGHT PROGRAMME

The launches in this programme were designed to qualify the Mercury Redstone combination for manned suborbital flights. It was originally intended that each of the seven Flight A astronauts would complete a Mercury Redstone suborbital flight before making a second orbital flight on a Mercury Atlas combination. This plan was changed in the wake of early Soviet successes.

When the Redstone launch vehicle shut down at the end of its normal powered phase, the Marman Clamp holding the LES to the neck of the spacecraft was released and both the Escape Motor and the Jettison Motor fired and pushed the LES clear of the spacecraft. On shutting down, the inert LES fell into the ocean below. The Mercury Spacecraft separated from the Redstone and followed a ballistic trajectory, deploying its ELS and recovery aids as it would in manned flight. One flight carried a chimpanzee and the last two Mercury Redstone flights carried human astronauts. (See Fig. 66.)

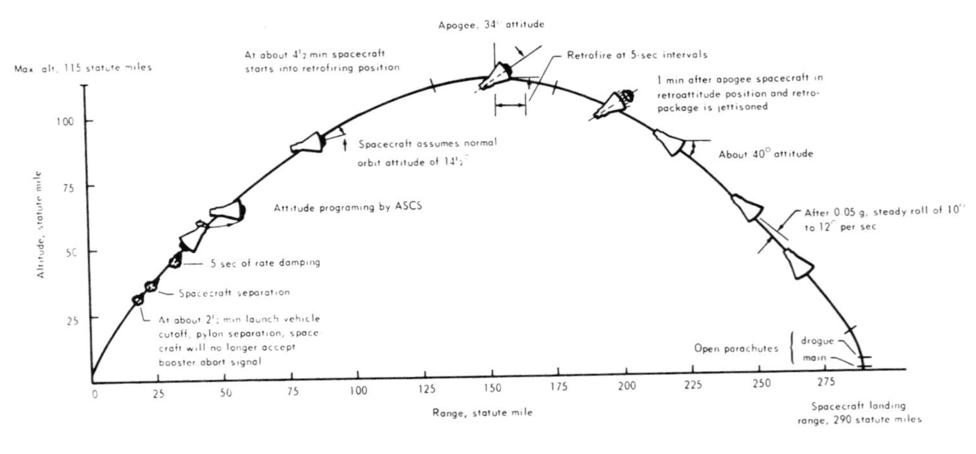

Fig. 66. Spacecraft flight events during a Mercury Redstone suborbital flight. (NASA)

19.21 MERCURY REDSTONE 1 (MR-1)

The Mercury Redstone 1 countdown to launch began at T – 10 h 40 min. The launch vehicle and spacecraft were powered up and all critical systems were checked. It was 6 November 1960 and the countdown proceeded well, until it entered the planned hold at T – 6 h 30 min, when everything stopped for the night.

The count was picked up at 0130 on 7 November. All proceeded well until T – 22 min, at 0738, when the launch was cancelled due to a pressure drop in the spacecraft RCS helium supply. The malfunction required the spacecraft to be de-mated from the Redstone, so that its heat shield could be removed, to allow the helium relief valve, the torodial propellant tank and the hydrogen peroxide tank to be replaced.

Wiring changes were also made to the two Marman Clamps on the spacecraft at the same time, to prevent a LJ-5 style failure. When the work was complete the spacecraft was re-mated with the launch vehicle and retested.

The second countdown commenced on 20 November and proceeded to T − 6 h 30 min, where it was held for the night. The following morning the count was resumed but a problem was discovered with the spacecraft's hydrogen peroxide system and the launch had to be held for one hour. At 0900 the Redstone's single R-7 rocket motor burst into life and MR-1 began to rise away from LC-5 AMR. With the vehicle just one inch above the launch table, two umbilicals connected to the base of fin II fell away. They were the launch vehicle power umbilical and the propulsion control umbilical and they should have fallen away simultaneously. In fact, at some point during an earlier Redstone missile launch from LC-5, not related to Project Mercury, the prongs on the plug at the end of the launch vehicle power umbilical had been filed down, making them slightly shorter than the prongs on the plug at the end of propulsion control umbilical. As a result, the power umbilical had disconnected 21 ms before the control umbilical. That time difference had allowed a 3-amp current from the control plug to pass to the Redstone's engine cut-off relay and its earth diode. If the umbilicals had disconnected simultaneously the current should have earthed harmlessly through the power plug and its umbilical. MR-1's onboard programmer received the nominal engine shutdown signal, as if the vehicle had completed its powered launch phase, and killed the R-7 rocket engine. The launch vehicle settled back on the launch table damaging its fins and creasing the skin of the oxygen tank. (See Fig. 67.)

Following the receipt of a nominal shutdown signal from the Redstone, the Marman Clamp at the neck of the spacecraft released and the LES Jettison Motor fired. The red tower climbed skyward, enveloping the launch complex in thick smoke. One second later the Jettison Motor shut down and, after reaching an apogee of 1.2 km, the LES began its fall earthwards. Observers on the beach dived for cover as the LES buried itself in the sand 365 m from the launch pad.

At the moment of Redstone engine shut down a 10-second timer was activated and ran its course. The spacecraft Marman Clamp should have released at the end of those 10 s, if the vehicle was experiencing less than 0.25*g*. Sensing the fact that the launch vehicle was subject to 1*g*, the signal to the Marman Clamp's limit switches was blocked and the clamp did not release. This was the correct action for the information received.

The barostat in the spacecraft's Antenna Canister correctly sensed that SC-2 was in thick atmosphere, deployed the radar chaff bomb and fired the drogue parachute mortar. The Antenna Canister was separated and pulled clear by the drogue parachute, which deployed and fell alongside the still vertical launch vehicle.

The lanyard attached to the Antenna Canister pulled the primary main parachute from its container. It too fell alongside the launch vehicle, before de-reefing as planned and beginning to fill with the breeze coming in off the ocean. When the strain gauges that were attached to the main parachute's harness registered no loads, the back-up main parachute was also deployed. It, too, fell alongside the launch vehicle, de-reefed and began filling with the sea breeze.

Figs 67. Mercury Redstone 1 rests at Launch Complex 5 while the Launch Escape System fires and the drogue parachute deploys. All actions were correct for the signals received by the launch vehicle and spacecraft's equipment. (NASA)

As the smoke cleared, MR-1 was still standing at LC-5, fully fuelled and surrounded by its parachutes, which were threatening to fill with wind and pull the vehicle over, thereby causing a huge fuel–air explosion. At the same time, engineers and flight managers ruled it unsafe to commence de-tanking the propellants until the vehicle's batteries had run down, rendering the numerous pyrotechnics on board safe. The batteries would not run down until the following morning.

In the launch blockhouse one of the German Redstone engineers who had come to America with von Braun, and now worked with him at MSFC, suggested that he be allowed to use his hunting rifle to shoot holes in the Redstone's liquid oxygen tank. This drastic action would have allowed the cryogenic liquid oxygen to come into contact with ambient air, warm up and turn to gas, thereby negating the possibility of an explosion. The engineer was refused permission to shoot the Redstone. Finally, when it was clear that MR-1 was not going to topple over, the blockhouse was evacuated and guards were placed around the launch complex. The vehicle's batteries were allowed to run down naturally.

Once the batteries were flat and the pyrotechnics rendered safe, the vent valves in the Redstone's liquid oxygen tank opened. Warm ambient air entered the tank and turned the cryogenic liquid to gas, which then escaped through the vent valves. With the danger of an explosion removed, the mobile launch structure was moved back around the launch vehicle. Engineers removed the pyrotechnics and cut away the parachutes before beginning the procedures to de-mate the spacecraft and return both it and its launch vehicle to Hangar S, where the investigation into what had happened would begin. That investigation would establish the above scenario.

In order to prevent a recurrence, all future MR launch vehicles were connected to the launch table by a new 30.5-cm long ground strap that ran between the launch table and fin I on the Redstone. It was protected from the launch vehicle's exhaust at lift-off and required 22.6 kg of pull to separate the plug from the base of the Redstone's fin. In addition, the normal shutdown signal in the MR launch vehicle was inhibited for the first 30 seconds of flight, to prevent re-contact with the launch table and the possibility of falling over during any similar event in the future. All future MR launch vehicles would also have their engine pressure monitored. Only if engine pressure were normal at T + 129.5 s, would the normal shutdown signal be armed. Prior to that time a shutdown of the Redstone's main propulsion system would result in an abort signal being sent to the LES.

19.22 KORABL SPUTNIK III

Korabl Sputnik III was launched at 1026 on 1 December 1960 with dogs Pchelka and Muska strapped to the ejection seat in a pressurised cabin. The initial orbit ranged between 166 × 232 km. Unlike earlier flight-tests of the Vostok Spacecraft, this vehicle was restricted to the low Earth orbit that later manned spacecraft would use. The orbital parameters ensured that the Re-entry Module re-entered the atmosphere before the Crew Compartment's eight days of air supply ran out.

After one day in orbit the spacecraft was aligned for retrofire. Once again the infrared sensor failed to align correctly with Earth's horizon and the spacecraft was in the wrong attitude when the retrograde rocket fired. When the Re-entry Module separated it entered the atmosphere on a trajectory that was too steep and burned up. Both dogs perished. The loss of this spacecraft led to the first manned flight of a Vostok Spacecraft being limited to a single orbit and a demand that it be preceded by two successful unmanned flight-tests of the entire flight programme.

19.23 MERCURY REDSTONE 1A (MR-1A)

Mercury Redstone 1A (Fig. 68) was an attempt to meet the goals of the MR-1 failed launch of 20 November. The countdown commenced on 18 December 1960 and was held overnight at T−6 h 30 min, before resuming the following morning, with launch planned for 1005. This flight re-used SC-2, the spacecraft that had failed to launch the previous month.

On 19 December strong winds in the jet stream led to the countdown being held for 40 min. A solenoid failure in SC-2's hydrogen peroxide system resulted in a further one-hour recycle. At 1145 MR-1A rose from the launch table and climbed into a clear Florida sky. During the launch vibration rates throughout the vehicle

Fig. 68. Mercury Redstone 1A undergoes liquid oxygen loading during a pre-launch flight readiness review. (NASA)

were telemetered to MCC. They reached unacceptably high levels in the Redstone's Adapter Section and Instrument Compartment. On future MR flights vibration-dampening material would be added to the Instrument Compartment. A second discrepancy noticed during the powered phase was the fact that MR-1A rolled about its long axis at almost twice the rate experienced on earlier Redstone missile launches. As the launch vehicle safety was not compromised, and at no time did the roll rate approach the 12-degree abort threshold, no corrective action related to this matter was taken after this flight.

Following a powered phase lasting 2 minutes 23 seconds the Redstone's single R-7 rocket motor shut down as planned. Momentum carried the vehicle to an altitude of 210.7 km before it was overcome by gravitational attraction and began its unpowered fall back towards the Atlantic Ocean.

At engine cut-off MR-1A was travelling at 2,170.2 m/s, 79.2 m/s greater than planned. The cause of the problem was later identified as excessive pivot torque on the launch vehicle's LEV-3 longitudinal integrating accelerometer. On later flights a softer wire was used, in conjunction with the re-routeing of other wires to the system. A back-up system was also developed where the Redstone's R-7 motor would be shut down by a time-based signal at T + 143 s. However, the rewiring of the accelerometer proved sufficiently reliable that the time-based motor cut-off signal was only used on the following two MR launches.

At shutdown the onboard programmer released the Marman Clamp holding the LES to the neck of the spacecraft. The Jettison Motor fired and pulled the LES clear of the spacecraft. When the motor stopped firing, the inert LES fell into the ocean below.

Ten seconds after Redstone shutdown the onboard timer released the Marman Clamp at the base of the spacecraft. The three posigrade rockets in the spacecraft's Retrograde Package fired, to ensure positive separation of the two vehicles in the microgravity environment. SC-2's ASCS carried out rate damping manoeuvres for 5 seconds before turning the spacecraft through 180 degrees and placing it in a 'heat shield forward, negative 14 degree pitch, Antenna Canister down' attitude.

Prior to entering the lower atmosphere the ASCS manoeuvred SC-2 to the correct 34.5-degree retrofire attitude. The spacecraft's three retrograde rockets fired at 5-second intervals, as planned. One minute after apogee the Retrograde Package was jettisoned, to fall away and burn up in the atmosphere. Retrofire was not necessary on suborbital MR flights, as gravitational attraction pulled the spacecraft back towards the ocean. On orbital flights, the retrograde manoeuvre was required to slow the spacecraft down, in order to allow gravitational attraction to pull it out of orbit. STG took the decision to fire the retrograde rockets on all MR and MA flight-tests, to ensure their reliability.

SC-2 maintained a stable 40-degree, heat shield forward, attitude throughout the re-entry heating phase. At the same time the spacecraft rolled around its long axis at a steady 10–12 degrees per second. The heavy beryllium heat shield used on SC-2 had been designed to survive re-entry from orbit. Its heat-sink capabilities coped easily with the suborbital trajectory of MA-1A. The Redstone's over-thrust resulted in the spacecraft experiencing 12.4*g* during re-entry, 1*g* greater than intended.

The spacecraft's barostat deployed the drogue parachute on time and the lanyard from the jettisoned Antenna Canister pulled the primary main parachute from its container. The main parachute de-reefed and lowered the SC-2 to a mid-ocean landing 378.1 km from the launch pad at Cape Canaveral. This was 28.9 km further than intended. The first MR flight had lasted 15 minutes 45 seconds.

The pilot of a search aircraft spotted SC-2 as it descended towards the water under its main parachute and observed it through landing. A Marine Corps helicopter was dispatched from the USS *Valley Forge* and the aircraft's winchman hooked the spacecraft as he had trained to do. The pilot then lifted it from the water and it was on the carrier's flight deck 12 minutes later. SC-2 was returned to Port Canaveral and transported to Hangar S, the following morning; it was found to be in excellent condition.

19.24 KORABL SPUTNIK

A new Korabl Sputnik was launched on 22 December 1960 carrying the dogs Shutka and Kometa. All went well with the launch itself, but during the latter half of the first-stage burn one of the R-7's four strap-on boosters shut down prematurely. The three still-functioning boosters and the second stage continued to fire and the climb into space continued. At T + 2 min the three boosters shut down and all four were jettisoned. The third stage continued the ascent until its propellant was depleted and it, too, shut down.

The premature shutdown of the first-stage booster meant that the third stage was well below orbital velocity when its single rocket motor stopped firing. The abort sequence was initiated and the Vostok Spacecraft separated from the third stage of its launch vehicle. The spacecraft manoeuvred through 180 degrees and fired its retrograde rocket against the direction of travel. The Re-entry Module separated from the Retrograde Module and both began the long fall back to Earth. Unprotected, the Retrograde Module burned up in the atmosphere. The Re-entry Module descended to the thick lower layer of atmosphere, where, at an altitude of 8,000 m, the side hatch was jettisoned and the ejection seat carried the two dogs clear of the descending spacecraft. The Re-entry Module and the pressurised container holding Shutka and Kometa landed in the winter landscape of Kungaska under separate parachutes. A search team from Baikonur Cosmodrome battled the atrocious weather to reach the landing site and recovered them.

3 January 1961: America severed all diplomatic relations with Cuba.

3 January 1961: STG was renamed the Manned Spacecraft Centre (MSC) and became a separate unit. MSC remained based at LaFRC although it became administratively responsible to NASA HQ, rather than GSFC. In February, Gilruth (Fig. 69) began the quest for a new location where NASA could construct a dedicated field centre to serve as the home of the MSC.

Fig. 69. Robert Gilruth. (NASA)

19.25 HANDING OVER POWER

President Eisenhower made his final budget speech on 16 January 1961. The text of his address included the sentence, *'Further testing and experimentation will be necessary to establish whether there are any valid scientific reasons for extending manned space flight beyond Project Mercury.'* The outgoing President had already refused funding for Mercury Mark II and manned circumlunar and lunar orbital flights in a new three-man spacecraft to be called Apollo. History would initially damn the outgoing President Eisenhower for failing to understand the propaganda potential of a dynamic space programme. But, as more of the information available to Eisenhower at the time was declassified and became available to historians, it was obvious that the President understood that the Soviet R-7 ICBM was an ineffective first-strike weapon and that the Soviet Union's missile and bomber forces were much smaller than the American Chiefs of Staff tried to convince him they were. History has proved the outgoing President to be correct.

On 19 January Kennedy and Eisenhower met at the White House in Washington DC for the official handover. Eisenhower advised Kennedy of his perceived weakness of the surrounding area if Laos fell to Communism. He also suggested that America should intervene militarily in Laos to prevent that happening. In the same region, Eisenhower advised of the danger that South Vietnam's anti-Communist leader Ngo Dinh Diem was unpopular at home and may be overthrown by a military coup.

Kennedy was also told that the Eisenhower Administration believed that one day East Germany would do something to stem the flow of East Germans fleeing to the West. The steady flow of people out of Soviet-occupied East Germany was draining the East's pool of professionally skilled personnel. Having previously received briefings on the training of anti-Castro guerrillas by the CIA in Guatemala, Kennedy was advised to support their operations *'to the utmost',* when he asked if he should do so. Finally, as the meeting ended, Eisenhower informed Kennedy that the 'Missile Gap' did not exist and that America led the 'Missile Race' due to its deployment of Polaris in submarines that patrolled just off of the Soviet coast.

19.26 PRESIDENT KENNEDY INAUGURATED

John F. Kennedy was inaugurated on 20 January and he reminded the nation that he was the first President born in the twentieth century. In a masterpiece of Cold War rhetoric he told them:

> Let the word go forth from this time and place, to friend and foe alike, that the torch has been passed to a new generation of Americans ... Let every nation know, whether it wishes us well or ill, that we shall pay any price, bear any burden, meet any hardship, support any friend, oppose any foe, to assure the survival and success of liberty ... Now the trumpet summons us again – not as a call to bear arms, though arms we need, not a call to battle, though embattled we are, but a call to bear the burden of a long twilight struggle year in year out ... And so my fellow Americans: ask not what your country can do for you, but rather, ask what can you do for your country.

Over the next few months the young President and his wife became the darlings of the Free World. They held Court like the Kings and Queens of Europe. *Life Magazine* named the American Court 'Camelot', after the Court of the legendary King Arthur.

On Inauguration Day both Glennan and Dryden tendered their resignation. Their posts were political. Eisenhower's Republican Administration had selected them. President Kennedy would select his own Administrator and Deputy Administrator to lead NASA. Dryden offered to remain in post as Acting Administrator until Glennan's replacement was found, and the Democrats accepted the offer.

Work on Project Mercury continued, but the apparent lack of interest from the new Kennedy Administration led to a fall in morale at NASA and within Project Mercury in particular. This was not helped when Kennedy initially tried to disband the National Space Council, NASA's direct-line to the President. This was only avoided when Kennedy passed the chairmanship of the Space Council to Lyndon Johnson, the Vice President. Many inside NASA felt that this was a good move, as Johnson had helped to establish NASA and was still its strong ally on Capitol Hill.

The most damaging problem was the amount of time that it took to find a replacement for Glennan. The Kennedy Administration was initially unable to find anyone that was prepared to take on the post. They had offered the post of NASA Administrator to almost twenty different individuals and every one of them had turned it down. The lack of visible movement in this area added to NASA's morale problem.

25 January 1961: Thirteen American women astronauts were selected by NASA to undergo the same selection process that had been used to the select the Flight A astronauts. The women became known as the Mercury 13, but this was not a serious attempt to select a group of female astronauts. They were given considerable media coverage at the time, but were then largely forgotten. It would be the mid-1980s before NASA launched its first female astronaut, Sally Ride. Eileen Collins became the first woman to command an American space flight in 1999. She invited several members of Mercury 13 to the launch. The women attracted considerable media coverage once more and their story returned to the public domain.

28 January 1961: Kennedy read a CIA report suggesting that Diem, the President of South Vietnam, be given better American support. The President replaced the staff at America's Embassy in Saigon. Earlier CIA reports submitted to the Eisenhower Administration, but shown to Kennedy, had suggested that Diem only be tolerated until a more suitable replacement could be found.

30 January 1961: In President Kennedy's first State of the Union Address he talked of 'national peril and national opportunity'. Then he suggested that the Russians were a major threat to America's national security. He told his audience: *'We cannot escape our dangers – neither must we let them drive us into panic and narrow isolation . . . There will be further setbacks before the tide is turned. But turn it must. The hopes of Mankind rest upon us.'*

February 1961: STG's James Chamberlin moved to McDonnell's factory where he began working with the contractor's engineers to define a manned spacecraft to follow on from Project Mercury. He called the spacecraft Mercury Mark II. On 12 April McDonnell agreed to undertake an engineering study of Chamberlin's numerous ideas and design a new spacecraft.

19.27 MERCURY REDSTONE 2 (MR-2)

It had long been a standing joke among some American test pilots that a chimpanzee would fly the Mercury Spacecraft into space before any of the Flight A astronauts. Chuck Yeager, the first man to fly an aircraft faster than Mach 1, had joked in public that he would not like to fly anything that required him 'to wipe the monkey shit off of the seat' before he could climb in. MR-2 was one of the flights that Yeager had been referring to. It would carry a chimpanzee on a suborbital space flight. If all went well the first Flight A astronaut would make an identical flight on MR-3.

A group of six chimpanzees had been trained to make the flight and were then moved to cages at the rear of Hangar S, Cape Canaveral. On 30 January two were chosen as the primary and back-up chimpanzee for Mercury Redstone 2. If all went according to plan, the primary chimpanzee would make the flight. If the primary animal became incapacitated, the back-up would fly. Even this mirrored the way in which the human astronauts had been selected for their flight.

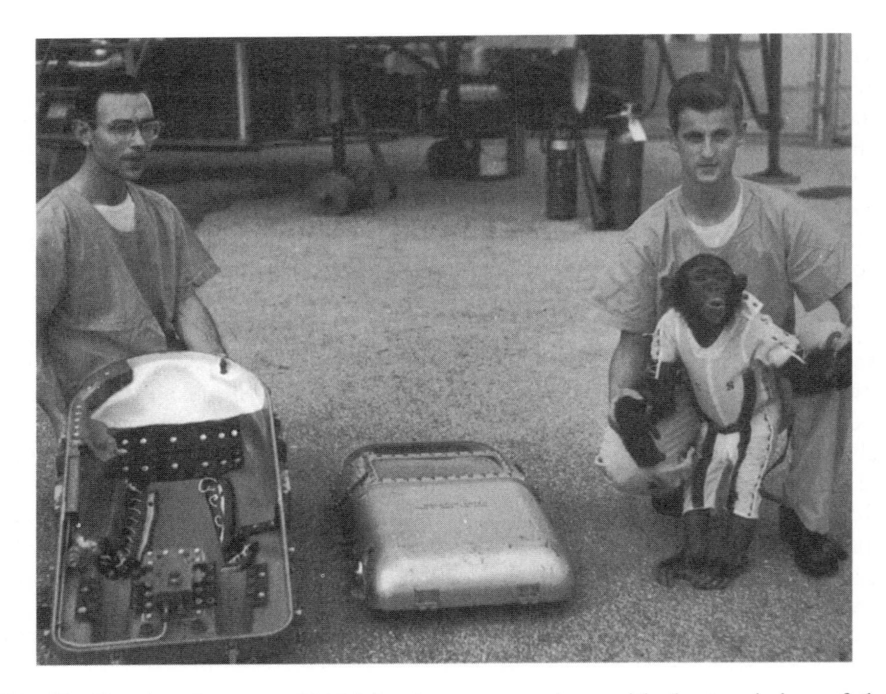

Fig. 70. Two handlers pose HAM for the cameras along with the two halves of the pressurised couch in which he will ride during the Mercury Redstone 2 flight. (NASA)

Because the chimpanzee had no way of understanding how to control the spacecraft in flight, it would be controlled by the ASCS. This was how all Mercury flights were originally intended to take place, with the astronaut along merely to read off the instruments, provide medical data and hopefully to make history, as the first man in space. This plan had provided the ammunition for Yeager's most enduring taunt – that the Flight A astronauts would be nothing more than the 'man in the can'. The taunt lost much of its original venom when the astronauts succeeded in having manual controls fitted in the Mercury Spacecraft.

As it was important to recover the chimpanzee as soon as possible after landing, the launch attempt would be cancelled if MR-2 were still on the launch table at 1200. A launch after that time might require a protracted search and recovery operation after nightfall.

At T – 19 hours the male primary and female back-up chimpanzees were placed on low residue diets in order to reduce their need to defecate over the coming hours. The procedure mirrored those that would be followed by the Flight A astronauts.

MR-2 began its 10 hour 40 minute countdown to launch on 30 January 1961. As usual it was held overnight. At T – 7 h 30 min, at 0030 on 31 January, the two chimpanzees were both given a final medical examination and tested on their training devices. The primary animal was keen and alert so he was selected to make the flight. The chimpanzee was named HAM, taking his name from the Holloman

Aerospace Medical Centre, where he had been trained. The countdown was resumed at 0130, T−6 h 30 min and proceeded towards launch. (See Fig. 70.)

At T−4 h HAM and his back-up were each fitted with three medical sensors and dressed in an Airtex body suit to stop them pulling the sensors away from their bodies. Once installed in the Mercury Spacecraft, the sensors would telemeter the animal's heart rate, breathing rate and body temperature to Mercury Control throughout the flight. The two chimpanzees were then placed in the contoured rear portion of the pressurised couches in which one of them would make the first ride into space in a Mercury Spacecraft. For their safety, their ankles were lightly laced into place and a body strap was secured across their torso. Finally, the upper portions of the couches were secured in place. Directly in front of the chimpanzee's face was a rectangular window. This not only allowed the animal to see out, but also allowed the movie camera mounted above the spacecraft's instrument panel to film the chimpanzee's reactions throughout the flight. A similar camera would record the reactions of the Flight A astronauts as they made their flights into space.

Mounted inside the upper portion of the couch, at the chimpanzee's waist level, was a psychomotor test that the animal would be expected to perform throughout the flight. The test consisted of two lights and two levers. When the blue right-hand light was illuminated the chimpanzee had 5 seconds to depress the right-hand lever. Likewise, when the white left-hand light came on the animal had 5 seconds to depress the left-hand lever. In both cases a correct action delayed the light coming back on for 15 seconds. An incorrect action resulted in a mild electric shock being applied to the soles of the animal's feet. Inside Ham's couch, placed between his feet, was a small white card, which carried the message, *'Have Missile, Will Travel'*.

With the upper portions of the two couches secured in place, the two chimpanzees were taken out to the waiting transfer van, where environmental control equipment was attached to both couches. Under a security escort the transfer van transported the two animals and their handlers from Hangar S to LC-5. At the launch complex only HAM's couch was removed from the transfer van and carried up the elevator to the white room, where SC-5 was waiting.

The countdown resumed at 0725, having been stopped to enable the vehicle's pyrotechnics to be installed. Twenty minutes later it was held, when an inverter in SC-5s ASCS began to overheat. Despite the delay, at 0753 Ham's couch was manhandled through the small side hatch and secured where the astronaut's prone couch would have been had this been a manned flight. The side hatch was lifted into place and the 40 torque bolts tightened.

When the faulty inverter had cooled down, the countdown was resumed, but was held again when the inverter overheated for a second time. That one inverter caused numerous small holds in the countdown. Throughout everything Craft, the flight director in charge of MCC questioned HAM's ability to withstand the delays. The representative of Holloman Aerospace Medical Centre in MCC monitored the data telemetered from the chimpanzee's three body sensors and assured Craft that all was under control.

The countdown resumed at 0945, but the inverter overheated once more and a new hold was called, at T−15 min. It was 1140 before the clock moved again, in a final attempt to launch before the day's deadline was reached. MR-2 lifted-off at 1155, 5 minutes short of the deadline.

At T+1 min the computers at GSFC showed that MR-2's trajectory was 1 degree higher than planned. They predicted that HAM would be subjected to 17*g* during re-entry, rather than the planned 12*g*. As with MR-1A, the present vehicle was rolling about its long axis at twice the rate of earlier Redstone missiles. The 8-degree roll did not endanger the launch and did not approach the 12-degree roll abort threshold. The high vibration rates experienced on MR-1A were also repeated on this flight. Before the next flight four stiffeners were added to the instrument compartment, along with 95.2 kg of ballast and insulating material.

Again, as with MR-1A, this launch vehicle produced more thrust than planned, causing the steeper trajectory. On both launches a servo control valve had failed to correctly control the flow of hydrogen peroxide to the steam generator and, on MR-2, this led to the early depletion of the liquid oxygen supply. The Redstone's AASS sensed the liquid oxygen depletion 0.5 second before the thrust chamber pressure abort switch was disarmed, the normal A-7 shut down command. Operating in a closed loop for the first time, the AASS released the spacecraft Marman Clamp and issued an abort signal to the onboard programmer. On future Redstone launches the thrust regulator and velocity integrator were modified. Also, the velocity cut-off arming switch activation was advanced to T−131 s and the combustion chamber pressure abort switch disarming was moved to T+135 s.

Having received an abort signal, the three bolts holding the retaining straps of the Retrograde Package to the base of the spacecraft fired, jettisoning the Retrograde Package. Meanwhile, MR-2's R-7 rocket motor shut down at T+2 min 17 s, as planned. The Escape Motor on the LES fired just 0.5 second after the R-7 shut down. The burn lasted for 1 second, pulling SC-5 clear of the Redstone and adding its own thrust to the launch vehicle's already extended apogee and predicted downrange distance. With the Escape Motor's propellant depleted, the Marman Clamp attaching the LES to the neck of the spacecraft was released and the Jettison Motor fired. The LES moved away from the spacecraft and, when the Jettison Motor shut down, fell to the ocean below.

During the unplanned abort the vibration shook loose the pin that held the inlet snorkel valve closed on the spacecraft. With the pin gone, the valve's spring-loaded flap moved to its open position and the Crew Compartment atmospheric pressure dropped from 351.5 to 70.3 g/cm^2 as the oxygen bled overboard through the snorkel valve. HAM was unaffected by the decompression as his couch contained its own ECS. The snorkel valve problem would be overcome on future flights by increasing the valve cover and its retaining pin's tolerance to vibration.

At apogee SC-5 was at an altitude of 252.6 km. This was 67.5 km higher and 77.2 km further downrange than intended for this flight. HAM experienced 6.6 minutes of microgravity, as opposed to the 4.9 minutes in the flight plan. During this time the camera recorded a fully relaxed chimpanzee as he went about his psychomotor tasks.

As gravitational attraction pulled the spacecraft back towards Earth it adopted a stable, heat shield forward attitude throughout re-entry. The barostat deployed the drogue parachute in the upper atmosphere and the lanyard from the Antenna

Canister pulled the main parachute clear of its container. The heavy beryllium heat shield dropped to deploy the landing skirt at the base of the spacecraft. The main parachute lowered SC-5 to the ocean. Landing occurred at 1212, after a flight lasting 20 minutes 12 seconds. On landing the heat shield contacted the base of the spacecraft, puncturing two holes in the Crew Compartments aft bulkhead.

Throughout the flight HAM performed faultlessly in reaction to the white light whenever it had been illuminated and had only reacted incorrectly to 2 out of 50 illuminations of the blue light. He had effectively proved that a human astronaut could not only expect to survive a flight into space, but could expect to perform all of his allocated tasks throughout that flight.

Landing took place 96.5 km from the nearest recovery vessel, but as the computer operators at GSFC had predicted the overshoot, recovery vessels were already steaming for the location. Meanwhile, SC-5 floated upright in mid-ocean. After 27 minutes in the water the crew of a search aircraft located the SC-5 and radioed their location to the primary recovery vessel.

The destroyer USS *Ellison* was the closest ship and still 2 hours away. Therefore, MCC had helicopters dispatched from the USS *Donner*, a landing craft-dock, to recover the spacecraft. When, at 1252, the first helicopter arrived over SC-5 it was no longer upright. Water entering through the punctured aft bulkhead had caused it to lie over on one side, where the open snorkel valve cover had allowed still more water to enter the Crew Compartment. Wave action had caused the landing skirt to rip around its circumference and the heat shield had been lost as a result. HAM was dry within his pressurised couch, but his spacecraft was at risk of sinking if it was not recovered soon. SC-5 was pulled from the ocean at the end of a helicopter winch line. The strain gauge on the winch showed that it was carrying 362.8 kg of seawater. It was returned to the USS *Donner* hanging beneath the helicopter and lowered onto the waiting dolly upon arrival.

Before the first manned Mercury Spacecraft flew, an additional fibreglass bulkhead had been placed inside the Crew Compartment aft bulkhead to prevent seawater entering the spacecraft if the aft bulkhead became punctured on landing. Cables and retention straps, designed to take the repeated shock of long-term exposure to wave movement, were added between the base of the spacecraft and the rear of the heat shield.

HAM's couch was removed from the spacecraft and the upper portion was removed. The chimpanzee took and consumed the apple and orange that he was given. In the ship's sick bay HAM was removed from the rear portion of the couch, undressed and his body sensors were removed. He was then subjected to a thorough medical examination, which revealed him to be fit and healthy. The only injury he had received during his 8 hour ordeal was a bruised nose. The procedures were all similar to those that would be followed after a manned flight into space.

HAM remained healthy in the weeks following his flight. Despite everything he had experienced that day, when placed on the training devices he had used before the flight, he performed well and enthusiastically. Many inside STG, including the Flight A astronauts, believed that Project Mercury was ready to proceed with the first manned MR suborbital flight. At Redstone Arsenal in Huntsville others disagreed.

Meanwhile, Alan Shepard had been selected as the prime astronaut for MR-3. Gus Grissom would support him throughout the training programme, while John Glenn would serve as the back-up astronaut. No public announcement was made at the time, to allow the three men to concentrate on their training without undue interference from the press and media.

19.28 NASA'S SECOND ADMINISTRATOR

In January 1961 Jerome Wiesner invited James E. Webb to Washington DC, where Vice President Johnson and then President Kennedy wished to discuss the administration of NASA with him. Webb travelled to the capital and familiarised himself with NASA's role and the Administrator's task. During this period Webb found out that he was far from the White House's first choice for the job. (See Fig. 71.)

Webb kept his appointment with the Vice President, but insisted that he would not take the post unless President Kennedy asked him to do so personally. An appointment was made for Webb to meet the President that afternoon. Kennedy told him that he wanted him in the Administrator's post because he had considerable experience of the political process in Washington. Webb accepted the post.

He would be expected to propose his own policy. The President also agreed to Webb's request that Dryden remain in place as NASA's Deputy Administrator. President Kennedy immediately took Webb to meet his Press Secretary, Pierre Salinger, who announced Webb's assignment to the NASA Administrator's position.

On taking up his new post Webb confirmed both Dryden and Seamans in their positions. The three men would lead NASA forward as a team, and all major policy decisions would be made between the three of them. Webb, Dryden and Seamans would prove to be one of the most effective senior management teams in NASA's history.

Fig. 71. James Webb, NASA's second Administrator. (NASA)

19.29 MERCURY ATLAS 2 (MA-2)

Mercury Atlas 2 would re-fly the suborbital flight-test that terminated so abruptly on MA-1, in July 1960. During the run up to launch George Low informed NASA Administrator James Webb.

> This will be the last of the thin-skinned Atlases to be used in the Mercury Programme. It differs from the booster used in the MA-1 test in that the upper part of the Atlas has been strengthened by the addition of an 8-inch (20.3-cm) wide stainless steel band. This band will markedly decrease the stresses on the weld located just below the adapter ring on top of the Atlas; the high stress region is shifted about 8 inches to a point where the allowable stresses are considerably higher. In addition to this strengthening of the top section of the Atlas, the bracing on the oxygen vent valve, which fits into the top of the Atlas tank, has been changed. The adapter between the Atlas and the spacecraft has also been strengthened.

MA-2 consisted of Atlas 67D, SC-6 and a production LES. The ASIS was flown in a closed loop and would activate the abort sequence in the event of a MA-1 style failure. The flight was intended to prove the spacecraft's ablative heat shield during a worst possible scenario re-entry, simulating a return from orbit. Media attention was firmly centred on the horse collar around the top of the Atlas and the possibility of another MA-1 style failure. (See Fig. 72.)

Countdown holds and recycles added 70 minutes to the planned launch time. It was 0910 on 21 February 1961 when MA-2 climbed away from LC-14, AMR. The sky was clear and the lighting conditions suitable for clear photography, as dictated by new rules governing weather conditions at launch, brought in after the loss of MA-1. Lift-off was perfect and the vehicle passed through max-*q* without breaking up. Two minutes after launch the two booster engines stopped thrusting and the booster skirt was released and jettisoned as planned. The Marman Clamp holding the LES to the neck of SC-6 released. Both LES motors fired and pushed the red tower away from the still ascending vehicle. The LES separated cleanly and shut down, allowing gravitational attraction to pull it down to the ocean below. The single Sustainer engine and the two steering verniers continued to push MA-2 towards space. Five minutes after launch the Sustainer and verniers shut down as planned.

A signal then separated the Marman Clamp holding the base of SC-6 to the top of the spacecraft adapter. The three posigrade rockets in the Retrograde Package fired to provide positive separation between the spacecraft and the now spent Sustainer. Following rate-damping manoeuvres, the spacecraft's onboard timer commanded the ASCS thrusters to turn the SC-6 through 180 degrees. Having reached an apogee of 183.5 km, the SC-6 took up the correct retrofire attitude. Retrofire occurred on time and the Retrograde Package was jettisoned. SC-6 began its re-entry.

Observers on the landing ship dock (LSD) USS *Donner* were able to see the spacecraft as it completed re-entry. They lost sight of it prior to parachute deployment, as it passed below their horizon. In the upper atmosphere the barostat

Fig. 72. The horse collar (directly beneath the dark-coloured launch vehicle spacecraft adapter) allowed Mercury Atlas 2 to use a 'thin skinned' Atlas despite the failure of Mercury Atlas 1. (NASA)

had deployed the drogue parachute, which pulled the Antenna Canister away from the neck of the spacecraft and deployed the main parachute. SC-6 landed in the ocean 2,303.5 km downrange from LC-14. The flight of MA-2 had lasted 17 minutes 56 seconds.

A clear signal was received from the spacecraft's Sarah beacon and helicopters were dispatched from the USS *Donner* to effect recovery. The spacecraft was only in the ocean for 24 minutes before being removed at the end of a helicopter's winch cable. Less than one hour after launch the recovered spacecraft had been secured aboard the USS *Donner*. The drogue parachute and Antenna Canister were found floating nearby and were also recovered. SC-6 proved to be in excellent condition.

During the post-recovery press conference, MSC Director Robert Gilruth was asked if a human astronaut would have survived the MA-2 flight. He replied with a cautious, *'Yes'*. He added that there were a number of critical items that still had to be completed before one of the Flight A astronauts would be asked to ride the Mercury Atlas combination. At the same press conference Gilruth informed the press and media that Flight A astronauts Glenn, Grissom and Shepard, in alphabetical order, had been named to begin training towards the first manned flight of Project Mercury.

19.30 KORABL SPUTNIK IV

At 0929 on 9 March 1961, the Soviets launched the human mannequin Ivan Ivanovich strapped to the ejection seat of the Korabl Sputnik IV Vostok Spacecraft. A pressurised container holding the dog Chenushka was strapped to the wall of the Crew Compartment. The flight progressed perfectly and the spacecraft was injected into a 173 × 239-km orbit. In a simulation of the planned first manned flight, as the spacecraft crossed Africa it was aligned for retrofire. The rocket burn passed without incident and the two modules separated. The spherical Re-entry Module pushed its way through the atmosphere, until the side hatch was jettisoned and the mannequin was ejected to land under its own parachute. The Re-entry Module's main parachute deployed and lowered the spherical module to a hard landing at 1116. Despite the severe landing forces, Chenushka was found to be fit and healthy when the recovery team let him out of his pressurised container. The Soviets had completed the first of the two unmanned precursor flights required before a cosmonaut would be allowed to ride the Vostok into orbit.

11 March 1961: President Kennedy was briefed at the White House on the CIA's training of anti-Castro Guerrillas. The Commander in Chief made it clear that there would be no American military involvement in any invasion of Cuba.

22 March 1961: President Kennedy informed NASA Administrator James Webb that he would not fund Project Apollo.

19.31 LITTLE JOE 5A (LJ-5A)

This flight was a repeat of the LJ-5 flight that had failed on 8 November. Production Spacecraft 14 would be used in the max-q abort. Although it was a production model, SC-14 contained only those systems required for this flight. It carried an impact skirt and a heavy beryllium heat-sink heat shield. It also contained a single retrorocket, to provide positive separation from the launch vehicle in the event of a repeat of the LJ-5 scenario, which was still not fully understood by March.

Following a 4-hour delay, lift-off occurred at 1149 on 18 March 1961. For the first time the launcher had been erected at LA-4, Wallops Island. LJ-5A's four Recruit and two Castor rocket motors performed flawlessly. At T + 22 s, with the final pair of Castor motors still burning, airflow once again deformed the spacecraft Marman Clamp and closed at least two of the limit switches in that area, resulting in an abort signal being sent to the LES. The Escape Motor fired 14 seconds earlier than the flight plan called for, but the spacecraft did not separate from the launch vehicle. The combination continued to arch through the sky belching smoke and flame from both ends.

At T + 35 s the spacecraft's onboard timer signalled the normal abort command, which released the spacecraft Marman Clamp. Aerodynamic pressure kept the now released spacecraft pressed firmly against the neck of its launch vehicle. At T + 43 s a signal was sent from the ground to fire the retrorocket and separate the spacecraft. Although the retrorocket did fire the vehicle had not reached its 12.43-km apogee and, therefore, the single rocket motor's thrust was insufficient to overcome the aerodynamic loads acting on the spacecraft and provide positive separation.

Finally, as the combination began to slow down and aerodynamic loads decreased, the spacecraft fell away from the launch vehicle. At separation SC-14 tumbled violently, narrowly avoiding a collision with the now inert LJ. The tumbling caused aerodynamic loads that ripped the retrorocket and the LES away from opposite ends of the spacecraft. At about this time the SC-14 also lost its Antenna Canister, complete with the drogue parachute and its mortar.

As the Antenna Canister fell away, at an altitude of 12,192 m, both the primary and reserve main parachutes deployed and filled with air, finally bringing the descent under control. Under its two parachutes the spacecraft drifted for 16.09 km. The barostat released the heat shield and deployed the landing skirt at an altitude of 3,048 m.

SC-14 finally splashed down 28.96 km downrange from LA-4, after a flight lasting 23 minutes 48 seconds. On impact with the water the heat shield re-contacted the base of the spacecraft, causing only superficial damage. Both parachutes fell over the neck of the spacecraft precluding the standard helicopter recovery. One-hour after landing, the spacecraft was recovered by the salvage vessel USS *Opportune*.

Once again corrections before the next flight concentrated on the two Marman Clamps. The spacecraft clamp was provided with a shield, to protect it from aerodynamic loads during flight. The limit switches in that area were changed so that they could not activate the abort signal unless the spacecraft Marman Clamp had been released first. Similar changes were made to the limit switches in the LES Marman Clamp at the neck of the spacecraft, to prevent them from activating the Escape Motor unless the Marman Clamp had been released.

19.32 MERCURY REDSTONE BOOSTER DEVELOPMENT (MRBD)

Following the discrepancies experienced during the MR-1A and MR-2 launches, Wernher von Braun, the director of MSFC, called for an additional flight to be added to the programme. The new flight would prove that the corrections that had been made to the MR launch vehicle worked. The additional flight was originally designated MR-2A, and ultimately flew under the designation Mercury Redstone Booster Development. The vehicle would consist of a live launch vehicle and the refurbished boilerplate spacecraft that had flown on LJ-1A with an inert LES. (See Fig. 73.)

On 24 March 1961 all countdown procedures passed as planned as launch time approached. Liquid oxygen loading commenced at T − 2 h. During loading high winds caused the vehicle to sway as it stood on the launch table and made the liquid oxygen slosh around in the vehicle's propellant tank. As the cryogenic propellant's temperature began to rise a boil-off valve opened and dumped liquid oxygen on the launch pad. The procedure was computer controlled. Loading continued and was completed on schedule.

Fig. 73. A boilerplate capsule is hoisted atop Mercury Redstone Booster Development. This additional flight, made at the demand of von Braun, cost Project Mercury the race to place the first man in space. (NASA)

At 1230 MRBD climbed into the Florida sky. Despite a 27 m/s over-thrust, it remained on its correct trajectory. The Redstone's rocket motor shut down on time and the combination coasted to an apogee of 182.62 km. No attempt was made to separate the spacecraft, or the inert LES. Having been overcome by gravitational attraction, the vehicle was pulled back towards the ocean below. Impact occurred 182.6 km downrange from LC-5, 8.5 km short of the target area. MRBD broke up on impact and all of its components sank. Although the boilerplate spacecraft's Sofar bomb detonated as the spacecraft sank no recovery efforts were made. The flight had lasted 8 minutes 23 seconds.

During a manned launch a cherry picker would be used to remove the astronaut from a malfunctioning vehicle in an emergency. To simulate this, an asbestos covered truck had been located 20 m from the launch table, at the location where the cherry picker would be located. This was an attempt to see what damage the cherry picker would suffer during a launch. The truck received no damage.

In the event of an emergency egress from a manned launch attempt, the M-113 would move up to the base of the cherry picker and serve as an escape vehicle for the astronaut. During the MRBD launch a M-113 armoured personnel carrier was parked 305 m from the launch table, in an attempt to see how close emergency crews could be placed to a launch. The M-113's crew had watched the launch without discomfort.

MRBD was successful. The difficulties of the previous two MR launches had been overcome and STG had qualified the MR launch vehicle for manned flights, but their caution would cost them dearly. Many people inside STG felt that MRBD had been unnecessary and that the 24 March launch should have carried NASA's first manned spacecraft. If MRBD had been manned, Alan Shepard would have been the first man in space.

19.33 DEATH STIKES

Valentin Bondarenko was the youngest of the twenty men selected for cosmonaut training in the Soviet Union. On 21 March he entered a pressure chamber to commence a test. Three days later, as he prepared for the end of the test, Bondarenko dropped a piece of cotton wool onto the grill and started a fire. It was 30 minutes before the pressure could be equalised and the door to the chamber opened. The trainee cosmonaut died 8 hours later. Manned space flight had its first casualty even before anyone had flown into space.

19.34 KORABL SPUTNIK V

The final unmanned flight-test of the Vostok Spacecraft lifted-off from Baikonur Cosmodrome at 0900 on 25 March 1961. The R-7 launch vehicle performed faultlessly and Korabl Sputnik V entered a 164 × 230-km orbit. The human mannequin, Ivan Ivanovich, was making its second flight into space, while the dog, Zvedochka, was in a pressurised compartment strapped to the wall of the Crew Compartment.

The spacecraft completed one orbit before aligning for retrofire, as it passed over Africa. Following the rocket burn the Re-entry Module separated from the Retrograde Module as planned and both modules were pulled towards the lower atmosphere. The Retrograde Module burned up while the Re-entry Module was protected by its heat shield. At 8,000 m the side hatch flew off and the mannequin was ejected to make a parachute landing. Zvedochka continued the descent in the Re-entry Module under its main parachute landing at 1047. The dog survived the severe landing loads and was found in a healthy state by the recovery crew. For the Soviets the flight-testing was over, the next Vostok Spacecraft would be manned.

5 April 1961: Despite the fact that a number of American newspapers and news magazines had recently run articles on the CIA giving military training to Cuban exiles, President Kennedy agreed to a date of 16 April for the invasion of Cuba by one group of those exiles.

19.35 VOSTOK I

Following a general training programme, in May 1960 Korolev chose Gagarin, Kartashov, Nikolayev, Popovich, Titov and Varlamov to commence training towards the first manned flight of the Vostok Spacecraft. Kartashov, however, injured his spine during a centrifuge ride and Varlamov injured himself while swimming as part of the training programme. Bykovsky and Nelyubov replaced them.

On 5 April the six cosmonauts were present at the vehicle assembly building at Baikonur Cosmodrome when Vostok I was mated with its R-7 launch vehicle as it rested horizontally on a special rail car. Three days later Korolev recommended to the State Commission that Gagarin make the first flight with Titov serving as his back-up.

At 0500 on 11 April the launch vehicle was moved out of the assembly building to the launch pad. There, it was raised to the vertical and surrounded by the launch structure. Final checks were completed and propellant loading began. Gagarin and Titov spent that night in the Cosmonaut's Cottage, a small wooden building at the Cosmodrome. They were woken up at 0530 and had medical sensors attached to their bodies before dressing in their pressure suits. A bright orange coverall formed the outer layer of the garment, to help locate an incapacitated cosmonaut in the event of his being injured on landing.

The cosmonauts were transported to the launch pad in a small coach. They embraced briefly at the base of the launch pad before Gagarin made his way to the elevator, where he made a speech to the gathered workers. He was then lifted to the level where his spacecraft waited. Entering the spacecraft, he was strapped tightly into his ejection seat before the circular hatch was bolted over his head. With no direct participation in the countdown he listened to music while the final checks were carried out around him. When everything was ready, the launch pad was evacuated and the service structure lowered to the ground. A single power umbilical remained connected to the launch vehicle.

Lift-off occurred at 0907 on 12 April 1961. Gagarin reported, *'Off we go. Everything is normal.'* Eight minutes later Gagarin was in a 169 × 315-km orbit. When the third-stage rocket motor shut down everything in the spacecraft entered a state of microgravity. Vostok was over Siberia and the Soviets had taken what the media and the public saw as the most important prize in the race for space. While Gagarin crossed the Pacific Ocean, Radio Moscow announced their achievement: *'Today, 12 April 1961, the first cosmic ship named Vostok, with a man on board, was orbited around the Earth from the Soviet Union.'*

In America President Kennedy had gone to bed the night before knowing that another R-7 stood on the launch pad at Baikonur Cosmodrome with a Vostok Spacecraft underneath the shroud and the rumours coming out of Moscow suggested that a human cosmonaut would be on board. Asked if he wanted to be woken up if the launch took place he said *'No'*. The following morning he was told that the first man in space was a Red Air Force Major and not an American military officer. Meanwhile, a journalist had telephoned Shorty Powers, the Project Mercury Public

Affairs Officer, at home in the middle of the night. Powers picked up the telephone and mumbled, *'We're asleep down here'*. The quote was applied to Project Mercury in general and printed in the morning newspaper.

Being only a passenger, Gagarin tried the food that was carried in his spacecraft. He squeezed a meal out of a tube directly into his mouth. Water escaped from another tube and formed droplets in the microgravity environment. The droplets drifted in the air until they struck the window and clung to it. The sun was extremely bright and Earth beneath him drew the comment *'How magnificent'*. The curve of the planet was obvious to him, as were the outlines of continents.

The spacecraft entered Earth's shadow over South America. As Vostok returned to sunlight half an orbit later the onboard timer aligned it for retrofire. The rocket motor fired against the direction of travel at 1025, slowing the spacecraft down and allowing gravitational attraction to pull it back to Earth. The straps holding the Re-entry Module in place separated, but the umbilical connector did not. Gagarin's Re-entry Module entered the atmosphere dragging the still attached Retrograde Module behind it. As the atmosphere thickened the umbilical burned through and the Retrograde Module drifted away and burned up. Gravitational attraction acted upon the contents of the Re-entry Module aligning it so that Gagarin was travelling with his back to the direction of travel.

At 8,000 m the hatch above his head jettisoned and Gagarin was blasted out of the spacecraft on his ejection seat. Separating from the seat, his personal parachute deployed and he watched the Re-entry Module drifting Earthward under its parachute. Beneath him the Volga River shone silver in the morning sunlight. The cosmonaut drifted to ground in a ploughed field close to a peasant woman and her daughter. It was 1055 and the previously unknown Red Air Force Major Yuri Gagarin was suddenly the most famous person in the world.

In the aftermath of Gagarin's flight Chairman Khrushchev challenged: *'Let the Capitalist countries catch up with our country.'* Webb swallowed his pride and sent NASA's congratulations to the Soviet Union on their achievement.

12 April 1961: In the wake of Gagarin's flight President Kennedy told a White House press conference:

> I do not regard the first man in space as a sign of the weakening of the free-world, but I do recognise the total mobilisation of man and things for the service of the Communist bloc over the last years as a source of great danger to us. And I would say that we are going to have to live with that danger and hazard through much of the rest of this century.

14 April 1961: Cuban exiles flying Liberator bombers carrying the markings of the Cuban Air Force, bombed military airfields in Cuba in an attempt to destroy the Cuban Air Force on the ground. The CIA had trained the exiles in South America. Their mission failed!

19.36 SHOOTING THE MOON

When Webb took the NASA Administrator's post, Project Apollo was scheduled to place a three-man spacecraft on a circumlunar trajectory in the late 1960s and to attempt a lunar landing in the early 1970s. Webb set NASA's engineers the task of reviewing Apollo to see if the lunar landing could be brought forward to 1969–70. He publicly supported Kennedy's reduction of his own request for an additional $308 million to NASA's budget, but had sound reasons for doing so. Out of the public arena the Kennedy Administration was seeking ways to take on the Soviet Union in space and prove the superiority of the American Capitalist system.

From the earliest days of his Administration President Kennedy had looked for a way to give the American people back their national pride in the wake of Sputnik, which had been launched three years before his election. In the intervening years he had made speeches supporting an accelerated missile programme. Kennedy viewed the space programme as a weapon in the Cold War. Following his inauguration he had delegated the chairmanship of the Space Council to the Vice President. Johnson had sought information from Webb and von Braun on the nation's ability to meet the challenge of a manned lunar landing programme and found both men in support of such a project. The Vice President then sought support on Capitol Hill for the Moon Landing. In May, Webb had provided Johnson with NASA's views on what an accelerated space programme should include. The list consisted of:

- The development of spacecraft and launch vehicles for a manned flight to the Moon.
- Scientific unmanned probes to survey the Moon.
- A nuclear rocket.
- Satellites for global communication.
- Satellites for weather observation and prediction.
- Scientific experiments to be carried out on the manned lunar landings.

On 14 April Kennedy held a meeting to discuss how America should react to Gagarin's flight on Vostok I. They discussed NASA's plans for Project Apollo and set a price tag of $40 million to place an American astronaut on the Moon by 1970. Kennedy was stunned by the projected cost but ended the meeting by stating that nothing was more important than America beating the Soviets to place a man on the Moon.

Following discussions with representatives of all interested parties, Kennedy sent Johnson a memorandum asking how America might compete with the Soviet Union's apparent lead in space flight. Johnson's reply suggested an accelerated space programme with a manned lunar landing as its goal. In a separate memorandum Webb informed the President that a manned lunar landing was technically feasible in 1967. Johnson and Webb argued that such a programme would require considerable technological advance beyond the present position and would provide America with a chance to prove her technological superiority. The extended timescale involved would negate the early Soviet lead in launch vehicle throw weight. The manned lunar landing, as part of an expanded Project Apollo, was officially accepted as the major American national goal in space beyond Project Mercury.

19.37 CUBA 'INVADED'

On 15 April President Kennedy gave the go-ahead for the invasion of Cuba by the 1,400 members of the Cuban Exiles Brigade 2506. The following day America suffered another public humiliation when a group of exiles landed at the Bay of Pigs in an attempt to inspire a public uprising against Castro. They were met with heavy resistance from the Cuban military forces. When it was obvious that the invasion was in trouble, Kennedy ordered the aircraft carrier USS *Essex*, which had been lying off the beachhead, to move beyond the horizon as seen from Cuba. He refused CIA calls for air strikes against the Cubans by the *Essex*'s aircraft. Without American air support, the Cuban Air Force easily sank the two freighter vessels carrying the invaders' reserve ammunition and equipment. The anticipated public uprising did not occur and the invaders either died on the beach, or were taken prisoner by the Communist Cuban army.

In the aftermath of the Gagarin flight and the Bay of Pigs defeat, Kennedy sought a way of recovering American leadership and prestige in the world. On 16 April he had addressed the space programme, and the manned Moon landing in particular, in conversation with Vice President Johnson. Johnson suggested that Kennedy send him a memorandum asking if there was any way in which America could show its superiority in space. The President sent the memorandum the next day. Johnson sought advice from numerous people and made his final decision following a meeting on 24 April, at which everyone present supported the call to place an American astronaut on the Moon. President Kennedy took Johnson's reply under consideration.

On 18 April Kennedy again refused CIA calls for air cover over the Bay of Pigs beachhead. He ultimately gave in, allowing one flight of aircraft from the USS *Essex* to escort the covert Liberator bombers flying in from South America. A failure to allow for the time difference between the bomber base and the aircraft carrier off Cuba meant that the two sets of aircraft never met.

President Kennedy met with Richard Nixon on 20 April, who agreed to back the President if he chose to mount a full-scale invasion of Cuba. On the same day, with the Cuban crisis still in full swing, Kennedy established a secret Task Force to deal with the growing problem of Vietnam.

At a White House press conference the following day the President took full responsibility for the Cuban invasion, telling the gathered journalists, *'There is an old saying that victory has many fathers, but failure is an orphan ... I am the responsible officer in this government.'*

The President signed the official report on the Cuban invasion on 27 April. With over 1,000 Cuban exiles taken prisoner on the beach, a Gallup Poll showed that 83 per cent of the American public supported the President over the Bay of Pigs action. As they saw it, President Kennedy was standing up to Communism, and that was the job of the American President during the Cold War.

19.38 MERCURY ATLAS 3 (MA-3)

When Kraft and his Operations Division personnel arrived at ETR in order to prepare for MA-3 they found that the Atlas launch facilities at Cape Canaveral had been commandeered by the USAF while the Bay of Pigs crisis was under way. As the invasion failed on the beach and President Kennedy made it clear that there would be no American invasion of the island, the Atlas facilities at LC-14 were returned to NASA's control.

Atlas 100D, Mercury Atlas 3, was intended to place SC-8 into orbit where a simulated man would test the ECS. After one orbit retrofire would slow the spacecraft down to a velocity that would allow gravitational attraction to pull it out of orbit. SC-8 was the last of McDonnell's first production run. (See Fig. 74.)

Following ignition on 25 April 1961, Atlas 100D made the transition to mainstage and the clamps holding it to LC-14 were released. MA-3 climbed off of the launch pad. At T + 40 s the launch vehicle should have begun a roll, to place it on the correct

Fig. 74. Mercury Atlas 3 at Launch Complex 14. (NASA)

trajectory. Inside Atlas 100D contamination on one of the flight programmer's pins, combined with in-flight vibrations caused the programmer to either fail to start, or to start and shut down again before it could issue the roll command. The roll manoeuvre did not occur.

In the blockhouse at LC-14 the Range Safety Officer noticed that the Atlas had failed to roll and pushed the button on his console to send the destruct signal. Ordnance devices built into the launch vehicle's propellant tanks exploded, destroying the Atlas. Sensing the Range Safety command, the ASIS issued the signal to release the Marman Clamp at the bottom of the spacecraft. The Escape Motor in the LES ignited and pushed SC-8 clear of the exploding Atlas. When the Escape Motor shut down the Marman Clamp at the neck of the spacecraft was released and the Jettison Motor fired to push the dead LES away from the spacecraft.

SC-8 reached an apogee of 7.24 km, before adopting the correct re-entry attitude and jettisoning its Retrograde Package. The ELS deployed as planned, lowering the spacecraft to the ocean 0.46 km from LC-14, after a flight lasting just 7 minutes 09 seconds. The Spacecraft's recovery aids deployed as intended.

Recovery helicopter crews based at Cape Canaveral reacted as if there was a human astronaut in SC-8. The spacecraft was pulled out of the ocean beneath a helicopter within minutes of landing. It was returned to the NASA facilities at Cape Canaveral and found to be in good condition. It would be reconditioned by McDonnell and re-launched on MA-4.

Once again divers searched for wreckage off LC-14. The flight programmer from Atlas 100D was recovered and close inspection of the unit led to the above scenario being given as the most likely cause of the vehicle's failure to commence the roll manoeuvre.

26 April 1961: President Diem was re-elected in a rigged election in South Vietnam. The President and his brother led their country with absolute power. Diem family members held all major posts in the government and military. The Diem family was unpopular with the South Vietnamese people and were only in power because they received American support to prevent a Communist takeover of the country.

19.39 LITTLE JOE 5B (LJ-5B)

The flight of the refurbished Little Joe factory test bed was originally going to be called LJ-7. Following the failure of both LJ-5 and LJ-5A to qualify the production spacecraft under Atlas max-*q* abort conditions, the airfame was designated LJ-5B. Spacecraft 14 had been returned to McDonnell and refurbished for this flight following the unsuccessful flight of LJ-5A. Redesignated SC-14A, it contained no landing skirt, but carried additional data recording equipment. It was returned to Wallops Island on 4 April 1961.

With everything ready the mated spacecraft and LES were bolted to the LJ-5B launch vehicle and lowered to the launch position. At 0903 on 28 April 1961, LJ-5B

Fig. 75. Little Joe 5B climbs into the morning sky at Wallops Island. (NAS)

climbed away from the launcher. All four Recruit rocket motors and one Castor motor had ignited at lift-off. The second Castor motor did not ignite until T + 5 s. The result was a lower apogee, at just 4.45 km. At that apogee aerodynamic pressure on the vehicle was approximately 937.4 g/cm^2, about twice that at an Atlas max-q abort. This flight would over-qualify the production spacecraft.

The four LJ propulsion units shut down as their propellant was consumed. The spacecraft's onboard timer initiated the abort signal at T + 33 s, as planned. The spacecraft Marman Clamp released and the Escape Motor fired. The LES pulled the spacecraft clear of the launch vehicle, before shutting down just 1 second after ignition. In its turn the LES Marman Clamp released and the Jettison Motor carried the LES away from the neck of the spacecraft. (See Fig. 75.)

The spacecraft's barostat fired the mortar at an altitude of 6,400 m and deployed the drogue parachute. The line from the base of the parachute harness pulled the Antenna Canister away from the top of the Recovery Section. A second line pulled the primary main parachute from its container in a reefed condition. When the reefing lines were cut the parachute deployed to its full extent. The line from the Antenna Canister was released, allowing the canister to descent under the drogue parachute. It was not recovered.

Spacecraft 14A made a perfect landing 14.48 km downrange from the launcher, after a flight lasting 5 minutes 25 seconds. It was recovered by helicopter and returned to Wallops Island just 30 minutes after launch.

With the exception of the late ignition of one Castor rocket motor, everything had worked as planned on this flight. STG engineers considered this 'twice as bad as planned' flight as proof that the production spacecraft and LES, as well as the reworked Marman Clamps and limit switches, could be trusted with the life of a human astronaut during a Mercury Atlas launch. The Mercury Little Joe programme had finally met its original objective. The flight-test programme had used seven Little Joe airframes, for just three successful flights.

April 1961: As the month ended Webb and Johnson reviewed the requirements of a manned Moon landing programme. Webb set the objective of landing a man on the Moon by 1967.

April 1961: Faced with the prospect of Chinese intervention in the civil war taking place in Laos, President Kennedy approached Chairman Khrushchev in Moscow with the proposition that Laos should remain neutral in South-East Asia. Despite sending troops to Thailand and basing the US Seventh Fleet off of that country, Kennedy did not want to fight in Laos. He was convinced that if America had to fight Communism in the region, they should do so in Vietnam. Khrushchev agreed with Kennedy's proposal and the Laos ceasefire document was signed on 5 May.

19.40 MERCURY REDSTONE 3 (MR-3)

On 30 April 1961 STG began the countdown for the first manned flight of Project Mercury. At T−6 h 30 min the count was held overnight, before being picked up again on 1 May.

On that date Gilruth confirmed his selection of Alan Shepard as the prime astronaut for MR-3 (see Fig. 76). The director of STG insisted that Shepard's name not be made public until the moment of launch. The continuing secrecy would allow Shepard to concentrate on his final pre-launch training without undue pressure from the media. In the event of Shepard becoming incapacitated, the secrecy would also allow his back-up, John Glenn, to make the first flight as if he had been intended to do so all along.

Shepard had spent the night in the astronaut quarters at Hangar S, Cape Canaveral. He was woken up at T−6 h and given time alone to complete his ablutions. Dressed in casual clothes, he enjoyed an informal breakfast with a few STG officials and invited friends. The astronaut's steak breakfast would become part of the Project Mercury launch day ritual. The high-protein, low-residue food consumed by the astronaut on launch day and for several days beforehand was carefully chosen in an attempt to reduce the astronaut's requirement to defecate during his flight.

After breakfast Shepard moved on to his pre-launch physical examination.

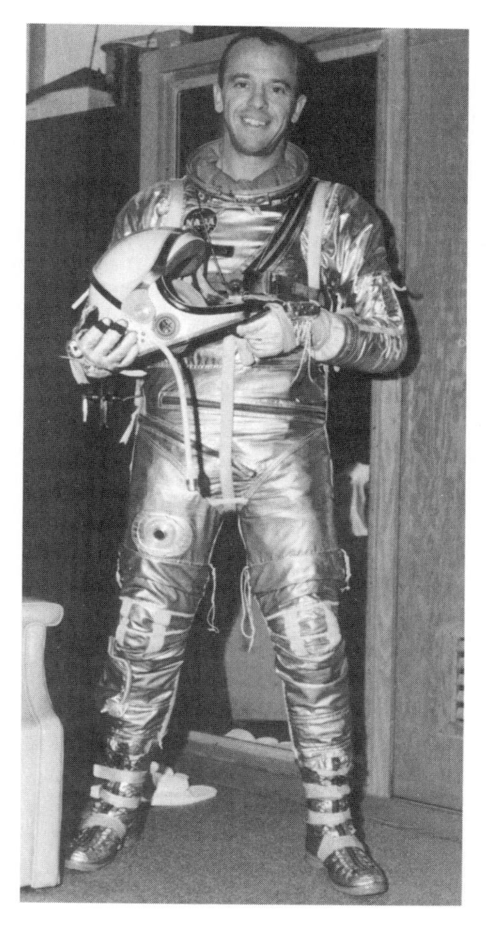

Fig. 76. Alan Shepard poses in his pressure suit. (NASA)

Having been pronounced fit to fly, he stood quietly while the physician strapped the biomedical harness around his waist and taped the sensors to his body. Next, pressure suit technician Joe Schmitt helped him dress in his long undergarment, silver pressure suit and white fibreglass helmet. The astronaut donned his pressure gloves and closed his faceplate while Schmitt inflated his suit to check its integrity.

Outside the stormy weather had caused concern in MCC throughout the previous night. At T − 4 h Shepard was instructed to remain at Hangar S, while MCC decided whether or not to commence the final portion of the countdown. When the storms did not clear the launch attempt was scrubbed at T − 2 h 20 min. Shepard returned to the suiting room, where Schmitt helped him out of his pressure suit. He then returned to his quarters. The late cancellation meant that the launch vehicle and spacecraft required a 48-hour recycle, after which the entire countdown would have to be repeated. The new launch date was set for 5 May with lift-off planned for 0700.

STG held a press conference at Cape Canaveral, during which they explained that the weather had made the launch untenable. In the course of the press conference STG public affairs officer Shorty Powers accidentally disclosed the fact that Shepard had been the prime astronaut for the launch attempt. Many years later it would become public knowledge that Gilruth had prepared three statements for the 1 May launch attempt. One would have been used if everything went well. The second would have been used if the launch was aborted, but Shepard had survived. The third would have been read at a press conference if the launch attempt had failed and Shepard had died.

The second countdown commenced on 4 May and was picked up after the standard hold, at 2000. The count stopped for the second planned hold at midnight while the pyrotechnics were installed in the launch vehicle.

Shepard was woken up at 0105 on 5 May. He showered and shaved before eating his second pre-launch breakfast. This time John Glenn joined the breakfast guests. He had spent the night in the spacecraft, setting switches to their launch positions and otherwise participating in the countdown while Shepard slept. Following breakfast Glenn returned to the LC-5

Shepard undertook his pre-flight physical and stood by while the sensors were taped to his body once again. Next he donned his pressure suit with Schmitt's assistance. With nothing left to do but wait until the countdown reached the point where he was called forward, Shepard rested in the suiting room.

At 0355, T − 3 h 5 min, Shepard left Hangar S. He entered the same transfer van that had carried HAM to the launch pad in January. He took his position in a recliner armchair, while Schmitt, physician Carmult Jackson and Gus Grissom occupied other seats in the van. Schmitt attached the umbilical from a portable life support system to the astronaut's pressure suit. He then fitted Shepard's gloves before closing and sealing his faceplate. From this point on Shepard breathed 100 per cent pure oxygen at 351 g/cm^2. The time left before launch was sufficient to purge his bloodstream of nitrogen before lift-off. While AMR security vehicles escorted the transfer van to LC-5, Shepard and Grissom relieved the tension by going through a routine from Bill Dana's cowardly Mexican astronaut, Jose Jimenez. As usual, Shepard took the role of Jimenez while Grissom played the television interviewer's role, normally taken by Dana.

When the transfer van arrived at LC-5, Gordon Cooper joined Shepard inside and briefed him on the weather and the countdown. After completing the briefing Cooper made his way to the launch blockhouse, where he occupied the 'capsule communicator's' (CapCom) position during the countdown. At T − 2 min responsibility for the launch passed from the blockhouse to MCC, where Deke Slayton was in the CapCom's seat.

Finally, Shepard left the transfer van and entered the pool of brilliant light surrounding MR-3. He walked slowly around the launch table and looked up at his vehicle, which was surrounded by white tendrils of evaporating liquid oxygen. Grissom, Schmitt and Jackson followed him at a discrete distance.

At 0715, dawn at Cape Canaveral, Shepard entered the elevator with Grissom and Schmitt and rode up to the white room surrounding the spacecraft that he had named Freedom 7. Jackson made his way to the blockhouse, from where he would monitor the telemetry from Shepard's medical sensors. In the white room, Glenn greeted Shepard with an OK hand signal. He was holding a rubber mat over the sill of the spacecraft's side hatch. The mat would protect Shepard's pressure suit as he entered Freedom 7. Holding on to a special handle installed above the side hatch, Shepard kicked off his plastic overshoes and swung his legs through the small square hatch. The launch crew, dressed in white smocks, supported the astronaut while he jack-knifed his body into his prone couch. Schmitt reached in and disconnected the portable life support system umbilicals from Shepard's pressure suit and replaced them with the umbilicals from the spacecraft's ECS, housed beneath the prone couch. He then fitted and tightened the restraint harness straps across Shepard's shoulders, chest, lap and knees. Schmitt, Grissom, Glenn and McDonnell's launch pad leader, Guenter Wendt, took turns to shake Shepard's hand and wish him good luck. The launch crew in the white room shouted '*Happy landings Commander*'.

Turning his attention to his instrument panel, Shepard read the message that Glenn had taped there. It read, '*No handball playing in this area*'. It was a reference to one of Shepard's favourite sports and mocked the tight confines of the Mercury Spacecraft's Crew Compartment. Glenn reached in and removed the tension breaker before the launch crew lifted the side hatch into position and began tightening the torque bolts that would hold it in place. Alone in his spacecraft, Shepard participated in the countdown confirming switch settings as Cooper read them off of the checklist over the radio.

The evacuation horn sounded and LC-5 was cleared of personnel. Blocks were established on surrounding roads to prevent anyone accidentally entering the area. The mobile launch gantry was rolled back to its parked position on the edge of the concrete apron. Shepard's only chance of a quick escape from a malfunctioning vehicle now lay with the yellow cherry picker parked alongside the side hatch. In an emergency he would fire the single chord of explosive to sever the bolts holding the side hatch in place, climb out onto the cherry picker and be lowered to the ground. He would then run to the waiting M-113 armoured personnel carrier and be driven to safety.

The countdown was picked up at 0600, T − 2 h, and the filling of the Redstone's liquid oxygen tank began. At 0625 Shepard began his official pre-breathing to remove the nitrogen from his bloodstream. At T − 55 min the cherry picker outside the spacecraft's side hatch was moved back to its secondary position, 7.6 m from the spacecraft's side hatch. Clouds had obscured the sky by the time the countdown approached its last few minutes. Following the loss of Mercury Atlas 1 Mission Rules had demanded clear skies at the moment of launch so that ground-based cameras could film the ascending vehicle. Meteorologists predicted that the sky would clear in the next 40 minutes, so the countdown was held at T − 15 min. Shepard spent the time looking through the spacecraft's periscope. When an inverter overheated in the launch vehicle, the hold was extended. It ultimately stretched to 52 minutes.

The countdown was recycled to T − 32 min and the clock began moving once more. At T − 15 min it was held for the second time, to correct a problem with one of the computers at GSFC. The hold lasted 1 h 42 min. Shepard returned to looking through the periscope. When the sunlight became too bright he placed a grey filter over the lens.

Mission Rules called for the launch to be cancelled if preparations required Shepard to wait more than 4 hours in his spacecraft. During the second hold Shepard passed the point where he had been in his pressure suit for almost 6 hours, without toilet facilities. When he told Cooper that he needed to urinate the initial reaction was to suggest cancelling the launch and removing him from the spacecraft. Shepard finally relieved the contents of his bladder into his pressure suit, where it formed a pool in the small of his back, but not before it had played havoc with the telemetry from his biomedical harness. Shepard told Cooper, '*I'm a real wetback now*'.

As the countdown began again and Cooper handed the microphone to Jackson, who asked Shepard, '*Are you ready?*', Shepard laughed and replied, '*Go*'. Jackson told him, '*Good luck old friend*', and let Cooper resume his role as CapCom.

Telemetry revealed that the pressure in the Redstone's fuel tank was higher than expected. The countdown was stopped. By that time Shepard was fed up waiting for his moment of glory; he called into his helmet microphones, '*Why don't you fix your little problem and light this candle?*' The countdown resumed and continued through launch without further delays.

Freedom 7's periscope retracted automatically and Shepard moved his left hand to remove the grey filter. As he did so the pressure gauge on his wrist brushed against the abort handle. If the handle had moved further it would have initiated the abort sequence and fired the LES Escape Motor. Freedom 7 would have been pulled into the sky and dropped unceremoniously in the ocean, while Mercury Redstone 3 remained on the launch table. The press and media would have had had a field day and the Soviets would have mocked America mercilessly. This was no time for a repeat of LJ-1 or MR-1. Shepard moved his arm away from the abort handle and the grey filter stayed on the periscope.

At T − 2 min the engineers in the blockhouse at LC-5 handed over their responsibilities to their counterparts in MCC. In Shepard's earphones Slayton's voice replaced Cooper's. Kraft was in charge of the flight, operating from the flight director's desk in the centre of the MCC. Mathews and other senior NASA managers were seated behind him. American television networks began broadcasting a live feed showing MR-3 standing on the launch pad. In the White House Oval Office President Kennedy took the time to watch the television coverage with his wife and the Vice President. Shorty Powers kept up a running commentary, explaining launch events as they happened to an eager American public. Across America people stopped working and turned their attention to the television. Cars pulled off the road so that their drivers could listen to the launch on their radios.

MR-3 transferred to internal power at T − 30 s. Through one of the small round portholes alongside his head, Shepard saw the electrical umbilical tower move backwards and fall to the ground as planned. Slayton gave him the final 10-second countdown over the radio. At his post-flight debriefing Shepard would admit that he could not remember hearing most of the numbers.

Fig. 77. Mercury Redstone 3 lifts off from Launch Complex 5 making Alan Shepard the second person and the first American in space. The shockwave from the Redstone's R-7 rocket motor was sufficient to knock a temporary building at the launch site over on its back, as is clearly shown in the film of the launch. (NASA)

Slayton told Shepard, '...*Zero. Lift-off.*' The Redstone ignited and climbed off of the launch table. Inside Freedom 7 Shepard reached up and started the mission clock. He reported, *'Roger. Lift-off, and the clock has started.*' Slayton came back with, '*You're on your way Jose.*' The shockwave from the launch vehicle's motor was sufficient to knock a small white building positioned alongside the launch pad over on to its back. (See Fig. 77.)

For Shepard the launch was considerably quieter than he had expected. In the centrifuge the noise of the Redstone motor was played directly into his earphones. On 5 May 1961 that noise was coming from 21 m below his back and he was dressed in a pressurised suit within a pressurised spacecraft. He felt no strong vibrations and had no difficulty hearing Slayton's voice in his headset.

At T + 30 s Shepard scanned the main instrument panel with his eyes and reported the readings to MCC, where his voice was played over loudspeakers so that the entire group of flight controllers could hear him. The astronaut was supposed to repeat the procedure every 30 seconds, but it proved unworkable during the busiest portions of the flight.

The smooth ride ended at T + 45 s. Thereafter, vibrations built up as MR-3 passed through the trans-sonic stage of the ascent. Gravitational forces built up steadily, but Shepard reported that they seemed less severe than they had been in the centrifuge. Max-q was at T + 1 min 28 s and saw the peak vibration and noise levels. Shepard's helmet was shaken severely within his shaped prone couch liner and his vision blurred slightly, but never to the point that he was unable to read the main instrument panel. Beyond max-q, Shepard reported, '*Okay, It's a lot smoother now. A lot smoother.*'

The Redstone stopped thrusting at T + 2 min 22 s, as planned. At shut down Shepard passed from 6g to microgravity in an instant. The Marman Clamp holding the LES to the neck of the spacecraft released and both motors fired together. The LES separated from the spacecraft and moved away at high speed. Shepard looked through the two small portholes, but did not observe the departing LES. On the instrument panel a green mission event light came on to confirm LES separation.

As he waited for spacecraft separation (the acronym for this action was CAPSEP, taken from the original term Capsule Separation) Shepard's heartbeat peeked at 138 beats per minute. Uniquely on this flight Shepard carried a personal parachute. In the event that CAPSEP failed, some people believed that Joe Kittinger's jump from Excelsior III had proved that Shepard might jump from the spacecraft and survive an extremely high-altitude parachute jump. More likely, he would have died as the still mated Redstone and spacecraft slammed into the ocean below.

The 10-second timer ran its course before the Marman Clamp at the base of the spacecraft was released and the three posigrade rockets in the Retrograde Package fired to provide positive separation in the microgravity environment. The green CAPSEP light lit up on the instrument panel. With Freedom 7 in free flight the periscope extended, but Shepard was too busy to look at its screen.

Freedom 7's automatic programmer fired the MPCS to carry out rate-damping manoeuvres before switching to the ASCS to commence the turnaround manoeuvre, and place Freedom 7 in a heat shield forward attitude, and a 14.5-degree negative pitch attitude. Shepard reported, '*CAPSEP. Periscope is coming out and turn around has started.*' He could not see Earth's surface, or the Redstone that had carried him into space, through either of his two portholes.

Shepard remained tightly strapped in his prone couch and could not float freely in the Crew Compartment. His most obvious sign that he was in microgravity was a single metal washer that had been left in the spacecraft by a careless engineer. The washer floated in mid-air, directly in front of his faceplate. He made a move to grab it out of the air with his left hand, but missed. The washer drifted out of his reach and he returned to his tasks.

As the spacecraft continued to climb towards apogee, Shepard unlocked the three-axis hand control and grasped it with his right hand. With his left hand he switched from ASCS to MPCS. He then switched to each axis in turn and reported the spacecraft's reaction to Slayton.

'*Okay. Switching to manual pitch.*'

'*Manual pitch.*'

He pushed the hand controller forward and then pulled it back. Thrusters fired and pitched the spacecraft's Antenna Canister up and then down again. Freedom 7 ended the manoeuvre in a 34-degree negative pitch attitude.

'*Pitch is okay.*'

While Shepard controlled pitch, the ASC remained in control of roll and yaw.

Next he took control of yaw, leaving the ASC to control pitch and roll.

'*Switching to manual yaw.*'

'*I understand. Manual yaw.*'

Twisting his wrist, Shepard found that the wrist ring of his pressure suit bumped against the emergency parachute package. He had to push hard to move the controller sufficiently to make the required manoeuvres. In reply to Shepard's movements of the hand controller a second set of thrusters fired and the spacecraft moved to left and right before returning to a central position. '*Yaw is okay.*'

Shepard returned control of yaw to the ASCS and reported, '*Switching to manual roll.*'

'*Manual roll.*'

'*Roll is okay.*'

'*Roll okay. Looks good here.*'

Finally, taking manual control of all three axes at the same time, America's first astronaut found that his spacecraft reacted in a similar manner to the ALFA trainer. He would later remark that the simulator was both valid and relevant. With the manoeuvres complete he returned control of Freedom 7 to the ASCS.

In accordance with the flight plan Shepard looked through the periscope and tried to identify any landmasses on Earth's surface below him. With the grey filter still in place, the view on the periscope screen between his knees looked like a black and white film. Even so, he told Slayton, '*On the periscope, what a beautiful view.*'

Slayton replied, '*I'll bet it is.*'

Shepard continued, '*Cloud over Florida, three to four tenths of the Eastern coast obscured, up through Hatteras. Can see Okeechobee. Identified Andrus Island. Identified the reefs.*'

As Freedom 7 reached the apogee of its trajectory, at 187.44 km, gravitational attraction overcame its upward momentum and the spacecraft began the long arching fall back to Earth. Inside the spacecraft Shepard activated the FBWS system. He noticed that the spacecraft was at 20–25 degrees negative pitch, rather than the intended 34 degrees. He began the manoeuvres to place his spacecraft in the correct re-entry attitude.

At the same time Slayton began reading the countdown to retrofire, '*Down to retro angle, 5, 4, 3, 2, 1, retro-angle.*'

Shepard replied, '*Tried retro sequence. In retro-attitude. Lights are green.*'

'*Roger.*'

'*Control is smooth.*'

'*Roger, understand. All is going smooth.*'

'*Retro-1. Very smooth.*'

'*Roger. Roger.*'

'*Retro-2. Retro-3. All retro rockets are fired.*'

'*All red on the button.*'

'*Okay. Three retros fired. Retro Jettison [switch] is back to arm.*'

Retrofire was not necessary on MR flights, but firing the retrorockets on those flights increased confidence in the automatic systems that would be employed on later manned orbital flights, when retrofire would be vital if the astronaut was to return to Earth.

Retrofire placed Shepard and his spacecraft under a retrograde acceleration of 5*g*, which stopped as soon as the third retrorocket had ceased firing. Shepard would later report, '*There is a smooth transition from zero gravity to the thrust of the retro rocket and back to weightless flying again.*' Following retrofire, the on board timer jettisoned the Retrograde Package. Shepard heard the straps separating from the wide end of his spacecraft. Looking out of the portholes he caught a glimpse of debris and a Retrograde Package securing strap as it drifted away. Returning his attention inside the spacecraft, he looked at the instrument panel, where the jettison light should have illuminated. It had not come on. He reported to Slayton, '*I do not have a light.*'

'*Understand. You do not have a light.*'

'*I do not have a light. I see the straps falling away. I heard a noise. I will use override.*'

He pushed a button to initiate the manual jettisoning of the Retrograde Package. The green light came on.

'*Override used. The light is green.*'

Slayton confirmed, '*Retro jet[tison].*'

'*Roger. Periscope is retracting.*'

Shepard returned Freedom 7 to FBWS. The spacecraft's thrusters were extremely responsive, so much so that he noticed a tendency to over steer. With acceleration forces beginning to build up the 0.5*g* light came on, Freedom 7 was commencing re-entry. Shepard turned his head to left and right in an unsuccessful attempt to observe stars and the coloured bands in Earth's atmosphere reported during the stratospheric balloon flights. His observation attempts caused him to fall behind in his tasks. He would later state that this was the one time in the flight that he did not feel that he was fully in control of the situation. Behind his back the ablative heat shield was melting away, taking the heat of re-entry with it. Ionised air molecules surrounded the spacecraft as a bow wave formed in front of it. Radio and telemetry communication was blacked out.

Shepard missed the planned altitude calls at 27,432 and 24,384 m. As the time for drogue parachute approached the astronaut became concerned that the automatic system would fail and he would have to make the deployment manually. The drogue parachute mortar fired at 6,400 m and Shepard watched the deployment through the periscope. The snorkel inlet valve opened at 4,572 m. At 3,048 m the Antenna Canister separated. It drifted away under the drogue parachute trailing the lanyard that pulled the main parachute from its canister. Shepard used the periscope to watch the main parachute deploy in the reefed condition and then de-reef and deploy to its full diameter. At main parachute deployment the on board programmer released the heat shield from the base of the spacecraft. It dropped to deploy the landing skirt. Separate green lights lit up on the main instrument panel to confirm the main parachute and landing skirt deployments.

The astronaut pressed a button to dump overboard the 70 per cent of his ASCS, and 90 per cent of his MPCS, propellant loads that remained in the tanks. As Freedom 7 drifted towards the ocean she passed below the horizon and radio communication with MCC was lost. Controllers were reduced to listening to the voices of the recovery forces.

The same crew that had recovered HAM at the end of the MR-2 flew the prime recovery helicopter for this flight. It seemed that Yeager's words would haunt Shepard to the very last. Having been plucked from the ocean and winched up to the helicopter, he would have to wipe the metaphorical monkey shit off of the seat before he could sit down.

The aircrew observed the descending spacecraft approximately 5 minutes before landing. In the final stages of the descent the helicopter pilot flew circles around the ballistic spacecraft as it hung beneath its parachute, waiting to be dumped into the ocean. In the spacecraft Shepard prepared himself for landing. He opened his faceplate, undid his chest and knee straps and removed the umbilical that supplied oxygen to the faceplate seal on his helmet. Then he lay back and waited for impact. Landing occurred at the end of a flight lasting 15 minutes 22 seconds. Shepard compared the landing forces to those experienced by a jet fighter pilot being catapulted from the flight deck of an aircraft carrier. On impact Freedom 7 listed 60 degrees from the vertical, Shepard noticed that one porthole was completely under water. He searched for any seawater inside the Crew Compartment, but found none. Relaxing, he removed his helmet.

The main parachute guillotine severed the parachute harness, to prevent the canopy filling with wind and dragging the spacecraft through the water. The disc of green fluorescent dye leaked its contents into the waves surrounding the small black cone as it bobbed in the water. Shepard observed the approaching recovery helicopter on the periscope screen as he waited for Freedom 7 to right herself.

Although the spacecraft was floating in an upright position after only 60 seconds on the ocean, US Marine Corps Airman George Cox had secured the helicopter's winch cable to the loop on the neck of the spacecraft by the time that it had righted itself. On Cox's orders, pilot US Marine Corps Airman Wayne Koons, took the strain on the winch line. At that moment the spacecraft's HF recovery antenna sprung out of its container and dented the underside of the aircraft. When Koons asked Shepard if he was ready to leave the spacecraft he reported that he could still see water against the portholes. Koons lifted the spacecraft until it was just out of the water.

Shepard detonated the explosive chord holding the side hatch in place. The hatch fell into the water and sank. He manoeuvred himself to a position where he could reach through the hatch to grab the waiting 'horse collar' at the end of the helicopter's second winch cable. When he was secure in the horse collar he let Cox winch him up to the waist door of the helicopter. Once in the aircraft he strapped himself into a seat. Koons lifted the aircraft and Freedom 7 and headed for the prime recovery vessel. After a short flight Koons lowered Freedom 7 on to a waiting dolly on the flight deck of the aircraft carrier USS *Lake Champlain*. When the winch line was released Cox reeled it in to its parked position. Koons then landed his aircraft on the carrier.

Still dressed in his silver pressure suit, minus his gloves and helmet, Shepard disembarked from the helicopter in front of a jubilant ship's company. He had ridden Freedom 7 487.2 km downrange from LC-5. He had experienced 6*g* during launch and 11*g* during re-entry. In the middle he had experienced microgravity for approximately 5 minutes. The only in-flight failure had been the bulb in the Retrograde Package jettison light and even that had come on when Shepard had pushed the manual jettison button. (See Fig. 78.)

Following recovery NASA physician Jerome Strong led Shepard below deck to the carrier's Sick Bay, where he gave the astronaut a post-flight physical examination. Shepard was pronounced fit and well. He showered before donning a bright orange astronaut's flight suit, with his silver pressure suit boots.

While his memories were fresh, Shepard began a personal debrief into a tape recorder. After 30 minutes he received a telephone call from President Kennedy. When he had completed his debrief he was flown to Grand Bahamas Island. On his arrival there he was greeted by Grissom and Slayton.

For the next three days Shepard was subjected to further physical examinations and debriefing sessions. The initial debriefings were in three separate areas: medical and life sciences, astronaut activities and performance and systems performance.

Shepard said that he was pleased with the performance of his pressure suit, although the pressure gauge on the left wrist had been hard to see during high acceleration periods. He suggested that it might be moved to the left knee, where it would be directly in front of the astronaut's face when he was in the prone position. He had had to continually withdraw his fingers from the rubber tips fitted to his gloves, as they had caused blood circulation difficulties. His helmet had been fitted with a larger faceplate prior to the flight and he felt that it had offered him a better view than the original faceplate.

During launch the restraint harness on his prone couch had seemed tight, but he had overcome the difficulty by shrugging his shoulders. The sensors on his

Fig. 78. Alan Shepard is recovered by helicopter after a flight lasting less than 16 minutes. A medical problem would ground him for many years and he would not fly in space again for another 10 years. (NASA)

biomedical harness had caused minor irritation and Shepard suggested that a different adhesive be identified and used. The pressure suit had maintained a comfortable temperature throughout the flight. The Crew Compartment temperature had also remained comfortable. Only one bottle of breathing oxygen had been used, the second bottle had remained untouched.

The astronaut said that the spacecraft had manoeuvred in a manner similar to the simulators that he had trained on. The one exception was the fact that the spacecraft had rolled slightly in flight. When engineers stripped the thrusters down they found debris in the hydrogen peroxide plumbing that may have caused it to leak. A continuous leak would have caused the spacecraft to roll in keeping with Newton's law of action and reaction. At one point, when he had activated FBWS, Shepard had forgotten to deactivate the ASCS. He stated that the rate changes had seemed high at the time. He had put the high rate changes down to micro-switch settings in the control system.

He reported that all pyrotechnics had fired on time. One back-up LES Jettison Rocket was later recovered and disassembled. It suggested that it had been ignited by manual control. Telemetry showed that Shepard had not jettisoned the LES manually. All other pyrotechnics had fired as planned and the Retrograde Package jettison light was the only indicator failure throughout the flight. Telemetry revealed that the Retrograde Package had been jettisoned by the primary system, but the electric current had failed to reach the indicator light. When Shepard had activated the back-up system the jettison light had come on.

All landing and recovery systems had worked as planned. The heat shield had recontacted the aft bulkhead after landing and one stud had pierced the fibreglass bulkhead installed after MR-2. Recovery was too rapid to test the seaworthiness of the landing skirt in its sea anchor role. Post-flight investigation revealed a number of tears in the landing skirt, but these were probably caused due to mishandling after recovery. All recovery operations went as planned.

On 7 May Shepard was flown to LaFRC and reunited with his family. The following day all seven Flight A astronauts and their wives attended a reception at the White House in Washington. In the Rose Garden President Kennedy presented Shepard with NASA's Distinguished Service Medal. Immediately prior to the presentation the President dropped the medal on the floor. Having stooped to pick it up he joked to Shepard, '*This medal has gone from the ground up*.'

Shepard and his wife were then driven to a dinner at The Capitol Building. Cheering crowds lined the streets. The following day Shepard received a tickertape parade along New York City's Broadway in an open top car. It did not seem to matter to the American people that Shepard had only made a suborbital space flight and that one-month after the Soviet cosmonaut Yuri Gagarin had made a complete orbit of Earth. Across the nation Alan B. Shepard junior was a hero.

In the aftermath of MR-3 President Kennedy began looking for ways to use the American public's new enthusiasm for space flight to improve American industry and education system. He wanted to dispel the doubts that the American people had suffered in the wake of Sputnik. Most of all, he wanted the American public and the entire world to be left in no doubt that the American free enterprise system was far superior to the Soviet's centralised Communist regime.

In the wake of the Bay of Pigs invasion and Gagarin's space flight, Kennedy had looked for a goal in which America had a chance to show its supremacy. Advisers had told him that sending men to the Moon was sufficiently long-range that new technology would have to be developed. That offered the American aerospace industry the opportunity to develop new, state-of the-art technologies – something they were good at. It also meant that the Soviets would have to develop comparable new technologies and would not be able to rely on their present lead in rocketry.

9 May 1961: Vice President Johnson travelled to Saigon, the capital city of South Vietnam. He carried a letter to President Diem from President Kennedy. The letter offered to change America's role in South Vietnam from that of a military advisory team working only in Saigon to that of sharing responsibility for the country's defence. Diem wrote back and accepted the offer on 19 May.

14 May 1961: A bus being used by Negro and white 'Freedom Riders' seeking basic Civil Rights for America's Negro population was firebombed in Annisan, Alabama.

17 May 1961: President Kennedy made a one-day visit to Canada. He failed to convince the Canadian President to allow him to base American nuclear-armed ICBMs on Canadian soil.

19 May 1961: The Freedom Riders continued their journey, but were attacked at Birmingham, Alabama. US Marshals escorted their bus beyond the Birmingham city limits.

20

Sustained effort

20.1 COMMITMENT

On May 25, Kennedy addressed Congress on what he termed, 'urgent national needs'. In what is regarded as a second State of the Nation address, Kennedy asked for additional spending on nuclear missiles and conventional weapons. He called for increased recruiting for the Marine Corps and a reorganisation of the Army to make it more flexible to respond to different international crises. He also called for the funding of a national civil defence programme to protect the American population from a Soviet nuclear attack.

With the advent of a successful first manned flight in Project Mercury President Kennedy made it clear that he was prepared to call for an expanded space programme to compete with and overtake Soviet achievements. Webb and McNamara cooperated on a memorandum for Kennedy that set out how they felt the future of America's space programme should develop. Project Apollo, now a manned lunar landing programme, would be the centre of a national space programme that also included the development of unmanned lunar and planetary probes, separate meteorological and communication satellite networks, a nuclear rocket and an increase in space science research.

Kennedy's speech covered everything in the Webb–McNamara memorandum, but is best known for the paragraph that committed America to an eight-year programme to land an astronaut on the lunar surface. While the nation's politicians listened to their President first hand millions of Americans watched on television as he told them:

> I believe that this nation should commit itself to achieving the goal, before this decade is out, of landing a man on the moon and returning him safely to the Earth. No single space project in this period will be so impressive to Mankind or more important for the long-range exploration of space. And none will be so difficult or expensive to accomplish.

Kennedy's speech was much more than a commitment to a greatly enhanced space programme. It also called for increased expenditure on the four armed services

Fig. 79. President John F. Kennedy addresses a joint session of Congress and commits America to landing a man on the Moon and returning him safely to Earth before the end of 1969. (NASA)

and increased recruitment to counteract the build-up of Soviet military forces. Congress debated the numerous recommendations, voted, and approved them. (See Fig. 79.)

As Eisenhower had seen Sputnik, so Kennedy saw the challenge of landing men on the Moon as a vehicle for advancing American education and the relationship between the country's universities, industry, the military complex and civilian government agencies such as NASA. He also saw the Moon landing as a direct Cold War challenge to the Soviet Union. The lunar landing goal came as a surprise to many of on NASA's payroll. Following President Kennedy's speech Project Apollo was redirected to meet the new goal.

Webb used Apollo funding to initiate an educational programme for engineers and scientists. NASA's Administrator worked hard to keep the nation's politicians behind Project Apollo even after the war in Vietnam began to demand more of their attention and funding. Contracts for Apollo hardware were used to gain political support from Congressmen and Senators who wanted money spent in their constituencies, thereby, providing work for their voters. As Project Apollo grew rapidly in size and complexity so did NASA.

May 1961: In Russia OKB-1 had commenced design studies for their N-1 launch vehicle in the early 1960s. When President Kennedy committed NASA to landing a man on the Moon OKB-1 engineers suggested using an N-class launch vehicle to send two cosmonauts around the Moon and back to Earth. An argument over what propellants to use in the N-1 launch vehicle led to a split between Korolev and Valentin Glushko, the principal designer of rocket motors in the Soviet missile programme. Korolev preferred to use the same propellants as in the R-7. Glushko wanted to use of cryogenic propellants and refused to participate in the N-1 programme. Korolev was forced to employ an engineer with only limited experience in developing rocket motors and employ numerous medium size motors in the N-1 rather than a few large motors, as the Americans had developed for their Saturn C5 launch vehicle.

Premier Khrushchev had previously assigned OKB-52, under the control of Vladimir Chelomei, the task of developing a launch vehicle and a manned spacecraft capable of supporting a circumlunar flight with a single cosmonaut. The three-stage vehicle, designated UR-500K, burned cryogenic propellants. Today it is better known by the name Proton. The new one-man spacecraft was designated LK-1. Flight tests of the LK-1 spacecraft would be carried out under the cover Zond.

25 May 1961: President Kennedy was warned of Soviet threats to sign a peace treaty with East Germany. Such a treaty would end the state of war between the Soviet Union and East Germany and give the German Democratic Republic control of its own country. With no existing state of war the East Germans would be able to demand the Americans, British and French occupation forces to leave West Berlin. That would result in the West Berlin population coming under control of the East Germans. Kennedy was not prepared to see that happen.

The Cold War conflict in Berlin revolved around two points:

- The Soviets needed to stop the flow of many thousands of professional East Germans defecting to the West.
- The Americans, British and French needed access to West Berlin in order to protect the West Berliners' freedom. In the wake of the Bay of Pigs, Laos and Gagarin's flight in space, Kennedy was not prepared to back down over Berlin.

31 May 1961: President Kennedy visited President De Gaulle in Paris, France. De Gaulle offered him advice on two subjects:

- Stand fast on West Berlin.
- Do not involve the US military in South Vietnam.

June 1962: James Chamberlin defined the Mercury Mark II spacecraft. It would be launched by a converted USAF Titan II ICBM and carry most of its equipment outside of the Crew Compartment. The single astronaut would use an aircraft-style ejection seat to escape from a launch failure and he would have manual control of the spacecraft throughout the flight. That flight would last up to 24 hours and would end with a land landing under a Rogello Flexible Wing. A large overhead hatch would give easy ingress and egress. The hatch also offered the opportunity to develop extravehicular activities. At a presentation to STG engineers on 27 July, Mercury Mark II was accepted. Future work on the concept would concentrate on a two-man spacecraft that McDonnell had proposed.

STG began negotiations with the USAF for the development of a man-rated Titan II. Meanwhile, Chamberlin commenced talks with Lockheed for the purchase of Agena rocket stages, which, with the addition of a Target Docking Adapter developed by McDonnell, would become the target for orbital rendezvous and docking practice.

Meanwhile, McDonnell was developing the Mercury Spacecraft for its own 22-orbit flight. This would become the final flight of Project Mercury.

1 June 1961: NASA announced that 10 Saturn C-1 flight-tests would be made using two-stage vehicles, the last few of which would carry boilerplate Apollo Spacecraft. Nineteen days later a committee was set up to review lunar flight modes for the various Saturn launch vehicles. The proposed Saturn C-3 and Saturn C-4 were both capable of completing a manned lunar landing mission by employing Earth Orbit Rendezvous methods. The committee also recommended the use of the J-2 rocket motor as a new second stage for the Saturn C-1 and the second stage of the Saturn C-3.

3–4 June 1961: Kennedy and Khrushchev meet in Vienna. In public they were friendly towards each other, but in private the Russian totally dominated the American. The Soviet Chairman refused to discuss a Nuclear Test Ban Treaty. Khrushchev was adamant that the Soviets would sign a peace treaty with East Germany by the end of the year, thereby destroying America's right to keep troops in West Berlin. The Soviet Premier told Kennedy that the Soviet Union would not resume atmospheric testing of nuclear weapons unless America did so first.

At the end of the second day of meetings Khrushchev informed Kennedy that he did not want to commit his country to a manned landing on the Moon as it would be too expensive. Kennedy offered him the opportunity for America and the Soviet Union to fly to the Moon together. Khrushchev refused the offer, saying that he would not consider such a thing until after their two countries had begun nuclear disarmament.

Kennedy was surprised by the way Khrushchev spoke to him throughout the two days of private meetings. He ended the final day warning the Soviet Chairman that he was prepared to go to war – all out nuclear war – with the Soviet Union over West Berlin.

5 June 1961: President Kennedy visited Prime Minister Macmillan in London, UK, on his way home and informed him of what had happened in Vienna.

6 June 1961: Kennedy returned to America and immediately began preparing for a war with the Soviet Union over Berlin. At that time the only plan America had to reply to a Soviet nuclear attack was all out retaliation against all Communist countries. This included launching nuclear weapons against China, which did not possess nuclear weapons.

20.2 MERCURY REDSTONE 4 (MR-4)

The countdown for Mercury Redstone 4 began on 17 July 1961. Although the weather forecast for Cape Canaveral was good, the weather for the landing area was predicted as cloudy for the next 48 hours. At T – 6 h 30 min the countdown was held overnight. The launch attempt was cancelled during the hold, in favour of waiting for better weather in the landing area. As propellant loading had not started, the countdown was recycled 24 hours. The new countdown began on 18 July and proceeded once again to T – 6 h 30 min, where it was stopped overnight prior to commencing propellant loading. Launch was planned for 0700 on 19 July.

The following morning Gus Grissom (Fig. 80) was awakened in the astronauts' quarters at Hangar S and allowed time to shower and shave. He then moved into his pre-launch breakfast with friends and Project Mercury managers. Next he undertook a pre-flight physical before joining Joe Schmitt to don his pressure suit. Following Shepard's 'wetback' incident, where he had been forced to urinate in his pressure suit, Grissom wore a makeshift urine reservoir beneath his undergarment. Other modifications to the pressure suit as a result of Shepard's experience included nylon ball-bearing joints in the wrist rings, new microphones in the helmet and a new design of parachute harness.

By 0500 the astronaut was in the white room at LC-5. On Wendt's signal Grissom moved forward and was helped into the spacecraft that he had named 'Liberty Bell 7'. The ECS umbilicals were attached to his suit and the restraint harness straps were tightened across his body. The new explosive hatch that the astronauts had demanded was bolted in place. When the hatch was secure the external lanyard that allowed it to be opened from the outside was securely stowed on the exterior of the spacecraft.

Fig. 80. Virgil Grissom. (NASA)

The countdown proceeded until T−10 min 30 s, when a hold was called to waitfor a gap in the solid cloud overhead. When no gap materialised the launch was cancelled and Grissom was removed from the spacecraft and returned to Hangar S. Because propellant loading had taken place the launch was recycled for 48 hours. A new launch target was set for 0600 on 21 July.

MR-4's third countdown began on 20 July and was held overnight at T−6 h 30 min. When it was picked up again in the early hours of the following morning the weather was still far from perfect, but the decision was taken to proceed with the launch attempt. Grissom was woken up at midnight and allowed to shower. He had shaved the previous night to save time. Following the now standard steak breakfast he took his pre-launch physical examination and then let Schmitt help him into his pressure suit.

Grissom left Hangar S at T−3 h 45 min and took his seat in the transfer van. During the drive out to the launch pad Slayton briefed Grissom on the weather and the condition of his spacecraft, the launch vehicle and the countdown, as he had on the previous launch attempt. When the transfer van drew up at LC-5 the countdown had been held pending Grissom's arrival. The astronaut rode the elevator up to the white room, where back-up astronaut John Glenn briefed him on the status of Liberty Bell 7. Glenn had spent much of the night on his back inside Liberty Bell 7, taking part in the countdown and ensuring that all instruments and switches were set to their correct launch positions.

Grissom was helped into his spacecraft and Schmitt connected his pressure suit to the ECS. Schmitt and Wendt then strapped Grissom tightly in place. The side hatch was bolted into place and its external lanyard secured. The launch structure was evacuated and rolled back to its parked position at the edge of the concrete apron. The evacuation horn sounded and LC-5 was emptied of personnel. Grissom was left alone at the top of MR-4 looking through Liberty Bell's periscope and the large 'heads up', or 'picture' window directly in front of him. At T−2 h, liquid oxygen began to flow into the Redstone's tank. Throughout propellant loading the cherry picker rested just outside the spacecraft's side hatch.

Cooper was flying a chase aircraft waiting to film the launch. Slayton was blockhouse CapCom, with Carpenter assisting him. Shepard was MCC CapCom, and would talk to Grissom throughout his flight. Schirra was in MCC as an observer. Glenn had been Grissom's back-up and was in the blockhouse at LC-5.

At T−45 min the countdown was held for a misaligned bolt holding the side hatch in place. After close inspection it was decided to launch with one less bolt holding the hatch. The misaligned bolt was removed and was presented to Grissom after the flight. The countdown resumed after 30 minutes. Fifteen minutes later it was held again, at T−30 min, while the launch pad searchlights were extinguished. They had been proven to interfere with the telemetry downlink on previous launches. The countdown resumed once more, but was stopped at T−15 min. The cloud cover had reduced the local lighting conditions to less than that required by the cameras that would record the launch, as demanded in the wake of the MA-1 loss. During the hold an inverter began to overheat. Grissom switched to a back-up converter, to allow the primary unit to cool down.

When the countdown resumed it continued without further difficulty. The cherry picker was lowered to the ground at T − 6 min. Everyone watched and waited, and at T − 0 the Redstone's single rocket motor burst into life. As thrust built up it finally overcame the weight of the fully loaded vehicle, and Mercury Redstone 4 climbed slowly away from the launch table. Grissom reported, '*Roger. This is Liberty Bell 7. The clock has started.*'

Shepard replied, '*Loud and clear Jose, don't cry too much.*'

Lift-off occurred at 0720. In his post-flight debriefing Grissom would admit to '*feeling a bit scared*' at lift-off. The feeling left him as the acceleration forces built up steadily, in a close approximation of the many simulated Redstone lift-off profiles that he had 'flown' in the Johnsville centrifuge. As acceleration forces increased, the spacecraft began to buffet, but additional foam placed in the prone couch headrest after Shepard's flight meant that Grissom's vision was never blurred as his predecessor's had been. MR-4 quickly passed through a layer of broken cloud and out of view of observers on the ground.

The Crew Compartment environment sealed off at 8,229 m as planned. Grissom felt a sense of elation that the ECS was working correctly. The Crew Compartment pressure was 97 °F, but his pressure suit remained at 57.5 °F. As Shepard had done before him, he kept up a running commentary of the readings on the instrument panel in front of him.

At T + 2 min 22 s the Redstone's A-7 rocket motor stopped thrusting. Grissom would report after the flight that he experienced a strong sensation of tumbling forward as he passed instantly from the high *g* forces of launch to free-fall. The feeling passed quickly. One second after shutdown he reported, '*There went the tower*'. Through his window he saw the LES Jettison Motor ignite and pull the red escape tower clear of the neck of his spacecraft. Nine seconds later the Marman Clamp at the base of the spacecraft fired, and was followed by the ignition of the three posigrade rockets in the Retrograde Package. The automatic systems then carried out rate dampening before commencing the turnaround manoeuvre. Grissom told Shepard, '*There went the posigrades, spacecraft has separated. We are go at zero-g and turning around and the sun is really bright.*'

In Mercury control the light had illuminated to confirm spacecraft separation from the Redstone. Shepard replied, '*Roger cap. sep. is green; turnaround has started, manual handle out.*' Grissom pulled the manual control handle out to its fully extended position and then came back with, '*Oh boy! Manual handle is out; the sky is very black; the spacecraft is coming around into orbit attitude; the roll is a bit slow.*' The slow turnaround was later found to have been caused by decomposed material in the roll thrusters.

Despite the large window, Grissom found nothing to look at to confirm that Liberty Bell 7 was turning around. He ended up relying on his instruments and the sunlight moving up his body confirm the manoeuvre. At no time during, or after the turnaround, did he see his Redstone launch vehicle.

Having activated the MPCS, Grissom took control of all three axes at once, unlike Shepard who had manoeuvred in one axis at a time. First he manoeuvred in pitch, but reported that the system was very sluggish. He pitched Liberty Bell 7 up, from negative 34 degrees to negative 24 degrees. He reported, '*I'm having a little trouble with rate ... ah ... with the manual control.*'

'*Roger.*'

'*If I can get her stabilised here, all axes are working alright.*'

'*Roger. Understand manual control is good.*'

'*Roger, it's, it's sort of sluggish, more than I expected.*'

'*Okay, I'm yawing.*'

'*Roger, yaw.*'

'*Left, ah ... Okay, coming back in yaw. I'm a little bit late there.*'

'*Roger. Reading you loud and clear Gus.*'

'*Lot of stuff ... There's a lot of stuff floating around up here.*'

'*Okay, I'm going to skip the yaw, ah, the roll, because I'm a little bit late and I'm going to try this rough yaw manoeuvre. About all I can really see is clouds. I haven't seen any land yet.*'

'*Roger. You're on the window. Are you trying a yaw manoeuvre?*'

'*I'm trying the yaw manoeuvre and I'm on the window. It's such a fascinating view out of the window you just can't help but look out that way.*'

The manoeuvres in pitch and yaw had both gone past their planned attitude and Grissom had had to make rate-damping manoeuvres to bring Liberty Bell 7 to a halt in the correct attitude. When time got the better of him he had cancelled the planned roll manoeuvre. Next he saw a cloud covered coastline that he was unable to identify. It was later identified as the Apalachicola region of Florida.

With the manoeuvres complete, Grissom returned Liberty Bell 7 to ASCS and was reminded by Shepard that, '*Retro-sequence has started, go to retro-attitude*'. Grissom replied, '*... I'm not in very good shape here.*'

At T + 5 min 11 s Grissom reported that the first retrograde rocket had fired. The remaining two fired at 5-second intervals and the onboard timer jettisoned the Retrograde Package. The appropriate lights illuminated on Grissom's instrument panel.

With retrofire complete Grissom activated the RSCS and then completed a quick communication check on HF, before returning to UHF. He reported that he could see Cape Canaveral through the window. The Earth observations and HF communications checks were written into the flight plan, but coupled with the difficulties he had had in manoeuvring the spacecraft manually, they caused him to fall behind schedule on his tasks. When the periscope retracted, at T + 6 min 41 s, it took Grissom by surprise. He quickly recovered and manoeuvred his spacecraft towards the correct re-entry attitude, which he achieved at T + 7 min. Liberty Bell 7 was at an altitude of 190.28 km.

Asked by Shepard what he could see out of his window, Grissom replied, '*Ah, the sun is coming in so all I can really see is just darkness. The sky is very black.*' The bright sunlight made it difficult for him to read his instruments.

The 0.5*g* light came on at T + 7 min 54 s, as Liberty Bell 7 began to feel the effects of gravitational attraction pulling it back towards the waiting ocean. Twenty-six seconds later Grissom reported that he was experiencing '*... about 10gs*'. Thereafter, the acceleration forces began to fall off again until they reached 1*g*.

At 6,400 m the drogue parachute deployed and the periscope extended. Grissom watched the parachute deploy through his window. The Antenna Canister was jettisoned and the lanyard pulled the main parachute from its canister, it deployed in a reefed condition. Liberty Bell 7 was at 3,749 m, several hundred metres higher than the main parachute was meant to deploy. The heat shield dropped away from the base of the spacecraft to deploy the landing skirt. All of the appropriate event lights illuminated on Grissom's instrument panel and in Mercury Control. The astronaut disconnected his oxygen umbilical from the torso of his pressure suit and opened the faceplate on his helmet. He did not seal the connector that the oxygen umbilical had been fixed to. He undid some of the straps of his restraint harness.

As Liberty Bell 7 dropped below MCC's horizon the Atlantic tracking ship *Coastal Sentry Quebec* began relaying communications back to Cape Canaveral. Grissom confirmed that his unused thruster propellant had dumped overboard as planned. He also explained that he had replaced one of the pins back into the main hatch jettison plunger to prevent it jettisoning on water impact. He was having difficulty relocating the second pin. He reported a triangular shaped rip in the main parachute, approximately 5 cm on each side. Although he continued to watch the hole, the tear-resistant parachute material prevented it from getting any bigger. He gave the tracking ship a report on the status of his spacecraft.

With the snorkel valve open the internal pressure in the Crew Compartment came up to that of the ambient air pressure outside. The ECS's pure oxygen was now mixed with air entering the spacecraft through the valve. As the spacecraft approached the ocean Grissom watched the waves on the periscope screen. He braced himself for impact. Liberty Bell 7 impacted the ocean at T + 15 min 37 s, after a flight that had taken it 486.07 km downrange from LC-5. It was 0735. The landing was 4.82 km downrange from the target area, where the aircraft carrier USS *Randolph* was waiting.

Landing was not as rough as Grissom expected it to be, but the spacecraft heeled over to one side before righting itself. He deployed the recovery aids and the whip antenna before jettisoning the reserve main parachute. The spacecraft moved continually up and down in the waves. The primary helicopter approached the spacecraft, circled above it, and the pilot began talking directly to Grissom, using the aircraft's radio call sign Hunt Club 1.

Having looked and found no water inside his spacecraft, Grissom disconnected the visor seal umbilical from his helmet. He unlocked his helmet neck ring but did not remove the helmet itself as it contained his radio microphones and earphones. Reaching through the open faceplate, he struggled to unroll the rubber neck-dam built into his pressure suit. The rubber collar kept rolling back-up each time he let go of it. He finally succeeded in deploying it correctly.

Talking to Hunt Club 1, Grissom asked the pilot to allow him a few minutes to note down the final instrument readings and switch positions in his spacecraft. Navy Lieutenant James Langley, the helicopter pilot, agreed and continued to fly circles around the spacecraft, while confirming that he was ready to move in and make the recovery as soon as Grissom told him to do so. The pilot of a second helicopter, with the call sign Card File 9, made contact with Langley and confirmed that his aircraft was in location to prevent any further aircraft approaching the recovery area. Once the recovery had been completed Card File 9 would escort Hunt Club 1 back to the USS *Randolph*.

Grissom found it difficult to record his instrument readings while wearing his pressure suit gloves. At the same time his body temperature began to rise, causing the rubber neck dam to balloon. He relieved the pressure by occasionally pulling the neck dam away from his neck with a finger. Finally, he was ready and called Hunt Club 1:

'*Okay, hunt Club, this is Liberty Bell. Are you ready for pick-up?*'

'*This is Hunt Club 1. This is affirmative.*'

'*Okay. Latch on, then give me a call and I'll power down and blow the hatch, Okay?*'

'*This is Hunt Club 1, will give you a call when we're ready for you to blow.*'

'*Roger. I've unplugged my suit so I'm kinda warm now.*'

'*One, roger.*'

'*Now, if you tell me to, ah, you're ready for me to blow, I'll have to take my helmet off, power down and then blow the hatch.*'

'*One, roger, and when you blow the hatch the collar will already be down there waiting for you, and we're turning base at this time.*'

As the helicopter began its approach, winchman Lieutenant John Reinhard reached out of the waist door with a long pole in his hand. On the end was an explosive cutter designed to severe the whip aerial that had deployed following landing. The aerial had to be cut away to clear the way for Reinhard to hook his winch line on to the recovery loop on the neck of the spacecraft. This was a new procedure brought in following Shepard's recovery at the end of the MR-3 flight, when the whip antenna had deployed and struck the recovery helicopter.

Reinhard was preparing to sever the antenna when he saw the side hatch come off of the spacecraft and skip across the water before sinking. Inside Liberty Bell 7 Grissom had disconnected the second umbilical from his pressure suit and was lying on his back waiting for Langley to tell him that everything was ready for the recovery. As instructed in his recovery checklist, he had removed the survival knife from its position above the side hatch and had placed it in the survival package. He was considering removing the knife from the survival package, so that he could keep it as a souvenir, when he heard 'a dull thud' above his head. He realised immediately that the side hatch had jettisoned.

Waves began lapping over the lower sill of the now open side hatch and water began entering the Crew Compartment. Grissom removed his helmet, grabbed the main instrument panel and pulled himself out through the side hatch. When he was in the open water he had to free himself from the chord holding the dye marker canister before swimming rapidly away from the spacecraft as it sank. When he stopped swimming he looked for Hunt Club 1's 'horse collar' and quickly realised that the recovery helicopter's crew were concentrating on trying to recover the spacecraft.

In Hunt Club 1, Reinhard cut the whip antenna and grabbed the tool that allowed him to guide the hook on the end of the extended winch line through the recovery loop on the spacecraft. He struggled with the task as the spacecraft was rapidly sinking. Grissom swam back to the spacecraft to see if he could assist Reinhard. When he reached Liberty Bell 7 only the spacecraft's neck was still above water, and Hunt Club 1's winch line was securely attached to the recovery loop, but Langley was flying his aircraft with three wheels in the water. Liberty Bell 7 sank out of sight but Reinhard's winch line snapped tight, the spacecraft was still attached to the end of it. Slowly, Langley began lifting his aircraft straight up. He succeeded in pulling Liberty Bell 7 clear of the waves but, with the spacecraft full of seawater, he was unable to make headway towards the USS *Randolph*. As he and Reinhard attempted to recover the spacecraft, Langley called another aircraft in to recover the astronaut.

Grissom was in difficulty. His pressure suit was taking in water through the oxygen valve that he had failed to close when he had unplugged the first umbilical. His pressure suit also contained a number of rolls of 25 cent coins that he had carried on the flight and intended to present to friends and colleagues as souvenirs. They were not heavy, but in his present situation they were unnecessary weight. His neck dam also leaked, allowing air to escape from his pressure suit. This not only retarded the suit's buoyancy but also allowed more water to be sucked in through the open umbilical connector. He struggled underwater to close the connector and finally succeeded, but his head and face were being pulled under the water. Each time he broke the surface again the waves broke over his head and he swallowed seawater, which made him cough.

When a second aircraft arrived Grissom began waving frantically to the aircrew. Rather than move in and recover him from the ocean, the aircraft began manoeuvring so that a cameraman on board could photograph the astronaut in the water. In the air the crew of the second helicopter misinterpreted Grissom's waving as a friendly gesture and waved back. Grissom would admit later that he started to panic at this point. He became annoyed and then scared as his pressure suit continued to take in water and the second aircraft made no attempt to drop a horse collar and recover him.

The spacecraft recovery attempt ended when a light illuminated on Langley's instrument panel, warning him that Hunt Club 1's engine was overheating. The safety of the aircraft and the lives of its two crewmen were more important that a spacecraft that had performed its only flight and was now little more than a museum piece. Langley instructed Reinhard to sever the winch line and let Liberty Bell 7 sink to the bottom of the Atlantic Ocean. The waist door camera recorded Liberty Bell 7 as it fell back into the water and began to sink. With the spacecraft gone, Langley was able to fly his aircraft back to the USS *Randolph*, where it would later be examined and the engine warning light determined to be a false indication. (See Fig. 81.)

With Hunt Club 1 now out of the area a third aircraft, piloted by the same crew that had recovered Alan Shepard, moved in to recover Grissom. Grissom found himself caught between the rotor wash of the aircraft that was photographing him and the aircraft that was attempting to recover him. As the recovery aircraft flew towards him its rotor wash pushed him away. The horse collar was 10 m away from him when it hit the surface of the ocean. Weary and frightened, Grissom summoned

Fig. 81. Water pours from Liberty Bell 7's side hatch as Hunt Club 1 pulls it from the water. A faulty indicator light in the helicopter led to the spacecraft being dropped and left to sink. (NASA)

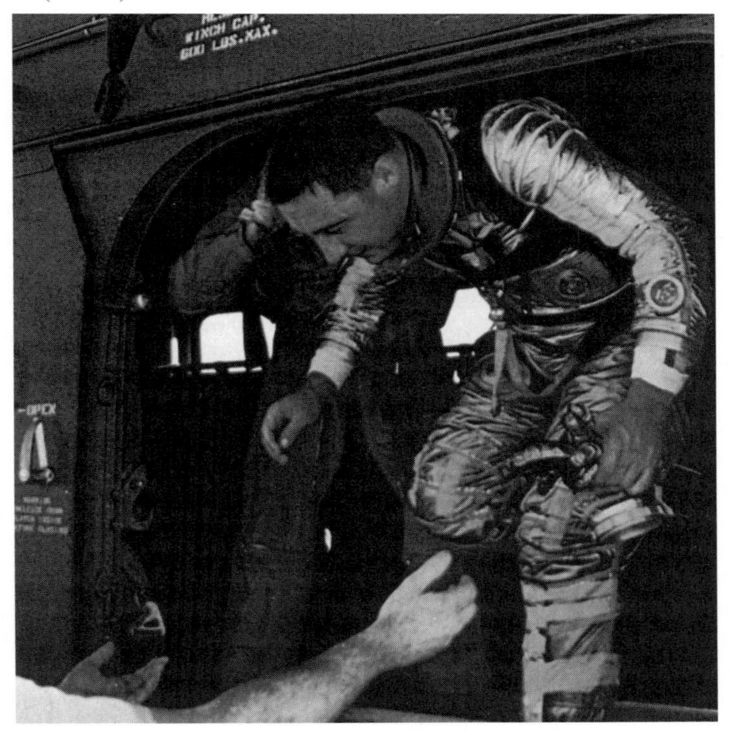

Fig. 82. Grissom leaves the recovery helicopter on the recovery vessel. (NASA)

up his energy and began swimming, in his restrictive, water-filled, pressure suit and against the aircraft's rotor wash in order to reach the horse collar. When he got there he allowed himself to flop backwards into the horse collar and be winched into the air. The winchman pulled him in through the waist door and removed the horse collar. In keeping with military instructions, as a passenger in an aircraft flying over open water Grissom donned a life preserver and strapped himself tightly into his seat. He had been in the water for approximately 4 minutes, but he would later say '*... It seemed like an eternity to me*'. (See Fig. 82.)

The astronaut was landed on the flight deck of the USS *Randolph* and hurriedly taken below deck to the sickbay. There he was examined for ill effects both of his space flight and of his unorthodox recovery. Dried off and dressed in civilian clothes he began his personal debriefing into a tape recorder. When his self-debrief was complete he was flown off the ship, to Grand Bahama Island where he underwent three days of technical debriefing, medical examinations and rest.

Grissom reported that he was happy with most aspects of his spacecraft's performance during the suborbital flight. He was unhappy with the slow turnaround rate following separation from the launch vehicle. He also felt that there was too much movement in the hand controller linkage when using the MPCS. The RCS had provided precise manoeuvring, but was extremely heavy on propellant. The astronaut also complained about dim instrument lights, too many restraint harness straps, a high oxygen consumption rate, the unsuccessful HF radio test and the makeshift urine collection device that he had had to wear.

On the subject of the loss of the side hatch Grissom originally said, '*I had the cap off and the safety pin out, but I don't think I hit the button. The spacecraft was rocking around a little, but there weren't any loose items in the spacecraft, so I don't see how I could have hit it, but possibly I did.*'

A similar production spacecraft was put through a series of tests at McDonnell under conditions far more severe than those experienced by Liberty Bell 7. The hatch did not blow off once. Grissom climbed inside a production spacecraft dressed in his pressure suit and attempted to retrace his actions. At no time did his arm or hand come close to the hatch detonator plunger. The hatch did not jettison. After an extended investigation the official STG report did not identify the reason why Liberty Bell 7's side hatch had jettisoned.

Grissom became America's second astronaut hero, and NASA Administrator James Webb personally pinned the NASA Distinguished Service Medal on his suit jacket when he returned to Patrick AFB after his debriefing at Grand Bahama Island. Realising that there was little chance of his receiving a second flight during Project Mercury, Grissom began to focus his attention on Chamberlin's two-man Mercury Mark II spacecraft.

Webb had cancelled MR-6 and the remaining MR flights in October 1960. John Glenn was expected to fly MR-5, which had been kept in the flight schedule pending the results of Grissom's flight. With MR-4 successfully completed Silverstein and Gilruth encouraged Webb to cancel MR-5. On August 1961 NASA Headquarters announced that all future Mercury Redstone flights had been cancelled. STG had stopped '*shooting young ladies from cannons*'.

25 July 1961: After many meetings with members of his government, his political advisers and the leaders of the US military, President Kennedy made a speech to the American population stating that he is prepared to risk nuclear war with the Soviet Union over the security of West Berlin.

20.3 VOSTOK II

Gherman Titov was launched on Vostok II at 0900 on 6 August 1961. The spacecraft entered a 166 × 232-km orbit. With the success of Gagarin's flight the Soviets reinstated their original goal of having a cosmonaut stay in space for a whole day. This spectacular leap in endurance was actually dictated by the secretive nature of the Vostok programme and the requirement to land within the Soviet Union, particularly in the primary landing area on the Kazakhstan Steppes. As the spacecraft orbited Earth, so the planet turned beneath it. As a result, each successive orbit flew over a different ground track to the previous one. If Titov did not land in the Soviet Union during the first three orbits then the huge country would pass beyond reach for the next 24 hours. The decision to fly for 17 orbits, 24 hours, was dictated by this fact rather than by any propaganda value of such a flight. Every 45 minutes, the cosmonaut saw the Sun set and was plunged into darkness as Vostok passed into Earth's shadow. Forty-five minutes later the Sun climbed above Earth's horizon once more and the cosmonaut observed a brilliant orbital sunrise.

During orbits 4 and 5 Titov became the first man to manually control the attitude of his spacecraft in orbit but, like the Mercury Spacecraft, the thrusters were too weak to change the spacecraft's orbital trajectory. He also became the first person to suffer from Space Adaptation Syndrome, which was referred to as Space Sickness in 1961. The vomiting was caused by the otoliths in his inner ear losing their gravitational reference in the microgravity environment. Although he had felt rough from the moment that he entered orbit, Titov put on a brave face. When he tried to sleep the nausea got worse and he had to lie completely still in order to overcome it. He ultimately overslept by 35 minutes.

The flight came to a close on orbit 17. As the spacecraft approached Africa it was aligned for retrofire. Following the rocket burn the two modules separated, but the electrical umbilical remained attached, as it had on Vostok I. As with Gagarin's flight the umbilical burned through during re-entry and the Re-entry Module turned under gravitational attraction to the correct attitude. Titov ejected from the spacecraft and landed under his personal parachute at 1018 on 7 August, after a flight lasting 25 hours 18 minutes.

Much of the Western media misunderstood the requirement to make 17 orbits before the Soviet Union passed back beneath the spacecraft. They heralded the 25-hour flight as a major advance in manned space flight and a massive propaganda coup for the Soviets, which it was.

13 August 1961: In the early hours of the morning East Berlin's authorities closed their borders with West Berlin. Four days later, while the world still marvelled at the

Soviet achievement of keeping a cosmonaut in orbit for over 24 hours, Khrushchev sent his soldiers into Berlin and divided the city by constructing the Berlin Wall. East German police initially constructed the 'wall' from barbed wire and left it to see how the Americans would react. When they did not react, the East Germans replaced the wire with prefabricated concrete sections a few days later. During that time almost 1,500 East Germans made successful attempts to get to the Western zone. Throughout the next few days the East Germans constructed the concrete monolith that would divide the city for the next three decades. On 22 August the East Germans declared a 100-m no-mans-land on either side of the wall. The Soviets had sealed their border and stopped the defection of professionals to the West.

30 August 1961: The Soviet Union detonated a thermonuclear bomb above ground in Siberia. They detonated two more in the next five days. Data from America's Corona satellites has shown that the Soviet Union actually had very few ICBMs, maybe as few as six R-7s. The Americans believed that the test was a show of strength to the Communist Chinese. For the first time the Americans realised that the Sino-Soviet Communist block was not in total agreement on all matters.

September 1961: President Kennedy commenced the reorganisation of America's military response to international incidents, making the change from Mutually Assured Destruction to 'Flexible Response'. The new strategy would allow America to tailor its response to the incident, rather than just launch nuclear weapons at everything in sight.

5 September 1961: President Kennedy announced that America would resume underground testing of thermonuclear warheads as a show of strength to the Soviets.

11 September 1961: North American aviation was selected to develop the S-II stage for the advanced Saturn launch vehicle models on this date. The S-II would serve as the second stage of the Saturn C-5. Stages for the various Saturn launch vehicles would be shipped to the Mississippi Test Facility where they would be test fired prior to delivery to Cape Canaveral. The Michoud Test Facility was under construction throughout 1961.

20.4 MERCURY ATLAS 4 (MA-4)

Cuba and Berlin were not the only trouble spots in the world. In Africa a number of small countries were seeking independence from the colonial powers that had ruled them. The Zanzibar tracking station was in an isolated location and many of the local African population did not want the Americans on their territory. Once in position for their training period during the run-up to MA-4 the Zanzibar crew found themselves surrounded by local nationalists. CapCom John Llewellyn gathered his team inside the tracking station and kept them inside until the exercise was over, including overnight. The exercise went well and everything proceeded towards launch.

Following a successful ignition and transition to mainstage, the unmanned MA-4 climbed away from LC-14 at 0904 on 13 September 1961. Telemetry recorders at MCC registered a vibration in the pitch plane from T + 15 s to T + 21 s. The launch vehicle passed safely through max-q, its thick-skinned liquid oxygen tank dome protecting it from the same fate as MA-1. An inverter, used to converting direct current, to alternating current, failed in the spacecraft at T + 52 s, but a back-up inverter automatically took on the role of the failed unit. During launch the spacecraft's ECS and the simulated pressure suit circuit both sealed off at 386.7 g/cm^2.

Vibration during launch dislodged the oxygen flow rate handle, allowing the oxygen flow valve to partially open, but not sufficiently to send the appropriate telemetry signal and inform MCC that it was open. The Crew Compartment and simulated pressure suit pressures rose to 421.8 g/cm^2. The open valve led to a higher than expected oxygen usage rate throughout the remainder of the flight. Post-flight review of the instrument panel camera film showed that the Oxygen Emergency Supply light had illuminated at the time. Had there been a human astronaut in Spacecraft 8A, the light coming on would have alerted him to the problem and caused him to take corrective action.

At T + 1 min 57.5 s, 2.5 seconds early, the Atlas's two booster engines stopped thrusting and the booster fairing was jettisoned to fall back into the ocean. The launch vehicle's velocity was 30.5 m/s high. Twenty seconds after Booster Engine Cut-Off (BECO) the Sustainer pitched down slightly and the Marman Clamp holding the LES to the neck of the spacecraft was released. The Escape and Jettison Motors fired and the LES moved rapidly away from the ascending vehicle. When the two motors shut down the LES fell into the ocean.

The onboard programmer then increased the Sustainer's pitch angle as planned in a manoeuvre that would result in the vehicle flying parallel to Earth's surface at Sustainer Engine Cut-Off (SECO). The Sustainer and the two vernier motors continued to fire for a further 2 minutes 50 seconds, before shutting down 10 seconds early. The Sustainer and SC-8A had become the first Project Mercury hardware to achieve orbit.

MA-4 entered orbit at a perigee of 159.13 km. One hundred and eighty degrees around the closed elliptical trajectory it would achieve an apogee of 228.63 km (such orbital parameters are commonly recorded as 159.13 × 228.63 km). The orbit was inclined at 32.8 degrees to Earth's equator and had a period of 88.19 minutes. The Sustainer's guidance system was fooled by light refraction as the vehicle approached the horizon resulting in an apogee that was 1.60 km and perigee 19.30 km lower than planned, the worst orbital injection parameters of the entire project. Even so, the orbital parameters achieved were within acceptable limits and the MA's guidance software was rewritten to avoid the problem on future flights.

With telemetry being relayed from the Bermuda tracking station, Mercury controllers watched and waited as the Marman Clamp at the base of the spacecraft was released and the three posigrade rockets mounted in SC-8A's Retrograde

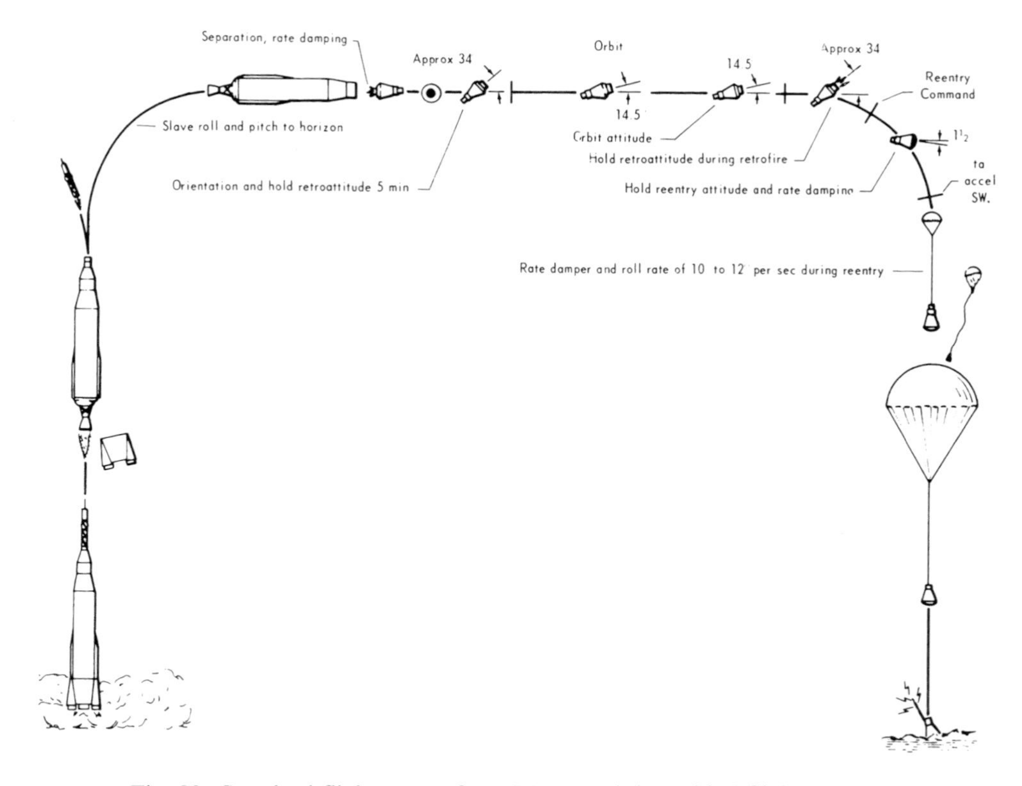

Fig. 83. Standard flight events for a Mercury Atlas orbital flight. (NASA)

Package fired to impart a positive separation force in microgravity. In the microgravity environment the Sustainer began to move in pitch, yaw and roll in reaction to the posigrade rocket exhaust impacting the Atlas's oxygen dome inside the confines of the spacecraft adapter. With separation complete the launch vehicle had performed its mission and would play no further role in the flight of MA-4. (See Fig. 83.)

After 5 seconds of rate-damping manoeuvres, SC-8A's onboard timer commanded the thrusters to turn the spacecraft through 180 degrees and adopt the correct heat shield forward orbital attitude, inclined at 34 degrees negative pitch to the orbital plane. The manoeuvre should have taken 20 seconds and consumed 0.99 kg of propellant. An open electrical connection in the spacecraft's pitch rate gyro led to large discrepancies during the turnaround and rate-damping manoeuvres, which took 50 seconds and 4.30 kg of propellant to achieve. With the spacecraft finally in the correct orbital attitude and steady, the decision was taken in MCC to let the flight continue. (See Fig. 84.)

SC-8A passed around the planet, its telemetry being relayed to each ground station in the Word-Wide Tracking Network as it passed overhead. After Bermuda

Fig. 84. The 'Mechanical Man' or crewman simulator flown on Mercury Atlas 4 released the same heat and gas output as a human astronaut into the spacecraft atmosphere to test the correct functioning of the Environmental Control System during an orbital flight. (NASA)

it passed over the Canary Islands, the Atlantic Ocean tracking ship, then Kano and Zanzibar in Africa. Information relayed to Zanzibar showed that the primary oxygen supply was down to 30 per cent of the amount carried at the moment of launch. The flight continued with the spacecraft passing over the Indian Ocean tracking ship, Muchea and Woomera in Australia and Canton Island. As the spacecraft approached retrofire, it twice dropped out of its correct orbital attitude. These discrepancies were caused by contamination in the metering orifices of some thruster units, which may have resulted in reduced thrust rates, or no thrust at all.

Approaching Hawaii, 1 hour 28 minutes and 59 seconds after launch, the first retrograde rocket fired to slow SC-8A down. The two remaining retrograde rockets fired at 5-second intervals as planned. The Guymas tracking station recorded the completion of retrofire and the jettisoning of the Retrograde Package. The re-entry logic activated and SC-8A began a slow roll around its long axis. Just before the moment when the 0.5*g* light was due to illuminate the spacecraft dropped out of orbital attitude for the third time. Passing over Texas, the atmospheric gases were slowly thickening around the vehicle as its onboard timer commanded it to take up the correct re-entry attitude. The manoeuvre was successfully completed and recorded in the telemetry received at MCC as the SC-8A approached Cape Canaveral at the end of its single orbit.

As it fell through the atmosphere spacecraft 8A compressed the ever-thickening gases into a bow wave in front of the heat shield. Ionised air glowed white-hot as it surrounded the small black-painted cone and temporarily blacked out all communication. The ceramic heat shield ablated as planned, carrying the incredible heat away from the spacecraft as it did so. Friction slowed the SC-8A from orbital velocity to less than Mach 1 in a just a few minutes. The ionised bow wave disappeared and communications were established once more.

The crew of a C-54 recovery aircraft watched the spacecraft's fiery trail as it passed through the atmosphere. Keeping it in their view they observed the drogue parachute deploy at 12,725 m. They also saw the Antenna Canister jettison and the main parachute deploy at 3,063 m. Spacecraft 8A hit the ocean at 1055, after a flight lasting 1 hour 49 minutes 20 seconds. The USS *Decatur* was the nearest recovery ship, some 54 km away, and she dispatched helicopters to effect recovery. The first Mercury Spacecraft to orbit Earth was secured onboard the USS *Decatur* 1 hour 22 minutes after landing. The ship carried the spacecraft to Bermuda, from where it was airlifted back to Hangar S, Cape Canaveral.

At the post-flight press conference STG Director Robert Gilruth confirmed that a human astronaut could have survived the MA-4 flight and could even have been expected to diagnose and overcome all of the problems encountered during the flight. Asked if a human astronaut would be on board the next Mercury Atlas flight, he explained that STG wanted a chimpanzee to make a successful three-orbit flight before one of the Flight A astronauts was asked to put his life on the line.

13 September 1961: President Kennedy admitted to American journalists that there was not now, and never had been, a 'Missile Gap' in the Soviet Union's favour. He blamed his campaign and post-election speeches on the misinterpretation of reconnaissance information by the country's military leaders.

19 September 1961: NASA Administrator James Webb announced that the MSC would be housed in a new NASA field centre to be built in Houston, Texas. The Manned Spacecraft Centre (MSC) would be the home of a new Mission Operations Control Room and the organisation that supported it. The centre would also be responsible for selecting and training all future groups of astronauts. Robert Gilruth was subsequently named as MSC's first Director.

24 September 1961: NASA HQ was reorganised on this date. The Offices of Advance Research Programs, Space Flight Programs, Launch Vehicle Programs and Life Science Programs were all closed. Four new Program Offices were opened. They were:

Office of Advanced Science and Research
Office of Space Sciences
Office of Manned Space Flight
Office of Space Applications

All of the new offices reported to Robert Seamans.

26 September 1961: Chrysler was named as prime contractor for the industrial production of the S-1 stage, the first stage of the Saturn C-1. The initial few Saturn C-1 stages had been constructed at MSFC, where they were designed. The Chrysler contract was subsequently changed to allow Chrysler to build the uprated S-1 stages employed on the Saturn C-1B launch vehicle.

30 September 1961: Chairman Khrushchev wrote to President Kennedy requiring that the two men commence a private correspondence, so that in times of crisis they might bypass the accepted political protocols.

11 October 1961: President Kennedy agreed to send US Special Forces into South Vietnam.

17 October 1961: At the 22nd Communist Party Conference, Chairman Khrushchev announced that the Soviet Union had set no date for signing the Peace Treaty with East Germany. He was backing down from confrontation with the Americans over West Berlin.

21 October 1961: US Deputy Defence Secretary Roswell Gilpatric gave a speech that had been edited by President Kennedy. The speech left no doubt over America's nuclear superiority over any other nation on Earth.

27 October 1961: The first unmanned Saturn C-1 flew a suborbital flight from the new LC-34 at AMR. Saturn was the first launch vehicle to be designed specifically for space flight. With only its first-stage live, the vehicle fell into the Atlantic Ocean once its eight first-stage motors shut down. The launch was near flawless.

20.5 MERCURY SCOUT FLIGHT PROGRAMME

Two Mercury Scout vehicles were intended to launch short-life satellites into Earth orbit, in order to test the World-Wide Tracking Network prior to commencing manned Mercury Atlas orbital flights. Only one launch was attempted.

20.6 MERCURY SCOUT 1 (MS-1)

On 31 October 1961 the countdown reached zero but Mercury Scout 1's first-stage rocket motors did not ignite. Having safed the launch vehicle, the launch crew set about testing the ignition circuits before making a second launch attempt. When the countdown reached zero for the second time MS-1 made the transition to mainstage and climbed away from the launch pad. Undetected by the men who had prepared the launch, inside MS-1 the pitch and yaw rate gyros had been wired in reverse. Errors in pitch were relayed to the yaw gyro, which initiated a yaw manoeuvre in an attempt to correct them. Meanwhile, errors in yaw were relayed to the pitch gyro, which tried to correct them with a pitch manoeuvre. The launch vehicle immediately began to make a series of erratic manoeuvres that subjected MS-1 to unacceptable aerodynamic loads and caused it to begin to break up. The Range Safety Officer sent the signal to destroy the vehicle by igniting the range safety ordnance packages built into the Scout launch vehicle. Much of the wreckage was recovered and an investigation identified the wrongly wired rate gyros. (See Fig. 85.)

Fig. 85. Mercury Scout 1 lifts off. (NASA)

With Mercury Atlas 4 having already tested the World-Wide Tracking Network in September and Mercury Atlas 5 planned for a three-orbit flight carrying a chimpanzee in November, the MS-2 launch attempt was cancelled on Low's recommendation. MS-1 proved to be the last major launch failure of Project Mercury.

November 1961: Milton Rosen reported to Seamans on the future of NASA's heavy launch vehicle programme. He recommended the development of a Saturn C-5, with a S-IB first stage with four F-1 rocket motors. This was later changed to five F-1 motors. Five J-2s motors propelled the S-II second stage and the S-IVB third stage contained a single J-2. The Boeing Company were named as prime contractor for the S-IB on 15 December 1961. The Boeing stage was later renamed S-IC and the S-IB designation was given to the first stage of the Saturn C-1B.

1 November 1961: Between 17 October and this date an American Commission had been visiting South Vietnam. Their report suggested that President Kennedy use the recent flooding in that country as an excuse to send in American military forces. They also suggested that Diem should be replaced as President.

29 November 1961: The prime contract for the Apollo Spacecraft was let to North American Aviation. The spacecraft consisted of a conical Command Module (CM), containing the Crew Compartment and recovery system. A cylindrical Service

Module (SM) contained the spacecraft's main rocket motor as well as manoeuvring thrusters, propellant tanks and high-pressure gas storage bottles. The two modules functioned as one Spacecraft, referred to as the Command and Service module (CSM) throughout the flight until just before re-entry. At that point the SM was jettisoned and the CM turned so that its high drag heat shield faced the atmosphere. Landing was to be in mid-ocean under a multiple parachute system.

20.7 MERCURY ATLAS 5 (MA-5)

If all went well, MA-5 would be the final flight-test in Project Mercury. SC-13 would carry a chimpanzee on a three-orbit precursor to the first American manned orbital flight. The countdown for Atlas 109D, Mercury Atlas 5, began on 28 November 1961 and was held overnight at T−6 h 30 min. Lift-off was planned for 0730 the following morning.

Eleven hours before launch the primary chimpanzee, Enos, passed his physical examination. When the time came for final preparations, veterinary specialists taped the biomedical sensors to Enos's body and dressed him in his Airtex flight suit. He was strapped into the lower half of his couch. The upper half of the couch, with its four-stage psychomotor test equipment, was bolted in place and the life support system was tested and found to be functioning correctly.

The couch was transported to LC-14 in the same transfer van that had taken Shepard and Grissom to the launch pad at the beginning of their Mercury Redstone flights. The pressurised couch was inserted into SC-13 approximately 5 hours before launch, after which Enos's health and performance were monitored in MCC. At T−1 h 30 min the spacecraft's side hatch was bolted in place. A hold was called when it was noticed that the launch crew had failed to install some hatch cover insulation. The task took almost an hour, after which the countdown resumed. With Enos acting calmly the launch gantry was rolled back to its parked position and LC-14 was evacuated. (See Fig. 86.)

Telemetry showed that a single switch in the spacecraft had been left in the wrong position. The countdown was held at T−30 min and the gantry was rolled back around the Atlas. The launch crew opened the spacecraft side hatch, an action that caused Enos to become agitated. With the switch reset the hatch was bolted back into place and Enos became calm once more. The gantry was returned to its parked position and the LC-14 was evacuated for the second time. The countdown was resumed from the T−30 min point, after a hold lasting 1 hour 25 minutes.

The clock stopped for the third time, at T−15 min. Four minutes were lost while the data link between the launch vehicle's guidance package and MCC was re-established after it had dropped out. A final 3-minute hold was called at T−7 min to correct a problem with a pulse beacon.

MA-5 finally left the launch pad at 1008 on 29 November. Feedback on the line caused the computers at NASA–Goddard to receive the lift-off signal 13 seconds before the event occurred. Shortly after lift-off MCC also received an erroneous signal to say that the Sustainer had shut down, when it had not. A second erroneous

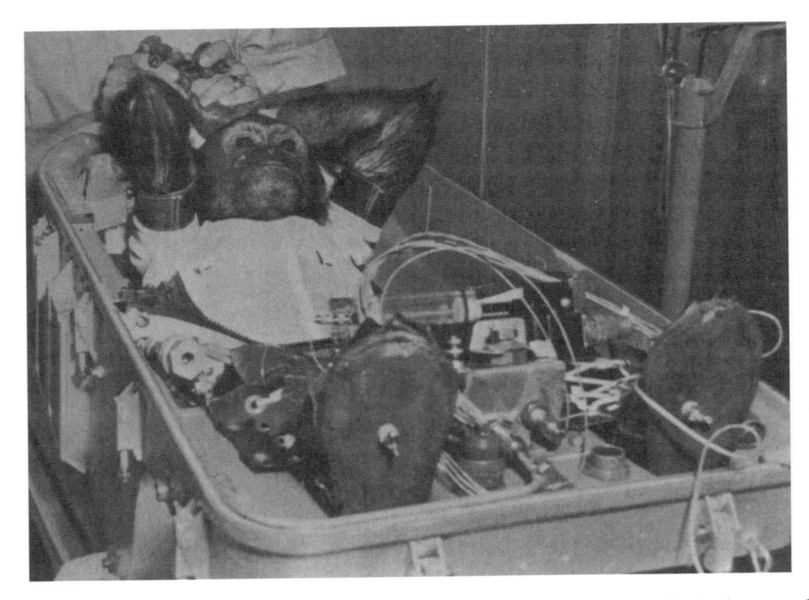

Fig. 86. A relaxed Enos lies in the rear portion of his pressurised couch during pre-flight preparations. (NASA)

Sustainer shutdown signal was received prior to BECO. Both signals were manually over ridden from MCC. During the first 2 minutes of flight the Atlas horizon scanner signal was lost briefly, but the drop out did not affect the ascent. At T+2 min the two booster engines shut down and the booster fairing was jettisoned. Twenty seconds later the LES was also jettisoned. With its single engine and two verniers burning at full thrust, the Sustainer continued the climb into orbit.

SECO occurred at T+5 min, as planned. The Sustainer/SC-13 was in a 160.9 × 237.1-km orbit, with an inclination of 32.6 degrees and an orbital period of 88.26 minutes. Spacecraft separation and rate damping occurred as planned and 30 seconds after separation SC-13 had turned through 180 degrees and adopted the correct 34-degree negative pitch orbital attitude. The spacecraft made very precise manoeuvres throughout the turnaround and propellant consumption was just 2.72 kg

Telemetry showed that SC-13 had performed flawlessly throughout the powered phase. The Crew Compartment atmospheric pressure had sealed off at 350.5 g/cm^2. For the first time on this flight the spacecraft primary and secondary oxygen bottles were at full pressure, rather than the half pressure that had been employed on all previous flights. The water separator in the ECS was carried for the first time on this flight and performed well.

Sealed inside his pressurised couch, Enos endured the launch well. He had begun his psychomotor tests at T−2 min and had performed well throughout the powered launch phase, despite undergoing maximum acceleration forces of 7.6*g*. During the flight he was expected to perform four separate test series using the psychomotor equipment mounted inside his pressurised couch. During the 6 minutes between each set of tests the chimpanzee rested, as he had been trained to do.

The first test consisted of continuous-avoidance and discrete-avoidance tests similar to those performed by HAM during the flight of MR-2. During four test sessions Enos would make 29 lever pulls and only received one mild electric shock to the sole of his feet. The second test required Enos to wait at least 20 seconds when a particular light was illuminated before pulling the lever beneath it in order to receive a drink of water. Throughout the flight Enos's average delay before pulling the lever was 33.8 seconds. The third test required 50 pumps of a lever in order to receive a banana-flavoured pellet. This was a purely voluntary test, with no electric shocks for incorrect actions. Enos participated eagerly and received 13 pellets during the four periods that the experiment was run. The fourth test required Enos to pull the lever beneath the odd one out among three displayed shapes. During the first session he made 18 correct and 10 incorrect pulls, receiving 10 mild electric shocks to the soles of his feet.

During the second test series the equipment failed, resulting in Enos receiving electric shocks to his feet regardless of whether he made the correct or incorrect lever pulls. The equipment continued to malfunction throughout the remainder of the flight and Enos received 36 shocks during the third session and 43 shocks during the fourth test session. Despite the shocks, Enos was so indoctrinated that he continued to pull the lever each time a shape was displayed, even when he was incorrectly punished for a correct lever pull.

During the first orbit, tracking and telemetry reception at the ground stations was excellent. UHF signals were received at each station for an average of 6 minutes, from horizon to horizon. HF communication even overlapped between some stations, while clear telemetry was received at all stations except Woomera.

The only problem arose as Enos approached Loss of Signal (LOS) from the Atlantic Ocean tracking ship, *Rose Knot Victor* (RKV). Enos fell behind on his psychomotor tasks and received a mild electric shock to his feet.

By the time the spacecraft came into range of Zanzibar, Africa, 10 minutes later, the flight surgeon saw a number of premature ventricular contractions on Enos's electrocardiogram readout. The readings continued and were observed and reported by the flight surgeon at each consecutive tracking station. Unknown to the majority of flight surgeons working the tracking stations, Enos had been fitted with an instrument to record central venous pressure. The catheter had been placed into a vein in his leg and then pushed up the vein, through the ventricles of his heart and into his lungs through the pulmonary artery, in order to prevent it coming out. The tube moved back and forth with each beat of Enos's heart, causing the premature ventricle contractions.

At the end of the first orbit telemetry showed that the spacecraft clock was running 18 seconds fast. As SC-13 passed over Cape Canaveral at the beginning of the second orbit a signal was sent from the ground to retard the clock by 18 seconds and resume correct time keeping.

Leaving Canaveral and Bermuda behind for the second time, SC-13 passed over the Atlantic Ocean tracking ship. Telemetry received on the ship showed that the temperature was rising in the spacecraft's ECS inverters. The information was relayed to MCC and was subsequently confirmed by the tracking station on the Canary Islands. Controllers were not too worried by this information as inverters had heated up on earlier flights and had either continued to perform their role, or had shut down and back-up converters had taken over.

Enos's electrocardiogram continued to show numerous premature ventricle contractions and several rows of four abnormal heartbeats in a row. The data also showed that the abnormal beats were coming from two separate areas of the chimpanzee's heart. In a human astronaut such readings would require immediate re-entry and emergency medical treatment, as they could be expected to lead to ventricular fibrillation and death. Unaware of the central venous pressure instrument, the flight surgeon on *Rose Knot Victor* recommended to MCC that the flight be terminated and Enos recovered. The decision was overridden by the flight surgeon in MCC. SC-13 and Enos passed on, towards Kano and the African tracking stations.

Approaching Kano, telemetry indicated that the Crew Compartment temperature had risen rapidly by 25 °F, indicating that the ECS heat exchanger had frozen. In his couch Enos's body temperature stabilised at 100.5 °F. When the primate's temperature stabilised it suggested that the heat exchanger was no longer frozen and the flight surgeon at MCC allowed the flight to continue. SC-13 passed across Africa and over the Indian Ocean, where a NASA tracking ship filled a large gap in the line of ground stations. Enos's electrocardiogram continued to show premature ventricle contractions to each flight surgeon as he passed over their station.

Muchea, in Australia, received conflicting telemetry signals. One set of signals suggested that the spacecraft was maintaining the correct heat shield forward, 34-degree negative pitch attitude. The second set suggested that it was manoeuvring erratically. The problem was unresolved before the spacecraft passed out of range. Telemetery received at Woomera was equally inconclusive.

The problem was not cleared up until SC-13 approached Canton Island. An obstructed metering orifice in one of the spacecraft's automatic roll thrusters was preventing that thruster from firing to maintain the spacecraft in the correct attitude. The spacecraft was drifting 30 degrees in negative roll, before reaching the limit programmed in to the ASCS. At that point the ASCS commanded thrusters to fire to halt the roll and place the spacecraft back into the correct 34-degree negative pitch orbital attitude. Over time the negative roll built up again until it reached 30 degrees and was corrected. The process was continually repeated as Spacecraft 13 completed its second pass across North America. The negative roll was corrected nine times, each requiring 0.45 kg of thruster propellant.

As Enos crossed America, concerns were being voiced in MCC were about the high propellant consumption rate caused by the negative roll. If the consumption continued at its present rate throughout the planned third orbit, there would be insufficient ASCS propellant left to control the spacecraft's attitude during re-entry. Flight Director Chris Kraft informed astronaut Gordon Cooper on duty at the Point Alluello tracking station to be ready to command retrofire at the end of the second orbit. Meanwhile Enos was approaching Point Alluello over Hawaii and California.

As the moment for the retrofire signal to be sent approached, communications between MCC and Point Alluello, dropped out. Arnold Aldrich, the leader of the team at Point Aguello sent the retrofire command as the spacecraft approached his station. In orbit SC-13 performed a flawless retrofire and the onboard timer jettisoned the spent Retrograde Package. The spacecraft rolled out of position one last time, before taking up the correct re-entry attitude. The 0.5*g* light illuminated on the instrument panel and the appropriate signal was telemetered to the ground. On board SC-13 the re-entry logic was activated and commanded the spacecraft to begin rolling around its long axis.

Re-entry passed without any problems and the ELS deployed as planned. Members of the recovery forces first observed the spacecraft at an altitude of 1,524 m, as it descended underneath its main parachute. The pilot of a P5M aircraft established his own orbit around the descending spacecraft and watched it until it hit the water and established an upright attitude in the ocean.

Water impact occurred at the end of a flight lasting 3 h 2 min 59 s. The heat shield had dropped away from the base of the spacecraft at main parachute deployment and deployed the landing skirt. When the heat shield struck the ocean it impacted the rear bulkhead of the spacecraft piercing it and compressing the shock-absorbing honeycomb in that area. A small amount of seawater entered the Crew Compartment through the punctures. Inside the landing skirt two straps were broken and two others bent, although this damage may have been caused by wave action as the heat shield and water-filled landing skirt performed their role of sea anchor.

The Destroyers USS *Stromes* and USS *Compton*, both of which carried Mercury Spacecraft recovery equipment, steamed towards the landing area, with the USS *Stromes* arriving first. When the spacecraft was recovered, after 1 hour 15 minutes in the water, it was noted that the centre plug of the heat shield had fallen out, either on water impact, or as the result of wave action.

With the spacecraft secure onboard the USS *Stromes*, the lanyard was pulled to explosively jettison the side hatch. Several of the hatch bolts were ripped out of their mountings in the process and the spacecraft's head's up window was cracked. Enos was removed from the spacecraft and taken below deck to the sickbay. There he was removed from his couch and subjected to a post-flight medical examination. He appeared fit and well after his flight.

At the post-flight press conference STG managers voiced the opinion that the loss of the final orbit was a small price to pay for the proof that the ASCS worked under flight conditions and the experience that the Flight Operations Division personnel gained in real-time planning during the flight. It was noted that an astronaut would have dealt with the negative roll by switching to a manual control system and keeping his spacecraft's attitude under control himself. It was suggested that an astronaut would not only have survived MA-5 as it was flown, but would probably have insisted on attempting, and would probably have completed, the third orbit.

At the same press conference Gilruth confirmed that John Glenn was the prime astronaut for MA-6, with Scott Carpenter serving as his back-up. MA-7 would fly with Deke Slayton as prime astronaut, or Wally Schirra as back-up astronaut. Both flights would attempt three orbits of Earth, but, as on MA-5, if circumstances dictated, they would be curtailed rather than risk the astronauts' lives unnecessarily.

Meanwhile, Enos was shipped back to Bermuda, where he was examined at Kindley AFB. He was then flown back to Cape Canaveral, arriving on 1 December and being subjected to a further week of post-flight medical and psychomotor tests. Finally, he was returned to Holloman AFB. He lived for one year, before contracting dysentery brought on by shingellosis, in September 1962. Despite two months of medical care, he died on 4 November 1962. His illness was in no way connected to his flight on MA-5.

In November 1961 Enos held a unique position in the annals of space flight. That position was recognised at the time by cartoonist John Fischetti. His cartoon shows two humans recovering the dripping SC-13 from the ocean in the background. In the foreground Enos and a second chimpanzee are walking on their hind legs. Both are wearing a pilot's flight suit and Enos is carrying a pilot's helmet complete with oxygen mask. In the caption Enos is telling his chimpanzee companion, '*We are a little behind the Russians and a little ahead of the Americans*'.

For the Flight A astronauts the time had finally come to wipe the last of the monkey shit off the seat and ride the Mercury Atlas combination into orbit.

December 1961: America sent 2,067 troops into South Vietnam.

7 December 1961: Seamans announced that NASA would fly a third series of manned space flights, between Project Mercury and Project Apollo. He referred to a '*two man Mercury*' spacecraft that would rendezvous and dock with an Agena target vehicle in Earth orbit. The information obtained on the flights would be transferred directly to Project Apollo and later manned space vehicles. The new spacecraft was Chamberlin's Mercury Mark II, developed by McDonnell. It would ultimately fly under the name Gemini. (See Fig. 87.)

20 December 1961: American troops in South Vietnam were given permission to use their weapons in self-defence.

21 December 1961: Douglas Aircraft was selected to modify the Saturn C-1's second stage, the S-IV into the S-IVB. The new stage would be powered by a single Rocketdyne J-2 rocket motor and would serve as the second stage of the Saturn C-1B and the third stage of the Saturn C-5.

22 December 1961: The first American soldier was killed in South Vietnam. By the end of the year 14 Americans had died in Vietnam.

9 January 1962: The Soviet Union and Cuba signed a trade agreement.

11 January 1962: President Kennedy made his second State of the Union Address. He concentrated on strengthening the economy at home.

Fig. 87. Full-scale mock-ups show both the obvious differences and similarities between the two-man Gemini Spacecraft and the one-man Mercury Spacecraft. (NASA)

18 January 1962: President Kennedy signed a National Security Action Memorandum authorising military counter-insurgency action in Laos, South Vietnam and Thailand. In South Vietnam American helicopters and pilots were being used to drop defoliant on the jungle surrounding major roads and on the rice paddies of villages thought to be cooperating with the North Vietnamese Communists. They also cleared the areas around South Vietnamese military bases.

25 January 1962: NASA agreed to develop the Saturn C-5 because it best met the requirements of the manned lunar landing mission. As a result of this decision the Saturn C-4 was cancelled.

The method by which Apollo should fly to the Moon was a complicated one. Most NASA engineers originally favoured Earth Orbital Rendezvous (EOR), where the various parts of the lunar spacecraft were launched on separate launch vehicles and assembled in Earth orbit. The launch vehicle's final stage would then be refuelled by an orbital tanker before commencing its lunar mission. The spacecraft would fly directly to the lunar surface before lifting off again and returning directly to Earth.

An alternative method was Lunar Orbital Rendezvous (LOR). This called for two smaller spacecraft to be launched on a single launch vehicle and placed into lunar orbit. From there two astronauts would descend to the lunar surface in one spacecraft while the third astronaut remained in lunar orbit in the second spacecraft. Following the surface exploration the landing vehicle would lift-off and return to lunar orbit. The two vehicles would then rendezvous and dock in lunar orbit and the two explorers would return to the orbiter. The landing vehicle would be jettisoned and the orbiter would return to Earth.

Throughout 1961–62 a group of LaFRC engineers fought for LOR to become the accepted route to the Moon in Project Apollo. They won their argument within NASA on 7 June. When President Kennedy's Scientific Adviser continued to argue against LOR, further debate took place but LOR was ultimately chosen as Apollo's route to the Moon.

10 February 1962: America swapped a Soviet spy for Francis Gary Powers, the CIA pilot of the U-2 reconnaissance aircraft shot down over the Soviet Union on 1 May 1960.

20.8 MERCURY ATLAS 6 (MA-6)

The first manned orbital flight of Project Mercury was originally set for 19 December 1961, just eight months after Gagarin's flight. To meet that schedule, primary astronaut John Glenn and his back-up Scott Carpenter moved into the astronaut quarters at Hangar S, Cape Canaveral, on 4 December.

On 7 December Gilruth announced that the flight was cancelled until 16 January 1962. He blamed '*minor problems dealing with the cooling system and positioning devices in the Mercury Spacecraft*'. Three days into 1962 difficulties with the propellant tanks in Atlas 109D caused a further delay, until 23 January 1962.

The pre-count began on 22 January and proceeded until T − 6 h 30 min, where it was held overnight. The following morning, Glenn was woken up and began his pre-launch routine. He shaved and showered before taking his breakfast. He then allowed the USAF physicians to carry out a pre-launch physical. At LC-14 the Atlas oxygen system was playing up. The launch was cancelled before Glenn began dressing in his pressure suit. A new launch attempt was scheduled for 27 January. Work continued to make Atlas 109D ready. (See Fig. 88.)

The second pre-count began on 26 January and was held overnight as usual. Glenn was woken up at 0233 on 27 January and followed the now standard pre-launch routine. He shaved and showered, dressed and ate breakfast with a few invited friends and programme managers. After his pre-launch physical he allowed Joe Schmitt to help him don his silver-coloured pressure suit. He left Hangar S at 0446 and entered the transfer van. Slayton briefed him on the launch day weather and the status of the countdown as they took the 21-minute ride to LC-14.

Leaving the van at the base of the launch pad, he rode the elevator up to the white room. There, Carpenter, who had spent the previous night in the spacecraft preparing it for launch, was waiting to brief him. Glenn entered his spacecraft at 0512 and Schmitt connected his pressure suit to the ECS. His restraint harness was pulled tight across his body and secured. The side hatch was bolted in place at 0624,

Fig. 88. John Glenn enters Friendship 7 at Launch Complex 14. (NASA)

but not before a problem with the hatch gasket had been overcome. Glenn was alone in the spacecraft that he had called 'Friendship 7'.

At 0700 the countdown was held, due to technical problems. The hold lasted 1 hour 45 minutes. At 0725 low clouds moved into the area, threatening the photographic limits placed on the launch attempt after the loss of MA-1. The countdown was restarted at 0845 and the launch gantry was rolled back to its parked position.

Finally, at 0911 (T − 19 min), the launch was cancelled due to low cloud causing unacceptable conditions for photography. The gantry was moved back alongside the launch vehicle and Glenn was removed from his spacecraft. He told the launch crew, '*There'll be another day*'. He was returned to Hangar S, where he removed his pressure suit. The new launch attempt was set for 1 February.

The third pre-count began on 30 January, but the weather was rough off of Bermuda, where the primary recovery site was located if the flight had to be cancelled after a single orbit. During the countdown procedures an engineer at LC-14 removed a drainage plug as part of the routine preparation of Atlas 109D. He discovered that propellant had leaked during the previous launch attempt and had flooded the insulation between the launch vehicle's two propellant tanks. To remove and replace the insulation would take 10 days. The launch attempt was cancelled and Glenn took the opportunity to visit his family. President Kennedy also invited him to the White House, to explain the continual delays to his launch.

Three weeks earlier a similar problem had been encountered on Atlas 121D, the launch vehicle for the unmanned Ranger 3 Moon probe. On that occasion the insulation and its mounting had been removed and the launch had been made without it. Convair decided that Atlas 109D could also be launched without the insulation between its propellant tanks.

The launch was rescheduled for 13 February. It was cancelled on 12 February due to bad weather both at Cape Canaveral and Bermuda and the next attempt was set for 14 February. The pre-count was completed on 13 February, but bad weather prevented the launch attempt the following day. The same occurred on 15 and 16 February. The launch attempt on 18 February also fell foul of the weather and the next attempt was set for 20 February.

The pre-count began on 19 February, as the bad weather began to dissipate. As usual, the countdown was held overnight at T−6 h 30 min. Despite a warning of patchy weather the countdown was resumed at 2330. Launch was set for 0730 the following morning.

Glenn's physician woke him at 0520 and gave him time to shower, shave and dress before attending the pre-launch breakfast. At 0305, Bill Douglas began Glenn's physical examination and taped the sensors from his biomedical harness in place. The astronaut laughed with Slayton while Joe Schmitt helped him into his pressure suit. He left Hangar S in the transfer van at 0501 and 16 minutes later arrived at the base of LC-14. Having ridden the elevator up the gantry, Glenn was greeted by Carpenter on his arrival in the white room. Carpenter, who had spent the night in Friendship 7 participating in the countdown, assured Glenn that all was well and that the one overnight problem, with the spacecraft's oxygen system telemetry link, had been corrected. Glenn entered the spacecraft and allowed Schmitt to connect the various pressure suit umbilicals and tighten the launch harness on his prone couch. It was 0603, 7 minutes after sunrise in Florida, T−1 h 27 min.

The countdown had been stopped at T−2 h due to a transponder in the launch vehicle that required replacement. When Glenn discovered that the respiration sensor attached to his microphone had moved from the position he had used in earlier simulations, he reported the fact. As moving it would require opening his faceplate it was decided to leave it where it was. Tracking stations were told to disregard the information from the sensor in question. The countdown resumed after a 50-minute hold, at 0625.

Cloud threatened the launch, but it was beginning to break up when the side hatch was bolted into place at 0700. Ten minutes later a hatch torque bolt was found to be broken. The countdown was held and the hatch unbolted so that the broken bolt could be replaced. Glenn passed his time watching the weather through Friendship 7's periscope. The countdown resumed at 0805, at T−60 min, but was held again 15 minutes later while 37.8 litres of RP-1 fuel was added to the Atlas's tank. Meanwhile, the gantry was rolled back to its parked position. Overhead the sky was now clear.

Liquid oxygen flowed into Atlas 109D at 0830, T−35 min. Thirteen minutes later a problem arose with the oxygen pump outlet and the repair required the countdown to be held for 15 minutes. During this hold, live television coverage of the launch commenced across America. The countdown resumed at 0925, T−10 min and was held again at T−6.5 min 30 s. The vital Bermuda tracking station had lost power to its mainframe computer. Two minutes later they were ready and the countdown continued through launch.

In New York 15,000 people crowded into Central Station to watch the launch on a huge screen. In the Oval Office at the White House in Washington, where President Kennedy was watching the launch, Vice President Johnson is reported to have quipped, '*If only he were a Negro*'.

The MA-6's three main rocket motors and two verniers burst into life at LC-14 AMR (see Fig. 89). Having achieved mainstage the vehicle was released to commence its climb into orbit. From MCC Carpenter came on the radio and sent his friend the message, '*May the good Lord ride all the way*'. Lift-off occurred at 0947 after a total of 6 weeks 2 hour 17 minutes of delays. Inside Friendship 7 Glenn acknowledged the lift-off and the fact that the primary mission timer on his instrument panel had started moving. Thirteen seconds later Mercury Atlas 6 was thundering through max-q. Considering the hours of testing and failure that had

Fig. 89. Mercury Atlas 6 climbs away from Launch Complex 14 carrying the first American into orbit. (NASA)

brought Project Mercury to that point, Glenn's call of '*Little bumpy along about here*' seemed somewhat nonchalant. Beyond max-*q* the ride smoothed out with Glenn pulling 6*g*.

At T + 2 min 14 s the two Booster engines stopped thrusting. The Booster fairing was released and slid along its guide rails before falling away from the base of the Sustainer. Glenn reported: '*BECO, BECO. I saw the smoke go by the window.*'

From MCC Shepard replied, '*Roger, we confirm staging on the TM [telemetry]*'.

With the Boosters gone the *g* forces dropped to just 1.5*g*. Twenty-two seconds later the Marman Clamp at the base of the LES released and the Escape and Jettison Motors fired, pushing the LES clear of the ascending launch vehicle. Glenn reported: '*There, the tower went right then. Have the tower in sight way out. Could see the tower go. Jettison tower [light] is green.*'

Once again the acceleration forces steadily built up to 6*g*. At T + 5 min 4 s the Sustainer shut down. Behind the astronaut's back the three posigrade rockets fired to push Friendship 7 away from its spent launch vehicle. Despite the fact that the rate-damping manoeuvre began 2.5 seconds late, a full 5 seconds of rate damping was completed before the ASCS commenced the turnaround manoeuvre. Glenn told Shepard, '*Zero-g and I feel fine. Spacecraft is turning around. Oh that view is tremendous*!' He added that he saw the Sustainer '*... just a couple of hundred yards behind me*'. Friendship 7 was in a 260.9 × 160.9-km orbit inclined at 32.54 degrees to Earth's Equator, with a period of 88.9 minutes. Just before MCC lost contact with Glenn, as he passed beyond their radio horizon, Shepard confirmed, '*Seven. You have a go for at least seven orbits.*' If all went well the flight would end after three orbits and Project Mercury would have met its primary goal.

As he passed into range of the Bermuda tracking station Glenn confirmed to Grissom that he had set his spacecraft up for orbital flight. His next task was to test the attitude controls. He manoeuvred in all three axes and found that he had no difficulty controlling Friendship 7's attitude. After just 4 minutes talking to Bermuda he was dealing with Canton Island. In 11 minutes of communication he managed to confirm the status of his spacecraft systems, telemeter his medical status to the ground and report seeing the Canary Islands, the African coast and dust storms blowing across the African desert.

At T + 24 min 43 s he switched to FBWS and yawed Friendship 7 through 180 degrees, so that he was facing the direction of travel. With the manoeuvre completed he found that he was not seeing instrument readings that he had expected. Engineers would put this down to the spacecraft's gyros labouring to keep up with the manoeuvring spacecraft. They would suggest that in future the gyros should be caged, or locked, before any spacecraft manoeuvres.

With the spacecraft back in orbit attitude and on ASCS, Glenn passed into range of Kano, in Zanzibar, off the east coast of Africa. Kano's controllers confirmed that they were in the centre of a huge dust storm.

He completed an exercise routine that required him to pull a bungee cord connected beneath the instrument panel up to his chin and release it slowly, once every second for 30 seconds. He then completed a series of planned head movements and reach-and-touch tests. All exercises were completed without difficulty. Before reading out a complete spacecraft status check (known to the astronauts as a 30-minute report) he increased the flow of cooling water into his pressure suit.

Then he was over the Indian Ocean and talking to the *Coastal Sentry Quebec* tracking ship in that location. By that time he had passed into Earth's shadow and reported, '*Had a beautiful sunset and can see the light way out, almost up to the northern horizon.*' Two minutes later he would add,

> ... The sunset was beautiful. It went down very rapidly. I still have a brilliant blue band clear across the horizon, almost covering my whole window. The redness of the sunset I can still see through some of the clouds way over to the left of my course ... The sky above is absolutely black. I can see stars ... I do not have any of the constellations identified as yet.

Later still he remarked,

> I am having no trouble at all seeing the night horizon. I think the Moon is probably coming up behind me. Yes, I can see it in the 'scope back here and it's making a very white light on the clouds.

A parachute flare was sent up from the tracking ship, but Glenn did not see it. Instead he observed a phenomenon that no one had expected:

> There appears to be a high layer, way up above the horizon; much higher than anything I saw on the daylight side. The stars seem to go through it and then go down towards the real horizon. It would appear to be possibly some 7 or 8 degrees wide. I can see clouds down below it; then a dark band, then a lighter band that the stars shine right through as they come down toward the horizon.

This would become known as the atmospheric haze layer and would be studied on later Mercury flights.

Next he was talking to Cooper at Muchea, Australia, who told him to look out for light patterns on the ground beneath him. A few minutes later Glenn reported: '*Just to my right, I can see a big pattern of lights apparently right on the coast. I can see the outline of a big city and a very bright light just to the south of it.*'

Cooper informed him, '*Perth and Rockingham, you're seeing there*'.

In the two communities concerned every available light had been turned on as Friendship 7 approached in a simple experiment to see how much an astronaut in orbit could observe on the Earth's surface. No one was even sure before the flight if Glenn would be able to see a fully lit city on the night side of the planet. Glenn ended the experiment with the words, '*The lights show up very well and thank everyone for turning them on, will you?*'

Then Muchea was behind him and he was transmitting a 30-minute report to Woomera. The communication session lasted 10 minutes, in the centre of which Friendship 7 passed the T + 1 h point. The conversation was all of a technical nature.

Canton Island was next in the chain of tracking stations. While talking to them Glenn opened his faceplate and consumed a single tube of applesauce. He reported no difficulty eating in microgravity. As he ate he reported, '*In the periscope, I can see a brilliant blue horizon coming up behind me; approaching sunrise*'. Having finished his meal he closed and resealed his faceplate. Behind him the Sun appeared from behind Earth's limb and Glenn had to put a dark filter on the periscope in order to watch it rise. At that moment he found himself by a mass of small particles. As he passed out of range of Canton Island he transmitted,

> This is Friendship 7 broadcasting in the blind. Sunrise has come up behind in the periscope. It was brilliant in the 'scope, a brilliant red as it approached the horizon and came up. And just as I looked back up out of the window, I had literally thousands of small, luminous particles swirling around the spacecraft and going away from me ... Now that I am out in the bright Sun they seem to have disappeared. It was just as the Sun was coming up. I can still see a few of them even now, even though the Sun is some 20 degrees above the horizon.

Throughout his 10-minute communication pass over Guaymas the astronaut kept returning to the subject of what became known 'Glenn's fireflies' after a description when he had used the analogy to explain what he was seeing. Among the more technical information exchanged in the pass, Guaymas reported that their telemetry showed Friendship 7's periscope was in the retracted position. Glenn confirmed that the periscope was extended and the reading was put down to an instrument failure.

While talking to Schirra at the California tracking station a problem arose. Glenn noticed that Friendship 7 had drifted out of the correct position, yawing to his right and had not been corrected by the ASCS low rate thruster. Rather, when the drift had reached 20 degrees, the high rate thruster bought the spacecraft back to the correct attitude. In time Friendship 7 began yawing right once more. The astronaut activated the FBWS system to correct the yaw drift in a manner that would consume less propellant that the high-thrust ASCS thruster. He would have to remain in FBWS and correct Friendship 7's attitude continually throughout the remainder of the flight. This was exactly the situation that Gilruth had predicted at the MA-5 post-flight press conference.

As Glenn approached the end of his first orbit the communications engineers at MCC were struggling to make the connections that would let President Kennedy in the White House talk directly to the astronaut in orbit. Then Glenn was back over Cape Canaveral and talking to Shepard at the start of his second orbit. It was T + 1 h 33 min. The remainder of the flight would continue in the same vein, a few minutes of snatched conversation with each tracking station in turn and where the stations' radio horizons did not overlap long minutes of no communication at all. During the communication blanks Glenn recorded his observations into a tape recorder.

When the White House telephone call reached MCC the line had not been established. Despite Shepard warning Glenn that '*The President will be talking to you*', the conversation never took place. At that time, as Glenn passed over Cape Canaveral and discussed Friendship 7's attitude control difficulties, a bent shaft on one of the spacecraft's rotary limit switches moved position and broke an electrical contact. As a result, a signal was transmitted to MCC that the heat shield had become detached from the wide end of the spacecraft and was only being held in place by the three restraining straps on the Retrograde Package. The problem had to be dealt with and it was more important than a telephone call from the President.

George Metcalf was left with the task of informing President Kennedy: '*Mr. President, we've gotten pretty busy down here now. I don't think we've got time to talk.*'

The men of the Operations Division were faced with their first real problem on a manned flight and had no idea how to deal with it. Despite many hours of in-flight abort simulation runs, no one had ever considered the possibility of the heat shield becoming detached while in orbit. The situation had never been practised and there were no written instructions to follow. In MCC Gene Kranz sent a teletype message instructing all remote sites to check their data relating to Segment 51 – the heat shield indicator.

Spacecraft experts were contacted at McDonnell and they in turn contacted their heat shield subcontractors for their advice. The spacecraft's designers were also brought into the loop at NASA–LaFRC. After long telephone conversations Walt Williams, Maxime Faget and John Yardley joined Flight Director Christopher Kraft at his console. Faget quickly decided that they had little choice but to re-enter Earth's atmosphere with the spent Retrograde Package still strapped to the centre of the heat shield. Hopefully the three titanium straps would keep a loose heat shield in place long enough for air pressure to take over the task when they finally burned through and the Retrograde Package fell away. If any of the retrograde rockets failed to fire Glenn would have no choice but to jettison the Retrograde Package as normal and take his chances with the loose heat shield. While the discussions took place and the decisions were made Kraft decided to keep Glenn oblivious of the signal suggesting that the heat shield was loose. Kraft himself firmly believed that the Segment 51 reading was a false telemetry reading from the spacecraft.

The second orbit was largely a repeat of the first, communication checks, status reports, medical reports and personal observations. The Canary Islands CapCom asked Glenn if he felt that the motions of the particles that he had seen at sunrise were in any way related to the firing of Friendship 7's thrusters. The astronaut replied that he had felt no movements and did not think that the particles were related to the thrusters firing.

During the second orbit he spoke to the *Rose Knot Victor* Atlantic Ocean tracking ship for the first time. He reported that his low rate left yaw thruster was now working, but the low rate right yaw thruster was now not working. Despite this new problem he managed to yaw Friendship 7 through 180 degrees and remarked that he liked being able to see where he was going. He also reported that he had a loose bolt floating around inside the periscope.

Over Kano Glenn returned Friendship 7 to its normal orbit attitude and returned control to the ASCS. He immediately noticed that the spacecraft was drifting in yaw. As he passed over the radio horizon and out of range of Kano the flight passed T + 2 h. Moving on to Zanzibar, he read down a status report in which he stated, '*All "T" handles are in except "manual" and I have pulled it so that I can immediately go to manual as a back-up in case the ASCS malfunctions further*'. As he made another yaw correction using FBWS, the Sun set behind his back and he watched it on the periscope screen between his knees. (See Fig. 90.)

In darkness he swept across the Indian Ocean. Talking to the *Coastal Sentry Quebec* he reported that he was on '*straight manual control at the moment*'. He added,

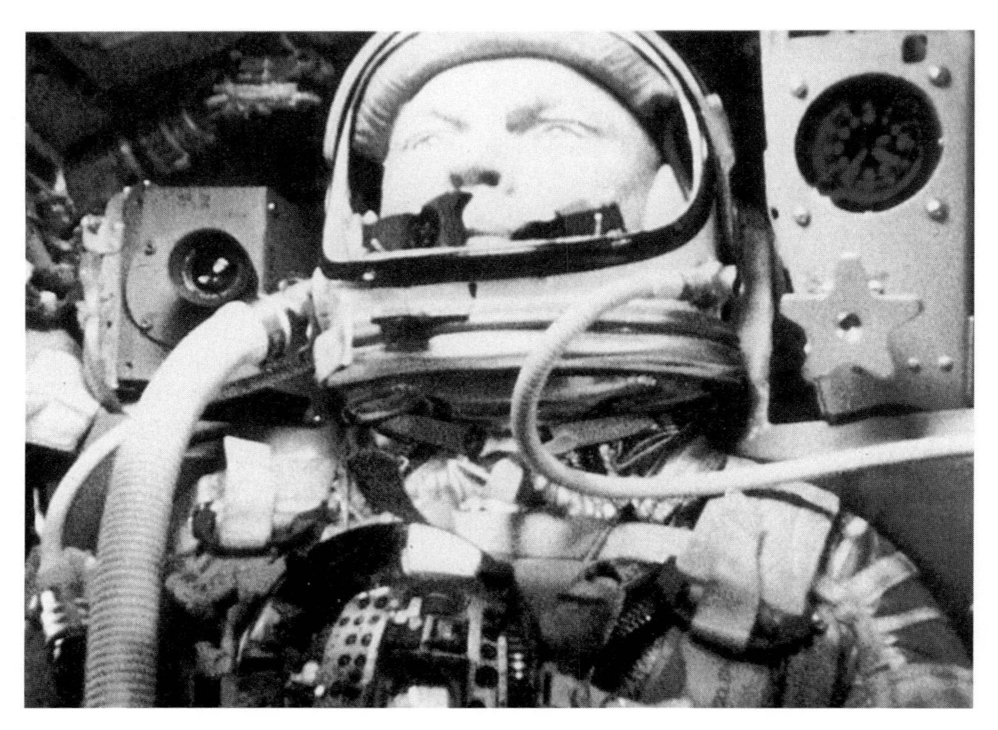

Fig. 90. One of two internal movie cameras films Glenn as he goes about his tasks in orbit.

'*The spacecraft dropped into orientation mode again on ASCS. I took over manually and am reorienting at the present time ... I have a fuel quantity light warning light on automatic. I am okay at present time. Have spacecraft under control and using manual. Over.*' The quantity warning light had come on when the automatic propellant supply had reached 60 per cent remaining, as that system would be needed to control Friendship 7's attitude during re-entry.

Glenn pitched Friendship 7 up, to look at the stars, in an attempt to identify the constellations. While he was looking for stars, a parachute flare was fired from the *Coastal Sentry Quebec*. Glenn did not see it. At T+ 2 h 19 min the *Coastal Sentry Quebec* CapCom told him, '*We have a message from MCC for you to keep your landing bag switch in the off position. Landing bag switch in the off position. Over.*'

Glenn replied, '*Roger. This is Friendship 7.*'

At the end of the communication session CapCom asked Glenn, '*Have you noticed any constellations yet? Over.*'

Glenn came back with, '*Negative. I have some problems with the ASCS. My attitudes are not matching what I see in the window. I've been paying pretty close attention to that; I've not been identifying stars. Over.*'

Communications with Cooper at Muchea were patchy, but Glenn managed to give more details of the problem he was having. He reported, '*I am getting some erratic indications in all axes. When I align everything on orbit attitude by the instruments I am considerably off where I should be. I am rolled some 20 degrees to the right; I'm also yawed to the right a little bit. Over.*' A few seconds later he added, '. . . *I think my* [horizon] *scanners are not working.*' In fact the horizon scanners were working correctly and the attitude indicator problem was not entirely unforeseen. Ground testing had shown that the attitude indicators gave false readings if the spacecraft spent any time manoeuvring outside of the normal orbital attitude. The problem would be overcome on later flights by allowing the astronaut to disconnect the horizon scanner slaving system and the programmed precession of the gyros during such manoeuvres.

Cooper asked him to confirm that his landing bag switch was in the centre, off position, and then asked, '*Have you heard any banging noises or anything like of this type at higher rates?*'

Glenn replied, '*Negative*'.

Over Woomera Glenn transmitted a 30-minute status report and a series of medical reports. The session ended before he could complete his planned exercise period using the bungee cord.

Just before leaving Earth's shadow Glenn aligned his spacecraft to the planet's horizon, caged the gyros and manoeuvred until the instruments showed no rates in all three axes and then uncaged the gyros once more in an attempt to reset them. The spacecraft instruments continued to be off 20 degrees right in roll and 11 degrees in negative pitch.

The Sun was rising as he approached Canton Island and Friendship 7 was once again surrounded by the tiny particles that he had reported during the first orbit. Canton CapCom told Glenn, '*We also have no indication that your landing bag might be deployed. Over.*'

Glenn came back with, '*Roger. Did someone report the landing bag could be down? Over.*'

Realising his mistake the CapCom came back with, '*Negative, we were asked to monitor this and ask if you heard any flapping when you had high Spacecraft rates*'.

Glenn replied, '*Negative*'. He then passed the whole incident off by saying that perhaps MCC thought that a deployed landing bag might be the source of the particles that he had seen at both sunrises.

The conversation over Hawaii concentrated on the attitude problems that Glenn was having and his capability to complete the third orbit as planned. Over California and talking to Schirra, Glenn reported that the attitude rates seemed to be under control and he returned Friendship 7 to the usual orbit attitude. He continued to exercise control with the MPCS. Passing over Texas tracking station Glenn manoeuvred to 60 degrees negative pitch and reported seeing El Paso through the clouds. Then he was talking to Shepard at Cape Canaveral. It was T + 3 h 7 min. His conversation with MCC was short and technical, with Shepard discussing the attitude control problem and recommending that he let the spacecraft's attitude drift for part of the third orbit, in order to conserve thruster propellant.

Over Bermuda the following conversation took place with Grissom:

'*I have the Cape in sight down there. It looks fine real fine from up here.*'

'Rog. Rog.'

'As you well know.'

'Yea, verily sonny.'

It was a conversation that could have only taken place between two of the five people who had seen Earth from space at that time. Glenn described how he could see the entire state of Florida *'laid out just like a map'*.

The communication with the Canary Islands was routine and technical. By the time Glenn was talking to *Rose Knot Victor* in the mid-Atlantic the conversation was back on the subject of attitude control. After passing over Cape Canaveral he had let the spacecraft drift through 180 degrees in yaw and had then reoriented it using MPCS. When he had established the correct orbit attitude using Earth's horizon and his periscope for visual cues his attitude indicators were all wrong. The roll indicator showed 35 degrees right and the pitch indicator showed plus 40 degrees. He caged the gyros, zeroed the indicators and switched control over to ASCS, which solved nothing.

There was no communication with Kano on the third orbit, and over Zanzibar the Sun set for the third time. He manoeuvred manually in yaw to observe the sunset then returned the controls to ASCS. He reported a huge storm over Africa, with lightning going off like firecrackers.

The CapCom on *Coastal Sentry Quebec* took a 30-minute report from Glenn in which the secondary oxygen tank, which had not been used, had dropped from 100 to 92 per cent oxygen remaining. Neither Glenn nor the controllers on the ship below could explain the drop. When the spacecraft drifted again Glenn switched to FBWS in order to control his attitude. He pitched up until he could see the constellation Orion in his window and used it thereafter to maintain his attitude.

Then he was talking to Cooper once more, at Muchea. At T + 4 h he began a 30-minute status check. As the 9-minute pass ended Glenn was trying to establish orbit attitude in the dark using FBWS.

At MCC Kraft had taken advice from all concerned and had made his decision: following retrofire Glenn would jettison the Retrograde Package and re-enter as normal.

As he approached Woomera, Glenn yawed Friendship 7 through 180 degrees, so that he could watch the last sunrise head on. He spotted particles outside his window, but hardly talked about them as he concentrated on maintaining his attitude. Then he was talking to Canton Island. With sunrise over, he manoeuvred back to orbital attitude. He began stowing everything for retrofire and re-entry. Meanwhile, Canton CapCom called for a status check. Following the 180-degree manoeuvre the attitude indicators once again showed erratic readings, until he caged and uncaged the gyros.

Finally, talking to Hawaii CapCom, Glenn was told of the heat shield problem. *'Friendship 7. We have been reading an indication on the ground of Segment 51. Which is Landing Bag Deploy. We suspect this is an erroneous signal. However, Cape would like you to check this by putting the landing bag switch in auto position, and see if you get a light. Do you concur with this? Over.'*

Glenn set the switch and reported, *'Negative. In automatic position and did not get a light and I'm back in off position now. Over.'*

The experiment appeared to confirm the suggestion that the signal on the ground was indeed an error. At T+4 h 23.5 min Glenn began his pre-retrofire checklist.

At MCC the arguments had raged long and hard. Kraft was still sure that the Segement 51problem was faulty telemetry, so much so that, based on the test over Hawaii, he informed Hawaii CapCom, '*Tell Glenn to go ahead with the normal re-entry sequence.*'

Schirra was next to talk to Glenn, from California. As retrofire approached the astronaut was unsure of the accuracy of his onboard clock. Schirra gave him a 45-second time mark. Glenn was flying Friendship 7 on ASCS but backing it up with FBWS.

With just 24 seconds to go before retrofire Schirra told Glenn, '*John, leave your Retrograde Package on through your pass over Texas. Do you read?*'

Glenn replied, '*Roger.*'

The first retrograde rocket fired at T+4 h 33 min 2 s. Glenn reported, '*Are they ever. Feels like I'm going back toward Hawaii.*'

Schirra joked, '*Don't do that. Go to the east coast.*'

In MCC Williams had been making phone calls ever since the Segment 51 light had come on at the beginning of the second orbit. Kraft confirmed with Schirra at the California tracking station that all three retrograde rockets had fired. When Schirra confirmed that they had, Williams told Kraft, '*That settles it, we're coming in with the pack on.*' Kraft had no choice but to concede to the deputy head of the Operations Division.

Over Texas Glenn was instructed to re-enter with the Retrograde Package in situ. When he asked if there was a reason for the instruction he was told, '*Not at this time, this is the judgement of Cape Flight [Director].*'

Six minutes after retrofire Glenn manoeuvred to re-entry attitude. Then he was talking to the MCC, where Shepard was CapCom. On Shepard's instruction Glenn retracted the periscope manually. While he was doing it Shepard informed him, '*We are not sure whether or not your landing bag has deployed. We feel it is possible to re-enter with the Retrograde Package on. We see no difficulty at this time with this type of re-entry. Over.*'

'*Roger. Understand,*' was all Glenn said.

Glenn controlled re-entry using FBWS. The spacecraft entered the upper atmosphere and began to establish a bow wave of ionised air in front of it. Outside the temperatures reached thousands of degrees as Friendship 7's forward velocity was transferred to heat. The astronaut heard a thump behind his back and one of the straps from the Retrograde Package flopped over the window. He watched the strap burn and then it was gone. He also watched two chunks of the Retrograde Package fly past his window consumed in flames.

With his velocity reduced below Mach 1 the bow wave dissipated and he was able to talk to Shepard once more. He told him, '*My condition is good, but that was a real fireball. I had great chunks of the Retrograde Package breaking off half way through.*'

Glenn had run out of propellant in both the automatic and manual control systems during re-entry. As a result he had no thrusters to control Friendship 7's oscillations as it fell through the atmosphere. At 10,668 m the spacecraft yawed through almost 90 degrees and Glenn deployed the drogue parachute manually, some 2,439 m higher than planned. The main parachute deployed 243 m higher than planned and Friendship 7 finally stopped oscillating. The high parachute deployments meant that the spacecraft drifted further uprange than planned. Friendship 7 finally hit the Atlantic Ocean within 20 minutes steaming time of the uprange Destroyer USS *Noa*, which was located 64.63 km from the original predicted impact point. America's first manned orbital flight had lasted 4 hours 55 minutes 23 seconds.

Seventeen minutes after landing USS *Noa* pulled alongside the spacecraft and a sailor cut the whip recovery antenna. Friendship 7, with Glenn still inside, was hoisted aboard using one of the ship's lifeboat davits, which had been converted for the purpose. Glenn began preparing to exit through the neck of the spacecraft, but ultimately warned everyone to stand clear and then blew the side hatch explosive chord. His first words to the recovery crew were, '*It was hot in there*'.

Fig. 91. Glenn's spacecraft is hoisted aboard the USS *Noa*. (NASA)

Fig. 92. Glenn carries out his initial personal debriefing into a tape recorder onboard the USS *Noa*. (NASA)

NASA physicians found Glenn to be hot, dehydrated and unusually quiet. During the flight he had sweated off 2.26 kg of body weight. After a cool shower and a glass of ice water he finally became more talkative about his experience. MCC routed a telephone call from President Kennedy through the Word-Wide Tracking Network to the USS *Noa*, where he could talk directly to Glenn and congratulate him on his flight. (See Fig. 91.)

Glenn found a secluded area on the ship's upper deck and recorded his self-debriefing into a tape recorder. (See Fig. 92.) Three hours after recovery he was airlifted by helicopter to the aircraft carrier USS *Randolph*. He remained on the carrier just long enough to receive a second physical examination and his first debriefing session from MSC experts. Then he was flown to Grand Turk Island. Friendship 7 remained on the destroyer and was returned directly to Port Canaveral, Florida.

At Grand Turk Glenn was greeted by Carpenter and Slayton. After a steak dinner he began three days of medical and technical debriefing sessions.

Vice President Johnson accompanied Glenn on 23 February 1961, when he returned to Cape Canaveral and was re-united with his family. At Patrick AFB the Vice President delivered a short speech and read a letter of congratulations from Soviet Premier Nikita Khrushchev. The group then left the USAF base in a motorcade that took them to Cape Canaveral. The streets were lined with cheering crowds.

John Glenn greeted President Kennedy as he left Air Force One, at 1022. The astronaut and Marine Corps officer led his Commander in Chief on a tour of the Project Mercury facilities. They visited the MCC, LC-14 and Hangar S. Outside Hangar S the President awarded Gilruth with NASA's Distinguished Service Medal. He then presented the NASA Distinguished Service Medal to Glenn, who accepted it on behalf of everyone who had been involved with his flight. Glenn showed the President his spacecraft, which was on display nearby. After an official lunch, NASA held a press conference during which Glenn answered reporters' questions on his flight. Following the press conference the Glenn family left for Key West Naval Base, where they spent the weekend away from the ever-present reporters.

On 26 February the Glenn family flew to Andrews AFB, Maryland. Prior to take-off the Glenns were introduced to Mrs Kennedy and her daughter Caroline. Following the introductions a disappointed Caroline asked, '*Where's the monkey?*' At Andrews AFB the Glenns joined President Kennedy for a motorcade through the District of Columbia, to the White House. Shepard and Grissom rode in the second car. The Glenn family then joined Vice President Johnson on a motorcade through Washington, DC.

At the Capitol Building Glenn was invited to address both Houses of Congress. All of the remaining Flight A astronauts, with the exception of Cooper who was still returning from Australia, were in the audience to hear him state:

> ... This has been a great experience for all of us present and for all Americans, of course, and I am certainly glad to see that pride in our country and its accomplishments is not a thing of the past. I still get a hard-to-define feeling inside when the flag goes by – and I know that all of you do too. Today as we rode up Pennsylvania Avenue from the White House and saw the tremendous outpouring of feeling on the part of so many thousands of our people I got this same feeling all over again. Let us hope that none of us ever loses it.

He related the story of his meeting with Caroline Kennedy earlier that morning and then joked, '*And I did not get a banana pellet on the whole ride.*' Next, he introduced his parents, his family the Flight A astronauts and the leading members of Project Mercury, so that he could share his moment of glory with them all. Finally he ended with:

> Today, I know that I seem to be standing alone on this great platform – just as I seemed to be alone in the cockpit of the Friendship 7 spacecraft. But I am not. There were with me then, and with me now, thousands of Americans and many hundreds of citizens of many countries around the world who contributed to this truly international undertaking voluntarily and in the spirit of cooperation and understanding. On behalf of all of those people, I would like to express my and their heartfelt thanks for the honours you have bestowed upon us here today. We are all proud to have been privileged to be part of this effort, to represent our country as we have. As our knowledge of the Universe in which we live increases may God grant us the wisdom and guidance to use it wisely.

The speech was pure John Glenn and the American journalists, their readers and even their politicians loved every word.

On 1 March Glenn led the seven Mercury astronauts on a motorcade through New York City. Each astronaut, with his family, had his own car. As they passed down Broadway they were treated to a tickertape parade. That evening the astronauts were put up at the Wardolf Astoria, purportedly the best hotel in America. Each family had its own suite.

On 2 March there was a reception at the United Nations' Building in Glenn's honour. The following day the Glenn family returned to Concord, New Hampshire, where John Glenn had been born. They were greeted with another parade. And so it went on. John Glenn was feted more than any other American astronaut, until the Apollo 8 crew flew around the moon at Christmas 1968. When the attention finally died down Glenn returned to LaFRC and took his place among the Flight A team. It was time to get ready for Mercury Atlas 7.

In the wake of Glenn's flight, Kraft had his controllers collected additional information on the spacecraft and launch vehicle systems. It was his intention that MCC would never again be caught unaware and unprepared, as it had been by the Segment 51 indication during MA-6. Never again should anyone have more information at their fingertips than the Flight Director and his immediate team of controllers.

February 1962: The Soviet Union announced that it was reserving the air corridors across East Germany that gave access to West Berlin for use by its own military aircraft. Although flights continued into West Berlin, the aircraft were frequently 'buzzed' by Soviet fighter aircraft. Soviet aircraft dropped 'Chaff' in an attempt to confuse Western radars tracking aircraft approaching West Berlin.

March 1962: Korolev signed a proposal that used a manoeuvrable Vostok Spacecraft to rendezvous and dock three separately launched rocket stages in Earth orbit, as the building blocks of a circumlunar spacecraft. This vehicle has become known as the Soyuz Complex. A new manned spacecraft, designated L-1, with three cosmonauts, would then be launched to dock with the orbiting Complex. When the Vostok Spacecraft had undocked and manoeuvred to a safe distance, the three rocket stages would be fired in sequence to place the L-1 on its circumlunar trajectory.

20.9 THEN THERE WERE SIX

When the seven Flight A astronauts were named to the public on 9 April 1959 it was expected that each one of them would make a Mercury Redstone suborbital flight before making a Mercury Atlas orbital flight. To that end, all seven astronauts began group training to prepare them for the manned flight-test portion of Project Mercury. In August 1959 the seven members of Flight A began centrifuge training at Johnsville Naval Air Station.

During the centrifuge training the astronauts' USAF physician, Bill Douglas, noticed that Slayton's heart occasionally displayed a condition called sinus arrhythmia. It was common among fit young men such as military pilots and often went away with exercise or exertion. On the day in question Slayton completed a centrifuge run at 1 negative *g* followed by another at 3 negative *g*. Finally he completed a run where the gondola spun on its axis and swapped him from 3 positive *g* to 3 negative *g*. When the third centrifuge run was complete the arrhythmia was still present on Slayton's heart trace. Douglas decided to take an EKG reading. Slayton's condition, identified as Idiopathic Atrial Fibrillation, consisted of his heart occasionally making an irregular beat, which reduced its capacity to pump blood around his body. The cause of the condition was unknown. Douglas informed Slayton of his condition and brought it to the attention of the Head of Cardiology at the Philadelphia Naval Hospital. Douglas was assured that the condition would not stop Slayton being selected to fly in space.

In October 1959, Douglas took Slayton to the USAF School of Aviation Medicine, Brooks AFB, San Antonio, for further medical examinations. He was checked over by Dr Kossman and Dr Lamb and once again he was assured that the condition would not stop him from being selected to fly as part of Project Mercury, although he was now sure that he had no chance of being selected to make the first flight.

In the third quarter of 1959, Douglas made Gilruth aware of Slayton's heart condition and that, in the opinions of two military medical centres, it should not stop him flying in space. In the meantime Douglas had also approached the Office of the USAF Surgeon General, which advised him to take no action to ground Slayton.

On 28 November 1961, MA-5 had carried the chimpanzee Enos on two orbits of Earth. At the press conference held the next day Deke Slayton was named as prime astronaut for MA-7. He would decide to call his spacecraft 'Delta 7' after the engineering term for the change in a number. Wally Schirra was named as Slayton's back-up. (See Fig. 93.)

In January 1962, while MSC was struggling to prepare John Glenn's MA-6 for launch, Webb was forced to order a review of Slayton's medical condition and his suitability to fly MA-7. Dr Lamb, who had examined Slayton in October 1959, had changed his mind and now felt that Slayton's condition should stop him flying in Project Mercury. He committed his new opinion to paper, in a letter addressed to Webb. Lamb was Vice President Johnson's cardiologist and may also have mentioned Slayton's condition within earshot of the White House staff, where it probably came to the attention of Jerome Wiesner, President Kennedy's scientific adviser who had lost his attempt to have the new President disassociate himself from Project Mercury and concentrate the nation's space efforts on unmanned space flight. Wiesner had informed Defence Secretary Robert McNamara who had passed it down through the USAF Chief of Staff until it came to the USAF doctors assigned to NASA-HQ. Although Webb had wanted the matter dealt with by civilian doctors, General Curtis LeMay insisted that Slayton was an USAF Major (he had been promoted in 1962) and should therefore be dealt with by USAF doctors. Slayton attended the office of the Surgeon General of the USAF on 13 March. There, eight doctors examined him and recommended that he fly MA-7. General LeMay accepted his doctors' recommendation.

Wiesner continued to argue against Slayton flying MA-7. Webb then insisted that three civilian doctors examine Slayton. The examination took place at NASA Headquarters on 15 March. When Slayton returned to Webb's office, NASA's Deputy Administrator Hugh Dryden informed him that he had been removed from

Fig. 93. Donald Slayton during training, before he was removed from flight status due to a perceived medical problem. (NASA)

MA-7. The following day NASA made Slayton attend a press conference to explain his condition and why he would not fly MA-7.

Walt Williams was given the task of selecting the new prime astronaut for MA-7. He argued that Carpenter had undergone the full training for a three-orbit flight as John Glenn's back-up on MA-6. Meanwhile, Mathews and Kraft both believed that Slayton's back-up, Wally Schirra, was better suited to the six-orbit flight that was due to be flown later in the year. Much to Slayton and Schirra's discontent, Williams named Carpenter as the new prime astronaut for MA-7, with Schirra continuing to serve in the back-up role. When the flight was made Slayton would be well away from the American press. He would serve as CapCom at Muchea, Australia.

Slayton continued to fight to be returned to flight status. When Schirra was assigned as prime astronaut for MA-8, Cooper was named as his back-up. By then plans were already underway to have Mercury Atlas 9 fly 18 orbits and Slayton fought to be assigned as the prime astronaut for that flight.

Douglas also continued to fight on Slayton's behalf and arranged for him to visit Dr Paul Dudley White in Boston. White had been President Eisenhower's personal physician and after the examination Slayton returned to work. Two weeks later Dudley White's report suggested that Slayton had no problem but, if NASA had astronauts who did not suffer with Idiopathic Atrial Fibrillation, then they should fly those astronauts rather than Slayton who did suffer from the condition. The recommendation grounded Slayton for the remainder of Project Mercury. In a private meeting Gilruth advised Slayton that he should not expect to fly on Project Gemini.

At that time NASA was considering a number of military personnel to come in and take charge of the Astronaut Office, which was about to be enlarged with the intake of a second group of astronauts required for Project Gemini. Shepard led the other five Flight A astronauts in a campaign to have Slayton assigned to the position, which would make him their first line manager. Slayton liked the idea and approached Gilruth with the suggestion personally. Gilruth agreed, and Slayton became the first Coordinator of Astronaut Activities. His first task was to oversee the selection of Flight B.

Thereafter, Slayton would be responsible for the selection of each subsequent astronaut flight crew, through the end of Project Skylab. He would also be responsible for flight crew selection during the same period. NASA management rejected only one crew that Slayton recommended and, ironically, Slayton's great personal friend Alan Shepard was the Commander of that crew.

18 April 1962: NASA announced that DX rating had been assigned to the Apollo Spacecraft and the Saturn C-1 and Saturn C-5 launch vehicles.

25 April 1962: The second Saturn C-1 was launched from LC-34, AMR. When the live first stage shut down at T+1 min 56 s the two dummy upper stages were exploded to release a large amount of water into the ionosphere. This was Project High Water. The flight was a success.

26 April 1962: America resumed atmospheric testing of nuclear weapons.

14 May 1962: Following Communist military action in Laos, President Kennedy positioned American troops training in Thailand along that country's border with Laos.

5 May 1962: The Communist Pathet Lao, with North Vietnamese backing, broke the ceasefire in Laos. President Kennedy sent the US Seventh Fleet to a position off Thailand.

22 May 1962: Following an exchange of private correspondence Khrushchev assured Kennedy that there would be no more large Communist offences in Laos. That country's neutrality in South-East Asia was re-established.

20.10 MERCURY ATLAS 7 (MA-7)

The pre-count for MA-7 began on 23 May 1962 and was held overnight, before being picked up on the morning of launch. Scott Carpenter (Fig. 94) was woken up

Fig. 94. Malcolm Scott Carpenter. (NASA)

at 0115 on 24 May. He completed his ablutions and joined a few Project Mercury personnel for breakfast. At 0215 he commenced his pre-flight physical examination and the sensors from his biomedical harness were taped to his body. He was fully dressed in his pressure suit by 0325.

After relaxing in an armchair for 20 minutes, Carpenter left Hangar S at 0345, with suit technician Joe Schmitt and astronaut John Glenn at his side. They entered the transfer van and took the ride out to LC-14. When they arrived at the base of the launch gantry the whole area was covered in an early morning mist. The astronaut remained in the transfer van while he was briefed on the weather both at Cape Canaveral and in his various predicted landing areas. For the first time Slayton did not present the launch day weather briefing. He was in Australia.

At 0435 Carpenter was instructed to leave the transfer van and ride the elevator up to the white room. Ten minutes later he was entering the spacecraft that Slayton had intended to call 'Delta 7'. Carpenter had named it 'Aurora 7'. There was a minor problem fitting the faceplate sealing ring inflation tube to his helmet, but otherwise the countdown proceeded smoothly. Carpenter shook hands with the launch team and watched while they bolted the side hatch into place and secured the explosive chord's lanyard. A few minutes later the evacuation horn sounded and Carpenter was left alone in Aurora 7 while the launch complex was evacuated. The mist had failed to clear and a 15-minute hold was called in the countdown at 0649, T – 11 min. The initial hold had to be twice extended by 15 minutes, before the countdown could be resumed. It was picked up at 0734 and continued through launch with no further difficulties.

Carpenter later reported that he did not hear any of the noises that Glenn had reported hearing prior to launch, as the Atlas was prepared for the task ahead. Rather, he reported that the first sign he had that anything was happening was the ignition of the main engines at the base of the launch vehicle. In his Astronaut's Report Carpenter would write, '*Lift-off is unmistakable*'. It occurred at 0745.

At T + 29 s Carpenter reported, '*A little rough through max-q*'. Prior to BECO, at T + 2 min 8 s acceleration forces had built up steadily from 1*g* to 6.5*g*. Carpenter heard and felt the Booster engines stop thrusting and the Booster skirt fall away. The Atlas pitched down slightly and the LES Escape and Jettison Motors fired at T + 2 min 30 s. Carpenter reported, '*The tower is way out. It's gone. The light is green.*' The Atlas pitched its nose up once more and the climb into orbit continued.

At T + 4 min 15 s hydraulic switch No. 2 on the Sustainer engine activated an abort signal. Telemetry from pressure transducer H52P, to which the switch was connected, showed a total loss of hydraulic pressure. This was in contradiction to the telemetry received from a second pressure transducer in the same circuit, which showed that the hydraulic pressure remained constant, at the correct level. As both transducers had to generate an abort signal to enable it to be acted upon by the ASIS, the flight continued and Carpenter was not told of the event.

During the Sustainer burn the astronaut experienced a smooth transition from 1*g* to 7.8*g* at SECO, T + 5 min 9 s. When the Sustainer stopped firing, Carpenter entered microgravity and he was awed by the silence of orbital space. The three posigrade rockets fired and separated Aurora 7 from its spent Sustainer. Carpenter then completed a manual turnaround, using the FBWS. With nothing but the black of space in his window, Carpenter had the sensation that the spacecraft was not turning but his attitude indicators told him differently. He saw Earth's horizon, then the Moon and finally the Sustainer and reported, '*A steady stream of gas, white gas, out of the Sustainer engine.*' He turned control of the spacecraft over to the ASCS while he photographed the Sustainer as it tumbled in the sunlight. Glenn's spacecraft had used 2.26 kg of propellant to complete the turnaround on ASCS. Carpenter had consumed just 0.72 kg of propellant on FBWS.

Aurora 7 was in a 160.7 × 268.4-km orbit, inclined at 32.5 degrees to Earth's equator. Each orbit had a period of 88.5 minutes. Grissom was CapCom at MCC, and talking through the antennae at Bermuda he told Carpenter that, '*We have a go, with a seven orbit capability.*' Like Glenn before him, Carpenter would complete three orbits if all went well. If not, there were numerous recovery zones along each orbit.

He freed the gyros and used FBWS again, in an attempt to keep the Sustainer centred in his window. Carpenter complained that their trajectories were too close for the exercise to be a good one and suggested that close-up tracking, such as he was doing, should only be completed using the low-rate thrusters, as the high-rate thrusters made the task too difficult.

Shortly after achieving orbit, telemetry received on the ground suggested that the pitch horizon scanner, mounted in the Antenna Canister at the narrow end of the spacecraft, was 18 degrees out of alignment. It would prove to be a random failure that would come back to distract Carpenter at retrofire, when he least needed it.

Passing out of range of the Bermuda antennae, the astronaut began talking to the controllers at the Canary Islands. He told them that he was adopting orbit attitude and returning the controls to ASCS. In so doing he complained that his attitude indicators did not agree with what he could see out of the window. He put it down to the time spent with the gyros free. Concentrating, as he was, on photographing the cloud patterns over the Canary Islands, Carpenter was not able to record the sunset and sunrise times when they were read up to him.

Over Kano he admitted, '*Roger. I am a little behind the flight plan at this moment.*' It was T+ 26 min 37 s. He photographed the Earth's limb using MIT film and filters, as one of the first experiments completed in space.

Then he was on FBWS and uncaging the gyros, over the Indian Ocean Ship, *Coastal Sentry Quebec*. As Carpenter opened his faceplate to observe his passage over the African coast, he glanced at his periscope screen but was unable to see anything. He asked, '*What in the world happened to the periscope*?' Then he answered himself, '*Oh, it's dark, that's what happened. It's facing a dark Earth.*' Two minutes later he was awed by sunset and struggled to stay focused on his attempts to observe Venus as it passed behind Earth's horizon. He noted that light leaking from around the clock on the instrument panel was making the cabin too light for his eyes to fully adapt to the darkness. He did not see the haze layer that Glenn had reported. At the end of the pass he noted, '*It's now nearly dark, and I can't believe where I am. Oh, dear, I've used too much fuel.*' His automatic propellant supply was down to 75 per cent.

Talking to Slayton at Muchea, Carpenter complained that his pressure suit temperature was too high and that he had increased the cooling water rate by two stops since lift-off. By the end of the pass the suit temperature had started to fall slightly. Post-flight investigation showed that the pressure suit cooling system was contaminated with dried silicone lubricant. Carpenter struggled throughout the flight to bring his pressure suit temperature under control. He admitted that he was behind the flight plan, before the rest of the pass was filled with technical details.

Over Woomera he switched back to FBWS, so that he could manoeuvre to plus 80 degrees yaw and minus 80 degrees pitch in an attempt to observe four flares launched from the tracking station. He did not see them due to the cloud coverage over the area. At T+1 h 2 min 41 s he caged the gyros, switched off power to the ASCS bus and entered drifting flight. Aurora 7's narrow end was free to describe arcs in the vacuum and, with the ASCS turned off, the spacecraft did not try to maintain the correct 34 degree negative pitch orbit attitude.

Carpenter attempted to eat at T+1 h 8 min. The bite size cubes of food were coated in glaze and packed in opaque packaging. The one that he opened had been crushed at some point and crumbs drifted into the Crew Compartment. The astronaut ate what he could but was unhappy about the free crumbs. This was the only meal that he ate in orbit, despite a second meal being scheduled. He would suggest later that similar bite size food cubes be packed in clear wrappings to enable the astronaut to see their condition before he opened the packet.

Canton CapCom asked him if he was comfortable, as they were showing his body temperature to be 102 °F. Carpenter stated that he was sure he was not as hot as that.

Talking to Hawaii he reported that he could see the constellation Cassiopeia in his window, just before he yawed through 180 degrees to watch the sunrise head on. The Sun came over Earth's horizon at T + 1 h 20 min 32 s and Carpenter reported that he could now see the particles that Glenn had reported during each sunrise on his flight. He reported that they were '*almost like a light snowflake particle caught in an eddy. They are not glowing with their own light at this time.*' He expressed the opinion that they might be frost from the spacecraft's thrusters.

Next he was talking to Cooper at Guaymas. Cooper confirmed that his instruments were also showing a body temperature of 102 °F and Carpenter stated once again that he believed it was false. The astronaut passed a message of welcome in Spanish to the Mexican people. At T + 1 h 29 min Cooper told Carpenter, '*It looks like we have a go for the second orbit as everything appears alright for you.*' He added, '*You start to conserve your fuel a bit and maybe, perhaps, use a little more of your manual fuel.*'

In his status report to Grissom at Cape Canaveral, Carpenter gave an automatic fuel reading of 62 per cent remaining and his manual fuel was at 68 per cent. The fuel quantity warning light and the excess cabin water warning light were both illuminated on his instrument panel.

At T + 1 h 38 min he pressed the button to deploy a tethered balloon from its container mounted in the Antenna Canister of his spacecraft. (See Figs 95 and 96.) He originally saw a piece of balsa wood packing and reported it as the balloon. Then he saw the balloon itself. He reported, '*The balloon is partially inflated. It's not tight*

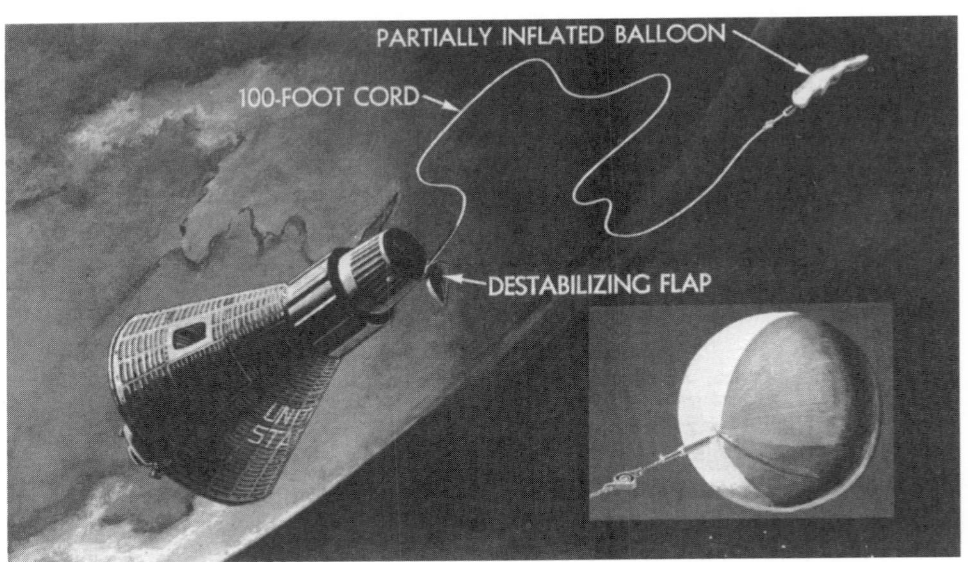

Fig. 95. Mercury Atlas 7 balloon experiment. (NASA)

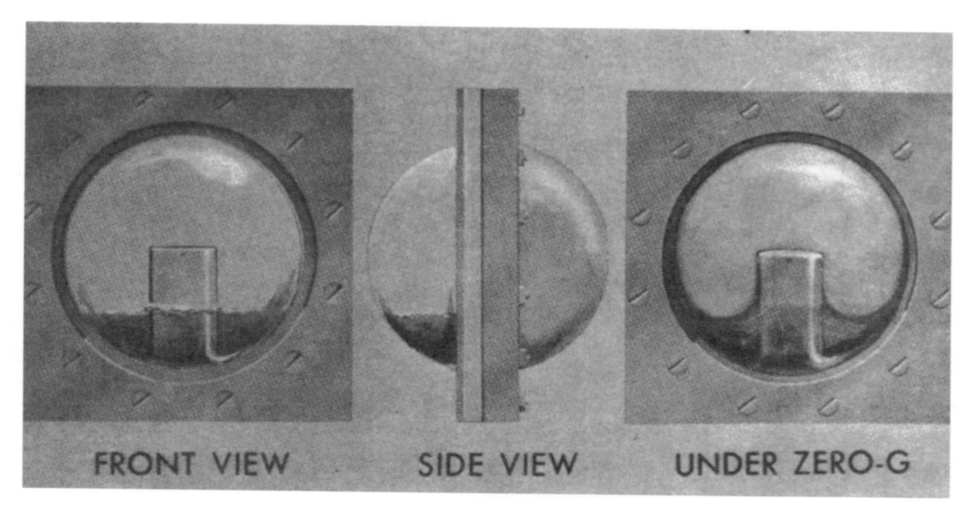

Fig. 96. Mercury Atlas 7 liquid behaviour in microgravity experiment. (NASA)

... There is an oscillation beginning ... The line is still not taut. I have some pictures of the line just waving out in back. I would say we have about a one cycle per minute oscillation. It's both in pitch and yaw.' Over Bermuda on his second orbit he reported that the balloon also oscillated in and out, towards the spacecraft and away from it. He stated that sometimes the line became taut and at other times it was so loose that it had loops in it. He also reported that he had turned his pressure suit cooling water valve off and then returned it to the first setting. He further stated that he was placing tape over the excess cabin water light and the fuel quantity warning light, in order to be able to obtain better dark adaptation during his next night pass.

The Canary Islands heard Carpenter talking about the haze layer as observed during his first night pass. The CapCom told him, '*Mercury Control Centre is worried about your auto fuel and manual fuel consumption. They recommend that you try to conserve your fuel.*' Carpenter replied, '*Roger. Tell them I am concerned also. I will try to conserve fuel.*'

As he crossed Africa for the second time he made a 30-minute report. Aurora 7's automatic fuel supply was down to 51 per cent. He was on ASCS in orbit attitude. Kano stated that they still had a body temperature reading of 102 °F. Carpenter repeated that he was sweating on his forehead, but he did not think he was that hot. His Crew Compartment temperature was 101 °F, but he his pressure suit exhaust valve was reading 74 °F and he turned the water flow setting up one notch in an attempt to bring that temperature down. He passed a greeting to the people of all Africa. The flight plan called for him to eat a second bite size food cube. He felt through the packet that it had crumbled and decided not to open it for fear that the crumbs might enter the spacecraft's systems and cause malfunctions.

Then he was talking to Zanzibar, as he turned his pressure suit cooling water flow from 4 to off and then back up to 3. The flight plan called for him to switch to FBWS, but he elected to stay in ASCS.

Passing over the *Coastal Sentry Quebec* in the Indian Ocean, the astronaut noticed that his attitude indicators did not agree with what he could see out of the window. He originally thought that he had an ASCS thruster malfunction, but having switched to FBWS and tested the thrusters he realised that they were all functioning properly. He manoeuvred Aurora 7 so that it was in orbit attitude. Returning the controls to ASCS he left them to see what happened. When the spacecraft began drifting in pitch and yaw he used FBWS to null the rates.

Meanwhile his Crew Compartment temperature was still high, as was his pressure suit temperature. He opened his faceplate because it felt cooler. The continuing high temperature in his pressure suit led to his electing not to complete the exercise session due at that time. At T + 2 h 16 min he admitted to his tape recorder, '*I have gotten badly behind the flight plan now*'. As he passed into Earth's shadow he noted that he did not see a haze layer.

Talking to Slayton in Muchea, Carpenter stated that his only problems were the Crew Compartment and his pressure suit temperatures. He stated that the Crew Compartment water valve was set '*past the marks*'. His suit flow valve was set on 4 and he increased it to 6 at that time. During his 30-minute report Carpenter read out an automatic fuel reading of 45 per cent; his manual fuel was at 68 per cent. When Slayton instructed him to go to manual control in order to conserve the automatic fuel supply for re-entry, Carpenter made the switch.

At MCC, Kraft was considering ending the flight after two orbits if Carpenter did not get control of his fuel consumption. A few minutes later Slayton warned Carpenter, '*For your information, Cape informs us that if we don't stay on manual for quite a spell here, we'll probably have to end this orbit.*'

'*I'll be sure to stay on manual.*'

'*Roger. You've got a lot of drift left here yet too.*'

'*Say again.*'

'*You've got drift capability left yet too.*'

'*Roger.*'

Carpenter closed his faceplate and noted that he now felt cooler than when he had the faceplate open. He decided to leave it closed for a time.

The planned flares were not launched as he passed over Woomera, because of the solid cloud cover. Rather, the conversation concentrated on the balloon experiment and the temperatures in the Crew Compartment and his pressure suit. The astronaut opened his faceplate long enough to consume a Xyclose pill and then closed it again. He noted that most of the uneaten food had melted in the heat.

Over Canton Island he was asked about the balloon. He reported that the day-glow orange panels were the easiest to see. Carpenter manoeuvred through 180 degrees to watch the sunrise head on. As he passed into range at Hawaii he tried to delay a short status report while he photographed the sunrise. He was instructed to make the status report first. He also reported a small number of particles outside the window at sunrise. He manoeuvred to within horizon scanner limits and uncaged the gyros as he passed out of range from Hawaii.

Shepard was CapCom at California, he passed on to Carpenter that MCC had cleared him for the third orbit. The message was followed by, '*General Kraft is still somewhat concerned about your auto fuel. Use as little auto; use no auto fuel unless you have to prior to retro-sequence time.*' The astronaut opened his faceplate and consumed 20 sips of water before resealing it.

Then he was talking to Grissom in MCC as his third orbit began. At T + 3 h 11 min he caged the gyros and began a second period of drifting flight in an attempt to conserve propellant. Grissom passed up a message from the flight surgeon that Carpenter should drink as much water as he could in order to stop the effects of dehydration in his warm pressure suit.

Carpenter manoeuvred so that the balloon was trailing out behind Aurora 7 before he pushed the button to jettison the spacecraft end of the tether. The tether did not jettison. The communication session ended with Grissom asking for and receiving a description of the fluid mechanics experiment mounted in the spacecraft. '*The standpipe is full of the fluid. The fluid is halfway up the outside of the standpipe. There is a rather large meniscus. I'd say about 60 degrees.*'

Passing the Canary Islands on his third orbit he reported that he could still see some particles outside the window, despite not having fired a thruster for some time. He reported that the Crew Compartment temperature had dropped and that his pressure suit cooling water was set on 7. He increased the latter setting to 8.

Approaching Kano in Africa he placed the MIT film back on his camera in preparation for more Earth limb photography. With the spacecraft in drifting flight he reported, '*The zero-g sensations are wonderful. This is the first time I've ever worn this suit and had it comfortable ... I don't know which way I'm pointed and I don't particularly care.*' At the end of the Kano pass he talked into his tape recorder, because he could not hear the Kano CapCom. He decided to finish the roll of film designated for the Earth limb photography, as he would have no time to do so as he approached re-entry. He described how he was '*... taking many MIT pictures*'.

Over the Indian Ocean ship he described sunset, which occurred at T + 3 h 41 min. '*The sunsets are most spectacular. The Earth is black after the Sun has set. The Earth is black; the first band close to the Earth is red; the next is yellow; the next is blue; the next is green; and the next is a sort of purple. It's almost like a very brilliant rainbow ...*'

After some discussion on the balloon, which was still attached to the Antenna Canister, CapCom informed Carpenter, '*Our medical monitor says that we are reading your respiration. I believe this is almost the first time it's come across.*' The astronaut replied, '*That's very good. I guarantee I'm breathing.*' He opened his faceplate to drink a large amount of water and left it open while he used the photometer to view stars through the atmospheric haze layer. He complained that the shiny finish on the star charts made them difficult to read when in Earth's shadow. He also began to report observing a comet, but stopped himself in mid-sentence when he realised that he was looking at the balloon attached to his spacecraft and lit by moonlight.

Back with Slayton in Muchea, Carpenter closed his faceplate before writing down the time for retrofire. On Slayton's instruction he completed a full 60-second exercise cycle before transmitting a blood pressure reading.

The 30-minute status report over Woomera included an automatic fuel quantity reading of 46 per cent and a manual fuel reading of 46 per cent. He remained in drifting flight with the gyros caged. The Crew Compartment temperature remained at 102 °F, while his pressure suit vent temperature was down to 60 °F. Carpenter had not yet noticed the fact, but his suit temperature had been high all the time that he had had his faceplate open to the 102 °F Crew Compartment. When he had begun passing longer periods with the faceplate closed his pressure suit circuit was isolated and the suit temperature had started to come down.

At T + 4 h 10 min he commented, '*I have 22 minutes and 20 seconds left to retrofire. I think I'll try to get some of this equipment stowed at this time ... There is the Moon [crossing his window} Looks no different here than it does on the ground.*' Once again he complained that it was impossible to make good observations as the light leaking from around the clock meant that his eyes could not become fully dark-adapted. He continued to try to make observations of the atmospheric haze layer.

The Sun rose behind his back at T + 4 h 19 min and he saw a myriad of Glenn's particles. He also made an observation relating to their origin. '*They are spacecraft emanating. I can rap the hatch and stir off hundreds of them. Rap the side of the spacecraft; huge streams come out. They – Some appear to glow. Let me yaw round the other way.*' He yawed Aurora 7 through 180 degrees to face the sunrise. '*That's where they come from. They are little tiny pieces of white frost. I judge from this that the whole side of the spacecraft must have frost on it.*'

At the beginning of his third pass over Hawaii he was yawing back though 180 degrees and taking up the standard orbit attitude. With less than 3 minutes of radio contact left he finally placed the spacecraft on ASCS and was about to let CapCom help him with the pre-retrofire checklist, when he noticed a new problem. '*Wait a minute, I have an ASCS problem. I think the ASCS is not operating properly. I have to evaluate this ASCS problem, Jim, before we go any further.*' The problem was caused by a random malfunction in the pitch horizon scanner as had been noted shortly after orbit insertion. He switched to FBWS, but forgot to turn off the MPCS that he had been using during the third orbit. For the next few minutes all manoeuvres were made on dual authority, with fuel being expended from both the automatic and manual propellant supplies. CapCom insisted that he complete the pre-retrofire checklist before the session ended, which he did. With less than 5 minutes to retrofire Carpenter told his tape recorder, '*I think we're in good shape. I'm not sure what the status of the ASCS is at this time.*' (See Fig. 97.)

Shepard was at California to talk him through retrofire. Carpenter was controlling on FBWS using visual cues from his window and the periscope. The MPCS was also still active. Shepard counted down the last few seconds to retrofire, but nothing happened. After waiting 3 seconds Carpenter initiated retrofire manually. All three retrograde rockets fired in sequence. Despite his best efforts to align the spacecraft manually, Aurora 7 was 27 degrees out in yaw at the start of retrofire. This fact meant that much of the retrograde rockets' thrust was lost firing in the wrong direction. The retrograde rockets also failed to thrust at full power. These two points and the late retrofire would result in the spacecraft overshooting its predicted landing site. One minute later the Retrograde Package was jettisoned and the periscope retracted automatically. His automatic fuel was down to 20 per cent and his manual fuel was showing 5 per cent. When he tried to put the spacecraft on

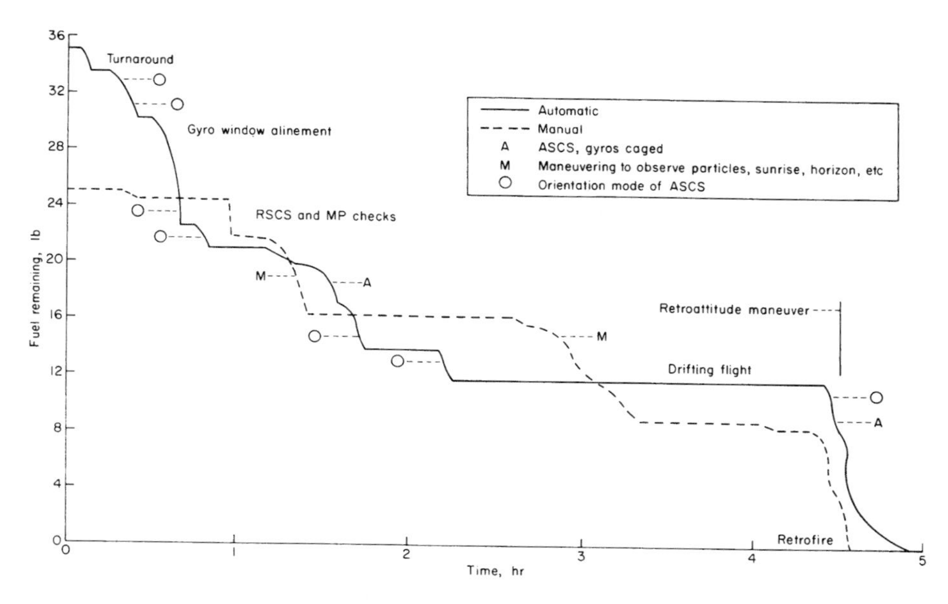

Fig. 97. Aurora 7 propellant usage. (NASA)

ASCS nothing happened, the automatic fuel tank was actually empty. He returned to FBWS for re-entry.

The spacecraft began tumbling in all three axes, and Carpenter nulled the rates each time before they reached 90 degrees in any one axis. As the communication session with California came to an end Shepard advised Carpenter to assume the correct re-entry attitude.

Over Cape Canaveral and talking to Grissom, Carpenter confirmed that he had completed his re-entry checklist and stowed all of his equipment. Grissom passed up a weather report for the primary recovery zone.

Aurora 7 entered radio blackout as a bow wave of ionised air surrounded the spacecraft. Carpenter described an astronaut's view of re-entry into his tape recorder: '*I've got an orange glow. I assume we're in blackout now ... There goes something tearing away. Okay. I'm getting a roll rate at this time. Going to Aux Damp. I hope we have enough fuel. I get the orange glow at this time ... bright orange glow. Picking up a little acceleration now.*' Over the next 3 minutes *g* forces climbed to 6.5*g*, at which point the orange glow disappeared. The forces on the astronaut's body slowly dropped to 1*g*.

The altimeter on the instrument panel began registering at 30,480 m. At that time Carpenter reported, '*smoke pouring out behind*'. At 21,336 m Aurora 7 began building up rates in all three axes. At the same time the automatic propellant supply ran out. With no way to counteract the rates Carpenter tried to use his view of the Sun out of his window to gauge how large they were. At 7,620 m, as the rates approached 45 degrees, Carpenter manually deployed the drogue parachute. He then manually deployed the air snorkel.

The main parachute failed to deploy automatically at 3,048 m, so Carpenter deployed it manually at 2,895 m. He watched as the striped parachute deployed in a reefed condition and then the reefing lines were released to allow it to inflate fully. The astronaut heard the heat shield detach from the base of the spacecraft and deploy the landing skirt. He opened his faceplate, removed his oxygen outlet hose and commenced the pre-landing checklist.

Grissom transmitted the prediction from NASA-Goddard that the spacecraft would land 321.8 km downrange from the primary landing zone and recovery personnel would not reach the area for at least one hour. Aurora 7 actually landed 402.25 km downrange from the primary recovery zone, after a flight lasting 4 hours 56 minutes 5 seconds.

Water impact was not as severe as Carpenter had been led to expect. His main cue was the noise of impact, which led him to suspect that the heat shield had re-contacted the aft bulkhead of the Crew Compartment. He was disconcerted to see water splash onto his tape recorder. The spacecraft listed heavily and did not right itself.

Carpenter was still hot and the humid sea air entering the Crew Compartment through the snorkel valve only added to his discomfort. Having removed his helmet he took the pliers from the survival kit and removed the right-hand side of the instrument panel. He then had a short struggle to remove one half of the forward bulkhead. Finally, he pushed the empty main parachute container out of the neck of the spacecraft and, taking his survival kit with him, he exited the spacecraft through the neck. He was the only Mercury astronaut to use this egress method at the end of a flight. Having inflated his life raft he tied it to the spacecraft and climbed aboard. When he realised that it was upside down he climbed back into the water, turned the raft over and reboarded it. He quickly realised that he had failed to seal the oxygen inlet on his pressure suit, or to deploy his rubber neck dam. He would describe his time in the raft as '*pleasant*'. He drank a lot of water from his survival kit during that period.

Six minutes after landing, one aircraft in the search pattern reported a possible UHF/DF contact with the downed spacecraft. Eighteen minutes after landing all aircraft in the search pattern had a positive UHF/DF contact with Aurora 7 and the Destroyer USS *Farragut* was steaming towards the landing area.

Meanwhile, Carpenter saw two P2V search aircraft and a single Piper Apache. The Apache pilot could be seen photographing Aurora 7 in the water. One of the P2V pilots reported identifying Aurora 7's green dye marker in the water and then located the spacecraft itself. Carpenter was seen sitting in his raft next to the spacecraft.

Fifty-three minutes after landing two SC-54 aircraft arrived over the area and dropped two USAF Para-Rescue swimmers into the ocean. They swam towards Carpenter, inflated their own rafts and tied them to Aurora 7. Carpenter offered the swimmers some of his survival rations, but they declined. In his Pilot's Report Carpenter would misidentify these men as 'Navy frogmen'.

The second aircraft dropped the Stullken Collar into the water. Despite having its own parachute the package impacted the water with sufficient force to burst one of the compressed-air inflation bottles. The two swimmers recovered the Stullken Collar, draped it around Aurora 7 and inflated the upper ring. The lower ring could not be inflated due to its gas bottle having burst. Both swimmers then returned to their rafts.

Two US Navy Seaking helicopters were dispatched from the prime recovery vessel, the aircraft carrier USS *Intrepid*, towards the recovery area. They arrived over the spacecraft 2 h 49 min after water impact. The aircraft with a Navy physician on board lowered a horse collar to Carpenter, who was then winched aboard. Carpenter would write in his report

> When the HSS-2 helicopter appeared it made a beautiful approach. One of the divers helped put me in the sling and I picked up my camera, which I had previously placed in the Recovery Compartment. I motioned to the helicopter pilot to take up the slack in the line, and I let go of the spacecraft expecting to be lifted up. Instead, I went down! The helicopter must have settled slightly, because I am sure there was a moment when nobody saw anything of me but a hand holding the camera clear of the water. A moment later however, I began to rise. I got into the helicopter with no difficulty and took off my gloves and boots. I poked a hole in the toe of my left sock and stuck my leg out of the window [the helicopter's waist door] to let the water drain out of my suit.

Carpenter was in the helicopter 2 hours 59 minutes after landing. The NASA physician described him as '*exhilarated*'. En route back to the USS *Intrepid* the physician subjected the astronaut to a brief physical examination. The helicopter landed on the aircraft carrier's flight deck 24 minutes after recovering the astronaut,

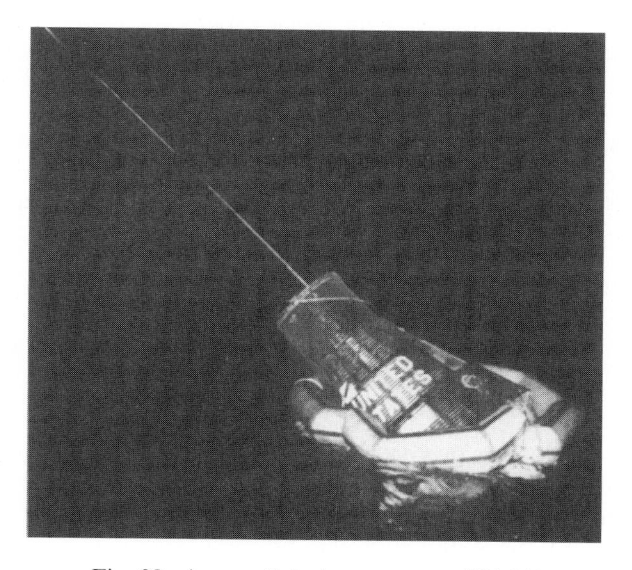

Fig. 98. Aurora 7 during recovery. (NASA)

at 1605 GMT. Carpenter was taken below deck to the sick bay, where he was subjected to a more detailed physical examination, after which he was flown directly to Grand Turk Island for three days of technical debriefings

The Destroyer USS *John Pierce*, recovered Aurora 7 at 1852, 6 hours 11 minutes after landing. Prior to recovery the spacecraft was listing at 45 degrees rather than the Mercury Spacecraft's usual flotation attitude of 15–20 degrees. At recovery the spacecraft contained 246 litres of seawater. The landing skirt was ripped in several places and most of its retaining straps were broken. The damage was probably caused by wave action after landing. (See Fig. 98.)

Following his time on Grand Turk Island Carpenter was flown back to Cape Canaveral. There he received the NASA Distinguished Service Medal from NASA Administrator James Webb. He received a number of parades in his hometown and other towns with which he had associations.

20.11 WAS CARPENTER GROUNDED?

Scott Carpenter's re-entry and recovery was unlike any that had gone before. In their search for a sensational headline the press and media at the time perpetrated the myth that Carpenter and Aurora-7 were lost and that MCC did not know where they were. That was not true. Search and recovery operations were carried out following standard military practices. The first aircraft picked up Aurora 7's recovery beacon only 6 minutes after the spacecraft hit the ocean. Twelve minutes later the entire recovery fleet knew exactly where the spacecraft was and its location was relayed to MCC.

A new myth was perpetrated in Tom Wolfe's book *The Right Stuff*, which claims that Kraft shouted, within the earshot of his controllers in MCC, that Carpenter would never fly for him again. Like the journalists' reports of 1962, *The Right Stuff* is written in a sensationalist way.

In 1990 the British magazine *Spaceflight News* carried two articles on Carpenter's flight written by Nigel MacKnight and Ed Pugh. Alongside the articles ran the transcript of a face-to-face interview between McKnight and Carpenter. The last part of that interview reads as follows:

> *MacKnight*: Some aspects of your flight caused controversy. When you got back ... did anyone take you to task or try to dress you down for those things?
>
> *Carpenter*: No. It wasn't necessary, you know. I think there was some undercover displeasure, but I really wasn't aware of that for quite some time afterwards. And, you know, no one wants that to happen to the space programme in the first place, so there's an honest effort to not make an issue of it – but it did get some noses out of joint.
>
> That is not uncommon in this business. Every flight is a compromise between conflicting purposes espoused by lots of different factions and organisations. The doctors want something, the scientists want something,

the flight-test people want something else, the guy who flies wants his own, and it's a big compromise. You could never please everybody.

MacKnight: I'd like you to comment, if you will, on what Tom Wolfe's book said, and that is – did Chris Kraft say of you, quote – unquote, 'That son of a bitch will never fly again!' – because of the feeling that he held you responsible for the splashdown overshoot. Did he say that to you?

Carpenter: He never said that to me, but I have heard that, of course. There's never been any animosity on my part towards him. I recognise that sort of thing as inevitable in the conduct of things: of experimental test flights like this. I don't . . . I feel charitable towards him for his views and I think that he probably – well, I don't know, I can't comment for him. I never heard that, but I have read it.

MacKnight: You've met him since. Have you ever had . . . did you get friendly with him and get on OK with him?

Carpenter: Of course. It was never mentioned and it is never . . . You know, I saw him lots and lots of times before I ever even heard that, so we had established a good relationship after the flight which existed for a long time. Then I heard that, and nobody likes to hear that, but it has never been an issue between us and I didn't let it affect my relationship with him at all. As far as I can see he has done the same with me.

While other Flight A astronauts took assignments in Project Gemini, Carpenter was assigned directly to the embryonic Project Apollo, where he oversaw the design and early construction of the Lunar Module at the Grumman plant in Bethpage. Photographs exists in the public domain of Carpenter on a jungle survival training course with the Flight A and Flight B astronauts as well as official photographs of both groups together. There are also photographs of him testing early Apollo hardware, including early models of a pressure suit designed for lunar surface extravehicular activities.

Scott Carpenter was officially removed from flight status for medical reasons in 1964, following a motor-scooter accident in Bermuda. Carpenter came off of his scooter in the accident and the injuries that he received left him with reduced mobility in his left elbow. In 1965 he took leave of absence from NASA to lead the diving team for the US Navy's Sealab II project. Carpenter and his team spent 30 days submerged at a depth of 62.5 m. He resigned from NASA in 1967.

July 1962: OKB-1 merged the original Soviet N-1 and N-2 launch vehicles into one giant vehicle designated N-1, with 24 rocket motors in its first stage. As with the early American Saturn designs, the N-1 was seen as the basis of a family of launch vehicles. A proposed N-11 would employ the N-1's second and third stages with a new upper stage, while the proposed N-111 used the N-1's third stage and the new upper stage.

11 July 1962: NASA announced that a new two-stage Saturn launch vehicle would be developed to support early manned flights in Project Apollo. The new vehicle was designated Saturn C-1B.

10 August 1962: Soviet missiles began to arrive in Cuba. Khrushchev undertook to install IRBMs on Cuba because he believed that Kennedy was weak and would not risk going to war over them.

20.12 VOSTOKS III AND IV

Andrian Nikolayev rode Vostok III into a 166 × 218-km orbit following lift-off at 1130 on 11 August 1962. His back-up cosmonaut was Valeri Bykovsky. Once again the flight passed the third orbit and the cosmonaut was committed to at least 24 hours in orbit if he was to land inside the Soviet Union. At T + 9 h, live television was transmitted to Soviet news networks. Nikolayev became the first person to unstrap himself from his ejection seat and float around his Crew Compartment in the microgravity environment.

After 23 hours 32 minutes Vostok III was approaching Baikonur Cosmodrome. At 1102 on 12 August 1961 Vostok IV, with cosmonaut Pavel Popovich, was launched and climbed into space. Popovich's back-up was Vladimir Komarov. As Vostok III passed overhead, Vostok IV separated from its spent third stage. The new spacecraft was in a 169 × 222-km orbit and was visible to Nikolayev in Vostok IV, just 6.5 km away.

The initial close approach was as much a matter of orbital mechanics as good engineering and launch preparation. It was not, as much of the Western media claimed at the time, the first rendezvous in space and a vital step on the road to the Moon. After the initial orbital insertion orbital mechanics quickly moved the two spacecraft away from each other. By Vostok III's 33rd orbit the two spacecraft were 850 km apart.

For the remainder of the flight the two cosmonauts performed individual flight activities although they spoke to each other every 30 minutes, taking it in turn to call the other cosmonaut. They followed similar flight schedules, eating and sleeping at the same time. Both men performed a number of scientific and physiological experiments. At one point Nikolayev sent a message to the American people. He said, '*Flying over your great country, I convey from the Soviet Spaceship, Vostok III, greetings to the gifted American people. I wish peace and happiness to the people of your country.*'

The double flight came to an end after three days. Vostok III performed retrofire on orbit 64, at which time Vostok IV was 2,850 km away. Nikolayev landed by personal parachute at 0954 on 15 August 1962, after a flight lasting 94 hours 22 minutes. Vostok IV performed retrofire only 6 minutes after Vostok III, on orbit 48. Popovich landed by personal parachute at 1009 on 15 August 1962 after 60 hours 57 minutes in flight.

13 August 1962: Violent demonstrations took place in West Berlin on the first Anniversary of the construction of the Berlin Wall.

29 August 1962: An American U-2 reconnaissance aircraft flown by a CIA pilot exposed photographs of Soviet SA-2 Surface to Air Missile (SAM) sites under construction on Cuba. The SA-2 was the same missile that downed Gary Power's U-2 over the Soviet Union in May 1960.

September 1962: AFRC began the development of the M2 prototype lifting body. It was the first in a series of vehicles designed to use the shape of its body to provide sufficient lift within the atmosphere to allow it to be flown to a standard landing without the use of wings.

9 September 1962: An American U-2 reconnaissance aircraft with a CIA pilot was shot down over Taiwan by Communist Chinese forces using a Soviet produced SA-2 missile.

9 September 1962: American forces in the Caribbean commenced manoeuvres off Cuba. The prospect of an American invasion of the island seemed extremely likely.

9 September 1962: After a week of riots and behind the scene negotiation President Kennedy called in the Army to ensure that Charles Meredith, a Negro student, could take his place at the university in Oxford, Alabama. Meredith took his place, but 23,000 troops were stationed around Oxford to keep the peace.

17 September 1962: Robert Gilruth announced the names of STG's nine Flight B astronauts. The nine new astronauts attended Cape Canaveral for the launch of Mercury Atlas 8 on 3 October.

20.13 MERCURY ATLAS 8 (MA-8)

Having lost the three-orbit flight of MA-7 to Scott Carpenter, Wally Schirra (Fig. 99) was named as prime astronaut for MA-8. His back-up was Gordon Cooper. Schirra intended to complete six orbits, flying almost the entire mission under manual control. With more time in orbit, the astronaut was expected to perform an enlarged scientific experiment programme.

Atlas 113D carried a number of updates on MA-8 that made it different from previous Mercury Atlas launch vehicles. These changes included:

1. New baffle injectors and hypergolic igniters replacing pyrotechnic igniters on earlier vehicles. This change meant that the launch vehicle would not be held down during the transition to mainstage.
2. The inter-propellant tank insulation was not fitted, following the success of MA-6 launch.

Fig. 99. Wally Schirra. (NASA)

3. Following the freezing of a pressure transducer on MA-7, leading to an abort signal being transmitted, the transducers on MA-8 were modified to prevent their freezing.

The pre-count for MA-8 was completed without difficulty on 2 October 1962. After the standard overnight hold the countdown was resumed the following morning. Having spent the night in the astronaut quarters at Hangar S, Schirra was awakened at 0140 on 3 October and completed his ablutions. He then took his pre-launch steak breakfast with guests that included Robert Gilruth, Walt Williams and Deke Slayton. Following breakfast he underwent his final physical examination and his biomedical harness sensors were taped to his body. Technician Joe Schmitt helped the astronaut to dress in his pressure suit and check that it was functioning correctly. Schirra left Hangar S at 0400 and entered the transfer van for the ride to LC-14.

On leaving the transfer van at the base of the gantry, Schirra received the good wishes of the Convair launch team. He was then greeted by Gordon Cooper, who had spent much of the previous night in the spacecraft, participating in the countdown. Schirra, Cooper and Schmitt entered the elevator and took the short ride up to the white room.

At 0441, Schirra was helped into the spacecraft that he had named 'Sigma 7'. Schmitt connected the various umbilicals between the spacecraft's ECS and the astronaut's pressure suit. With his body harness pulled tight, Schirra began participating in the countdown. The McDonnell launch team had hung a car ignition key on the three-axis hand controller safety catch. There was also a steak sandwich wrapped in plastic in one of the spacecraft's storage compartments. The close out crew bolted the side hatch in place and secured its external lanyard. The launch complex was evacuated and the gantry rolled back to its parked position.

The countdown continued without any problems until T−45 min at 0615, when a driver unit failure took the Canary Islands radar off line. The radar in question would provide early tracking information following orbital insertion, information that would allow Kraft to make the Go/No Go decision on continuing the flight, or bringing the spacecraft down at the first opportunity. A hold was called while the problem was diagnosed and corrected. When the countdown resumed, 15 minutes later, it continued through launch without any further stoppages.

Inside Sigma 7, Schirra heard the vernier engines on the Atlas fire, followed by the three main propulsion units. Although lift-off occurred slightly earlier than Schirra had expected, he would later write that there was no doubt when first movement occurred. At T+10 s the vehicle rolled to within 20 degrees of the abort threshold. The manoeuvre was monitored at MCC, but Schirra was not informed how close he had come to ending his flight being dropped into the ocean from beneath the LES.

Throughout the launch, communications with Slayton in MCC were good, with the exception of the period as the launch vehicle passed through max-q. At that time vibrations from the launch activated Schirra's microphone, making it impossible for him to receive Slayton's transmissions. The problem passed as the vibrations reduced following max-q.

At T+2 min 5 s Slayton warned Schirra, '*Stand by for staging.*' Unknown to Schirra his Atlas was burning propellant faster than planned and was subsequently flying higher and faster than it should have been. BECO occurred 2 seconds earlier than planned. Schirra confirmed that the two Booster engines had shut down and staged as planned. The Sustainer pitched down and the LES Escape and Jettison Motors fired. Schirra reported, *There goes the tower ... This tower is really sayonara.*' The exhaust plume from the LES motors left a thin film on the exterior of the spacecraft window. As programmed, the Sustainer pitched back-up and the climb into orbit continued. The Sustainer burn struck Schirra as slower and longer than expected. SECO occurred 10 seconds later than planned.

The 10-second timer ran its course before the Marman Clamp released and the posigrade rockets fired. The Auxiliary Damping System damped the rates set up during separation. Sigma 7 was in a 161.3 × 289.9-km orbit inclined at 32.5 degrees. Each orbit had a period of 88.5 minutes. Slayton confirmed, '*You have a go. Seven orbit capability.*'

Schirra confirmed that his spacecraft had separated from the launch vehicle and switched to FBWS. He then tested the thrusters in all three axes, before yawing Sigma 7 through 180 degrees. Throughout the turnaround Schirra forced himself to concentrate on his instruments and not look out of the window. When the manoeuvre was complete he looked up to see the Sustainer trailing along behind him, apparently hanging above the blue and white horizon of Earth. Having placed the spacecraft in retro-attitude and switched to ASCS Schirra told Slayton, '*I'm in chimp mode now and she is flying beautifully*'.

Looking at his Sustainer once more, Schirra noticed that he was staring down the rocket engine. The Sustainer had also turned through 180 degrees. He saw none of the white crystals that Carpenter had seen, but he did see a band of ice around the cryogenic LOX tank. Where Glenn and Carpenter had reported that their Sustainers had appeared silver, Schirra said that his looked black. He manoeuvred to keep the Sustainer in the centre of his window in order to practise attitude control in both FBWS and MPCS, before returning Sigma 7 to ASCS.

Moving into range of the Canary Islands tracking station, Schirra reported that his pressure suit temperature was climbing and was presently reading 75 °F. The cooling water circuit was set at 4.5. The astronaut had arranged with the pressure suit technicians that he would increase the cooling water setting in 0.5 unit increments every 10 minutes until he had his suit temperature under control, before it threatened to bring his flight to an early close. As on MA-7, Schirra's pressure suit cooling system was contaminated with dried silicone lubricant, but this would not be established until the system was stripped down during the post-flight investigation of the temperature control problem.

Passing over Kano in Africa the astronaut confirmed that all was well with his spacecraft and that he had raised the cooling water setting on his pressure suit to 5. Schirra yawed his spacecraft 5 degrees to the right, but noticed that the yaw indicator on the main instrument panel showed 10 degrees. His final comment as he passed out of range from the Kano tracking station was: '*Now this may be a problem; we will have to observe it.*'

Two minutes later he was talking to Zanzibar and confirmed that he had increased his pressure suit water flow rate by another 0.5 increment. Schirra let the ASCS reduce the yaw rates to zero. He reported that it was possible to gauge yaw rates by using the edge of the window against the position of clouds, or even gaps in the clouds. Ten minutes after the previous water flow rate increase he increased the setting to 6 and 10 minutes later still he increased it by a further 0.25 of an increment.

Approaching sunset, he observed the particles that Glenn and Carpenter had seen. He banged the side of the spacecraft with his hand and a shower of particles passed in front of the window. The astronaut switched the cabin lights from white to red, but like Carpenter he found that light leaking from around the time correlation clock stopped his eyes becoming fully dark-adapted. In changing the lights over, he almost missed the sunset completely.

Crossing the Indian Ocean, he talked to Muchea and confirmed that he had set his suit cooling water flow rate to 6.5. At the same time he had decreased the Crew Compartment cooling water flow rate from 4 to 3.5. During the communication session Schirra admitted that he was sweating slightly, but did not want to open his faceplate to take a drink of water until he had the pressure suit temperature under control.

Approaching Woomera, he switched to FBWS and pitched the spacecraft down to 50 degrees negative pitch, in order to observe a number of flares that would be set off at the tracking station. He initially mistook lightning in the clouds below him for the flares. Despite being informed exactly when the flares had been lit, he did not see them. Instead, he described the lightning in the thick cloud cover. As he pitched up again he observed the city of Brisbane. During the manoeuvre he confirmed that he had increased his pressure suit cooling water flow rate to 7.

As he passed over Woomera, the tracking station reported a telemetry dropout. The link was re-established almost immediately. Two minutes later Schirra pitched up to re-entry attitude and switched to ASCS. The automatic systems made the last few adjustments to Sigma 7's attitude using only the low-rate thrusters.

The session ended with Schirra moaning about the intensity of the red light in the Crew Compartment. He acknowledged that he could turn the lights down, but admitted that he was reluctant to do so while he was still having difficulty bringing his pressure suit temperature under control. He also suggested that later spacecraft be fitted with a small bag to hold those items that drifted around the Crew Compartment once the spacecraft entered microgravity.

As he lost contact with Woomera, he switched back to FBWS and manoeuvred to retrofire attitude as he began his first pass over the Pacific Ocean. He then returned the spacecraft to ASCS. Two minutes later he was talking to Canton Island, but communications were poor. He passed a 30 minutes status report and the news that he had increased the pressure suit cooling water flow to 7.5. He reported that he was feeling '*slightly warm*'. He still refused to open his faceplate and drink some water.

Over Hawaii he watched the sunrise on his periscope screen. He remarked that it was the first light that he had seen in the periscope throughout the entire passage through Earth's shadow. He continued, '*It is obvious that the periscope has no function whatsoever in retro-attitude on the night side.*' He saw more particles and made showers of them appear by banging on the spacecraft walls. When the sunlight through the periscope became blinding, he put the dark grey filter in place.

Speaking from Guaymas, Carpenter confirmed that Schirra was cleared for the second orbit. The astronaut relayed a 30-minute status report, beginning: '*I'm in chimp configuration* [ASCS] *the spacecraft is flying beautifully. All thrusters are working well.*' He still had 100 per cent of his automatic and 95 per cent of his manual propellant supply. His clock was one second fast. He transmitted a blood pressure reading, but spoilt it when he moved his arm. The communication with Carpenter ended with Schirra switching to FBWS to complete the first of two daylight yaw checks over Cape Canaveral. (See Fig. 100.)

For the yaw checks he first covered the yaw attitude indicator and then manoeuvred in yaw using just an external marker viewed through the window. When he had manoeuvred back to orbit attitude he uncovered the attitude indicator to see how accurate he had been. He then repeated the experiment using the periscope screen rather than the window. Both attempts were then repeated when Sigma 7 was in Earth's shadow.

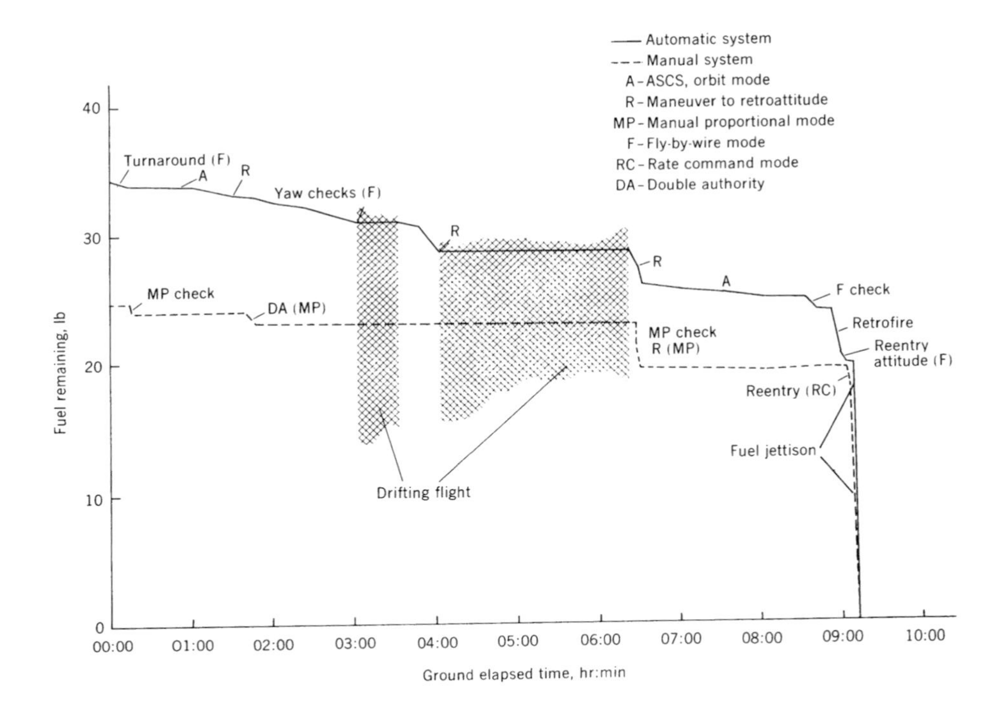

Fig. 100. Sigma 7 propellant usage. (NASA)

Over Cape Canaveral he was talking to Slayton when he returned to ASCS and increased the suit cooling water flow rate to 8. On Slayton's suggestion he opened his faceplate and took a drink of water. He talked his way out of a recommendation that he return the suit cooling water rate back to 3.

At T + 1 h 41 min he switched to FBWS and began his first yaw manoeuvre using the low rate thrusters. The Bermuda tracking station reported telemetry suggesting that the high rate thrusters had fired, but Schirra confirmed that they had not. He yawed the spacecraft 8 to 10 degrees left. Two minutes after starting the experiment he zeroed his rates visually and uncovered the yaw indicator. It showed 4 degrees left yaw. The astronaut returned the spacecraft to ASCS to allow the gyros to reset.

Over the Canary Islands Schirra took Kraft's recommendation and turned his pressure suit cooling water flow rate back to 3. He voiced his opinion that if the suit temperature did not continue to fall, as it had begun to do, then he would return the flow rate to 8 after a few minutes.

At T + 1 h 50 min, he covered the yaw indicator once more and switched to FBWS low. He yawed 25–30 degrees to the right using visual cues on his periscope screen. He zeroed his rates visually and uncovered the yaw indicator, which showed one-degree right yaw. Schirra returned the spacecraft to ASCS.

Communications with Kano were poor. Schirra reported that his suit temperature was climbing once more. He returned the cooling water flow rate setting to 7.5. The temperature began falling again immediately, so he turned the water flow rate up to 8 to help the temperature fall more quickly. At T + 2 h 20 min he reported, '*I feel very comfortable. I'm cooling off at last.*' The session ended with Schirra switching to MPCS.

The session over Zanzibar only lasted 4 minutes. Schirra admitted that he had had a brief period of double authority when he switched back from MPCS to FBWS. The result of this error was a single high-rate thruster burst in pitch, before the astronaut turned the rate command switch to automatic.

Talking to *Coastal Sentry Quebec* in the Indian Ocean, the astronaut reported that he felt that his pressure suit temperature was now under control. As the spacecraft passed into Earth's shadow he said, '*Sunset is rather striking. I don't think I need waste much time looking at them.*' One view out of the window over Australia did allow him to see the lights of Perth. He reported, '*Oh, what a beautiful yaw check this is, and its approximately 10 degrees left of path.*' Always the engineer and test pilot, Schirra had decided to concentrate his efforts inside the spacecraft, rather than allow the view through the window to distract him from his tasks.

At T + 2 h 25 min, over Muchea, he made his first night-side yaw check using the window. He yawed 10 degrees left and then zeroed his rates. The yaw indicator showed 3 degrees left at the end of the manoeuvre. The fourth manoeuvre should have been made using the periscope screen, but it was too dark. At T + 2 h 28 min he yawed left and back once more, using the Moon and Venus through the window as his visual cues. When he uncovered the yaw indicator it showed 1 degree left yaw. Switching to ASCS he reported, '*I'm going to set up and uncover the instruments for chimpanzee configuration. Dammit, I'm sorry, auto mode.*'

Woomera was under a large thunderstorm as Schirra transmitted a 30-minute report. The engineers at Woomera did not attempt to launch a flare on this pass.

Talking to Canton Island he reported that he was ready to receive instruction for retro-sequence at the end of his second orbit. He quickly added, '*And you'd better not give me one.*' Describing Sigma 7, he added, '*She is really performing like a little jewel right now.*'

Talking to Grissom at Hawaii he described the sunrise as '*rather disappointing because it is just about socked in completely*'. Then he was talking to Glenn at California. He told him, '*I do see fireflies.*' The two astronauts discussed the particles that Schirra could see outside his window. Below the spacecraft the cloud cover gave way to a clear atmosphere as he crossed the coast. The view remained clear as he passed across the continental United States.

The communication session with Carpenter at Guaymas lasted just 29 seconds, and then Schirra was talking to Slayton at MCC once more. Slayton advised him to turn off the electrical power to the ASCS and go to drifting flight. Schirra did as he was told at T + 3 h 10 min. A few minutes later the astronaut reported that Sigma 7 had begun a slow left roll. With the ASCS turned off the spacecraft's thrusters did not fire to bring it back to orbital attitude. The period of drifting flight continued over the Canary Islands and Kano.

At T + 3 h 44 min Schirra powered up the ASCS and switched to FBWS before he manoeuvred to zero the rates and then switched to ASCS. There was a single burst from one of the high-rate thrusters. By that time he was talking to *Coastal Sentry Quebec* located in darkness in the Indian Ocean.

Over Muchea and Woomera Schirra performed a number of star sightings. Back over Hawaii, Grissom passed up the message that MCC had cleared him to complete six orbits. Schirra replied, '*Hallelujah*.' Grissom told him, '*You're looking real good down here Wally. We can see nothing wrong*.' Schirra took a large sip of water and consumed a tube of peaches and several bite-size food cubes. The session ended with a hint of sarcasm between the two friends when Schirra remarked, '*Well, Gus. At least we got some fuel coming over here this time*.' It was a jibe at Carpenter's MA-7 flight.

Having passed through sunrise without even mentioning it, Schirra discussed the particles outside his spacecraft briefly with Glenn at California. At T + 4 h 34 min he powered down the ASCS once more and entered a second period of drifting flight. The spacecraft began a slow right roll. The flight plan called for Schirra to complete experiments and photographic tasks.

Slayton took over the communication from MCC. During the pass over Cape Canaveral Kraft came on the radio and told Schirra, '*Been a real good show up there. I think we are proving our point, old buddy*' Schirra replied, '*I hope so Chris. I am enjoying it*.'

At the start of the fourth orbit 89 per cent of the automatic and 90 per cent of the manual propellant supply still remained in the tanks. Two minutes later Slayton passed up an instruction for Schirra to advance the spacecraft clock by 1 minute, to bring it in line with the clock in MCC.

Talking to Ascension Island, Schirra described the usual shower of particles at sunset, but not the sunset itself. After an orientation test Schirra turned out the cabin lights and began taking photometer readings of the stars that he could see through his window as he flew with his head down towards Earth.

As his trajectory was taking him further away from the WWTN stations, he employed the Pacific Command Ship for the first time. Shepard was CapCom but communications were poor. Schirra compensated by switching his communications through a relay aircraft to Hawaii, where Grissom was CapCom. He joked that he was '*sure breaking tradition on this pass*'. It was a comment on the fact that his was the first Mercury Spacecraft to go beyond three orbits.

Over California Schirra's voice was relayed live on American television for 2 minutes, commencing at T + 6 h 8 min. He told the audience, '*I suppose the old song Drifting and Dreaming is approprios at this point, but at this point I don't have a chance to dream. I'm enjoying it too much*.'

With the broadcast over Glenn and Schirra discussed the atmospheric haze layer. A brief communication session with Carpenter at Guaymas was followed by a renewed link with Slayton in MCC. Sigma 7 was approaching the beginning of its fifth orbit.

Schirra told Slayton that he had powered up the ASCS over California, but was still drifting in FBWS low. He explained, '*No problem, just, the bird's flying beautifully ... give her a break*.' When Slayton asked how he was feeling Schirra replied, '*Very fine Deke. It's the first time I've had to relax since last December*.'

Over the Gulf of Mexico he told his tape recorder that he had switched to ASCS at T + 6 h 24 min. His attitude control on FBWS was so good that there was no high-rate thruster activation at switch over. As he began orbit 5 his automatic and manual propellant supplies stood at 85 per cent and 90 per cent respectively.

Flying across South America he switched to MPCS and tested the thrusters in that mode. He switched back to FBWS having re-established retro-attitude. As he did so he operated the two systems on duel authority for a single thruster burst. He admonished himself with '*That was stupid*'. With Sigma 7 back on ASCS the astronaut concentrated on his cloud photography tasks. He dictated details of everything that he did into the spacecraft's tape recorder.

Passing into Earth's shadow once more Schirra easily identified the lights of Port Elizabeth in South Africa, off to his left. Over the Indian Ocean ship he confirmed that he was in ASCS and using the Moon and Venus in his window as visual cues to any build up of rates. Following a full 30-minute report he praised Sigma 7, saying, '*This is as tight a ship as anyone can imagine.*'

Then he was talking to Alan Shepard on the Pacific Command Ship. The following conversation took place: '*Well, I would say you were definitely go. We are out of contact with the Cape at the moment, but looks like you are go for the full route.*'

'*Right you are.*'

With no access to the flight director, two experienced test pilots had reviewed the available information and decided that Schirra should make the sixth orbit and complete his flight plan. As one Naval officer to another, Shepard joked, '*We recommend that you drain the bilges before re-entry.*'

The next links in the extended communication network were the tracking ships *Huntsville* and *Watertown*, from the Pacific Missile Range. Communications were poor and most of the session was taken up with unsuccessful radio checks. Communications were clearer when he began talking to Grissom at Hawaii. Schirra repeated the 30-minute report that he had transmitted to the two tracking ships. Outside, the Sun rose without comment from the astronaut.

Grissom relayed the end of mission retrofire time, but forgot to allow for the 5-second discrepancy shown on Sigma 7's clock. Glenn corrected the error as Schirra passed over California. A power failure on California's receiver meant that Glenn had to transmit '*in the blind*', and could not know if Schirra received his message. Carpenter confirmed the retro-sequence time from Guaymas, while Schirra consumed a tube of beef and vegetable paste.

Switching to FBWS, to correct a 4-degree roll, Schirra tested Sigma 7's thrusters before beginning his sixth and final orbit. His orbit was now so far south that he could not talk to MCC without relaying his communications through the tracking station at Quito, South America. He photographed the continent and dictated the details into his tape recorder.

At T + 8 h 2 min he selected FBWS low, to zero the rates before switching to ASCS. He also turned his pressure suit cooling water flow rate up to 8, '*to increase the cooling for re-entry*'.

The Sun set once more and Schirra found himself talking to the *Coastal Sentry Quebec*. He reported that he could see lights beneath the clouds, but could not confirm if he was looking at the city of Durban. He did not see the flare launched from the tracking ship located off Madagascar. CapCom on the ship had Schirra turn his attention to the pre-retrofire checklists. The flight was coming to an end and the astronaut complained, '*Seems so sad, just a little less than half an hour left to play with this.*'

In the final status check before retrofire Schirra confirmed that both his automatic and manual propellant loads stood at 78 per cent. Everything was working well with the exception of the finger light on the index finger of his left hand, which had stopped working. Talking to Shepard on the Pacific Command Ship, Schirra confirmed that he intended to let the RSCS control the spacecraft through retrofire. In the event of a malfunction he would use MPCS as a back-up

Shepard counted down the final few seconds, then, at T + 8 h 52 min 4 s Schirra reported, '*I've got one and she's holding* [attitude] *real tight. Very tight. I got two. My attitude is right on the money. I've got three.*' All three retrograde rockets had fired as planned. He switched to FBWS and manoeuvred to re-entry attitude. The Retrograde Package jettisoned and he reported the fact. Having set the correct attitude he switched back to ASCS. Prior to re-entry Shepard read up the spacecraft's post retrofire propellant rates as automatic 65 per cent and manual 78 per cent. The periscope retracted automatically and its door latched shut.

Talking to the *Watertown*, Schirra confirmed that his faceplate was closed and sealed. As the 0.5*g* light illuminated on his instrument panel he removed his hand from the controller. The spacecraft established the normal re-entry roll. When the sheath of ionised air cut off his communication link he dictated what he saw and felt into the tape recorder.

The RSCS controlled the spacecraft throughout the *g*-force build up, while Schirra described the green and orange hues that he could see out of the window. After peaking at 7.5*g* the deceleration forces began to ease off. As the bow wave in front of the spacecraft disappeared Schirra was able to talk to the *Watertown* once more. He would write in his post-flight report:

> There were two occasions when I nearly switched from RSCS to auxiliary damping mode. One was while I was monitoring the fuel gauge; it looked like a flow meter. The indicator for the manual tank was visibly dropping. Yet I continued with the RSCS because I wanted to give it every chance to complete the re-entry control task in order to evaluate it sufficiently. The second time that I thought of going to auxiliary damping mode was when the yaw mode left the nominal 2.5 to 3 degrees per second and went off the scale (6 degrees per second) to the left. Soon after this occurrence it held at 5 degrees per second ... Since it had started to hold again I did not switch to auxiliary damping because I still wanted to allow the RSCS a full demonstration.

Then he was talking to Grissom at Hawaii. Schirra deployed the drogue parachute at an altitude of 12,496 m; it was T + 9 h 6 min. He reported, '*I can see the drogue now. Drogue looks very good.*' He switched to ASCS and let the remaining propellant dump overboard while he manually deployed the snorkel to let ambient air into the Crew Compartment. The main parachute deployed at 3,200 m and

Fig. 101. A whaler from the USS *Kearsarge* draws alongside Sigma 7. (NASA)

Schirra told Grissom, '*There goes the drogue and the main 'chute is out. She's out, beautiful. Bright blue sky. And it's de-reefed and looks like a real sweetie pie.*'

As he opened his faceplate and removed the seal, Schirra elected to remain in Sigma 7 until she had been hoisted aboard the recovery vessel. On impact with the ocean he felt that the spacecraft '*seemed to sink way down in the water ... The spacecraft seemed to take a long time to right itself, but time is merely relative and in actuality the spacecraft righted itself in less than one minute.*'

Landing had occurred north-west of the Hawaiian Islands, after a flight lasting 9 h 13 min. The spacecraft was 7.40 km from the prime recovery vessel, the aircraft carrier USS *Kearsarge*. Some of the carrier ship's company had first observed the spacecraft following drogue parachute deployment. Many of them had observed the landing from the flight deck. (See Fig. 101.)

The astronaut guillotined the parachute lines, deployed the recovery aids and began running through his post-landing routine. Grissom told him, '*The carrier has you in sight and the helos* [helicopters] *are on their way.*' Those helicopters dropped three recovery swimmers into the water to deploy the Stullken Collar around the base of the spacecraft. Schirra told the pilot of the primary recovery helicopter, '*Tell the fellows I'm perfectly comfortable. I can wait as long as they want ... Don't let them get their hands cut on something on here, go at it casually.*'

When the USS *Kearsarge* was positioned alongside the spacecraft a motorised whaler was lowered into the water and made its way to Sigma 7. The sailors in the whaler attached a lifting line to the spacecraft's recovery loop. The spacecraft was then hauled to a position beneath the ship's hoist. Finally, Sigma 7, with Schirra inside, was lifted onto one of the carrier's aircraft elevators. Recovery was complete 40 minutes after landing (Fig. 102). Five minutes later Schirra stuck the plunger to jettison the spacecraft's side hatch. Like Glenn, he suffered a bruise on his hand from this action, despite the fact that he was still wearing the glove to his pressure suit. The astronaut was escorted below deck to the officers' sick room. Still dressed in his pressure suit he received telephone calls from President Kennedy, his wife Jo and Vice President Johnson.

Fig. 102. Schirra egresses Sigma 7 on board the USS *Kearsarge*. (NASA)

Unlike earlier Mercury astronauts, Schirra remained on the USS *Kearsarge* for the next 72 hours and began his debriefing sessions while on the carrier. Sigma 7 was offloaded at Midway Island and returned to Hangar S, Cape Canaveral. It was found to be in a similar condition to the two Mercury Spacecraft that had made orbital flights before it.

After a session on a tilt table, Schirra's heart started showing different rates of beating depending on whether he was lying down, sitting or standing up. Also, after 9 hours of microgravity the return to one-gravity resulted in his blood pooling in his lower limbs, turning them purple. These proved to be the only effects of his flight. The astronaut began a sleep-period 6 hours after the flight had ended. When he awoke the following day his heartbeat was back to normal and there was no pooling of blood in his lower limbs.

Schirra was landed at Hawaii, where he received an enthusiastic welcome from the local population. He flew back to Texas, where MSC was now based. A press conference at Rice University was well received. The astronaut and his family were given a parade in his hometown of Oradell, New Jersey. They then travelled to Washington DC, where President Kennedy presented Schirra with the NASA

Distinguished Service Medal and the Naval Chief of Operations presented him with the US Navy's Astronaut Wings.

14 October 1962: An American U-2 reconnaissance aircraft photographed Soviet IRBMs in Cuba. The launch sites were still under construction and, while the missiles were on the island, the Americans never did establish if their nuclear warheads were also on Cuba, or still en-route from the Soviet Union. They were in fact already under secure storage on the island, one of the first items to be delivered.

President Kennedy and Chairman Khrushchev had been carrying out a private correspondence on the matter of Soviet missiles in Cuba since the first photographs of the SA-2 launch sites had been taken. Khrushchev had assured Kennedy that all Soviet missiles in Cuba were of a defensive nature and that America was not under threat from those missiles. This had now been proven to be a lie.

America was considering six possible actions to remove the missiles:

- Negotiation with Khrushchev.
- Surgical bombing of missile sites.
- Continuous bombing of missile sites to ensure they were destroyed and not replaced.
- A naval quarantine of Cuba.
- A full-scale military invasion of Cuba to depose Castro.
- Remove obsolete Jupiter IRBMs from Turkey in return for removal of Soviet missiles in Cuba.

20 October 1961: China attacked India across their mutual border.

22 October 1962: The MA-8 post-flight celebrations were overshadowed by President Kennedy's public announcement that the Soviet Union had established a number of nuclear-tipped missiles on the Island of Cuba, from where they could reach most major American cities and military bases. The missiles had been identified by their standard deployment patterns, which had shown up on reconnaissance photographs taken of the island.

President Kennedy went on television and explained his plan to place a naval quarantine around Cuba and turn back any Soviet ships carrying missiles or any other items necessary for their operation. He explained that the commanding officer any Soviet ships found carrying a cargo of nuclear missiles bound for Cuba would be extended the opportunity to turn their vessels around. The President had given orders that no Soviet ships were to be fired upon without his personal order to do so. Such action might lead to war with the Soviet Union and such a war might include the exchange of nuclear weapons between the two super powers. He told the astounded American public:

> I have directed that the following initial steps be taken immediately:
>
> First: To halt this offensive build up, a strict quarantine on all offensive military equipment under shipment to Cuba ...

> Second: ... Should these offensive military preparations continue, thus increasing the threat to this hemisphere, further action will be justified. I have directed the armed forces to prepare for any eventuality ...
>
> Third: It shall be the policy of this nation to regard any nuclear missile launched from Cuba against any nation in the Western Hemisphere as an attack by the Soviet Union on the United States, requiring a full retaliatory strike against the Soviet Union.

The fear was that, if American aircraft attacked the missile sites, the Cubans, or their Soviet advisers might fire a missile with a nuclear warhead at the American mainland. In that case American military planning at the time meant that every American nuclear missile would be launched in reply. The American leaders had no doubt that the Soviet leaders knew that and would feel obliged to launch a full pre-emptive strike as soon as the Cuban missile had been launched. The world was closer than it had ever been to a nuclear conflagration.

American military forces around the world were put on full alert and the US military began a 10-day preparation period, leading to a full military invasion of Cuba. Meanwhile, President Kennedy continued to exchange personal correspondence on the Cuban crisis with Chairman Khrushchev.

26 October 1962: The first Soviet ship was stopped in US Naval quarantine of Cuba. By that time Khrushchev had already ordered the ships that were en-route to Cuba with offensive military equipment to turn around and return to the Soviet Union.

27 October 1962: Chairman Khrushchev made a public announcement on Radio Moscow that he was prepared to remove the Soviet missiles from Cuba, if America would remove its Jupiter IRBMs from Turkey. Although the Kennedy Administration had been considering this option from the outset they believed that their Western allies would view it as America selling out their allies for their own security. The President feared that such a move might lead to the break-up of the North Atlantic Treaty Organisation.

On the same day a U-2 reconnaissance aircraft was shot down over Cuba using one of the Soviet SA-2 SAMs recently installed on the island. The pilot, who had taken the first photographs of missiles on Cuba, was killed.

President Kennedy decided to answer both Khrushchev's radio broadcast and his last private correspondence. Using the Voice Of America radio station he offered to remove the US Naval quarantine, promise not to invade Cuba and arrange to remove the Jupiter IRBMs from Turkey at a later date, when it would not be viewed by the American press and media as a conciliatory measure.

28 October 1962: Radio Moscow carried Chairman Khrushchev's reply to President Kennedy's broadcast. He agreed to remove all Soviet missiles from Cuba. For one week the world had lived in fear of a nuclear war, but Chairman Khrushchev had to admit that his country could not win an all-out nuclear war with America and he had backed down.

28 October 1961: America agreed to send arms to India to help in its border dispute with China.

2 November 1962: President Kennedy announced that the Soviets had dismantled their nuclear missile sites in Cuba and returned the missiles to the Soviet Union.

7 November 1962: The Apollo Lunar Excursion Module (LEM) prime contract went to Grumman Aircraft Engineering Corporation. The final public announcement that Project Apollo would land men on the Moon using the LOR method was made on the same day. The Apollo Spacecraft now consisted of two vehicles. The North American Aviation CSM would remain in lunar orbit while Grumman's LEM landed on the lunar surface. The crew would travel to and from the Moon in the CM.

21 November 1962: China and India declared a ceasefire.

19 December 1962: Kennedy offered to sell the Polaris submarine-launched nuclear missile to Britain in place of the Skybolt air to ground missile, which the Americans wanted to cancel. The offer was accepted, but Britain would supply its own warheads for the missiles, thereby retaining control of when the missiles were used. Polaris was also offered to France, but President De Gaulle refused the offer.

December 1962: In return for $53 million worth of goods from America that were unavailable due to the American trade embargo, Castro released over 1,000 members of the Cuban Exile Brigade 2506 that had been captured during the Bay of Pigs invasion attempt.

As the year ended America had 11,500 military personnel in South Vietnam and the surrounding area. One hundred and nine Americans had been killed in Vietnam during 1962.

January 1963: The governments of Turkey and Italy announced that America was removing the Jupiter IRBMs from their countries. America explained that the Jupiters were obsolete, while American Polaris submarines were now able to protect Turkey and Italy.

In the American Government's Fiscal Year 1964 budget, NASA's allotment rose from $2.4 billion in 1963 to $4.5 billion in order to finance a manned lunar landing by 1967.

January 1963: The year began with the first launch attempts with a new generation of Soviet Luna probes. Luna's 4 to 8 all failed in their primary objectives, before Luna 9 completed the first survivable landing on the lunar surface in March 1966.

14 January 1963: President De Gaulle announced that France would develop its own nuclear strike capability outside of NATO. In America President Kennedy viewed the decision as against American policy to keep all nuclear weapons in the Western Hemisphere under American control.

7 February 1963: The C designation was dropped from the various Saturn launch vehicles and their numerical designations were changed from Arabic to Roman numerals. The three Saturn launch vehicles selected for development became the Saturn I, Saturn IB and Saturn V.

19 February 1963: The Soviet Union removed its military forces from Cuba.

March 1963: STG announced the completion of a study into manned Earth orbital science stations. They concluded that Apollo Saturn hardware and facilities could be used in the development of such a vehicle.

21 March 1963: STG engineers established baseline requirements for project Gemini extravehicular activity. In the same month NASA HQ accepted STG's plans to rearrange Project Gemini's flight schedule taking into account the cuts in NASA FY'63 budget request. STG had originally wanted to share the cuts between Project Gemini and Project Apollo, but were instructed that the entire budget cut had to be carried by Project Gemini.

4 April 1963: President Kennedy was advised to return 1,000 troops from South Vietnam before the end of 1963. He was told that such a move would prove to outside observers that America believed she was winning the fight against the Communist forces in North Vietnam.

20 April 1963: President Kennedy advised his joint Chiefs of Staff to begin planning for direct military action against North Vietnam. At the same time he was hoping to be able to begin withdrawing American troops from South Vietnam by the end of the year.

23 April 1963: The last Jupiter IRBM left Turkey. President Kennedy had kept the promise that he made to Chairman Khrushchev to remove these missiles in return for the removal of Soviet missiles from Cuba.

28 April 1963: Fidel Castro visited the Soviet Union.

April 1963: Dr Martin Luther King led a series of Civil Rights actions in Birmingham, Alabama. They led to a month of rioting in the town. Dr King was arrested several times. Thousands of State Troopers were called to Birmingham on the same day that two bombs exploded, at Dr King's motel and his brother's home. The white Troopers turned on the Negroes in an orgy of violence.

May 1963: Korolev redefined the Soviet manned lunar flight. An Escape Stage would now use its propellant to place itself into orbit and then be refuelled by a number of smaller tanker stages that would each be launched on a separate R-7 launch vehicle. When the Escape Stage was fully loaded and the final tanker had departed a new spacecraft designated 7K, the precursor to the Soyuz Spacecraft, would be launched to rendezvous and dock with the Escape Stage, which would be used to place the 7K onto a circumlunar trajectory.

8 May 1963: President Kennedy announced that agreement had been reached in Birmingham Alabama. Negroes would cease their demonstrations and the local industrialists and business owners would oversee desegregation in their companies. The peace was short-lived and the riots continued.

8 May 1963: Nine Buddhists were shot in South Vietnam when they demonstrated after not being allowed to celebrate the birthday of their movement's founder. President Diem, a Catholic, blamed the shootings on the North Vietnamese Communists, but his Buddhist population did not believe him.

20.14 MERCURY ATLAS 9 (MA-9)

Gordon Cooper was named as prime astronaut for MA-9 on 13 November 1962 (Fig. 103). His back-up was Alan Shepard, who had been pushing for an orbital flight since the Mercury Redstone programme had been cancelled after just two manned flights.

During his time as an astronaut Cooper had come to the notice of NASA and MSC management as a bit of a rebel. It had been Cooper that had made the group's discontent over the lack of high performance training aircraft known to the press. He raced fast cars and boats and flew fast aircraft. He also picked up a high number of speeding tickets from the local police departments both in Virginia and Florida.

Just days before the Mercury Atlas 9 launch he had buzzed Hangar S, roaring in fast and low, in a jet fighter. Cooper had taken the aircraft up to release his frustration. An alteration had been made to his flight suit, involving passing a medical sensor wire through the pressure bladder, without first seeking his permission, as protocol demanded. Certain that Cooper had been the astronaut responsible, Williams telephoned Slayton, the astronauts' line manager, and threatened to ground Cooper and give the flight to Shepard. Slayton called all 15 astronauts together and threatened to ground the astronaut responsible for buzzing the MSC offices if his identity could be confirmed. Gordon kept his place on MA-9. (See Fig. 104.)

The countdown for the final manned flight in Project Mercury began on 13 May 1963 and proceeded as planned until the routine overnight hold. It was restarted at 0200 on 14 May and continued throughout the early morning hours. Cooper was woken up in the astronaut quarters at Hangar S at 0500. He completed his ablutions before moving on to the pre-flight steak breakfast. After the meal he underwent his pre-flight medical and stood patiently while the sensors from his biomedical harness were taped to his body. Moving across the corridor, he let Joe Schmitt help him into his pressure suit and check it for leaks.

The astronaut left Hangar S and was transported to LC-14. There, he rode the elevator up the gantry and was greeted in the white room by Shepard, who had spent

Fig. 103. Leroy Gordon Cooper. (NASA)

the night in the spacecraft participating in the countdown. Schmitt assisted Cooper as he climbed into the spacecraft that he had named 'Faith 7'. The pressure suit technician then connected the umbilicals between Cooper's suit and the spacecraft's ECS. Cooper's parachute harness was pulled tight across his body. Left alone in his spacecraft the astronaut found a plumber's suction plunger, a comment from Shepard on the confined conditions inside Faith 7. The wooden handle bore the words 'remove before launch' so Cooper passed the plunger out through the open hatch. The side hatch was bolted in place at 0636.

At Bermuda members of the Operations Division ran a series of slew tests on the FPS-16 radar. The system failed the test in both azimuth and range. It was estimated that the problem would take one hour to locate and correct. At Cape Canaveral the countdown continued. At T − 60 min the gantry should have been rolled back to its parked position, but the diesel generator that drove motors, that in their turn drove the wheels, failed to start. At 0700 the first in a series of holds, that would finally total 2 hours 9 minutes, was called. The entire fuel system in the gantry's generator

Fig. 104. Mercury Atlas 9 completes the final launch of Project Mercury. (NASA)

was stripped down and replaced. The work revealed a fault in the fuel injection system. Lying on his back inside Faith 7, Cooper fell asleep while the work went on below him. The countdown resumed at 0909.

A number of tests were run on the malfunctioning Bermuda radar, but at T−20 min its data was still unacceptable. The countdown was cancelled at 1000, T−13 min, and Cooper was removed from the spacecraft. His words on egress were, '*It was just getting to the fun part.*' He returned to Hangar S, where his pressure suit was removed, checked and stored. Cooper spent the afternoon fishing off of the beach, within the confines of Cape Canaveral.

Tests of the Bermuda radar continued and the faults in the preamplifier in the azimuth digital data channel and the shift register in the range digital data channel were finally corrected at 1800. The radar was declared ready to support the new launch attempt, which had been recycled 24 hours, to 15 May.

The new countdown was picked up at 0200 on 15 May. Cooper repeated the same pre-launch routine as he had performed the previous day. The astronaut entered the spacecraft and was secured in place. The side hatch was bolted in place, LC-14 was evacuated and the gantry rolled back to its parked position. At T−11.5 min a hold was called to investigate a signal fluctuation in the launch vehicle's Mod III guidance system. An outside source of radiation was cited as the cause. The hold lasted 4 minutes.

Lift-off occurred at 0804. Talking to Schirra in MCC, Cooper reported, '*Sigma 7, Faith 7 on the way.*' MA-9 rolled 0.3 degrees anticlockwise before the autopilot stopped the roll and corrected it after 101 cm of travel. Although he experienced the same vibrations as all of the other astronauts to fly a Mercury Atlas, Cooper did not comment on them.

BECO took place at T + 2 min 12.5 s. Staging occurred 3 seconds later. When the LES was jettisoned, at T + 2 min 38 s, Cooper reported, '*And there goes the tower. Does she take off.*' The single Sustainer engine continued the push into orbit, before it too shut down, at T + 5 min 4 s. Schirra informed Cooper, '*Seven, we're right smack dab in the middle of the plot.*'

Faith 7 separated from the Sustainer and the auxiliary damping system nulled all attitude rates resulting from the manoeuvre. Cooper selected FBWS and manually controlled his spacecraft throughout the 180-degree turnaround. He was in an orbit with a perigee of 161.38 km and an apogee of 269. 93 km. The trajectory was inclined at 32.5 degrees to the equator and each orbit would take 88.45 minutes to complete. If all went well Cooper would complete 22 orbits on a flight lasting 34.5 hours.

The astronaut reported, '*She's yawing around very nicely. What a view, boy oh boy. And there's the booster ... Boy oh boy, it is very close.*'

Schirra enquired, '*Fun isn't it?*'

Cooper replied with a simple, '*Yeah.*' He added, '*Fly-by-wire is working just like advertised ... Booster is still smoking. It looks silver Wally.*' Schirra had reported that his jettisoned sustainer had appeared black.

At T + 8 min 35 s Cooper switched Faith 7 to ASCS. The spacecraft was in orbit attitude, 34 degrees negative pitch. With both the Crew Compartment and his pressure suit displaying temperatures in the 60s he increased the latter's cooling water flow rate and then returned it to its original setting only 4 minutes later. As the spacecraft reached the limit of the MCC-Bermuda communication session Schirra told Cooper to '*Have a good flight, boy*'. Cooper replied, '*Thank you, buddy.*'

The passes over the Canary Islands and Kano tracking stations were taken up with technical conversations regarding the Crew Compartment and pressure suit temperatures. Cooper confirmed that he had zeroed both cooling water systems and then returned them to their normal launch level. The Crew Compartment was at 52 °F and the pressure suit in the low 40s. He reported that he had turned the water flow rate on his pressure suit down once more, in an attempt to increase the suit's temperature. Zanzibar CapCom relayed a message from MCC that Faith 7's trajectory was good for the following 20 to 25 orbits.

As he passed over the Indian Ocean Cooper transmitted a short status report into the spacecraft's tape recorder before observing his first sunset. He remarked on the wide blue band surrounding Earth and observed the atmospheric haze layer. Beneath him lightning flashed in Earth's shadow. Talking with Muchea CapCom he reported that the Crew Compartment and pressure suit temperatures were still fluctuating. He confirmed that he had turned the pressure suit cooling water flow rate back to 1.

The citizens of Perth had turned their lights on, as they had for John Glenn's flight. Cooper reported that he could see the city. He could also see the flames from the oil refinery to the south of Perth.

The Sun rose at T + 1 h 22 min, as he was passing over the Pacific Ocean and talking to Canton Island. Cooper observed particles similar to those seen by each of the Mercury astronauts who had made an orbital flight.

Gus Grissom, serving as Guaymas CapCom, passed up the news that Cooper was go for a further seven orbits on top of the 25 that he had already been cleared for, taking his total to 32 orbits. He also passed up the good wishes to the staff of the Gemini Project Office at MSC. The flight so far was summed up when Grissom requested, '*Will you give me a short report.*'

Cooper replied, '*Roger. It's great.*'

Grissom came back with, '*That's good enough.*'

Then Cooper was talking to Schirra at MCC. As he started his second orbit he turned on the television camera mounted in the spacecraft. The picture was displayed on the large screen at the front of MCC. While the camera was on, Cooper caged the spacecraft's gyros, switched to FBWS and turned off the electrical power to the ASCS, thereby entering drifting mode. As Faith 7 passed out of range Schirra reminded him to turn off the television camera. The broadcast had lasted 5 minutes.

Schirra joked, '*... you can stop holding your breath any time and use some oxygen if you'd like*'. Cooper replied, '*Okay. You set such a good example; I've got to equal you here.*' Schirra came back with, '*Yeah, you son of a gun. I'm still higher and faster, but I have an idea you're going to go farther.*' In drifting flight, the spacecraft began to develop slow rates in all three axes, but no thrusters fired to correct them.

The second pass over Zanzibar concentrated on the Crew Compartment temperature. When the conversation dried up Cooper fell asleep for almost 4 minutes before waking up again of his own accord. Faith 7 passed into Earth's shadow without Cooper even mentioning the fact, until he reported that he was drifting in an inverted attitude with the Moon in his window and lightning in the clouds beneath him. Each time the lightning flashed he heard static in his headset. Over Muchea he reported that, due to his inverted attitude, he could not see the lights of Perth on this orbit.

The Sun rose again as he was talking to Canton Island. The spacecraft was surrounded by the usual particles. By the time he was talking to California he had applied power to the ASCS, but was flying in FBWS. He uncaged the gyros and switched to ASCS at T + 3 h 7 min, as the pass over California was drawing to a close and he was approaching Cape Canaveral and the start of the third orbit.

Talking to MCC, he transmitted a further 3 minutes of television before turning the camera off. The pictures were recorded and broadcast by some of the American networks. After switching back to FBWS he pitched up to 20 degrees negative pitch before deploying a flashing light experiment from its position on the Retrograde Package at T + 3 h 25 min. Having returned to orbit attitude he caged the gyros and switched off the power to the ASCS to enter a second period of drifting flight 1 minute later. Using FBWS he yawed Faith 7 through 180 degrees in order to observe the sunset head on. Despite the manoeuvre, he did not observe the flashing light that he had just deployed.

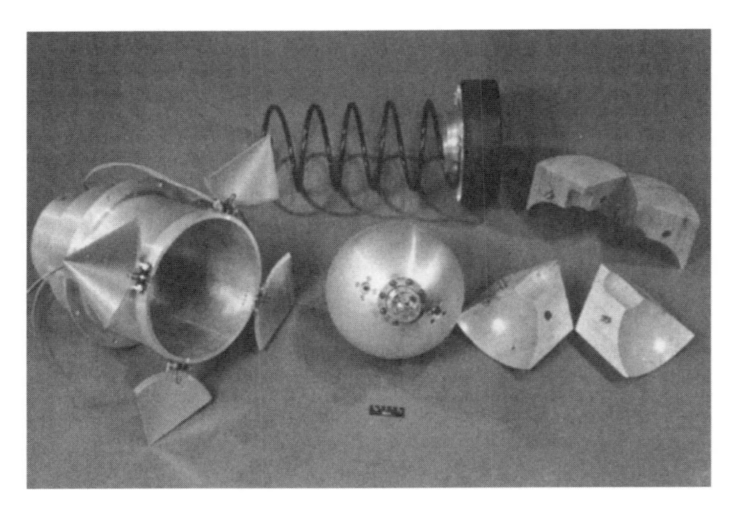

Fig. 105. Flashing beacon experiment hardware. (NASA)

The next tracking station on this orbit was Muchea, in Australia, and then Hawaii. Talking to Hawaii on the third orbit Cooper confirmed that he still had 96 per cent manual and 102 per cent automatic propellant remaining. His primary oxygen supply was reading 90 per cent and the secondary tank was still untouched.

A short communications pass over California was followed by contact with MCC at Cape Canaveral at the beginning of the fourth orbit. He reported that he had a clear view of the entire state of Florida. Schirra passed up the message, '*We are very impressed with the work you're doing ... We lay a pat on the back from Walt Williams.*' This was praise indeed from the man who had wanted to take the flight away from Cooper just a few days before.

Due to the precession of his orbit, the next tracking station was Hawaii. In the 60 minutes between the two stations he described what was happening into the tape recorder. At T + 5 h 5 min Faith 7 passed into Earth's shadow and Cooper reported that he could see the flashing light experiment several kilometres away. (See Fig. 105.) The '*reddish brown*' light was clearly visible, even when there were stars, or thunderstorms behind it. Cooper used the '*Little Rascal*' as a visual cue for a yaw manoeuvre. He managed to centre the light in his window without difficulty at the end of the manoeuvre. The light remained in sight throughout the remainder of the pass through Earth's shadow. (See Fig. 106.)

The communication session with Hawaii was followed by contact with California, before Cooper passed across the Gulf of Mexico and the northern edge of the South American continent at the start of orbit 5. He pointed the television camera out of the window as he passed Florida. Meanwhile, he discussed his observations of the flashing light experiment with Schirra.

From losing contact with Schirra, at T + 6 h 18 min, he was again reduced to talking into his tape recorder. He reported eating some food and taking a drink of water at T + 6 h 25 min. The spacecraft crossed the Atlantic Ocean and the southern

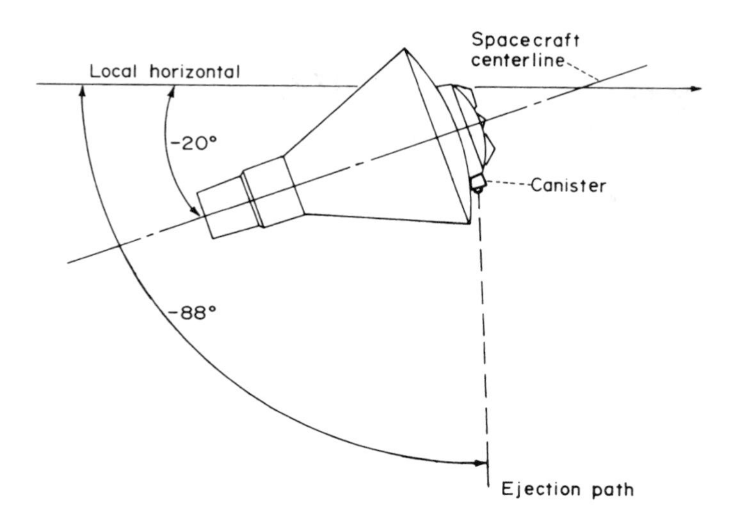

Fig. 106. Mercury Atlas 9 flashing beacon launch trajectory. (NASA)

tip of Africa, before sweeping across the Indian Ocean. Passing into Earth's shadow once more he saw the flashing light experiment and described it as, '... *just barely discernible*'. Of the view through his window he reported, '*The moon is out and the water is very, very bright below. It's quite a lovely Moonlit night.*'

His trajectory took him between the Hawaii and California tracking stations. Talking to Hawaii, he was instructed to switch Faith 7's telemetry transmitters off during all periods where he was not in contact with a tracking station for greater than 30 minutes. Hawaii CapCom told him, '*You're looking good on the ground Gordo.*' While in communication with California he exercised by using the bungee cord attached to the bottom of the instrument panel.

Grissom was the CapCom at Guaymas who instructed Cooper to power up the ASCS and let it take control of the spacecraft as he began orbit 6. With no direct communication with MCC at the end of the pass Cooper explained his actions into the tape recorder as he passed over South America and the Atlantic Ocean. As Faith 7 crossed South Africa, where it was night, he observed a flare that had been fired outside of the town of Bloemfontein. He also saw the distinctive horseshoe shape of the town itself.

Faith 7 had no periscope; in its place the spacecraft had been fitted with a folddown desk. Cooper found that in microgravity his legs occupied a position that made it impossible to push the desk down far enough to lock it into position. He reported, '*The damn desk is unusable.*'

Coastal Sentry Quebec was slightly further north-west of its usual position, off the west coast of China. Cooper has spoken to the ship very briefly on orbit 5. On orbit 6 the ship played its role as a tracking station and link in the communication network to the full. At T + 9 h Cooper was talking to John Glenn, the *Coastal Sentry Quebec* CapCom, as he pressed the button to deploy the balloon experiment, similar to that carried on Mercury Atlas 7. Using FBWS he pitched down to observe the balloon – but saw nothing. He came to the conclusion that it had not deployed. He admitted to the tape recorder, '. . . *I'm doing a rather sloppy job of flying now, trying to look for the balloon.*' As Cooper approached Hawaii the CapCom at that location relayed a request from MCC that he try to deploy the balloon a second time. He did so at T + 9 h 7 min. When the balloon failed to deploy he reported, '*No joy.*'

Talking into the tape recorder once more, he reported that he had manoeuvred from re-entry attitude to orbit attitude. When he reached the latter he had activated the ASCS to maintain Faith 7's attitude. Faith 7 passed south across South America and the Atlantic Ocean, before heading north across Africa, where the spacecraft passed south of Zanzibar tracking station. Zanzibar CapCom passed two messages from MCC. The first confirmed that Faith 7 was Go for the following 17 orbits. The second instructed Cooper not to jettison the balloon experiment as MCC was trying to work out an alternative method of deploying it. Meanwhile, he increased the cooling water flow rate to his pressure suit from 1.5 to 2.7. Faith 7 crossed the southern tip of Madagascar and passed on over the Indian Ocean.

His next radio contact was with Glenn on *Coastal Sentry Quebec*, south of Japan. Cooper transmitted television pictures but technical difficulties prevented their being displayed on the ship. At T + 10 h 26 min he turned off the electrical power to the ASCS and entered a second period of drifting flight. Faith 7 was flying north across the Pacific Ocean.

Apogee was north of Hawaii before he began a long sweep south across the Pacific Ocean at the start of orbit 7. Talking into the tape recorder, Cooper noted, '*It is a rather strange feeling being able to place objects out into the cabin and let go of them and they'll stay in relatively the same position. This is worrisome as well as an odd sensation. Handy sometimes.*'

Landfall lay midway up the western coast of South America and perigee over the Atlantic Ocean. Travelling north once more, he crossed Africa and the northern tip of Madagascar before passing over the Indian Ocean, India and Asia. With apogee north-east of Hawaii, over the Pacific Ocean, Faith 7 turned south and passed directly over Hawaii before crossing South America and the Atlantic Ocean.

Cooper's next communication session was with Zanzibar on orbit 8. As he crossed India and China he was over land that no American spacecraft had crossed before. He took photographs through the window and explained everything that he did to the tape recorder.

The long pass over the Pacific Ocean was broken by communication sessions with *Coastal Sentry Quebec*, where Glenn informed him, '*You're sure looking good. Everything couldn't be finer on this pass.*' The television camera worked on this pass and Glenn was able to see Cooper's face on the monitor.

A brief period talking to Hawaii was followed by the first communication session with *Rose Knot Victor* in the southern Pacific Ocean. Cooper made a 30-minute report and added that his pressure suit cooling water flow rate was set at 2.5. An instruction was passed on from MCC to switch the spacecraft telemetry system to continuous transmit, then Cooper settled down to become the first American astronaut to take a scheduled rest period in space.

He passed around the Earth in near silence, making just the odd comment to the tape recorder. Over Africa he reported, '*I can see roads and rivers and some small towns here on the ground. Small villages are pronounced. Can almost make out the individual houses.*' He took photographs as he passed over Africa, Arabia, India and the Himalayan mountain range. He also checked his control in all three axes using both manual control systems.

During the pass over *Coastal Sentry Quebec* on orbit 9 Glenn promised to '*tell everyone to go away and leave you alone now ... You're looking real good Gordo. Everything is real fine, boy.*' It was T + 13 h 35 min.

Cooper slept soundly, remarking to the tape recorder that he was initially not aware of where he was each time he woke up. Once on each of the next four orbits he made a full 30-minute report to the tape recorder each time he woke up. During orbit 12 he commented,

> Have a note to be added for the head shrinkers. Enjoy the full drifting flights most of all ... you aren't worried about the systems fouling up. You have everything turned off and just drifting along lazily. However, I haven't encountered any of this so-called split off phenomena. Still, note that I am thinking very much about returning to Earth at the proper time and safely. Over.

Over Muchea in Australia on orbit 14 Cooper reported, '*I'm awake now. Just thought I'd check in with you.*' He reported that he had slept well and shared a joke over how he wanted his coffee. Cooper said that he would prefer tea and then added, '*In fact, hot black tea would go down very well right now.*' In the 30-minute report that followed he reported that he had 69 per cent of his automatic and 95 per cent of his manual propellant left. Outside the window it was dark and the astronaut observed the atmospheric haze layer and a number of small towns on Australia's eastern coast.

Travelling north across the Pacific Ocean and the Gulf of Mexico, he was able to talk to MCC at the start of orbit 15. He transmitted television pictures of himself that were displayed on the screen at the front of MCC. At apogee, off the north-eastern coast of Africa he made a second television transmission as he was talking to the CapCom at the Canary Islands. As he passed diagonally south down the African continent he was able to hold brief communication sessions with both Kano and Zanzibar. The flight surgeon at the latter station requested that Cooper report how he felt. He replied, '*Fine. Excellent.*'

Then it was south across the Pacific Ocean, to perigee over Australia. Over Muchea he powered up the ASCS and tested the thrusters used by the two automatic systems. It was T + 23 h 30 min. With the spacecraft in Earth's shadow he was able to tell Guaymas CapCom, '*Checking Fly-by-wire now. Man do those ever throw out fire at night ... you can really see the sparks from the thrusters at night.*'

Over Cape Canaveral at the start of orbit 16 Cooper nulled the rates in all three axes and switched control of the spacecraft over to the ASCS just as the communication session ended. The trajectory had precessed around the planet by this time and orbit 16 would follow the same ground track as the first orbit. Likewise, each subsequent orbit would follow the same ground track as each subsequent orbit at the start of the flight.

The pass over the Canary Islands station was occupied with medical experiments. Then he was talking to Kano in Africa. As that session ended Faith 7 and its astronaut passed the 24-hour point in their flight. Seven minutes later Zanzibar CapCom asked, '*How does it feel on the second day, Gordo?*' He replied, '*Fine. I may get used to this thing yet.*' Switching to FBWS he performed a 34-degree right yaw before returning control to the ASCS. The manoeuvre allowed him to observe and photograph the various dim light phenomena visible when the spacecraft was in Earth's shadow. The passes over the Atlantic Ocean, Africa, the Indian Ocean, Australia and the Pacific Ocean were occupied with taking the relevant photographs.

Back over Cape Canaveral Cooper complained light-heartedly to Schirra, '*Man, all I do is take pictures, pictures, pictures.*' As he passed overhead at the start of orbit 17 he transmitted a few minutes of television. At the same time he powered down the ASCS and entered drifting flight. Schirra told him, '*You sure are a miser on the control fuel.*'

The communication passes over the Canary Islands and Kano were short, the latter lasting just 3 minutes. Over Zanzibar he transmitted a 30-minute report before beginning the long pass over the Pacific Ocean. The session with Muchea lasted 7 minutes, while Canton Island handled just 3 seconds of conversation. Hawaii confirmed that everything was going well and then he was talking to California, where he reported, '*I can make out individual fields. Smoke from a smokestack down there. There's some roads, houses, a little airstrip. There's a dry lake.*' On the ground some people doubted that Cooper could actually see as much detail as he was claiming but the photographs that he exposed during his flight supported his claims.

Crossing Cape Canaveral as orbit 18 commenced, he began a series of Geiger counter radiation experiments. These were followed by Earth photography experiments as he crossed the coast of Africa. He recorded his actions on the tape recorder as he was out of range of all tracking stations until he reached Muchea in Australia. There he complained that his pressure suit temperature was too cold and that he had shut off the cooling water flow completely in an attempt to increase the temperature. Meanwhile, one of the needles on the condensate trap attached to the front of his pressure suit had broken and he was unable to remove any more water from the suit oxygen circuit as both holding tanks were full. He complained into the tape recorder, '*I wish some of you guys who try to stick in some of this plumbing ... would sit in here and try and use the stuff.*' Talking to California he reported that he had clear views of Baja, California and Houston, Texas, as he passed over each of them.

Once more over Cape Canaveral, he reported details of the damage to the condensate trap and water removal system. He attempted to transmit television pictures, but turned the camera off after MCC informed him that the picture was too dark to resolve any detail. He reported that he was comfortable, was in drifting flight and had a clear view of Miami Beach as he passed over it.

At T + 28 h 59 min the perfect flight began to go wrong. The green 0.5g warning light came on and Cooper turned off the two 0.5g fuses. Out of communication, he reported the details to the tape recorder. He also stopped the radiation experiment that he was performing and dictated a short status report. Thirty-four minutes later

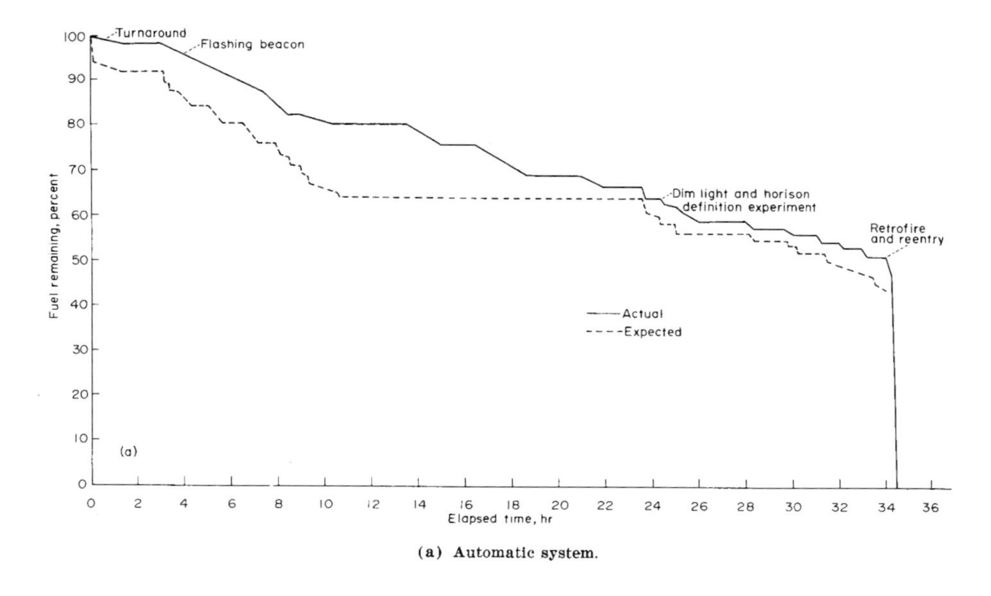

(a) Automatic system.

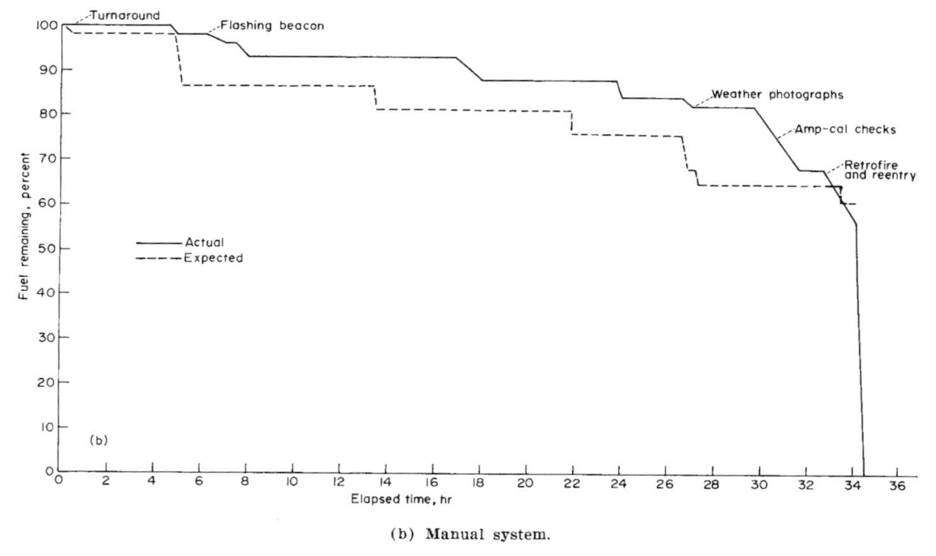

(b) Manual system.

Fig. 107. Faith 7 propellant usage. (NASA)

he was able to report the situation to Hawaii and ask them to relay the details to MCC. Meanwhile he had crossed the Atlantic Ocean and South Africa and the Indian Ocean without communications. Faith 7 had passed north of Australia out of range of both Muchea and Woomera and north across the Pacific before he could talk to Hawaii. (See Fig. 107.)

After Hawaii, he had a brief conversation with California before Guaymas passed up instructions on how to test to see if the spacecraft had switched to post-0.5*g* re-entry logic. Despite everything, Cooper completed the high-frequency communication test planned for the Guaymas tracking station.

Orbit 20 began over Mexico; with MCC passing up details of the re-entry logic test that they wanted Cooper to perform. The test called for him to power up the ASCS bus, slave the gyros and initiate a manoeuvre in any axis. If the attitude indicators did not follow the manoeuvre then the spacecraft was on re-entry logic. Cooper completed the test during the 48 minutes that he had no communications and in which time he passed most of the way around the planet. He powered up the ASCS bus and slaved the gyros. When he saw that he had no attitude indication on the instrument panel he returned to drifting flight in the assumption that the spacecraft had switched to re-entry logic. His trajectory took him south over Columbia and Brazil across the Atlantic before turning north above the southern tip of South Africa, crossing the Indian Ocean and Indonesia. As he crossed the Pacific Ocean he spoke to Glenn on *Coastal Sentry Quebec* and explained what had happened during the test.

The next voice that he heard was Hawaii CapCom, who soon gave the session over to MCC CapCom relayed through the Hawaii tracking station. Cooper was instructed to turn on the 0.5*g* ASCS fuses individually and report which, if any, indicator lights illuminated on his instrument panel. When he activated the primary fuse the green 0.5*g* light came on. He then turned off that fuse and activated the 0.5*g* emergency fuse. No light came on. The communication ended with Cooper giving a 30-minute status report.

California CapCom talked him through the next phase of a new test. He was instructed to power up the ASCS bus and switch control of Faith 7 to ASCS. At T + 31 h 18 min Cooper switched to ASCS and Faith 7 began a slow roll, indicating that the re-entry logic had been activated. The roll did not register on the attitude indicator in front of him. Two minutes later he followed instructions and powered down the ASCS bus. MCC now knew that Cooper would have to perform a manual retrofire burn, but that the ASCS could be trusted to control the spacecraft once it was past the initial stage of re-entry.

Grissom spoke to him from Guaymas and reminded him that it was the last time they would speak to each other during the flight. Grissom told him, '*You're doing an outstanding job. I'm proud of you.*' Cooper joked, '*Roger. Muchas Gracias. Muchas Gracias. That's French for thank you.*'

Orbit 21 began high over the Pacific Ocean and carried him south across central South America. Perigee was east of South Africa and he continued north across the Indian Ocean and Vietnam. At T + 32 h 21 min Cooper told the tape recorder, '*I'm observing some cities through the clouds ... Seeing out over Laos.*'

Glenn on *Coastal Sentry Quebec* read up a series of procedures for re-entry. Cooper would have to fire the first retrograde rocket and jettison the Retrograde Package manually. Glenn cautioned him, '*Be sure that you do not arm the retro-jettison switch until after the rockets have fired.*' An attempt to advance Faith 7's clock by one hour by command from the ship was unsuccessful. Cooper made the correction manually before the session ended.

As his spacecraft reached apogee and turned south once more, Hawaii took over communications. They advised Cooper, '*Faith 7, Hawaii Cap Com recommend take a green as for go now and go over your stowage checklist now. Did you copy?*' The first part of the message, '*take a green as for go*', was a coded message advising him to take a Dexadrine stimulant pill. The flight surgeon at MCC had doubts that Cooper was fully rested from his sleep period earlier in the flight and, although there was no concern over his condition, had wanted him to take the pill to ensure that he was alert during the manual retrofire manoeuvre.

At T+33 h 3 min a warning tone sounded in Cooper's headset. He reported, '*Well, things are beginning to stack up a little. ASCS inverter is acting up and my* CO_2 *is building up in the suit. Partial pressure is decreasing in the cabin. Standby inverter won't come on line. Other than that things are fine.*'

The 250-volt-amp main inverter failed to operate and Cooper noticed two small fluctuations in the ammeter. He checked the entire electrical system, but everything appeared to be in working order. At T+33 h 3 min the light came on to indicate that the spacecraft had automatically switched from the primary to the back-up inverter. After a second check of his instruments Cooper realised that the back-up inverter had failed to start. He turned both inverters off, thereby turning off the ASCS and disabling the re-entry control mode. He would have to make not only the retrofire burn, but the entire re-entry under manual control using FBWS.

Free moisture, either from the nearby condensate trap mechanical sponge, or from the astronaut's exhaled breath in the Crew Compartment atmosphere, had caused a short circuit between three pins on two electrical plugs associated with the ASCS. Post-flight x-rays revealed that the current had travelled through the insulation material surrounding the pins and free moisture was identified as the most likely cause of the short circuits that resulted. This had happened to two independent plugs in the electrical system. One effected plug caused the 0.5*g* light to illuminate and initiated the re-entry logic. The second plug resulted in the loss of the two inverters. With electric current no longer flowing to the ASCS Cooper could not use that system to control re-entry and would have to set up the re-entry roll manually. The failure meant a change in only one switch setting on the re-entry checklist that he had written down over *Coastal Sentry Quebec* on the previous orbit.

The partial pressure of oxygen in the Crew Compartment had been steadily decreasing throughout the flight until it was down to 246 g/cm^2, while the partial pressure of the carbon dioxide in Cooper's pressure suit had risen to the same level. As a temporary measure he turned his oxygen flow rate to high, but switched back to a lower rate when he realised that his breathing had become more rapid and deeper.

Faith 7 crossed South America and the Atlantic Ocean. Zanzibar CapCom talked Cooper through his pre-re-entry checklist as he crossed Africa and the Indian Ocean. He passed over southern China and was then talking to Glenn on *Coastal Sentry Quebec*. He explained his new difficulties and informed Glenn that he would control re-entry on FBWS.

Glenn provided a countdown to retrofire and Cooper pressed the manual override button. The first retrograde rocket fired at T + 34 h and the remaining two rockets ignited as planned. Throughout retrofire he maintained the correct position using MPCS. All three axes remained within 12 degrees of the correct attitude. One minute after the third rocket stopped thrusting Cooper jettisoned the Retrograde Package manually and then switched to FBWS. Glenn signed off with, '*It's been a real fine flight Gordo. Real beautiful all the way. You have a cool re-entry, will you?*' Cooper replied, '*Roger John. Thank you.*'

Re-entry occurred without further incident. Cooper kept Faith 7 under control using FBWS. Following radio blackout he established voice contact with Carpenter at Hawaii. Tumbling rates reached their peak at 15,240 m. Cooper deployed the drogue parachute at 12,801 m and the spacecraft stabilised. The main parachute and landing skirt were deployed at 3,828 m. The astronaut spent the next few minutes preparing for an emergency egress if the spacecraft sank on landing.

By the time Faith 7 was at 152 m, recovery helicopters were circling around the descending spacecraft, which was visible from the flight deck of the USS *Kearsarge*. Landing occurred in the Pacific Ocean north-east of Hawaii, at T + 34 h 19 min 19 s. Faith 7 settled on one side and then righted itself. Cooper jettisoned the main parachute manually. Overhead, recovery swimmers were entering the water to install the Stullken Collar around the base of the spacecraft. (See Fig. 108.)

Four minutes after landing, with one swimmer already climbing on the exterior of his spacecraft, Cooper elected to remain inside Faith 7 until they could both be lifted

Fig. 108. Cooper is helped from Faith 7 onboard the USS *Kearsarge*. (NASA)

directly on to the USS *Kearsarge*. He was informed that there would be a 45-minute delay while the aircraft carrier steamed alongside the spacecraft. As an USAF officer, Cooper made a formal request and was granted permission to come aboard the Naval vessel.

Despite having landed just 2,133 m from the prime recovery vessel, Cooper apologised to her Captain for having missed the ship's third elevator. He received the reply, '*I think it's a quite acceptable shot Major*.'

The USS *Kearsarge* drew close to the spacecraft and launched a whaler to carry a securing line to Faith 7's recovery loop. When the line had been secured the spacecraft was hoisted onto one of the aircraft carrier's side elevators. With the spacecraft secured on the elevator Cooper jettisoned the side hatch and received the same bruise to his wrist as Glenn, Carpenter and Schirra. NASA physicians gave Cooper an 8-minute physical examination as he lay on his prone couch. They found him to be dehydrated.

Finally, Cooper was assisted out of his spacecraft and he quickly found that he felt dizzy. He was taken below deck to the officer's sick room, where he drank several glasses of milk and orange juice to overcome his dehydration and dizziness. While in sickbay he received telephone calls from his wife and from President Kennedy. Like Schirra before him, Cooper remained on the USS *Kearsarge* and was debriefed while the ship returned to Hawaii.

The final astronaut to fly in Project Mercury received a tumultuous welcome when he landed at Honolulu, where he was given his first parade. From Hawaii he was flown to Cape Canaveral. On 19 May, he received a second parade through the town of Cocoa Beach, where he gave his personal impressions of his flight at a press conference. There was another parade in Houston, the home of MSC.

On 21 May Cooper and his family attended the White House in Washington, where President Kennedy presented him with the NASA Distinguished Service Medal. Five other members of the Project Mercury management team were also decorated. After the presentation Cooper was given a parade through the city before following in Glenn's footsteps and addressing a joint session of Congress. One week later he received the USAF's astronaut's wings for flying higher than 50 km. The tickertape parade that Cooper received in New York was one of the largest parades seen in the city up to that date. Finally, he received a parade in his hometown of Shanwee, Oklahoma.

28 May 1963: Race riots began in Jackson, Mississippi. Dr Martin Luther king called for a 'national event', a march on Washington DC to make the President act on desegregation. Throughout the summer the southern half of North America was swept with Civil Rights demonstrations and the violence that accompanied them.

10 June 1963: Following an exchange of private correspondence between himself, British Prime Minister Macmillan and Chairman Khrushchev, President Kennedy called for better understanding of the Soviets by the American people. He announced new negotiations were about to commence on a Nuclear Test Ban Treaty. He also announced that America would meanwhile resume atmospheric testing of nuclear weapons. Radio Moscow carried a full translation of the speech across the Soviet Union.

11 June 1963: A Buddhist monk burned himself to death in the centre of Saigon. President Diem's Catholic sister in law offered to provide gasoline and matches to any other Buddhist monk, or American reporter, who wanted to follow the monk's example.

11 June 1963: President Kennedy Federalised the Alabama National Guard to protect two Negro students as they registered at the Alabama State University, the last segregated university in America.

11 June 1963: President Kennedy went on television to call for the enactment of Civil Rights legislation to make segregation illegal in America. Over the next few days the Southern states erupted with racial hatred.

20.15 VOSTOKS V AND VI

The 13 June attempt to launch Vostok V was cancelled due to bad weather and was rescheduled for the following day. Valeri Bykovsky was the prime cosmonaut when the launch occurred at 1500 on 14 June 1963. His back-up cosmonaut was Boris Volynov. Vostok V entered a 175 × 221-km orbit but attracted only cursory attention. Moscow was alive with rumours that a woman would be launched into space soon. The Soviet media did nothing to dispel the rumours. Despite those rumours nothing happened when Vostok V passed over Baikonur Cosmodrome at the beginning of its second day in space.

Unspecified technical difficulties are believed to have delayed the launch of Vostok VI, which was rescheduled for the following day and launched at 1230 on 16 June 1963. The prime cosmonaut was Valentina Tereshkova and her back-up was Irina Solovyova. The spacecraft was injected into a 181 × 231-km orbit, just 5,000 m away from Vostok V. The two cosmonauts were able to see each other's spacecraft and communicate by radio. They also repeated the actions of the previous Vostok 'twins', in that they followed the same flight regime. Due to their different launch times they ate, slept and performed experiments to different schedules.

During her first afternoon in orbit, Tereshkova began her sleep period. She did not wake up at the scheduled time and repeated calls from Mission Control in Moscow failed to wake her. She finally woke up 35 minutes late. She promised, '*I shall do better. I feel fine.*' Despite the incident Soviet mission planners agreed to extend her flight beyond the first 24 hours, first to 48 hours and then to 72 hours. On 18 June Bykovsky broke Nikolyev's duration record. The now bearded cosmonaut remained cheerful as he completed his tasks and reported the status of his spacecraft to the ground.

Tershkova's spacecraft was aligned for retrofire during its 48th orbit. Re-entry followed without incident and the first woman in space ejected from her spacecraft and landed under her personal parachute at 1120 on 19 June 1963, after a flight lasting 70 hours 50 minutes. Tereshkova had orbited Earth 14 times more than all four Flight A astronauts put together.

Bykovsky returned to Earth on his 81st orbit, landing under his personal parachute at 1400 on 19 June 1963. The final Vostok flight had lasted 119 hours 6 minutes. In the year 2000 this still stood as the longest flight by a one-man spacecraft.

In the years that followed the flight of Vostok VI there were persistent rumours that Tereshkova had spent a miserable three days in flight pleading to be allowed to return to Earth. The rumours insist that she was continually vomiting. She may well have been the second person to suffer from Space Adaptation Syndrome, which later experience would show could take several days to get over. That would certainly explain the reports of her vomiting.

Whatever the truth of the matter, Tereshkova married cosmonaut Adrian Nikolayev, the pilot of Vostok III, in 1964, and gave birth to a healthy daughter seven months later. The couple were subsequently divorced and Tereshkova jealously guarded her daughter's privacy, only allowing her to be photographed shortly after her birth and not again until she came of age.

20.16 MERCURY ATLAS 10 (MA-10)

In January 1962 Seamans had informed Congress that the 24-hour manned Mercury flight should be completed during fiscal year 1963. At that time he made it clear that a back-up spacecraft and launch vehicle were available if MA-9 failed to meet its objectives.

SC-15 was originally delivered to Hangar S, Cape Canaveral, on 13 August 1961. It was sent back to McDonnell for reconfiguration in support of a 24-hour manned flight and returned to Cape Canaveral on 16 November 1962. By that time it had been redesignated SC-15A and assigned to a flight that had been tentatively designated MA-10. The flight was seen as a potential 72-hour manned flight with additional consumables such as oxygen and water being carried in a container attached to the wide end of the spacecraft and jettisoned prior to retrofire. Alan Shepard was the astronaut most frequently associated with the flight.

On 14 January 1963 SC-15A was redesignated SC-15B and named as the back-up to SC-20, which was the prime vehicle for MA-9. On 11 May 1963, three days before the first MA-9 launch attempt, Julian Scheer, NASA deputy administrator for public affairs, announced: '*It is absolutely beyond question that if this flight* [MA-9] *is successful there will be no MA-10*.'

MA-9 was successfully flown on 15–16 May 1963. At the post-flight press conference held at Cocoa Beach on 19 May, Seamans was asked if there would be any further manned flights as part of Project Mercury. He replied, '*It is very unlikely*.' President Kennedy was asked a similar question on 22 May. He replied that the final decision regarding any additional flights would rest with NASA's management. Nine days later DoD announced that they were ready to support MA-10 if NASA decided to make the flight.

On 6–7 June 1963 the Office of Manned Spaceflight made a two-day presentation to NASA Administrator James Webb in favour of flying MA-10. They argued that SC-15B was capable of supporting a 72-hour manned flight, the information from which could be applied to Project Gemini and Project Apollo. They added that both spacecraft and launch vehicle were available and approaching flight-readiness.

Five days later James Webb testified before the Senate Space Committee. He told the committee, '*There will be no further Mercury shots*.' It would be a further 22 months before American astronauts returned to space, at which time they would be flying the two-man Gemini Spacecraft.

Having lost MA-10, Slayton named Alan Shepard as command pilot for the first manned flight in Project Gemini, with Flight B astronaut Thomas Stafford as his pilot. While he was training for the flight Shepard was grounded with a medical condition called Menniers Disease, which affected his balance. When Shepard was removed from the first manned Gemini flight Stafford also lost his place on the crew. Slayton's second choice for the crew was Flight A astronaut Gus Grissom as command pilot, with John Young as pilot. Grissom would become the first person to fly in space twice, while Young would become the first member of Flight B to make a space flight.

20 June 1963: President Kennedy and Chairman Khrushchev agreed to establish a direct secure telephone line between the White House in Washington DC and the Kremlin in Moscow, for use in times of crisis. The telephone link quickly acquired the name of 'The Hot Line'.

22 June 1963: President Kennedy visited West Berlin. During his motorcade through the city he made two stops to allow him look over the Berlin Wall into East Berlin. Outside the city's Government Building he told the 150,000 Berliners who had come to hear him speak, '*Today in the world the proud boast is "Ich bin ein Berliner"*.' No one told him that 'ein Berliner' was the name that locals gave to a sweet doughnut-like cake.

14 July 1963: Negotiations began in Moscow between America, Britain and the Soviet Union, aimed at achieving a Nuclear Test Ban Treaty. At the same time Kennedy went so far as to suggest joint American–Soviet action to prevent China becoming a nuclear military power. The draft of the Test Ban Treaty was signed on 25 July and banned nuclear testing in the atmosphere, in the sea and in space. The treaty was officially signed on 5 August. China accused the Soviets of selling out Communism and France insisted that she retained the right to continue atmospheric testing of her own nuclear weapons.

28 August 1963: Two hundred thousand Civil Rights supporters from across the country, both Negro and white, marched on Washington DC. The march ended with a meeting at the Lincoln Memorial, where Civil Rights leader Dr Martin Luther King addressed the crowd. He told them that, '*I have a dream*'. That dream was that one-day Negroes and whites would live in an America where they shared equal rights.

September 1963: Four black Negro children were killed in a bomb blast at a Negro church in Birmingham, Alabama. Two Negro boys were shot later in the day, one by police and one by a white segregationist.

1 November 1963: A military coup occurred in South Vietnam, overthrowing Diem's government. Having found it increasingly difficult to make Diem and his brother do as they wished, American ambassadors had been in collusion with the South Vietnamese military Generals for some weeks. Although American forces played no part in the coup it had President Kennedy's full backing. Diem and his brother were captured by the Army and shot in the back of the head. Their bodies were mutilated.

8 November 1963: America officially recognised the new military government in South Vietnam.

20.17 PRESIDENT KENNEDY ASSASSINATED

On 22 November 1963 President Kennedy visited Dallas, Texas. Having begun the day in Fort Worth, the Presidential group flew to Love Field, Dallas. Leaving the aircraft, the President and his wife spoke briefly to the people that had come to greet him. They then took their places in the back seat of an open top car and the motorcade made its way to the city's business centre.

As the motorcade drove through the city centre three shots rang out. One hit the President in the neck, another hit him in the head. Mrs Kennedy, the President's wife climbed on to the trunk of the car and had to be helped back into her seat. The President's car left the motorcade and followed a prearranged route to Parkland hospital, while Mrs Kennedy cradled her husband in her arms. On arrival at the hospital the President was rushed into an operating theatre. He died without regaining consciousness, 30 minutes after being shot.

Witnesses pointed police in the direction of the Dallas Book Depository, where a rifle was later found next to a window and behind a wall of boxes. Three blank cartridges were found lying on the floor. The window in question offered a clear view of the Presidential motorcade at the moment when the shots were heard.

Two hours after the shooting Vice President Johnson was sworn in as America's 35th President. The ceremony took place at Love field, on the aircraft that would carry President Kennedy's body back to Washington. Mrs Kennedy stood alongside the new President as he took the oath of office.

That evening Dallas Police arrested Lee Harvey Oswald, an employee at the Dallas Book Depository, for the President's assassination. Two days later Jack Ruby shot Oswald as he was being escorted across the basement car park at Dallas Police Headquarters. Ruby was arrested for Oswald's murder.

President Kennedy was given a state funeral in Washington, DC, on 25 November.

20.18 FLY ME TO THE MOON

As Project Mercury ended, NASA was busy with a multitude of programmes. Unmanned spacecraft were flying to the Moon, Venus and Mars. Early meteorological satellites were proving invaluable to the science of weather prediction, and communications satellites were already beginning to change the way people and businesses talked to each other. Project Gemini was in the final stages of hardware development. A number of Flight A and Flight B astronauts were already training for Project Gemini. The first of two unmanned flights was scheduled to occur in December 1963. Ten manned Gemini flights would be flown during 1965–66. NASA used Project Gemini to give their astronauts practice in rendezvous and docking techniques and extravehicular activity in orbit while the hardware for Project Apollo was being developed.

When America's Project Mercury ended, the situation in the Soviet Union was complicated. The manned space programme had become politically motivated and driven by Khrushchev, who demanded a series of propaganda 'firsts', to give the appearance of Soviet supremacy, while not allowing his designers to concentrate fully on follow-on vehicles. Soyuz, the spacecraft designed to follow Vostok, was delayed while Korolev was forced to risk his cosmonauts' lives in a converted Vostok Spacecraft called Voskhod. Voskhod offered no means of escape in the event of a launch failure. Even so, two manned Voskhod flights allowed the Soviets to claim the first multi-man crew and the first extravehicular activity. Nikita Khrushchev fell from power on 15 October 1964 and was written out of the Soviet Union's official history. General Secretary Leonid Brezhnev and Premier Alexei Kosygin replaced him and were far less well disposed towards America and her Western allies.

Apollo was still in the early stages of hardware development. The Command and Service Module had largely been defined, but the Lunar Excursion Module was still in its infancy. The Apollo Saturn IB and Saturn V launch vehicles were also under development, as were the new Apollo facilities at Merritt Island, Cape Canaveral. Following President Kennedy's assassination the NASA facilities at AMR were named the John F. Kennedy Space Centre and the whole geographical area was renamed Cape Kennedy. When Project Apollo finished the geographical region returned to the name Cape Canaveral. The NASA facilities retained the dead President's name.

Meanwhile, the Soviets had developed two manned lunar programmes. Korolev's N-1 programme was directed towards a two-man lunar landing programme, while Chelomei's UR-500K programme concentrated on sending a single cosmonaut around the Moon on a circumlunar trajectory. This splitting of the Soviet lunar programme and the deep personal rivalries between the leaders of the various design bureaux resulted in the failure of both programmes. They were both abandoned in the wake of Project Apollo's success. The Soviet government refused to acknowledge the existence of either manned lunar programme until the after the collapse of the Soviet Union.

By the end of 1965 Project Gemini had destroyed the impression of Soviet supremacy in space.

Part V: AFTERMATH

21

Scientific experiments

With the advent of manned orbital flight many people inside NASA and the American scientific community wanted to fly scientific experiments on Project Mercury. An experiment programme was therefore commenced as part of a greater overall national space science programme (see Table 9).

Table 9: Project Mercury in-flight experiments (taken from a NASA original)

Experiment	MA-6	MA-7	MA-8	MA-9
Known earth lights recognition	S	S	S	S
Ground-launched flares	U	U	U	—
Ground-based Xenon light	—	—	U	S
Flashing light beacon	—	—	—	S
Horizon definition photography	—	S	—	P
Weather photography	N	N	S	S
Terrain photography	+	+	S	+
Zodiacal light photography	—	—	—	P
Air-glow photography	—	—	—	S
Radiation detection (film pack)	—	—	S	S
Radiation detection (emulsion pack)	—	—	S	S
Radiation detection (ionisation chamber)	—	—	S	S
Radiation detection (Geiger–Mueller)	—	—	S	S
Tethered balloon	—	P	—	U
Liquid behaviour in microgravity	—	S	—	—
Ablative materials study	—	—	S	—
Micrometeorite impact study	S*	S*	S*	S*

S = Experiment carried out successfully.
P = Experiment partially successfully.
U = Experiment attempted but unsuccessful.
N = Experiment carried but not completed.
+ = Terrain photography completed as part of general photographic tasks.
* = Study of last four spacecraft revealed only one possible impact crater, in the window of the MA-9 spacecraft.

In-flight experiments were limited by the amount of free space in the Mercury Spacecraft and the severe weight restrictions dictated by the lift capability of the Atlas launch vehicle. The limited electrical power available and equally limited availability of attitude manoeuvring propellant further restricted the number of acceptable experiments. Experiments had to be designed so that their failure to perform would not endanger the astronaut, or the spacecraft.

The photographic experiments carried out by MA-6 were simply suggested by NASA personnel and accepted by mission planners in STG's Flight Operations Division. Glenn purchased his own auto-focus camera in a store in Cocoa Beach. For MA-7 and MA-8 all experiments were submitted to the Mercury Scientific Experiment Panel (MSEP), which was formed in April 1962. The MSEP brought together representative of all the interested parties involved in Project Mercury and NASA's space science programme to encourage the development of suitable experiments, vet them and recommend an order of priority in which they should be flown. In October 1962 the MSEP was surpassed by the MSC In-flight Experiments Panel (IFEP), which continued the role of the original panel, for MA-9, while also beginning the planning of experiments for Project Gemini and Project Apollo.

21.1 LUMINOUS PARTICLES

On MA-6, astronaut Glenn observed that his spacecraft was surrounded by luminous particles at orbital sunset and sunrise, and the astronauts on MA-7, MA-8 and MA-9 also observed the phenomenon. Carpenter identified the fact that the particles originated from the spacecraft when he discovered that he could induce showers of particles by hitting the inside of the spacecraft hatch with his hand. This gave rise to the early theory was that the particles were ice particles that had formed on the exterior of the spacecraft. Later observations suggested that the particles were associated with thruster firings. The final conclusion was that the particles were extremely small amounts of thruster catalyst, or steam, which left the thruster at relatively slow velocities at the end of a thruster firing and cooled rapidly in the space environment.

21.2 KNOWN EARTH LIGHTS RECOGNITION

On all four manned orbital flights the astronauts attempted to identify cities and other known light targets on the night-side of Earth. The most famous of these experiments took place on MA-6 when astronaut Glenn successfully identified the lights of the Australian city of Perth, after the occupants had been requested to turn on as many lights as possible as MA-6 passed overhead. Astronauts Carpenter, Schirra and Cooper were also able to identify known targets from orbit during the flights of MA-7, MA-8 and MA-9 respectively.

In general, the Flight A astronauts reported that Earth from the Mercury

Spacecraft's orbit looked much like it did from a high-altitude aircraft except that a much greater field of view was visible from orbit. When looking straight down, colours and small detail were easily visible but they became lost in a blue haze as the angle increased. This was to be expected as the observer looked along a longer visual path through the atmosphere. Smoke, dust particles and other aerosols caused haze within the troposphere to the extent that whole cities could not be seen, even when the spacecraft was passing directly above them and the astronaut was well aware of their existence.

Clouds were clearly visible on the sunlit side of Earth and the astronauts could determine different cloud layers and different types of cloud formation. The ground was visible in hues of brown, and green foliage was best viewed from directly above. Water was easily distinguished from orbit, as all astronauts stated that they could identify roads and rivers within a given landmass.

The astronauts reported that stars were visible in twilight conditions. During orbital sunset the astronauts reported three distinct colour bands surrounding Earth's horizon – a lower band of orange, a centre band of white and an upper band of blue. These light effects were caused by the atmosphere acting upon the light passing through it. In the upper atmosphere the light scattered in the normal way, acting upon the blue wavelengths the most, as observed by the astronauts. In the middle atmospheric layers the light was scattered to saturation point in all wavelengths, thus causing a visible white band. In the lower layers the astronauts saw the red light that had passed through the atmosphere rather than the light-scattering effects as seen in the other two bands.

On the night portion of each orbit Earth appeared darker than the space behind it. Clouds were just visible to the dark-adapted observer. If moonlight was striking Earth then clouds and their movement within the atmosphere were easily discernible. City lights were identifiable, even through thin cloud layers. Space appeared deep black and stars appeared as pinpoints of light that did not twinkle as they were no longer being observed through the ever-changing layers of Earth's atmosphere. On the other hand, astronaut Cooper stated that when he looked down through the atmosphere from his MA-9 some lights on Earth's surface did appear to twinkle.

21.3 GROUND-LAUNCHED FLARES

During the first and second orbital passes of MA-6 Glenn attempted to observe flares fired from mortars mounted on the Indian Ocean tracking ship. On both passes cloud cover prevented the astronaut from seeing the flares despite their being launched as planned. Ground flares of 1,000,000-candle power intensity were launched from the tracking station at Woomera, Australia, during the flights of MA-7 and MA-8. On all occasions the flares were launched as scheduled, but cloud cover prevented astronauts Carpenter and Schirra from observing them.

21.4 GROUND-BASED XENON LIGHT

A ground-based bank of 18 Xenon arc lights mounted in a highly polished reflector was located at Durban, South Africa, during the flight of MA-8. The lights flashed at 150 flashes per second and should have appeared to the astronaut as a solid light of between 30,000 and 35,000 candle power intensity. The light should have been visible from a slant-distance of 592.5 km, but cloud and rain in Durban prevented astronaut Schirra from observing it.

The Xenon light experiment was moved to Bloemfontein, South Africa, for MA-9 and was successfully observed by astronaut Cooper on the sixth orbit. The MA-9 experiment was considered successful, but flashing lights, or distinctive patterns of lights were considered necessary if such light banks were to be used as ground-based navigation beacons on future space flights. The spacecraft's rapid angular pass over the light bank was considered a problem in the future use of such beacons, as was the unpredictability of atmospheric weather conditions.

21.5 FLASHING LIGHT BEACON

A flashing light beacon experiment was carried into orbit mounted on the Retrograde Package of the MA-9 spacecraft. The experiment was an attempt to see if astronaut Cooper could locate and identify the beacon in orbit. It was a forerunner to the use of navigation lights on the Gemini Agena Target Vehicle, Apollo and future generations of spacecraft, which could be expected to perform orbital rendezvous and docking.

The experiment was manufactured at LaFRC. The beacon was a 14.6-cm sphere with a flashing light at opposite ends of one axis. The mercury battery powered lamps flashed simultaneously, at a rate of one flash per second. They were expected to be as bright as a second magnitude star at a distance of 15 km. Cooper used markings on his spacecraft window to manoeuvre Faith 7 to a minus 20-degree attitude, which resulted in the beacon being released on a minus 88-degree trajectory. When the canister lid was jettisoned a compressed spring pushed the beacon from its canister at a speed of 3 m/s.

The beacon was ejected 15 minutes before sunset on the third orbit. Cooper was unable to locate the beacon (which he referred to as the 'Little Rascal') during the night pass of the third orbit because his spacecraft was incorrectly aligned in pitch. He acquired the beacon visually during the night-side on the fourth orbit. He was able to manoeuvre his spacecraft's attitude away from the beacon and then manoeuvre to visually reacquire it. Cooper estimated that the beacon was one magnitude dimmer than expected. He also saw the beacon on the fifth orbit night pass. The light was 'barely discernible' due to the distance between Faith 7 and the beacon. At no time did Cooper see the flashing light during a pass over the sunlit side of the planet.

21.6 HORIZON DEFINITION PHOTOGRAPHY

Astronaut Carpenter on MA-7 and astronaut Cooper on MA-9 carried out horizon definition photography in an MIT experiment to determine if Earth's sunlit limb might be used as a navigational aid on future space flights. Carpenter employed a 35-mm robot camera with a split blue and red filter placed in front of the film plane. The filter represented the two extremes of the visible spectrum. When Earth's limb was photographed through the blue filter it appeared to have a higher elevation than when photographed through the red filter. The image though the blue filter appeared more stable, while the image through the red filter was subject to interference effects from clouds and the atmosphere.

Cooper used a 70-mm Hasselblad with a filter split into four exposure areas. The two outer areas contained blue filters while the two inner areas held red filters. The MA-9 experiment was also intended to define the limb's radiance and the effects of variations of scattering angle of incident light on limb height, and to establish the height of the limb above Earth's surface. The flight plan called for Cooper to photograph Earth's limb during the sunlit pass on most orbits. He was also tasked with photographing the setting Moon in the four quadrantal directions relative to the Sun.

The sunlit photographs were not obtained due to operational difficulties at the time when Cooper was supposed to expose them. The setting Moon photographs were exposed, but the resulting images were too indistinct to allow the limb's height above Earth's surface to be measured. One possible cause of the indistinct image was the film of debris that formed on the exterior of the spacecraft window during launch and when the LES was jettisoned. When the four quadrantal photographs were compared no difference in limb height was measured. The experiment was considered a success, but more work was considered necessary on future flights to determine if Earth's limb was a viable navigational aid.

21.7 WEATHER PHOTOGRAPHY

The weather photography performed during Project Mercury was intended to augment similar work throughout America's space programme. The astronauts on all four orbital flights performed weather photography. Glenn on MA-6 and Carpenter on MA-7 both carried cameras and the appropriate filters for photographing areas of special meteorological interest viewed through their spacecraft windows. Operational difficulties on both flights meant that no such photographs were exposed. Meteorological data was obtained from some Earth photographs exposed on these two flights as part of the astronauts' general photographic tasks.

Weather photography experiments carried out by Schirra on MA-8 and Cooper on MA-9 were performed for the National Weather Satellite Centre. They were intended to examine some of the spectral reflectance characteristics of clouds, waters and land. The same 70-mm Hasselblad camera was used on both flights.

Schirra employed a six-element filter covering a range of spectral transmission from red to blue in the wavelength range 3,700–7,200 ångströms. Neutral density was added to each filter to give each image the same density. Schirra's 13 black-and-white photographs revealed that contrast increased with wavelength in the visible spectrum. They suggested that future meteorological satellites should image Earth's surface in the near infrared. The photographs also showed a 16-km thick bright band on Earth's horizon. This was identified as the troposphere, with the band being made visible by the scattering of light by dust particles, water droplets, etc.

Cooper's photographs were exposed using infrared film with a three-element filter, covering ranges 6,600–9,000, 7,300–9,000 and 8,000–9,000 ångströms. Water was shown to have a low reflectance in the infrared and appeared very dark in the images. Land and cloud were both shown to have a very high reflectance. As a result, coastlines and clouds were easily identified next to water. On the other hand, clouds over a vegetation-covered landmass were much more difficult to differentiate. The experiment suggested that future meteorological satellites should image Earth in the range 5,000–7,500 ångströms.

21.8 TERRAIN PHOTOGRAPHY

Terrain photography was performed as part of the Mercury astronauts' general photographic tasks on all four manned orbital flights. Glenn on MA-6, Carpenter on MA-7 and Cooper on MA-9 all took photographs of interesting features in the terrain of the planet beneath them, but these were always targets of opportunity and not prearranged targets of specific scientific interest. Uniquely, on MA-8 astronaut Schirra was given a series of specific targets to photograph as he passed over them. The photographs taken on the first three orbital flights suffered from poor weather conditions above the few landmasses that those spacecraft flew over.

Weather conditions were better during the flight of MA-9. In addition, the extended duration of that flight meant that it flew over many more landmasses than the previous three flights. Cooper was able to obtain excellent photographic coverage of many areas, including the African and Asian deserts and the Himalayas mountain range. The terrain photographs were used to compile a catalogue of space images showing geological features on the surface of the planet below.

21.9 ZODIACAL LIGHT PHOTOGRAPHY

Zodiacal light is a dim-light phenomenon. Astronaut Cooper was tasked with obtaining photographs of the Zodiacal Light from orbit on MA-9, in order to establish its origin, distribution and association with solar radiation and solar flares. Cooper observed the Zodiacal Light, which he described as a faint band concentrated along the ecliptic. He took photographs with exposure times ranging between 1 and 30 seconds, commencing immediately after orbital sunset. The photographs were underexposed and yielded no useful data.

21.10 AIRGLOW PHOTOGRAPHY

Airglow was another dim-light phenomenon that astronaut Cooper was tasked to photograph in the 5,577 ångström line during the flight of MA-9. The phenomenon is caused by the emission of light by gases in the upper atmosphere. Glenn, Carpenter and Schirra had all commented on a layer centred approximately 6–10 degrees above Earth's visible horizon. Setting stars were seen to disappear as they entered the layer and then reappear below it, before passing behind the planet's horizon.

Cooper's photographs were taken in an attempt to better understand the solar energy conversion processes that take place in Earth's upper atmosphere. Cooper took photographs using exposure times ranging from 10 to 120 seconds periodically throughout an entire orbital night-side pass. Using the filter resulted in the airglow layer appearing in the photographs while all other details of Earth's horizon were filtered out. The 15 resulting usable images were employed to determine the surface brightness of the airglow layer and the height of the layer. Cooper described the airglow layer as being whitish-green in colour and having the brightness of clouds illuminated by a quarter full Moon.

21.11 RADIATION DETECTION (FILM PACK)

Schirra on MA-8 and Cooper on MA-9 both wore badges containing film and lithium-fluoride thermoluminescent detectors to measure the electron dosage that they received during their flights. The detectors were mounted on the helmet of their pressure suits and at the chest and thigh of their undergarments. The badges revealed that the astronauts received less exposure during their flights than they did from background radiation during two weeks of their lives on Earth's surface.

21.12 RADIATION DETECTION (EMULSION PACK)

The MA-8 and MA-9 spacecraft carried emulsion packs at several internal locations. The packs produced data suggesting that a spacecraft was subjected to only very low levels of radiation.

21.13 RADIATION DETECTION (IONISATION CHAMBER)

An ionisation chamber was mounted on the inside of the side hatch of both the MA-8 and MA-9 spacecraft. This produced data suggesting low radiation levels.

21.14 RADIATION CHAMBER (GEIGER–MUELLER)

On 8 July 1962 the United States exploded a 1.4-megaton nuclear device in the upper atmosphere. Electrons from the explosion were trapped in Earth's magnetic field and formed an artificial radiation belt around the planet. The explosion was one of a series called Fishbowl, and the 8 July test was code-named Starfish Prime. By the time Schirra flew MA-8, on 3 October 1962, the radiation belt had disappeared.

The two Geiger–Mueller tubes mounted on the Retrograde Package of Cooper's MA-9 spacecraft were used to measure the electron flux in an anomaly in Earth's magnetic field over the Atlantic Ocean. One tube, which viewed along the spacecraft's roll axis, obtained good data. The second tube viewed a hemispherical area 40 degrees below the roll axis. The combination of the fact that this tube had been set to reject energy levels below 2.5 meV and the fact that the radiation belt from the Starfish Prime explosion had deteriorated resulted in this tube failing to be activated. Data was obtained during the spacecraft's seventh orbit. The radiation level was shown to have degraded significantly since the flight of MA-8.

21.15 TETHERED BALLOON

Carpenter's MA-7 spacecraft and Cooper's MA-9 spacecraft both carried a 76-cm balloon. The folded balloons were carried into orbit mounted in the Antenna Canister of both spacecraft. The experiment was initiated by jettisoning a cover by firing two small explosive charges. The balloon was then forced out of its housing by a compressed spring. The inflated balloon was attached to the spacecraft by a 30.4-m tether. The tether was attached to a strain gauge that would measure the drag on the balloon caused by the molecules in the thin upper atmosphere. The experiment failed on both occasions. A split seam meant that the MA-7 balloon failed to inflate fully, while the MA-9 balloon failed to deploy from its canister.

The gores on the balloons were coloured black, white, gold and silver, and the astronaut was intended to make observations on which colours were easiest to see at close quarters in the space environment. Carpenter stated that the gold and silver gores were the most visible. It may be coincidence, but the Apollo LM was coloured gold and silver.

21.16 LIQUID BEHAVIOUR IN MICROGRAVITY

MSC and LaFRC co-sponsored the liquid behaviour experiment. The experiment was carried on Carpenter's MA-7 spacecraft and was related to the possible refuelling of spacecraft in the future. The liquid was held in a spherical glass chamber and rested in the bottom under 1*g* conditions and acceleration forces. In the microgravity environment of orbital flight the liquid clung to the sides of the chamber. This was the expected result.

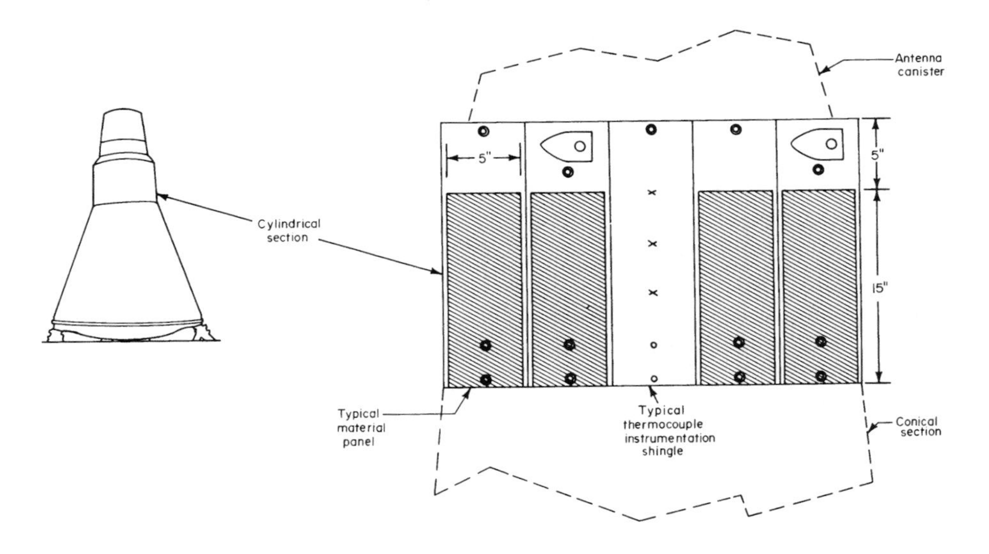

Fig. 109. Mercury Atlas 8 Ablative Materials Study. (NASA)

21.17 ABLATIVE MATERIALS STUDY

A selection of advanced ablative materials were flown on Schirra's MA-8 spacecraft. The materials were centred on 12.5 × 51-cm beryllium shingles mounted around the cylindrical neck of the spacecraft. The samples themselves were 38 cm long. Heat-sensitive paints were applied to the rear of the shingles as a means of measuring the temperatures that each sample was exposed to during re-entry. The shingles were removed after the flight and the ablative materials examined to see how they had withstood the flight. All samples showed increased charring towards the Antenna Canister. No one material proved superior to any other, but elastromeric materials did prove better than hard ablation materials. (See Fig. 109.)

21.18 MICROMETEORITE IMPACT STUDY

The MA-6, MA-7, MA-8 and MA-9 spacecraft were all examined after their flights for signs of micrometeorite impacts. No impacts were discovered on the first three spacecraft. A small crater was discovered during the post-flight inspection of the window on the MA-9 spacecraft. Although the crater displayed the characteristics of a hypervelocity impact it was impossible to determine if the window had been struck by a micrometeorite or by a piece of spacecraft debris during re-entry.

21.19 AEROMEDICAL CONSIDERATIONS

During the selection of an astronaut as the prime or back-up pilot on any space flight, the individual's medical records were thoroughly reviewed. Thereafter, the astronaut was subject to physical examinations prior to a number of training exercises in order to build up a baseline database against which medical readings taken during his flight could be compared (see Table 10). Both the prime and back-up pilot underwent thorough physical examinations at 10 and 3 days before launch. For one week prior to launch the astronauts ate only a specially controlled diet and 3 days before launch, and including any launch delays, they ate a low-residue diet in an attempt to reduce their need to defecate during the flight.

On launch day the primary astronaut underwent a brief physical examination to ensure his readiness to fly. He was then fitted with the biomedical sensors that would relay his physical condition back to Earth. Sensors relayed information on the astronaut's respiration rate, body temperature, chest movement and heart action (ECG). The data received was used by Flight Surgeons at the tracking stations to evaluate the pilot's condition and his ability to continue the flight beyond that point. The Flight Surgeon could then pass his recommendations to the Flight Director at either MCC, or Bermuda. Along with the Capsule Communicator (always an astronaut) and the Flight Director, the Flight Surgeon at any of the tracking stations retained the right to talk directly to the astronaut in flight. This right was very rarely exercised during Project Mercury.

Respiration rate was measured by means of a thermistor on the helmet microphone pedestal on early flights. When this proved unreliable it was replaced by an impedance pneumograph on the last two flights. This instrument gave excellent data.

Body temperature was originally registered by a rectal thermistor, which was shown to give reliable information. On the longer flight of MA-9 the rectal thermistor was replaced by an oral thermistor. Two electrocardiographic sensors stuck to the astronaut's chest recorded his heart rate. Pulse rates were recorded throughout each flight as a physiological response to the conditions of that flight. Blood pressure was not recorded in-flight until MA-6, at which time the equipment consisted of an un-directional microphone and an inflatable arm cuff. Following

Table 10: Project Mercury astronaut's in-flight pulse rates

Flight	Launch	Microgravity	Re-entry
MA-3	138	108–125	132
MA-4	162	150–160	171
MA-6	114	88–114	134
MA-7	96	60–94	104
MA-8	112	56–121	104
MA-9	144	50–60 (sleep) 80–100 (awake)	184

elevated readings on MA-7, as a result of badly calibrated equipment, excellent results were achieved on MA-8 and MA-9.

Voice communications, both in real time and recorded on magnetic tape, were used to evaluate the pilot's physiological response to the conditions of space flight.

Project Mercury proved that trained individuals could endure the many different conditions experienced during space flight. These included high acceleration forces during launch, microgravity and high deceleration forces during re-entry. Contrary to many early fears, all six Project Mercury pilots found that they were able to perform in space as well as they were in high-performance aeronautical flight. Astronauts were able to make observations, monitor and even override the systems of their spacecraft. They also proved that the more-outlandish worries of some physicians were unfounded; sleeping, eating, and urinating were all performed in the microgravity environment as they are on Earth. None of the physiological effects highlighted during Project Mercury threatened the 8–14 flights planned for Project Gemini and Project Apollo.

22

Recovering Liberty Bell 7

Liberty Bell 7, Mercury Spacecraft 11, was Gus Grissom's spacecraft. He flew it on the second manned suborbital flight of Project Mercury, Mercury Redstone 4, on 21 July 1961. Following a successful flight the spacecraft landed in the Atlantic Ocean. As Grissom recorded the final instrument readings the side hatch prematurely jettisoned and the spacecraft began filling with seawater. The astronaut had to evacuate the spacecraft and swim for his life. Although the prime recovery helicopter succeeded in lifting the water-filled spacecraft clear of the ocean, a faulty instrument reading on the flight deck resulted in the spacecraft being dropped back into the water and left to sink. A second helicopter recovered Grissom and carried him to the recovery vessel, USS *Randolph*.

On 19 April 1999 an expedition led by Curt Newport and financed by the Discovery Channel, sailed from Port Canaveral, Florida, with the intention of locating and recovering Liberty Bell 7. Newport was a highly experienced diver who had previously worked on the recovery of wreckage from the STS-51L Challenger explosion in 1986 and the recovery of artefacts from the Royal Mail Steamer *Titanic*. He had made two earlier attempts to locate Liberty Bell 7, in 1992 and 1993, both of which were unsuccessful. In January 1999 Newport had formed his own company, Liberty Bell Incorporated, with the sole aim of locating and recovering the spacecraft.

Newport's reasons for wanting to recover the lost spacecraft were simple. He told a press conference, '*Outside of Challenger* [STS-51L], *this is the only one we haven't gotten back. It's the right thing to do. It's patriotic. It was one of ours.*' At the time of the third recovery attempt interested parties were able to e-mail Newport's vessel *Needham Tide*. When one person asked if Newport intended to recover Liberty Bell 7, even if the spacecraft was damaged in the process, he showed little concern for preserving a national historical artefact intact, rather than preserving his own reputation. Newport replied, '*Yes, if we find the capsule, it is coming to the surface no matter what!*'

Following his Mercury flight Grissom went on to command Gemini Titan III before being assigned as Commander of Saturn Apollo 204, Apollo 1 and dying alongside Ed White and Roger Chaffee in the Apollo fire on 27 January 1967. NASA had put the Gemini III Re-entry Module on permanent display in the town of Mitchell, Indiana, where Grissom and his wife Betty had grown up and courted, before getting married. Betty Grissom successfully fought an attempt by the Kansas Cosmosphere and Space Centre to obtain the Gemini III spacecraft and it remains on display in Mitchell. A few days before *Needham Tide* sailed from Port Canaveral, Betty Grissom told a journalist from the *Houston Chronicle* newspaper, '*I hope that they do not find it. I'm keeping my fingers crossed that something happens that they don't.*'

Needham Tide arrived at the search area on 21 April but the robotic equipment that would be used for the search presented a series of problems, and it was 25 April before the Ocean Explorer 2000 'towfish' was lowered the 4.8 km to the floor of the Atlantic Ocean. Connected to *Needham Tide* by a steel wrapped communication umbilical, the towfish contained a side scanning radar. As *Needham Tide* travelled up and down the search area in parallel strips, the radar sent out regular signals. The ocean floor absorbed much of the signal, but metal objects did not; therefore, metal objects reflected a stronger signal than the seabed. Signals were displayed on monitors in an operations room that had been established in a cargo container bolted to the deck on *Needham Tide*'s fantail.

After a week the towfish was recovered and replaced in the water by the submersible 'Magellan'. A pilot in the operations room 'flew' Magellan by using the pictures transmitted by its cameras. Signals were transmitted to and from Magellan via its communication umbilical. Using the radar data supplied by the towfish, the pilot could guide Magellan to each of the 88 'targets' that the towfish had identified in turn, in an attempt to locate Liberty Bell 7.

By 1 May the weather had deteriorated to the point that the ocean's surface was too rough to recover the spacecraft. Five kilometres down the water was smoother, so it was decided fly Magellan to the first target. The pilot picked up a debris trail and many on *Needham Tide* thought that the target would turn out to be the wreckage of a light aircraft that was known to have crashed inside the search area. To everyone's surprise Magellan's cameras showed something sitting upright on the ocean floor. Closer inspection showed it to be Grissom's lost spacecraft.

The small black Mercury Spacecraft was sitting on the sea floor, its wide end surrounded by the remains of its heat shield, which had disintegrated since it had sunk 38 years previously. The straps that had held the heat shield in place when the landing skirt was deployed were clearly visible. Nothing appeared to remain of the canvas landing skirt itself. The periscope door was open and the periscope was deployed. White upper-case letters spelled out the words 'UNITED STATES'. The white representation of the crack in the original Liberty Bell was also visible. The titanium airframe had not corroded, but the aluminium around the neck of the Recovery Section had. Views through the open side hatch showed the instrument panel in its correct position, but covered in crustacean. The name 'LIBERTY BELL' was clearly visible next to the side hatch and above it the picture window was unbroken. (See Fig. 110.)

Success was quickly followed by disaster. The large waves at the ocean's surface caused Magellan's umbilical to twist. Ultimately the umbilical broke and Magellan

Fig. 110. Thirty-eight years after it sank Liberty Bell 7 is returned to Port Canaveral. Guenter Wendt, McDonnell's pad leader, addresses the assembled audience. (Joel Powell)

sank to the ocean floor. Newport summed up the day in just six words: '*We went from triumph to tragedy.*' *Needham Tide* returned to Port Canaveral on 4 May. Newport was optimistic that he would be able to return and recover Liberty Bell 7, but not for several weeks. He told a press conference, '*I don't think this is going to drag into months.*' Meanwhile, Betty Grissom had told a *Florida Today* journalist, '*It's okay to bring it up, but it should be displayed whatever it looks like. If there's a dent, I want to see a dent.*'

Just five weeks later Newport was ready to try again. The vessel *Ocean Project* sailed from Port Canaveral on 5 July. In the intervening period a replacement for Magellan had been constructed and named 'Ocean Discovery'. It was now lashed to the fantail of *Ocean Project*. The vessel also carried three special guests: Guenter Wendt, the pad leader at Launch Complex 5, who had overseen Grissom's launch; Jim Lewis, who had been the pilot of Hunt Club 1, the Marine Corps helicopter that had tried to recover Liberty Bell 7; and was Max Ary, the Director of the Kansas Cosmosphere, where the spacecraft would be preserved and displayed if it was recovered.

Newport elected to search for Magellan first. If it could be recovered, its lights would be used during the attempt to recover Liberty Bell 7, to allow the recovery attempt to be televised live. The search for Magellan would be carried out within a defined time, after which it would be abandoned and the attempt to recover Liberty Bell 7 would begin. Between 6 and 9 July, Ocean Discovery made the 5-hour journey to the seabed twice and was lowered into the water on a third occasion, only to suffer technical difficulties that required its recovery without any attempt being made to search for Magellan. On the fourth attempt the search began, but there were only 11 hours before Newport's self-imposed deadline. Magellan had not been located when the search was abandoned at midnight on 9 July. Ocean Discovery was winched aboard *Ocean Project* and the latter moved to a new location above Liberty Bell 7.

Despite *Ocean Project*'s location being set by Global Positioning Satellite data it took a further 20 hours of searching before Liberty Bell 7 was relocated on the ocean floor. Ocean Discovery carried out a complete photographic survey of the spacecraft before screwing the first of three clamps to the top of the Recovery Section. When the data link to Ocean Discovery was lost, the remote-operated vehicle was winched back to the surface and recovered. Over the next 48 hours Ocean Discovery was repaired and made two further 12-hour trips to the ocean floor and back. On the first dive two more clamps were screwed to the Recovery Section at intervals of approximately 120 degrees around its circumference. During the second dive a three-point lifting harness was attached to the three clamps. Before returning to the surface at the close of the second dive Ocean Discovery's mechanical 'hand' was used to grasp a black metal cylinder lying in the mud next to the spacecraft. This was the container that had held the green dye that had been released into the sea following landing on 21 July 1961. Ocean Discovery brought the dye float to the surface, where it was placed in a container full of seawater to preserve it.

At the end of the 12 July dive, Ocean Discovery suffered further technical difficulties that delayed the next dive until 14 July. On that date the 5-hour descent to the seabed was followed by another malfunction and an abandoned dive. Finally on the night of 18–19 July Ocean Discovery carried the Kevlar recovery line down to the ocean floor and was then used to attach the line to the three-point harness that had previously been attached to Liberty Bell 7's Recovery Section. When Ocean Discovery was recovered it brought the other end of the Kevlar line back to the surface with it.

The line was fitted to *Ocean Project*'s fantail winch and Liberty Bell 7 began the 6-hour journey back to the surface. The spacecraft was pulled up through the water at a speed that was sufficiently slow to prevent objects falling out through the open side hatch and being lost. The spacecraft was lifted on to *Ocean Project*'s fantail, but only after most of the crew had been evacuated from that area. Two pyrotechnics experts were the first to examine Liberty Bell 7, locating an unexploded Sofar bomb in the Recovery Section and throwing it back into the ocean.

Ocean Project returned to Port Canaveral on 21 July. The spacecraft, in its white protective container, was hoisted off the ship and onto a flatbed truck. The top section of the container was removed to show the spacecraft and Newport addressed the crowd that had turned out to welcome Liberty Bell 7 home. He told them, '*I feel somewhat vindicated. A lot of people told me this couldn't be done.*' Grissom's brother Lowell was present and described the recovery as '*tremendous*'. When the press conference was over, Liberty Bell 7 began her overland journey to the Kansas Cosmosphere and Space Centre.

At the Kansas Cosmosphere Liberty Bell 7 was carefully emptied of 265 litres of silt and mud accumulated over its 38 years on the ocean floor. Fifty-two dimes were found in the spacecraft, a reminder of the rolls of coins that Grissom had carried on his suborbital flight in order to give away to friends and space programme employees as souvenirs. Some of them had initials scratched on them suggesting that some at least had been handed to Grissom by Project Mercury workers who had asked him to carry them. The survival knife that Grissom had been thinking about taking as a souvenir was among the items removed from the silt, as was the notepad on which he had been recording the final instrument readings and switch settings when the side hatch blew off and the spacecraft began taking on water. The notations that he made on the pad with a grease pencil in those final moments were still readable.

Grissom's life raft was where he left it when he evacuated the sinking spacecraft. It was found to be in good condition and even held air when it was pressurised. The astronaut's survival kit was found to still contain shark repellent and a vacuum-packed bar of soap. The tape that had recorded Grissom's voice communication was still on the reel, which was still in place on the tape recorder. A plastic cup and the remains of a cigarette filter showed that some McDonnell workers were not as careful as they might have been when working inside the spacecraft

With the silt removed, Liberty Bell 7 was constantly sprayed with fresh water for two months, in an attempt to wash away the salt that would corrode it now that it was no longer in the ocean. The spacecraft's outer shingles were then removed and hung on the wall in the order in which they had been fitted. One shingle revealed a patch welded into place over a hole that must have been made in it accidentally by a McDonnell employee. The muslin thermal insulation between the spacecraft's inner and outer shells was found to be in excellent condition with its stitches still tight.

The titanium strip along one side of the side hatch opening was buckled. The hatch sill, on the lower edge as Liberty Bell 7 rested in the ocean, was bowed outward by 1.2 cm. Staff at the Kansas Cosmosphere suggested the theory that the impact forces of landing in the ocean may have popped free a few of the bolts holding the hatch in place along its lower edge. The bolts on the remaining three sides may have popped free a few at a time as Liberty Bell 7 rode the ocean swell, until the hatch finally fell off. They stressed that it was only a theory and could not be proved. On the other hand, they pointed to the fact that there were no burn marks from the explosive cord that should have been fired to jettison the hatch under normal recovery conditions, suggesting that, had the explosive cord been ignited, the titanium strip should have been deformed around all four sides of the hatch, rather than just along the lower edge. The hatch was not located on the ocean floor during Newport's recovery effort.

The spacecraft's interior was stripped and the individual parts were placed in plastic boxes full of fresh water until they could be cleaned and stabilised. When the spacecraft was reassembled, corroded or missing parts were replaced by Plexiglas. Following the restoration of its individual parts Liberty Bell 7 was reconstructed. It then began a tour of North America, after which it was placed on permanent public display in the Kansas Cosmosphere.

23

Project Mercury in retrospect

From its official authorisation, on 7 October 1958, to the recovery of the Mercury Spacecraft Faith 7 on 16 May 1963, Project Mercury lasted 4 years 7 months and 9 days. Conceived as it was in the period of American self-recrimination in the wake of Sputnik, Project Mercury was a weapon in the Cold War. Having been beaten to place the first satellite into orbit by their political and military enemy, the Soviet Union, America aimed to place the first man in orbit and thereby reinstate their technological supremacy in the eyes of the non-aligned nations of the world.

From the beginning, neither President Eisenhower nor T. Keith Glennan, the first NASA Administrator, believed that America should be drawn into a 'Space Race' with the Soviet Union. Both men believed that the Soviets would stop their efforts to explore space once America placed the first man into orbit. America's own manned space programme would then be cancelled, when the Soviets conceded the race. It was not to be.

Eisenhower appeared to fail to understand the propaganda value of the space programme. Nor did he appear to understand the need felt by the American people to be shown that their education system, scientists, engineers, technology and especially their ICBMs, were at least equal to and preferably better than those possessed by the Soviets. In failing to react immediately to the Sputnik launch by greatly enlarging America's efforts to explore space, Eisenhower's voters thought that the old man had failed to understand the threat that the large Soviet missiles posed to their country.

In fact Eisenhower had U-2 spy plane reconnaissance that showed the R-7 missile to be a ponderous machine that required a large concrete launch complex and several hours to prepare for launch. His intelligence also showed that the Soviets did not have R-7s in large numbers. The secret nature of the information meant that he could not tell the American public why he was so sure that Sputnik did not represent a threat to their country and why, therefore, there was no need for a largely accelerated space programme. The President was also aware of Project Corona, an attempt to build a reconnaissance satellite that would photograph the Soviet Union in detail from orbit. He did increase funding to Project Corona following the launch of Sputnik, but the classified nature of the project meant that he could not make the American people aware of its existence.

When public and political pressure demanded that the American effort in space be accelerated the Commander in Chief insisted that space exploration be taken out of the hands of the military and given to a civil administration that would attract the participation of the world's best scientists. The American civil space programme was to be carried out in full public view while drawing attention away from the military, which then left free to concentrate on passive information gathering in its own secret space programme.

The 1961 Presidential election was partly fought on the false premise that a 'missile gap' existed between Soviet and American missile capabilities. Even so, the election result was very close. Kennedy, the new President, initially distanced himself from Project Mercury, but his attitude changed when the first man in space was a Soviet cosmonaut and not an American astronaut. The Soviet cosmonaut's launch vehicle was more powerful than the missiles available to the Americans for Project Mercury. His Vostok Spacecraft was heavier than the Mercury Spacecraft, if not as sophisticated. It was also capable of carrying more consumables and therefore capable of longer flights than its American counterpart. Having taken advice and sought alternatives, Kennedy committed America to a manned landing on the moon before the end of 1969.

Despite denials from the Political Administration and the new NASA Administrator that Project Mercury would be accelerated in the wake of Soviet achievements, such acceleration did take place. Originally, all seven Flight A astronauts were scheduled to make a suborbital Mercury Redstone flight before they progressed to making an orbital flight on a Mercury Atlas, 14 manned flights in all. Following only two manned suborbital flights the Mercury Redstone programme was cancelled. The two pilots who had made those flights did not make orbital flights during Project Mercury.

Compared at face value to the Vostok programme, Project Mercury does not look very spectacular. But while NASA was left to run Project Mercury largely unmolested by the White House, Korolev's Vostok programme was subject to continual demands for new propaganda 'firsts' from Soviet Premier Nikita Khrushchev. Perhaps Khrushchev's ultimate propaganda coup, after Gagarin's flight, was the flight of Valentina Tereshkova, the first woman in space. The young woman was in space 17 hours longer than the total flight time accumulated by the six American astronauts who flew in Project Mercury. At the end of the Vostok programme the Soviets had 13.5 man-days more experience in space than the Americans.

Within its own technological limitations Project Mercury was a great success. In 55 months manned space flight had gone from theoretically possible to a technology with which some elements of the general public were already becoming blasé. Mercury Atlas 6 achieved project Mercury's original goal on 20 February 1962, that was 40 months after the Project had begun and 22 months behind its original schedule. A second astronaut repeated the flight in May of the same year. In October 1962 the flight duration was doubled and May 1963 saw a 34-hour, 22-orbit flight that had not been part of the original flight schedule. Although each manned Mercury flight set new records, at the close of Project Mercury the Soviet Union's Vostok cosmonauts held all of the internationally recognized manned space flight records.

The four manned orbital flights in Project Mercury carried a total of 17 scientific experiments, some of which were carried on more than one flight. The experiments were split into four groups: Earth Observations, Photographic, Radiation and Miscellaneous Studies. Twenty-five experiments were completed successfully, three were partially successful, five were attempted but were unsuccessful and two were not attempted due to unexpected flight events.

Many of Project Mercury's achievements took place on the ground, rather than in space. The design of a high-drag low-lift spacecraft with an ablative heat shield, along with the astronaut's prone couch, and a rocket-powered Launch Escape System, were all achieved by NACA engineers before NASA was formed. The man-rating of two military missiles was the direct result of NASA–military cooperation, as was much on the operational side of the project.

The WWTN and the MSC's Operations Division were established to support Project Mercury. The tracking network would continue to support American manned space flight through Project Gemini, Project Apollo, Project Skylab, the Apollo–Soyuz Test Project and even the early flights of the Space Shuttle. In the 1980s the Tracking and Data Relay Satellite System replaced the WWTN. The Flight Operations Division continues to support Shuttle flights from the ground and still applies the principles that were established by Charles Mathews, Walt Williams, Christopher Kraft and the many dedicated engineers that worked under them.

Lessons were learned from the design of the Mercury Spacecraft that were applied to all later NASA spacecraft. Those lessons made the new spacecraft easier to service and prepare for launch.

In flight, the six Flight A astronauts who flew Mercury Spacecraft proved that a human astronaut could succeed where an automatic machine would have failed. On all four manned orbital flights the astronaut had to use manual control to overcome malfunctions in the ASCS. Only one of those four astronauts was able to rely on the automatic system to control his spacecraft throughout re-entry. Yet, it was only the astronauts' belief in their own capabilities when they joined Project Mercury that led to the manual flight control systems being installed in the Mercury Spacecraft at all. The STG/MSC and McDonnell engineers had originally wanted the astronauts to rely completely on the automatic systems. The success of the astronauts' operations during Project Mercury led to all future American manned spacecraft being fitted with a manual flight control system.

Project Mercury's true legacy lies not in its 53 hours 55 minutes 27 seconds of manned space flight, but in the infrastructure that it constructed and on which NASA has established the past four decades of America's civil manned space programme.

24

Project Mercury as a first step

The Mercury Spacecraft was a ballistic vehicle; that is to say, its external shape generated practically no lift as it passed through the atmosphere. Like a missile warhead the spacecraft was aerodynamically shaped to help it pass out of the atmosphere. It was not capable of leaving the ground without the thrust produced by its launch vehicle, and in the event of a launch failure an escape rocket would pull the spacecraft to an altitude where it could be safely dumped in the ocean. During such an abort the astronaut could expect to be subjected to an incapacitating 20*g*.

Throughout the flight the astronaut lay strapped tightly in his prone couch dressed in a pressure suit and surrounded by the spacecraft's instrument panels at less than an arm's length away from his face. During launch and re-entry he was subjected to high acceleration and deceleration forces. He was forced to perform all bodily functions within his pressure suit, because there was insufficient room to allow him to take it off. Most of his food was puréed and packaged in tubes, from which it was squeezed into his mouth. Alternatively food was in small bite-size chunks. Drinking water was taken directly from the hose through which it was pumped from the storage tank.

The astronaut breathed 100 per cent oxygen because it was easier to handle than a mixed gas atmosphere. The risks of a fatal fire happening as a result of this decision were assessed and considered to be worth taking. So was the fact that the spacecraft was fitted with a side hatch that contained a round of explosive cord that was designed to blow it off during recovery operations and might therefore be blown off by accident while the spacecraft was in orbit.

A high-drag low-lift design slowed the spacecraft down during re-entry and caused a bow wave to form in front of it which carried away most of the heat produced. A heat shield on the spacecraft's wide end ablated to carry away the heat produced by the friction caused by entry through the atmosphere. (See Fig. 111.) Landing occurred in the oceans of the world because they were vast and easy to hit. The water also served to cool the spacecraft quickly after re-entry. The mid-ocean landing meant, however, that large recovery fleets had to be available to retrieve the spacecraft and its pilot wherever they landed.

The launch vehicles and spacecraft flown during Project Mercury were used only

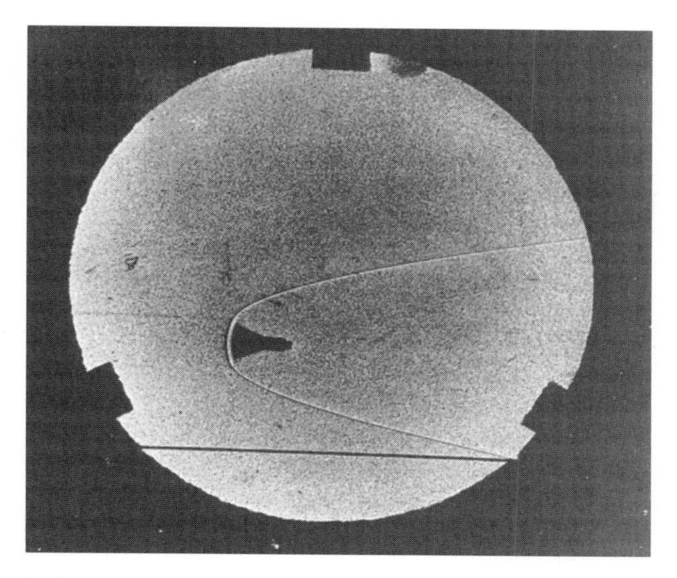

Fig. 111. A shadowgraph image of a Mercury Spacecraft model shows the bow wave formed by its high drag shape. (NASA)

once and then became museum pieces. Each flight required new hardware. In short, Project Mercury represented the 'quick and dirty' approach to manned space flight made possible by the advent of the first ICBMs. The first manned space flights came about as a result of the political situation of their era. The International Geophysical Year presented the opportunity to launch the first Earth satellite. When the Soviets did so before the Americans the race was on to place the first human being into space. The Soviets did that first too. They also used a ballistic spacecraft, which they called Vostok, although its design was considerably different to that of the Mercury Spacecraft. The paranoia of their political leaders dictated that they land their spacecraft on land, inside the Soviet Union. The loads involved in such a landing demanded that the pilot eject from the spacecraft and land under a personal parachute.

24.1 SMALL STEPS AND GIANT LEAPS

Having lost the race to place the first man in space President Kennedy committed NASA to an 8-year programme to land men on the Moon. NASA had already commenced a programme to define a three-man spacecraft capable of circumlunar flight, which they had called Apollo. Project Apollo was redefined to meet Kennedy's goal. In time Project Gemini was introduced to fly after Project Mercury, but before Project Apollo, with the goal of exploring orbital rendezvous and extravehicular activity.

Both the Gemini Re-entry Module and the Apollo CM were ballistic vehicles, although a certain amount of lift was designed into their external shape. Both were conical in shape, like the Mercury Spacecraft, with an ablative heat shield on their wide end. Both vehicles used 100 per cent oxygen in their Crew Compartments and both were dropped into the ocean under parachutes at the end of their flights and were recovered using principles established for Project Mercury. Each Gemini and Apollo Spacecraft made only one flight and both used one-shot launch vehicles.

Gemini and Apollo were second-generation spacecraft and contained modular design principles. The re-entry vehicle contained everything necessary to support the crew and allow them to complete their flight. Lessons learned on the Mercury Spacecraft ensured that as much equipment as possible was placed outside the Crew Compartment and was accessible through external access panels. Anything not required for re-entry was placed outside the re-entry vehicle.

A two-piece Adapter Section was mounted behind the Gemini Re-entry Module. The Equipment Module held the main power supply (batteries or fuel cells depending on the flight), propulsion systems and high-pressure gas bottles. The Equipment Module was jettisoned at the end of the flight to expose the four solid propellant retrograde rockets housed in a separate Retrograde Module. Following retrofire the Retrograde Module was jettisoned to expose the Re-entry Module's heat shield. The experiences that the Flight A astronauts had had in manually controlling the attitude of their spacecraft led to the astronauts being given manual control of both the Gemini and Apollo Spacecraft.

The Gemini Spacecraft's rendezvous radar interrogated a transponder in the Gemini Agena Target Vehicle (GATV) and displayed range and rate data to the crew. The astronauts then manually fired the spacecraft's Orbital Attitude Manoeuvring System (OAMS) thrusters to change their orbit. (See Fig. 112.) Gemini docked to the GATV by driving its Re-entry and Recovery Section into the Target Docking Adapter (TDA) on the GATV. The GATV was an unmanned rocket stage with the TDA attached to the opposite end of the Primary Propulsion System. An umbilical was then extended from the Gemini Spacecraft that allowed the crew to send commands to the GATV's propulsion systems. Undocking was performed by first retracting the umbilical and the docking latches before manoeuvring the Gemini Spacecraft away from the GATV to provide positive separation. (See Fig. 113.)

Gemini astronauts performed extravehicular activity (EVA) by first preparing their pressure suits and associated EVA equipment, before closing off the oxygen supply to the Crew Compartment and dumping the oxygen already in the Crew Compartment overboard. The pilot then opened the hatch above his head and drifted out of the spacecraft. He remained attached to the spacecraft by an umbilical that carried his oxygen, cooling water, radio communication and a safety tether. Having completed his EVA tasks the process was reversed.

The first manned Gemini flight flew in March 1965. On the third and final orbit Command Pilot Gus Grissom became the first person to change his spacecraft's orbital trajectory. Gemini IV flew in June and included the first failed attempt at orbital station keeping and the first American EVA. The flight lasted 4 days and was the first to be controlled from the Mission Operations Control Room at MSC in

Fig. 112. Gemini Spacecraft 7 in orbit, taken from Gemini 6. (NASA)

Fig. 113. A Gemini Agena Target Vehicle in orbit. (NASA)

Houston, Texas. Gordon Cooper commanded the Gemini V flight in August. Although the planned rendezvous experiment had to be considerably changed to meet real-time conditions, it was successful. The 8-day flight duration was considered the minimum required for an Apollo lunar landing flight. Gemini VI was to have made the first attempt to rendezvous and dock with a GATV, but was cancelled when the GATV failed to achieve orbit.

The decision was made to rendezvous Gemini VI with Gemini VII and the flight was redesignated Gemini VIA. Gemini VII was launched on a 14-day medical flight in December. Eleven days later Gemini VIA was launched, after a second aborted launch attempt a few days earlier. Wally Schirra flew his spacecraft to the first orbital rendezvous, while Gemini VII was a passive target. Gemini VIA landed the next day. The 14-day duration of the Gemini VII flight was the same as that proposed for later lunar landing flights of Project Apollo.

The first manned flight of 1966 was Gemini VIII, in March. The crew completed the first rendezvous and docking with a GATV, but had to abort their flight when a manoeuvring thruster stuck on and the crew had to initiate an emergency procedure to bring the spacecraft under control. The pilot's EVA was not completed. Gemini IX through Gemini XI all included rendezvous and docking attempts, and EVA, although the latter proved far more difficult than many NASA engineers had anticipated. The final Gemini flight, Gemini XII, completed a series of rendezvous and dockings with its GATV in December. The pilot used new techniques to conquer orbital EVA and set a new EVA endurance record.

In the Apollo Spacecraft the Service Module (SM) contained the main power supply (fuel cells), propulsion system and pressurised gas bottles. The Command and Service Modules functioned as one vehicle throughout the flight until a point just prior to re-entry. At that point the SM was jettisoned and burned up in the atmosphere. The CM turned heat shield forward and was protected as it passed through the atmosphere. Original plans called for the CSM to land on the Moon and support the surface activities before lifting off again for the return to Earth.

When the Lunar Module (LM) (the word Excursion had been dropped from the name) was introduced into Project Apollo it was designed to fly only in the vacuum and free-fall of space and in the Moon's weak gravitational field. It was protected as it passed out of Earth's atmosphere and was not required to survive re-entry. The spacecraft that resulted from that unique programme was designed from the inside out. It carried no aerodynamic shaping, no heat shield and was incapable of supporting its own weight on Earth.

The two Apollo Spacecraft performed rendezvous using similar methods to those employed in Project Gemini. The docking system consisted of a probe at the apex of the CM and a drogue in the roof of the LM. The probe was extended prior to docking. The Command Module pilot then manoeuvred the CSM so that the probe slipped into the drogue and activated the spring-loaded capture latches in the head of the probe. By retracting the probe the LM was pulled towards toward the CM until contact was made. An airtight seal then formed and a series of docking latches closed to secure the two spacecraft together with an internal tunnel between them. Access between the two spacecraft required the LM and the tunnel to be pressurised. The crew then opened a hatch in the apex of the CM Crew Compartment, removed the probe and drogue and opened a second hatch in the roof of the LM Crew Compartment. In order to undock the LM, the probe and drogue was reinstalled and both tunnel hatches sealed. The probe was then extended and the CSM was manoeuvred away from the LM to provide positive separation.

When 1967 dawned the first manned flight of Project Apollo was set for 28 February. It was to be an orbital flight-test of the CSM (see Fig. 114). On 27 January Gus Grissom and his crew entered their spacecraft as it stood on its Saturn IB launch vehicle at Launch Complex 34, Cape Kennedy. After a day of continuous difficulties, during the simulated launch a fire broke out in the Crew Compartment's pure oxygen atmosphere. The crew was asphyxiated within 14 seconds. Project Apollo was delayed for 18 months while the cause of the fire was investigated and the CM was redesigned. Meanwhile, unmanned flight-tests of the Saturn IB and Saturn V launch vehicles continued. The first of two unmanned Saturn V flight-tests was launched on 27 November 1967.

The first manned Apollo flight, Apollo VII, finally took place in October 1968, with Wally Schirra as Commander. The second manned flight was the first manned launch of the Saturn V. With the LM behind schedule, the flight of Apollo VIII took

Fig. 114. Apollo Command and Service Module in lunar orbit. (NASA)

a manned CSM into lunar orbit over Christmas 1968. The flight was a great success and America reaped the propaganda return.

Apollo IX flight-tested the LM in Earth orbit, in March 1969, before Apollo X carried out similar flight-tests in lunar orbit two months later. Finally, in July 1969, five months before President Kennedy's 'decade' ran out, Apollo XI carried the first two men to the surface of the Moon and returned them safely to Earth. Apollo XII made a second lunar landing in November.

Apollo XIII's Service Module exploded on the way to the Moon. The third manned lunar landing was abandoned and the LM became a 'lifeboat', providing everything vital to bringing the three astronauts back to Earth. The LM was jettisoned as Apollo XIII approached the planet's atmosphere and the CM made a normal re-entry. The project was delayed again while the cause of the explosion was investigated. Many people consider the safe recovery of the Apollo XIII crew to be the highlight of the project – NASA's finest hour.

Apollo IV was commanded by Alan Shepard, the only Flight A astronaut to reach the surface of the Moon (Fig. 115). The flight was flown over the end of January and the beginning of February 1971. The crew made two lunar surface EVAs. Apollo XV followed in July. This was the first of three J-flights, with extended consumables, a Manned Lunar Roving Vehicle and three 7-hour lunar surface EVAs.

Fig. 115. In February 1971 Alan Shepard became the only Flight A astronaut to walk on the Moon during Project Apollo. (NASA)

The SM also contained a Scientific Instrument Module to study the Moon from orbit, which required the Command Module Pilot to make an EVA, to recover the film cassettes from two cameras, during the flight back to Earth. The last two Apollo flights, XVI and XVII, flew in April and December 1972 respectively. Both were J-flights and followed the same flight profile as Apollo XV.

With the end of Project Apollo NASA turned its attention to Project Skylab, a prototype space station built inside a converted S-IVB rocket stage. The station was launched in June 1973 and visited by three three-man crews in Apollo CSMs launched by Saturn IB launch vehicles throughout 1973–74.

In July 1975 the last Apollo CSM was launched by a Saturn IB to rendezvous and dock with the Soviet Soyuz 19 in Earth orbit. After years of selecting other astronaut crews and watching them fly in space, Flight A astronaut Deke Slayton finally got his first flight, as Docking Module pilot on the American ASTP crew. The two crews worked together for several days before returning to Earth. This was America's final manned ballistic spacecraft.

The ballistic spacecraft and the 'quick and dirty' approach to manned space flight pioneered by Project Mercury had served NASA well and carried American astronauts to the Moon and back. They also gave America its first prototype space station in Earth orbit, before bringing the first round of the 'Space Race' to a close in a joint flight with a Soviet Soyuz Spacecraft, in 1975.

25

Alternative paths

25.1 VON BRAUN'S GRAND SCHEME

In 1952 the German rocket pioneer Werner von Braun published two books that explained his vision of how space would be explored. In *Across the Space Frontier* and *Man on the Moon* there was no place for a one-man ballistic spacecraft and the 'quick and dirty' approach to space flight.

Von Braun believed that the first stage of space exploration would be the development of a three-stage launch vehicle, to launch 'instrument heads' (satellites) into orbit. The second stage of the exploration would involve the development of a seven-man, winged spacecraft that would lift-off vertically as the third stage of a rocket launch vehicle, fly in orbit and land like an aircraft. Having expended its propellant supply achieving orbit, the manned spacecraft would rendezvous with an unmanned 'tanker vehicle' which would follow it into space. Von Braun believed that rendezvous would be a simple matter of pointing the active vehicle at its target and thrusting towards it. He was wrong! With rendezvous complete and docking not planned, two astronauts would perform an EVA to transfer propellants from the tanker vehicle to the manned spacecraft. The astronauts would return to their spacecraft and the empty tanker vehicle would be left drifting in space. Following the refuelling operation the astronauts would perform the tasks for which they had been sent into orbit. For re-entry the crew would perform retrofire and then turn their spacecraft upside-down so that its wings would produce 'negative lift' as it entered the atmosphere. The crew would turn the spacecraft right-side-up for the final landing, which would be on a standard runway in the style of a large aircraft.

After initial flights the manned spacecraft would be used to construct a huge circular space station, with a central hub and three spokes, in orbit 1,609 km above Earth. The manned spacecraft would rendezvous with 'cargo rockets' containing the materials required for constructing the space station. In an alternative design the tanker and cargo rockets were combined to provide a joint vehicle that could carry both propellant for the manned spacecraft and cargo for the space station into orbit. Von Braun described how the first two stages of each launch vehicle would fall into the ocean, where they would be recovered by specially designed ships. The stages were to be refurbished and relaunched in 10 days.

In von Braun's view the construction of the 76.2-m diameter space station would require a 10-year programme of manned flights into space. The station's outer ring would be constructed from 20 sections of 'flexible nylon plastic'. The sections would be transported into orbit in a collapsed state, inflated and pressurised and connected to the station. The hub and three spokes, each with an elevator, would be constructed within the outer circle. Spinning the station about its axis would induce centrifugal force in the outer ring, while the central hub remained subject to microgravity. Counter-rotating airlocks in the central hub would allow for the docking of spacecraft to the space station.

In the Cold War atmosphere of the time, von Braun presumed that his space station, with its 80 staff, would be used as a military observation platform, a weapons platform, a scientific laboratory and a staging post for longer journeys, e.g. to the Moon. He also presumed that the space station would be completed by 1967 and that the first manned lunar landing flight would take place by 1978.

The rocket pioneer saw the lunar spacecraft being constructed in orbit alongside the space station. The lunar landing expedition would be preceded by a manned 'survey vehicle' to photograph potential landing sites. Starting from the space station's orbit, the survey vehicle would be launched on a circumlunar flight, behind the Moon and back to rendezvous with the space station 10 days later.

For the lunar landing three vehicles would be constructed. The refurbished survey vehicle and its twin would each be manned with a crew of 25 people. The third spacecraft, a cargo vehicle, would be unmanned. All three spacecraft would be of skeletal design with the Crew Compartments at one end. The spacecraft were 48.6 m tall and 33.5 m across their extended landing gear. Each one would have 36 individual 567-newton thrust engines at its base and 12 'hinged rockets', three on each side for attitude control.

Each manned vehicle would have four large spherical propellant tanks to fuel the trans-lunar burn. Once the spacecraft were on their way to the Moon astronauts performing an extravehicular activity would manually jettison the propellant tanks. Ten further tanks would hold the propellant for the lunar landing, lift-off and braking back into the space station's orbit at the end of the flight. The crews would live in their five-storey Crew Compartment. In the event of one vehicle malfunctioning, either one could house all 50 men. The cargo vehicle would house a 23 × 11-m cylinder beneath its four large propellant tanks. The cylinder would contain the equipment that the astronauts would require on the lunar surface, including three vehicles and six trailers.

The flight trajectory would be guided by computers, which in 1952 used punched tape, with manual control as a back-up option during vital portions of the mission. Five days after leaving Earth the spacecraft would approach the Moon. The 'guidance tapes' in the computers would control the retrograde burns that would land the spacecraft on the lunar surface within a few hundred metres of each other.

Having checked the integrity of their vehicles the crews would don their pressure suits and make their way out onto the lunar surface. They would unload the cargo vehicle and set out in one of two tracked vehicles to find a narrow, steep-walled 'crevice' in which to build their base camp. Meanwhile, the remainder of the crew would continue to unload the cargo vehicle. Finally, the two halves of the main cargo container would be dismantled and transported to the selected crevice, where it would form two Quonset hut type buildings as the centre of the base camp. The buildings would offer a pressurised shirtsleeve environment for the astronauts. With the base camp set up, the astronauts would begin their scientific work.

During the lunar 'day' the astronauts would collect samples, take photographs and perform scientific experiments. During the lunar 'night' the samples would be examined and the photographs processed in the buildings at the base camp. Recordings of conversations held on the surface would be transmitted to Earth and new instructions received. Instructions for the next 'day' would be radioed from Earth.

Two weeks after landing the Sun would set over the base camp. Although work would continue out on the surface, preparations would also be made for the journey to Harpalus Crater, some 402 km away. Two tractors, each with two trailers full of equipment and each with five astronauts, would traverse the lunar terrain using headlights and spotlights. In the event that one vehicle malfunctioned the other would be capable of carrying all 10 astronauts and pulling all of the equipment. Five days would be set aside for the outward trip with geology and other science being performed en-route. The majority of samples collected would be examined inside the tractor and then discarded. Only the most interesting samples would be retained for return to Earth. By the time the expedition reached the crater the Sun would be above the horizon once more.

If possible the tractors would be taken right up to the crater rim. If that proved impossible they would be taken as close to the rim as possible and the astronauts would manhandle their equipment the rest of the way. Geology would be a prime concern at the rim and one of the astronauts would be lowered to the crater floor by rope to collect samples and expose photographs. After the exploration of the crater was complete the expedition would return to base camp by following the tracks left in the dust during their outward journey.

During the six weeks of surface exploration the Moon would complete 1.5 orbits around Earth and would be in the correct position for the return trip to commence. The astronauts' final act on the lunar surface would be to deploy three stations of solar-powered experiments that would transmit their data to Earth through the antennae on the abandoned cargo ship.

The two manned spacecraft would lift-off under the guidance of their computers and enter a direct ascent return trajectory. Thirty hours and 210,779 km from the space station's orbit the two spacecraft would turn around and perform a retrograde burn to allow them to rendezvous with the space station 5 days after leaving the Moon. Small orbital ferry vehicles would then transfer the lunar explorers and their equipment to the space station. From there the standard winged manned spacecraft would be used to return them to Earth, where they would doubtless receive a hero's welcome.

25.2 THE REALITY

In 1952 von Braun could not foresee that the Soviet Union would launch the first satellite into orbit and that his adopted country would enter a political race to place the first man in space. In order to save time he would have to abandon his grand dream. America would elect to develop small ballistic spacecraft and launch vehicles that could only be used once. When they lost the race to place the first man in space they would by-pass the development of the space station and continue to use ballistic spacecraft and one-shot launch vehicles to fly to the Moon.

Project Apollo flew to the Moon for the wrong reasons and employed the wrong methods. The Apollo Moon landings were a spectacular achievement, but their 'footprints and flags' approach was a dead end. Walking on the Moon before a Soviet cosmonaut was the sole reason for Project Apollo. There was never a long-term lunar science programme, or even a long-term lunar exploration programme, to follow on from the initial landings. Apollo, like Project Mercury, was a child of its times. The political situation in the early 1960s were right for a crash programme to land men on the Moon and technology had reached a stage where such a programme was possible.

Today the Moon landings are history. They are not taught in school as a great leap forward for Mankind, because they were not. They represent little more than the American nation flexing its muscles in front of the unaligned nations of the world, in order to prove its technological supremacy and the supremacy of its free market and its political system over the tightly controlled technology and paranoid political system of the Soviet Union. The Cold War is now over and consigned to history. The early 'Space Race', up to the close of Project Apollo, has also been consigned to history, as a part of that Cold War.

But not everyone abandoned the dream of exploring space in vehicles that were under the control of the pilot, were landed like aircraft and were re-usable.

25.3 X-15: THE HYPERSONIC AIRCRAFT

The X-15 hypersonic research aircraft was developed as a join US military–NACA project in the late 1950s. (See Fig. 116.) When NACA became the basis for NASA and the ballistic satellite vehicle programme began, the X-15's development continued unabated, and three aircraft were produced.

X-15-1 made its first solo glide-flight on 8 June 1959, dropping from the wing pylon of a NASA B-52 bomber. A few minutes later the aircraft was safely on the ground. X-15-1 was used in a series of similar glide tests, to define the aircraft's handling and landing characteristics. Following just four powered flight-tests, X-15-1 was handed over to NASA on 23 January 1960.

X-15-2 came on line in May 1959 and on 17 September it made the programme's first powered flight using two XRL-11 rocket motors. It was a success. Test pilot Scott Crossfield even performed a barrel roll in flight, despite not being able to take the aircraft under Mach 2 in powered flight, even with the airbrakes deployed.

Fig. 116. The X-15 hypersonic aircraft lands at Edward Air Force Base, California. (NASA)

X-15-2 made its third powered flight on 5 November 1959, with Crossfield at the controls. Shortly after the drop from the B-52 a chase plane pilot informed Crossfield that he had a fire in one of the X-15's rocket motors. The test pilot shut down both engines and jettisoned as much propellant as he could. He then brought the aircraft in to land. Upon contact with the dry lakebed the aircraft buckled between the rear of the cockpit and the front of the propellant tanks. The buckling was caused by a structural weakness that been in the aircraft from the beginning. X-15-2 was returned to North American Aviation for a rebuild.

In April 1960 the first two XLR-99 rocket motors were delivered and were subsequently fitted to X-15-2 (which had made four flights following its rebuild) and X-15-3. On 8 June X-15-3 exploded during a static firing of the new motor. The wreckage was returned to NAA for the aircraft to be rebuilt. X-15-3 did not return to the flight programme until December 1961.

On 15 November 1960 Crossfield flew X-15-2 on the first powered flight employing the XLR-99 rocket motor. He made his last two X-15 flights on 22 November and 6 December 1960. Having gambled on leaving NASA to join North American Aviation as a test pilot, Crossfield was removed from the X-15 programme when the second aircraft was handed over to NASA in December 1960. In March 1961 Bob White flew the X-15-2 to Mach 4.4.

In April 1961 the Soviet Union orbited Vostok I, containing cosmonaut Yuri Gagarin. The X-15 programme and the test pilots at Edward's AFB lost much of their audience as the public turned its attention to the manned space programme. On 5 May 1961 Alan Shepard carried out Project Mercury's first manned suborbital flight. The X-15 programme continued, and on 25 May 1961 Joe walker piloted the X-15-2 to Mach 4.95. That evening President Kennedy addressed the nation and committed America to landing a man on the Moon. On 9 November 1961 Joe White flew the X-15-2 at Mach 6.4 and an altitude of 30,967 m. The X-15 had reached its design speed.

X-15-3 came on line, with an XLR-99 rocket motor, on 20 December 1961. It was the 46th flight in the X-15 flight programme. This was the third X-15 flight for NACA test pilot Neil Armstrong, and being the first flight of the new aircraft, the speed was restricted to Mach 3.76.

The USAF had set the definition of space at 80.45 km and had announced that any USAF pilot that crossed that threshold would receive a set of 'Astronaut's Wings'. However, the Federation International Aeronautique (FIA), based in Paris, France, was the official body responsible for recording aviation record attempts and set the rules by which all such attempts had to abide. When the FIA extended its domain into space it established the threshold at 100 km.

On 17 June 1962 Bob white flew the first X-15 flight into 'space' as defined by the USAF. He took the X-15-3 to Mach 5.4 and an altitude of 91.1 km. At the apogee of his trajectory he could see the curvature of the planet with its clearly defined layer of thick lower atmosphere. During the descent White held his aircraft in a nose high attitude to present its flat underside to the ever-thickening atmosphere. The effects of atmospheric heating were clearly visible to White, who received his 'Astronauts Wings' at a ceremony in Washington two days after the flight.

On 9 November 1962 Jack McKay made an emergency landing after his X-15-2 had suffered a propulsion system malfunction in flight. The nose wheel collapsed on landing and the aircraft tumbled numerous times, ending upside down on the desert floor. The seriously injured McKay lay trapped in his cockpit for 4 hours before the rescue team could remove him. The X-15-2 was returned to North American Aviation for its second rebuild. In time, plans would be made to modify the X-15-2 to make an attempt on reaching Mach 8. The modified aircraft was designated the X-15-A2, with the 'A' standing for 'Advanced'.

On 17 January 1963 Joe Walker flew the X-15-3 to Mach 5.4 and an altitude of 82 km. Robert Rushworth made the X-15's third 'space flight' on 27 June 1963, when he took the X-15-3 to 86.9 km at Mach 4.8.

On 19 July 1963 Joe Walker flew the X-15-3 to 105 km at Mach 5.5 and became the only pilot to fly an X-15 past the official French definition of space. During the flight Walker was to deploy a nitrogen-filled balloon and tow it behind the aircraft to measure air density at that altitude. This was similar to the balloon experiments carried on Scott Carpenter's MA-7 and Gordon Cooper's MA-9 flights in Project Mercury. As had happened on Cooper's flight, Walker's balloon failed to deploy from its canister. Walker passed the FIA's 100 km threshold a second time on 22 August 1963, when he flew the X-15-3 to 107 km at Mach 5.5. This was the highest altitude that any X-15 flight reached.

The X-15-A2 re-entered the programme in February 1964. The aircraft included two large external propellant tanks slung beneath the wings. When their propellants were depleted the tanks would be jettisoned and recovered by parachute. The landing gear was longer and it was even possible to mount a ramjet engine beneath the ventral fin. Initial flight-tests, without the external propellant tanks, began on 25 June 1964. Following the flight-tests the X-15-A2's fuselage was covered in a white ablative material to protect it during the attempt on Mach 8.

On 29 June 1964, Joe Engle made his 14th X-15 flight in the X-15-3. He reached an altitude of 85.5 km and a speed of Mach 4.9. Engle made a second 'space' flight on 10 August 1965, reaching 8.32 km and a speed of Mach 5.2.

By late 1964 Jack McKay had made 15 additional flights in the X-15 after his return to flight duties, following his 1962 crash in the X-15-2. On 28 September 1964 he flew the X-15-3 to an altitude of 90 km at a speed of Mach 5.3.

Engle flew the X-15-1 to 80.7 km at Mach 5 on 14 October 1965. It was his third flight above 50 km and his last X-15 flight. In April 1966 he was selected as a NASA astronaut.

On 3 November 1965 the X-15-A2 made its first flight with its two external propellant tanks. On this flight the tanks were empty. The relatively low-speed flight-test, with Rushworth at the controls, was a success, but only one propellant tank was successfully recovered. The other was destroyed on impact with the ground after its recovery parachute failed to open. The flight was Rushworth's last in the X-15.

William Dana joined the programme and on his fifth flight he flew the X-15-3 to an altitude of 93.3 km and a speed of Mach 5.4. It was 1 November 1966.

The first flight-test of the X-15-A2 with full external propellant tanks was flown by William Knight on 18 November 1966. Knight made another flight-test in the X-15-A2 on 8 May 1967, after which the aircraft was returned to North American Aviation to have its white ablative heat shield material applied, in preparation for the attempt to fly a piloted aircraft to Mach 8. The aircraft was ready in July 1967 but the additional weight meant that the aircraft was no longer capable of reaching Mach 8. NASA set a new goal: Mach 7.2. Knight made one low-speed flight-test of the heavier aircraft before beginning the attack on Mach 7.2. On 3 October 1967 Knight was dropped in the X-15-A2 for the high-speed attempt. The mock-up of a ramjet engine mounted on the ventral fin was ripped off while the aircraft was flying at Mach 6. The flight peaked at Mach 6.7 but was seriously damaged in the attempt as the unprotected wreckage of the ramjet mount was heated by impact with the surrounding air. Knight landed safely and the X-15-A2 was returned to North American Aviation for repair, but insufficient funds remained in the programme budget. The X-15-A2 had made its last flight. Two weeks later Knight flew the X-15-3 to an altitude of 85.9 km and a speed of Mach 5.3, making him the seventh X-15 pilot to pass 50 km.

Michael Adams was dropped in the X-15-3 on 15 November 1967. Throughout the climb to altitude a new experiment was causing radio interference in the voice and telemetry links. The interference then began affecting other systems and caused the Reaction Control System to be shut down. Just as the XLR-99 rocket motor shut down the electrical fault shut down the aircraft's computer and it did not restart as planned. A warning light on the main instrument panel told the pilot that his displayed attitude indications were subject to error. The aircraft climbed to 81 km at a speed of Mach 5.2. As he approached apogee Adams performed a number of planned experiments. As the series of malfunctions built up they caused Adams to lose control of his aircraft during re-entry through the atmosphere. The aircraft was subjected to unacceptable stresses and broke up. Adams was killed.

The programme continued with only the X-15-1 in a flyable condition. Bill Dana flew the X-15-1 to 81 km and Mach 5 on 21 August 1968, and that would be the programme's last flight above 50 km. Dana flew the 199th and final X-15 flight on 24 October 1968. Numerous attempts were made to launch the planned 200th flight before the end of 1968, but it was ultimately cancelled when technical difficulties and bad weather threatened to push it into 1969.

25.4 X-20 DYNA-SOAR

Taking the rocket-propelled aircraft concept one stage further, the USAF's Air Research and Development Command (ARDC) began the definition of a winged vehicle that was capable of orbital flight and a standard landing under the control of the pilot. The new vehicle would be launched vertically on a converted USAF Titan I ICBM (this was subsequently changed for the Titan II) and would land on the dry lakebeds at Edwards AFB, California, as the X-15 did. The new programme was given the designation X-20.

In October 1957, the month that the Soviet Union launched Sputnik, the X-20 picked up the name 'Dyna-Soar', for its dynamic soaring capabilities. Like many military space programmes, the X-20 benefited financially from the public outcry that followed the launch of Sputnik. USAF Headquarters approved the programme in November 1957, the month that the Soviet Union sent the first living creature, the smooth-haired terrier bitch, Laika, into orbit.

Study contracts for the X-20 spacecraft were issued to aerospace companies on 1 January 1958. Eight companies submitted their proposals and in June USAF selected two teams, one headed by Boeing and the other by the Martin Company to compete against each other in the initial definition of the X-20 spacecraft. The two proposals were subsequently reviewed during the 'Phase Alpha' Configuration and Verification Study, which ran between December 1958 and April 1960. On 25 April 1960 the prime contract was issued to Boeing. Martin would provide the Titan launch vehicle

After 'Phase Alpha', planning began and resulted in the following programme, in three stages:

- *Phase One*
  1962: Scout launched suborbital tests of X-20 scale models.
  1963: Unpowered drop-tests from beneath a B-52 bomber at Edwards AFB.
  1964: Unmanned launches on a Titan II launch vehicle.
  1965: Manned suborbital flight on Titan II launch vehicle.
- *Phase Two*
  Manned operational orbital flights on Titan II launch vehicle.
- *Phase Three*
  Manned orbital flights, with weapons systems, on Saturn I launch vehicle.

The Boeing X-20 was a manned, delta-winged, hypersonic vehicle capable of launching vertically on a large rocket, flying in space, re-entering the atmosphere and landing on a runway under the control of its USAF pilot. It was designed to give the

USAF its own military manned space flight capability. A single pilot would be seated in a pressurised compartment, but would wear a full pressure suit to guard against a failure of the ECS. The front windshields would be protected during launch. The protective plates would be jettisoned once the spacecraft achieved orbit.

The vehicle had short stubby wings. The underside of the fuselage was flat, to provide the high-drag required for re-entry. A controlled 30-minute re-entry phase would restrict deceleration forces to those experienced in a commercial airliner. Three landing skids, similar to those used at the rear end of the X-15, would be used to support the landing on the dry lakebeds of Edwards AFB. The two rear skids contained a series of wire brushes on the underside, to increase friction with the lakebed and bring the spacecraft to a gradual stop. The smaller, circular front skid had no wire brushes. The X-20 would carry 456 kg of payload, or four men in a pressurised container. Later plans would be developed to use an undefined Service Module to carry extra cargo or additional LSS gases to extend the spacecraft's orbital life.

Meanwhile, NASA had replaced NACA and the new Administration had accepted the ballistic spacecraft designed by Gilruth's PARD engineers as America's first manned spacecraft. In the Soviet Union Korolev's engineers were also designing a ballistic spacecraft. On 12 April 1961 Korolev's engineers launched Vostok I and made Red Air Force Major Yuri Gagarin the first man in space. Alan Shepard flew Project Mercury's first manned suborbital flight in May. In 1961 Congress passed a FY'62 budget of $100 million for the X-20. Gagarin's launch resulted in an additional supplement of $85.5 million being assigned to the programme, with the demand that it be advanced at greater speed.

President Kennedy's Secretary of Defence, Robert McNamara, cancelled the supplement and demanded that the USAF justify the X-20 programme to him personally, on purely operational military grounds. The grounds quoted to McNamara were: manned reconnaissance, satellite repair and retrieval, an orbital bomber, the ability to carry four crew members in a pressurised compartment. With the two-man NASA Gemini Spacecraft already in definition McNamara was convinced that Air Force astronauts in 'Blue Gemini' spacecraft could perform each of the tasks that the USAF had assigned to the X-20, with the exception of the four-man crew. Convinced that there was no military justification for the X-20 the Defence Secretary began his attempt to militarise Project Gemini.

McNamara suggested that USAF pilots be allowed to fly on NASA Gemini flights and even on All-USAF 'Blue Gemini' flights. He even suggested that the USAF and NASA space programmes be amalgamated under USAF control. NASA refused, although they did agree to fly military officers on their Gemini flights (at the time all of NASA's pilot astronauts were serving military officers, or civilian test pilots that had previously served in the military). Ultimately, no non-NASA astronauts would fly during Project Gemini.

NASA did agree to a joint USAF–NASA Experiment Selection Board. The Board acted in an advisory role, recommending experiments to the Gemini Project Office for inclusion on the 10 manned NASA Gemini flights. One USAF experiment that was accepted for flight was the Astronaut Manoeuvring Unit, the predecessor of the Manned Manoeuvring Units that were carried on some Space Shuttle flights in the 1980s. The concessions were few and small, but McNamara turned his attention from Project Gemini.

Within the X-20 programme Boeing undertook 'Project Streamline', which resulted in a programme that would progress directly from hardware development to unmanned and manned orbital flights on a NASA developed Saturn I launch vehicle. Early flight-tests would launch out of Cape Canaveral, with operational military flights taking place from Vandenburg AFB, California. Landings would take place at Edwards AFB. The plan stated that the X-20 would be flying operational military flights by the last quarter of 1962. The new flight programme did not include a flight schedule. A review of the X-20–Saturn I flight programme revealed that the vehicle would be torn apart as a result of aerodynamic stresses during the launch. The Saturn I was dropped from the X-20 programme.

In August 1961 Gherman Titov spent 25 hours in orbit aboard Vostok II.

Secretary McNamara insisted that the USAF accept the Project Streamline proposals in order to prevent the X-20 being cancelled. Meanwhile, McNamara recommended to President Kennedy that all military aspects of the X-20 programme be abandoned, thereby allowing the spacecraft itself to be developed more quickly. He felt that the concept of a high-lift vehicle capable of manned orbital flight was too important to be delayed by excessive military demands. At this time Boeing commenced the manufacture of the first X-20 airframes at their Seattle plant.

In November Gus Grissom flew Project Mercury's second manned suborbital flight and in February 1962 John Glenn flew three orbits of Earth during the MA-6 flight.

During the same month the Martin Company commenced work on the Titan IIIC heavy lift space launch vehicle, with a Titan II central core and two solid boosters to give additional thrust. The Titan IIIC was assigned as the new launch vehicle for the X-20. McNamara ordered the X-20 programme to be subjected to a second 'Phase Alpha' study centred on the X-20–Titan IIIC configuration.

On 15 March 1962 a group of six USAF and NASA civilian test pilots were named as 'Pilot Engineer Consultants' for the X-20. They would undergo an 8-month training programme to prepare them to fly the X-20. The six pilots were:

N. A. Armstrong (NASA)	H.C. Gordon (USAF)	W.J. Knight (USAF)
R.L. Rogers (USAF)	M.L. Thompson (NASA)	J.W. Wood (USAF)

A second group of all USAF Pilot Engineer Consultants were named on 20 April:

C.C. Book	A.H. Crews	L.N. Hoover	B.F. Knolle
R.H. McIntosh	R.W. Smith	D.L. Sorlie	T.W. Twinting

Scott Carpenter flew MA-7 into space in May 1962 and orbited Earth three times. In August McNamara approved the X-20–Titan IIIC programme.

In September, Gordon, Knight, Rogers, Thompson and Wood were named as the prime pilots for the X-20. The six prime pilots attended the Air Force Association Convention and posed for publicity photographs with a full-scale mock-up of the X-20. Neil Armstrong was originally named as one of the X-20 prime pilots. Three days before the USAF announced the names of the pilots in the group Armstrong was selected as one of the nine astronauts in NASA's Flight B.

In October Wally Schirra completed six Earth orbits during the flight of MA-8.

The final group of Pilot Engineer Consultants were named on 22 October. Once again all 10 men were USAF pilots of the highest standard:

A.L. Atwell	C.A. Bassett	T.D. Benefield	M. Collins
J.H. Engle	N.R. Garland	E.G. Givers	F.G. Newbeck
J.A. Roman	A.H. Uhalt		

Of this group, Charles Bassett and Michael Collins were selected as part of NASA's Flight C in October 1963. Joseph Engle was selected as a NASA Flight E astronaut in April 1966.

In March 1963, McNamara ordered the X-20 programme to undergo another full review of its objectives and uses. In May 1963 Gordon Cooper spent 34 hours in orbit during the last flight of Project Mercury.

The only X-20 related launch took place in September 1963. A Thor launch vehicle carried a scale model of the X-20 to an apogee of 61.87 km. Following separation, gas thrusters on the model aligned it for re-entry into the atmosphere. The model made its final descent under a parachute and landed in the ocean off Ascension Island. Following landing the flotation bag failed to inflate and the model sank.

On 22 November 1963 President Kennedy was assassinated in Dallas, Texas, and Vice President Lyndon Johnson was sworn in as President on the aircraft that carried Kennedy's body back to Washington. McNamara continued to serve the new President as Secretary of Defence.

On 10 December 1963 McNamara cancelled the X-20 programme. He demanded that the partially built airframes be destroyed, that the three groups of Pilot Engineer Consultants be disbanded and the Air Force pilots be returned to operational service. X-20 funds were redirected towards the Air Force's other manned space programme, the Manned Orbiting Laboratory.

25.5 LIFTING BODIES

At the NACA 1958 Conference on High-Speed Aerodynamic a group of engineers from Ames had presented a paper on the potential of the high-lift, high-drag half-cone as a manned satellite vehicle. The Ames engineers did not push their lifting body concept, as they were less enthusiastic than the engineers at LaFRC to become involved in the manned satellite vehicle programme. A lifting body was a vehicle that was shaped to provide sufficient lift, without the use of wings, as it fell through the thick lower atmosphere to allow it to be flown and landed by its pilot in a controlled manner. Although Faget's ballistic capsule design was chosen for the Mercury Spacecraft, the desire to develop and flight-test the lifting body concept did not die in 1958.

M2F1

NASA engineers began developing lifting bodies in 1963, as Project Mercury drew to a close. The one-man M2F1 (Fig. 117) was constructed from plywood and stood on a tricycle undercarriage. On 5 April the new aircraft flew for the first time, having been towed across the dry lakebed at Edwards AFB by a car. The M2F1 completed 60 further tow tests, including some that used a solid rocket motor to increase velocity. On 16 August the M2F1 completed its first free flight, having been towed into the air by a NASA C-47 aircraft. The M2F1 flight-test programme ultimately included 90 flights.

M2F2

In 1964 Northrop was selected to develop the metal M2F2 (Fig. 117) and the HL-10, a lifting body of a different design. Both vehicles would employ off-the-shelf technology and the M2F2 was designed so that it could be retrofitted with an XLR-11 rocket motor. Both lifting bodies would be dropped from beneath the wing pylon of a NASA B-52 bomber in a similar manner to the X-15.

The first M2F2 free flight was made on 12 July 1966, after two years of wind tunnel testing and four months of captive flights. On 10 May 1967 the M2F2 crashed on landing at the end of its 16th glide flight and was severely damaged. The airframe began a rebuild in 1968, having spent the intervening period in storage.

Fig. 117. (Left to right) NASA's M2F1 and M2F2 lifting bodies at Edwards Air Force Base. (NASA)

HL-10

The HL-10 flew free for the first time on 22 December 1966, but proved difficult to control. It was removed from the flight programme and subjected to a new series of wind tunnel tests. The HL-10 returned to flight on 15 March 1968. After eight glide-flights the HL-10 was removed from the lifting body flight programme so that it could be fitted with a XLR-11 rocket motor. The aircraft resumed glide-flights on 24 September. On 3 October the HL-10 made its first powered flight. It was only partially successful, but four low-speed powered flights followed in quick succession. On 9 May 1969 the HL-10 made its first powered supersonic flight. When the HL-10 made its final flight, on 17 July 1970, it had made 37 flights in total. Fifteen of those were gliding flights, 6 were low-speed powered flights and 16 were made at supersonic velocities up to Mach 1.86.

X-24A

With the cancellation of the X-20 the USAF contracted Martin Marietta to develop the X-24A lifting body, with an XLR-11 rocket motor and two smaller liquid propellant rocket motors. The X-24A made its first glide-flight on 17 April 1969. Eight further glide-flights were followed by the first low-speed powered flight on 19 March 1970. The vehicle performed its first supersonic flight on 14 October 1970. The programme ended with the 28th flight, on 4 June 1971. It had consisted of 10 glide-flights, 9 low-speed flights and 9 supersonic flights up to Mach 1.39.

M2F3

The rebuilt M2F2 was designated M2F3 and made its first glide-flight on 2 June 1970. The first powered flight, employing a four-chamber XLR-11 rocket motor, occurred on 25 November 1970, with the first supersonic flight taking place on 25 August 1971. M2F3 completed its 27th flight on 20 December 1972, bringing to a close a programme of 7 glide-flights, 8 low-speed flights and 12 supersonic flights up to a velocity of Mach 1.44.

X-24B

The last of the first-generation American lifting bodies was the X-24B, which used the same propulsion units as the X-24A. The new machine's first glide-flight took place on 1 August 1973, with the first powered flight following on 15 November 1973 and the first supersonic flight on 15 February 1974. The programme ended with flight 36, on 26 November 1975 after 12 glide-flights, 3 low-speed flights and 21 supersonic flights up to velocities of Mach 1.76.

In July 1975 America's last ballistic spacecraft had flown as one-half of the Apollo–Soyuz Test Project. Meanwhile, a totally new type of spacecraft was being slowly brought to fruition.

25.6 SPACE SHUTTLE

Today the Space Shuttle is the backbone of the American piloted spaceflight programme. Developed in the 1970s, it has been flying since 1981. The Shuttle is a combined launch vehicle and piloted spacecraft that consists of three major parts. Its design draws heavily on the data collected during the X-15 flight-test programme.

When a Shuttle lifts off from Launch Complex 39, Cape Canaveral, two Solid Rocket Boosters (SRB) provide the majority of thrust. These are each 45.46 m long and 3.70 m in diameter. The two SRBs burn solid rocket propellant to provide 1,315,440 kg of thrust. This represents some 71 per cent of the total thrust produced at launch.

The SRBs are connected on either side of the External Tank (ET), which holds the liquid oxygen and liquid hydrogen propellants for the three Space Shuttle Main

Engines (SSME). The ET is 14.63 m long and 8.53 m in diameter and carries attachment points for the Orbiter, the piloted portion of the vehicle, which has the superficial appearance of an aircraft. At the rear of the Orbiter are three SSMEs, each producing 170,100 kg of thrust at launch. Above the SSMEs are two Orbit Manoeuvring System (OMS) pods, each containing a single 2,722-kg thrust OMS engine and 12 individual 395-kg Orbiter Reaction Control System (ORCS) thrusters. These engines are used to perform orbital manoeuvres, along with other manoeuvring thrusters mounted at the front of the Orbiter. Each pod also contains two individual 11.3-kg vernier thrusters.

The main body of the Orbiter, between the wings, holds the payload bay. As the name suggests, this 18.29 m by 5.18 m by 3.96 m area is where the Orbiter carries its payload. On flights to the International Space Station (ISS) this may include the Orbiter Docking Module, ISS Modules, Spacehab Modules, Multi-Purpose Logistics Modules or other as yet undefined payloads. The Orbiter can carry a Remote Manipulator System (RMS) mounted down one side of the payload bay. At launch, the payload bay is enclosed by a set of payload bay doors.

The forward part of the Orbiter contains the Flight Deck and, beneath it, the Mid-Deck. The Flight Deck is where the Commander and Pilot control the Orbiter. Mission Specialists operate the RMS from the Aft Flight Station on the Flight Deck.

The Mid-Deck provides additional seating for crew members during launch and re-entry as well as sleeping quarters, galley and personal hygiene facilities. The Mid-Deck also contains the hatch giving access to the internal airlock, if it is carried. Alternatively that hatch can give access to pressurised modules in the payload bay, or to ISS via the Orbiter Docking System mounted in the payload bay.

The Shuttle is launched vertically. The SSMEs are ignited first, and when they achieve mainstage the SRBs are ignited to provide sufficient additional thrust to raise the Shuttle beyond the thickest portion of Earth's atmosphere in less than 2 minutes. After burnout the SRBs are jettisoned, deploy parachutes, and land in the ocean, from where they are recovered and returned to Cape Canaveral. They are then returned to the manufacturer for breakdown, cleaning, refurbishment and preparation for relaunch on another Shuttle flight. (See Fig. 118.)

The Shuttle continues to climb, the SSMEs continuing to take propellant from the External Tank. When the correct altitude and velocity are reached the three SSMEs are shut down and the ET is jettisoned. The OMS thrusters are then fired to push the Orbiter into orbit.

Once in orbit the payload bay doors must be opened to expose the radiators mounted on their interior. If the doors fail to open the Orbiter must return to Earth before its electrical equipment overheats. With the payload bay doors open the Orbiter is ready to perform the on-orbit portion of its mission.

The payload bay doors must be closed before the de-orbit burn, which is performed using the OMS thrusters. The Orbiter re-enters Earth's atmosphere in a belly-first, nose-high attitude. As it flies through the lowest, thickest portion of the atmosphere the aerodynamic surfaces on its wings and tailplane become operational. A microwave landing system guides it towards the landing strip, and the three-point

Fig. 118. The Space Shuttle lifts off from Launch Complex 39 at Cape Canaveral. (NASA)

undercarriage is lowered 30 seconds before landing. Mechanical braking and a brake parachute bring it to wheel-stop at the end of its post-landing rollout.

Having won the race to the Moon, NASA seems to have returned to the programme in von Braun's Great Scheme. Since 1981 America has been flying the re-usable spacecraft that von Braun had dreamed about in 1952. The Shuttle launches like a rocket, but lands like an aircraft. In order to gain political support it was designed to be all things to all people. It was a satellite launch vehicle and an orbiting laboratory, but from the outset it was always intended as the crew and cargo ferry for an orbital space station. In the late 1980s it proved its worth in that role during the Shuttle–Mir programme. It is presently involved in the construction of the International Space Station (ISS) – a giant space platform being built as a cooperative effort between 16 nations.

25.7 X-38 CREW RETURN VEHICLE

The X-38 is an American vehicle being developed to serve as a Crew Return Vehicle for the ISS. Its shape is based on the US Air Force's X-24A experimental lifting body tested in the 1960s. Wherever possible, the X-38 will employ off-the-shelf technology, rather than develop new items from scratch. The CRV's computer is a commercial product that is currently being used in aircraft. The vehicle's software is also a commercial package.

Following separation from the ISS in an emergency the X-38, with up to seven astronauts onboard, would perform a de-orbit rocket burn and then jettison the module containing the de-orbit rocket engine. Following entry through the atmosphere the vehicle would glide to an unpowered landing under a steerable parafoil. The X-38 will deploy skids to absorb final landing loads.

The 9.1-m long by 4.4-m wide vehicle employs a nitrogen attitude control system and a bank of batteries to provide power for up to 9 hours. All de-orbit and landing procedures will be automated, although the crew will have the option to use back-up systems to control attitude, select a de-orbit site and steer the parafoil.

The X-38 programme began as an in-house study at NASA's JSC, in early 1995. Throughout the summer of 1995 the parafoil concept was tested by dropping pallets, with parafoils attached, over the US Army's Yuma Proving Ground in Arizona. In early 1996 a contract was awarded to Scaled Composites Ltd, Mojave, California, for three full-scale atmospheric demonstration vehicles. The first vehicle, known as Vehicle 131, was delivered to JSC in September 1996. It was out fitted with avionics, computer systems and other hardware before being shipped to the NASA Dryden Flight Research Centre, California, on 4 June 1997. The vehicle was used in a series of captive flights, slung under the wing of Dryden's B-52 aircraft throughout July and August 1997. The first glide flight was made on 12 March 1998. It was the beginning of an ongoing series using two test vehicles. A third vehicle, V201, will be carried into orbit in a test-flight, and make a full return to Earth. If it is developed the Crew Return Vehicle is due to become operational in its ISS CRV role in 2003.

25.8 VENTURESTAR AND THE X-33

On 2 July 1996 NASA Administrator Dan Goldin announced that a consortium led by the Lockheed Martin 'Skunk Works' had been selected to develop a possible replacement for the Space Shuttle. The new Reusable Launch Vehicle (RLV) would be called VentureStar. Using new technologies VentureStar would demonstrate a Single Stage To Orbit (SSTO) launch concept in a vehicle based on the M2F1, the original wooden NASA lifting body. It would be triangular in shape with a flat upper surface and rounded underside. The vehicle would employ linear aerospike rocket motors, an internal payload bay, and a three-point undercarriage. It would be unmanned, with a capability to carry astronauts in a pressurised container. Lockheed Martin would develop and operate the spacecraft, which NASA would rent as a contractor.

Before building VentureStar the consortium was to construct a half-scale X-33 Advance Technology Demonstrator (ATD) to prove the concept in suborbital flight, and this would bring together a number of cutting edge technologies for the first time in the new vehicle. Vertical launches would take place from a new facility constructed near Haystack Butte, in the California desert close to Edwards AFB. The unmanned X-33 would follow the flight plan written into its computer software and land at one of two military air fields, depending on the flight path followed.

By the end of 1999 Lockheed Martin had decided that VentureStar's internal components could be better fitted within the airframe if the payload bay was removed, and for this reason VentureStar's design was changed to facilitate the use of an external payload canister carried on the upper surface of the airframe. However, the contractor failed to attract finance from the private sector to pay for VentureStar's development and flight-testing.

The X-33 ATD programme was delayed when the composite materials used in the multi-lobe propellant tanks proved difficult to secure in place and it was decided that the composite material tanks should be replaced with heavier aluminium tanks. The programme was effectively cancelled when the NASA/Lockheed Martin agreement ran out on 31 March 2001. This followed NASA's refusal to allow the X-33 to compete for funds as part of the Space Launch Initiative.

Part VI: APPENDICES

Appendix A

Astronauts' biographies

A.1 MALCOLM SCOTT CARPENTER

Scott Carpenter was born on 1 May 1925, in Boulder, Colorado. His parents separated when he was 3, and following his mother's contracting tuberculosis he was raised by a family friend. After graduating from high school in 1943, he joined the US Navy as part of the V-5 flight-training programme at the University of Colorado. With the end of World War II the V-5 programme was terminated. He entered the University of Colorado, studying aeronautical engineering, and received his degree in 1949.

He married Rene Price on 9 September 1948 and they subsequently had four children, Marc, Kristen, Candace and Robyn. He then joined the Navy and received flight training between November 1949 and April 1951. Following further training he was assigned to Patrol Squadron 6 at Barbers Point, Hawaii. During the Korean War his squadron carried out anti-submarine patrols, shipping surveillance and aerial mining operations in the Yellow Sea, South China Sea and the Formosa Straits.

He entered the Navy Test Pilot School at Naval Air Test Centre (NATC), Patuxent River, Maryland, in 1954. After completing his test pilot training he was assigned to the Electronics Test Division at NATC, where he flew a variety of single and multi-engine aircraft with both propeller and jet propulsion. In 1957 he spent 10 months at Naval General Line School, Monterey, followed by 8 months at Naval Air Intelligence School, Washington, DC. He was assigned to the USS *Hornet* anti-submarine aircraft carrier as Air Intelligence Officer in August 1958.

He was serving on USS *Hornet* when he received orders to attend Washington, DC, for a briefing on a new government project. He had been selected as one of 110 military test pilots that met the initial selection criteria for NASA's Research Astronaut Candidate programme. On 9 April 1959 he was named as one of the Flight A astronauts for Project Mercury and reported for duty with the Space Task Group at Langley. His experience in communication and navigation aids led to his being assigned that area of speciality within the project.

Carpenter served as John Glenn's back-up for MA-6. When Deke Slayton was medically grounded from MA-7 Carpenter was named as the new prime pilot. Walter Schirra served as his back-up. On 24 May 1962 Carpenter orbited Earth three times in the Mercury Spacecraft that he had named Aurora 7. In 1963 he monitored the design and development of the Apollo LM. He also served as Executive Assistant of the Director of the MSC, Houston.

Taking leave from NASA in 1965 he took part in the US Navy's Sealab II project as training officer and officer-in-charge of the submerged diving teams. He spent 30 days living on the seafloor as part of Sealab II, earning himself the Navy's Legion of Merit award. During his time on the seafloor he was able to talk to fellow Flight A astronaut Gordon Cooper and Flight B astronaut 'Pete' Conrad while they were in orbit on Gemini V.

Returning to NASA Carpenter served in a liaison role with the Navy for the astronauts' neutral buoyancy training programme, which was being introduced at that time. On 16 July 1964 he had a motorcycle accident and broke his arm, which led to his being removed from flight status. He retired from NASA on 10 August 1967.

Carpenter continued to serve with the Navy's Sealab III project, pioneering saturation diving and deepwater search and rescue capabilities, before retiring from the service as a Commander on 1 July 1969. He has since served as an engineer consultant, and is a wasp breeder and novelist. Following the collapse of his first marriage, he married Maria Roach in 1972. The couple have two children, Mathew and Nicholas.

A.2 LEROY GORDON COOPER

Gordon Cooper was born on 6 March 1927 in Shawnee, Oklahoma, and graduated from high school in 1945. As the Army and Navy flying schools were not recruiting at the time, he joined the US Marine Corps. World War II ended before he saw combat and he was assigned to guarding duties in Washington, DC. He was serving on the Presidential Honour Guard when, as a Marine reservist, he was released from service.

Living with his parents in Hawaii, he attended the University of Hawaii. He married Trudy Olson on 29 August 1947 and they lived in Honolulu while Cooper continued his studies. At University he received a commission in the US Army ROTC. While married to Trudy Olson, they had two children, Camilla and Janet.

Cooper transferred to the USAF and was called to active duty for flight training in 1949. Receiving his wings in 1950, he was assigned to the 86th Fighter-Bomber Group, Landstuhl, West Germany, where he flew F-84 and F-8 jets for the next four years. He later became Flight Commander of the 525th Fighter-Bomber Squadron.

Returning to America in 1954, he attended the Air Force Institute of Technology and gained his degree in aeronautical engineering in 1956. His next assignment was to the Experimental Test School, Edwards AFB, California, until 1957. Thereafter, he was transferred to the Fighter Section of the Flight Test Engineering Division at

Edwards AFB, where he worked on the F-102A and F-106B. This was his assignment when he was called to Washington, DC, to receive a briefing on Project Mercury. He subsequently underwent the Mercury selection process and was one of the Flight A astronauts named to the public on 9 April 1958.

His area of specialisation was the Mercury Redstone launch vehicle. He also developed the knife carried in the astronauts' survival pack and served on the committee responsible for defining the emergency egress procedures to allow an astronaut to leave his Mercury Spacecraft rapidly but safely during an emergency on the launch pad. Cooper served as Capsule Communicator during both Glenn's MA-6 flight and Carpenter's MA-7 mission, before serving as back-up pilot to Wally Schirra on MA-8. He was then assigned as prime pilot for MA-9, with Alan Shepard serving as his back-up. He called his spacecraft Faith 7. MA-9 was the last flight of Project Mercury and Cooper completed 22 orbits in 34 hours over 15–16 May 1963. He was the first American astronaut to sleep in space.

Cooper made his second flight into space as Command Pilot on Gemini Titan V, with Charles 'Pete' Conrad as his Pilot. The flight began on 21 August 1965 and lasted 7 days 22 h 55 min 14 s, the expected length of the initial lunar flights of Project Apollo. The astronauts practised terminal rendezvous techniques. On 29 August Cooper was able to talk to Scott Carpenter on the seafloor in the US Navy's Sealab II.

Cooper served as back-up Command Pilot for Gemini Titan XII, before being assigned to Project Apollo and the Apollo Applications Programme (ultimately re-named Skylab) as assistant to Deke Slayton. After serving as back-up Commander for Apollo X, he was named as Commander of the prime crew for Apollo XIII but was replaced by Alan Shepard when he returned to flight status after being grounded for 4 years for medical reasons. Shepard's crew was ultimately reassigned to Apollo XIV.

On 31 July 1970 Cooper resigned from both NASA and the USAF, where he held the rank of Colonel. He set up Gordon Cooper Associates Inc. and has served in senior management positions in a number of aviation-based companies. Following the collapse of his first marriage, Cooper married Susan Taylor in May 1972. They have two children, Elizabeth and Colleen.

A.3 JOHN HERSCHEL GLENN

John Glenn was born in Cambridge, Ohio, on 18 July 1921. He graduated from New Concorde High School and then obtained his Bachelor of Science in engineering from Muskingum College, New Concorde. While still under full-time education he took advantage of a government civilian training programme to learn to fly.

He passed the Army Air Corps physical examination and was sworn in. When he received no orders he took the Navy physical examination, passed it, and was sworn in as a Naval Aviation Cadet. He completed his flight training at Corpus Christi, Texas. While there he learned that he could volunteer for the US Marine Corps (USMC) and took a commission in that service. He received his aviator's wings and the rank of Lieutenant in 1943. He married Anne Castor in the same year. They have two children, John and Carolyn.

He joined Marine Fighter Squadron 155 and flew 59 combat missions in the Pacific theatre, in F4Us, before being stationed to Patuxent River where he performed as a test pilot. With the end of World War II Glenn was stationed with Fighter Squadron 218 on North China patrol. From mid-1948 until the end of 1950 he served as a flight trainer at Corpus Christi Naval Air Station and then attended Marine Amphibious Warfare Training in Quatnitco, Virginia.

With the beginning of the Korean War Glenn requested combat duty. During the hostilities he flew F9F jets on 63 ground support flights with Fighter Squadrons 311 and 27. During an exchange with a USAF pilot he served with an F-86 squadron over the Yalu River. During the last 9 days of the war he shot down three Communist MIG fighters.

When the war ended Glenn applied for and received assignment to the Navy Test Pilot School, Patuxent River. Following his graduation, on 23 July 1954, he was assigned to the Fighter Design Branch of the Navy Bureau of Aeronautics as a test pilot of Navy and Marine Corps aircraft in Washington, DC, from November 1956 to April 1959. At the same time he attended the University of Maryland. In July 1957 he flew an F8U on the first transcontinental flight from Los Angeles to New York to average supersonic speeds.

While at Patuxent River Glenn attended the NACA Langley Flight Research Laboratory, where he took place in simulation runs as part of NACA's manned satellite programme. He also made simulated re-entry rides on the Johnsville centrifuge as part of the same assignment. It was during this work that Glenn first met Scott Carpenter and the two men became good friends. He was then assigned as one of the military service advisers on the NACA–McDonnell manned satellite mock-up then being defined.

In 1958 Glenn found that he was one of 110 military test pilots qualified to undergo screening for selection as a NASA Research Astronaut Candidate and agreed to undergo the selection process. On 9 April 1958 he was named as one of seven Flight A astronauts. Glenn was the oldest of the group and had flown more different types of aircraft than any of the others. He did not have a degree, however, and admitted many years later that he had used influential friends in the Marine Corps to fight his case and claim that his education was the equivalent of the required qualification. With his wide experience in flying military aircraft he was given the specialist responsibility for the development of the Mercury Spacecraft's instrument and display panel, Crew Compartment layout and control functions.

Glenn served as back-up pilot to Alan Shepard on MR-3, the first American manned space flight, and to Gus Grissom on MR-4. Glenn was originally intended to fly as prime pilot on MR-5, but after MR-4 all further manned Mercury Redstone flights were cancelled. Glenn was assigned as prime pilot on MA-6, with Scott Carpenter as his back-up. Glenn named his Mercury spacecraft Friendship 7.

MA-6 was launched on 20 February 1962 and Glenn became the first American astronaut to go into orbit. Friendship 7 made three orbits of Earth and Glenn returned to an international hero's welcome.

Glenn resigned from NASA on 16 January 1964, when it became obvious that his age would prevent him flying in space again. The next day he announced that he was running for the Democratic nomination to become Senator for Ohio. On 26 February he fell in his bathroom, struck his head and suffered a concussion that affected his balance and on 3 March 1964 he withdrew from the election campaign. Glenn was promoted to Colonel USMC on 27 October 1964. He resigned from the Marine Corps on 1 January 1965 and served as a consultant to NASA Administrator James Webb.

For five years he worked with Royal Crown International, a soft drinks company. He made a second run for the Ohio Senate position in 1970 but was defeated and remained in his job at Royal Crown International. Finally, in 1974, he was successful in seeking election as Senator of Ohio, and was re-elected in 1980. In 1984 he ran for the Democratic Presidential nomination, but withdrew before the convention when it became obvious that he would not win. He was returned as the Senator of Ohio in 1986 and 1992.

In 1997 he announced his plan to retire from politics and said that he was ready to return to space if he could be offered the chance to carry out experiments into ageing while in orbit. In the years since Project Mercury the physical effects of long-term microgravity had been found to be similar to those experienced during the ageing process.

On 16 January 1998 NASA Administrator Dan Goldin announced that Glenn would fly as a Payload Specialist on the crew of STS-95. At 77 years of age when the space shuttle Discovery lifted-off, on 29 September 1998, Glenn was the oldest person to fly in space to that date. The first American astronaut to take photographs in space was in charge of STS-95's video and still photography. The flight was given wide media coverage and Glenn's return to space was the subject of several books and updated autobiographies.

A.4 VIRGIL IVAN GRISSOM

'Gus' Grissom was born in Mitchell, Indiana, on 3 April 1926. He graduated from Mitchell High School in 1944. Having enrolled as an aviation cadet in his senior year, he reported for duty with the Air Force in August 1944. In July 1945 he took leave and married Betty Moore, his High School sweetheart. The couple would have two children, Scott and Mark. When Japan surrendered and World War II ended Grissom had still not received flight training. After a series of administrative postings he took the opportunity to resign from the Air Force in November 1945.

After working in a bus manufacturing plant, in 1946 Grissom enrolled at Purdue University in order to obtain a bachelor's degree in mechanical engineering. He graduated in 1950 and re-enlisted in the USAF, underwent flight training and earned his wings. Less than one year later he was shipped to Korea, where he flew 100 combat missions in six months with 334th Fighter Interceptor Squadron. His request to fly a further 25 missions was denied and he was posted back to America. He ended up training cadets but used every opportunity to increase his flying hours and flight skills. After instruction at the Institute of Technology, Wright-Paterson AFB, Grissom attended test pilot school at Edwards AFB, California. He graduated as a test pilot in 1957 and began testing jet aircraft for the USAF.

While at Edwards AFB he received orders to attend Washington, DC, where he was briefed on Project Mercury. He agreed to proceed with the selection process, and on 9 April 1959 he was named as one of the seven Flight A astronauts.

Grissom served in support of Alan Shepard's MR-3 flight, while John Glenn served as Shepard's immediate back-up. Grissom was then named as prime pilot for MR-4, America's second manned suborbital space flight. He named his Mercury Spacecraft Liberty Bell 7. Glenn served in the back-up role for the second time. The flight took place on 21 July 1961. Liberty Bell 7 sank on landing and was not recovered for a further 38 years. MR-4 was the final suborbital flight in Project Mercury.

Unlike Shepard, Grissom did not seek an orbital flight within Project Mercury. Rather, he concentrated his efforts on the new Project Gemini, a two-man spacecraft being developed to investigate orbital rendezvous and docking. Grissom participated in the development of the Gemini Spacecraft's Crew Compartment, which was designed around him. Grissom's small stature led to a Crew Compartment that larger astronauts found cramped. The others referred to the Gemini Spacecraft as the 'Gusmobile'.

The Command Pilot's position on first manned Gemini flight went to Alan Shepard with Thomas Stafford, a Flight B astronaut, as his Pilot. When Shepard was medically grounded Stafford was also removed from the crew. The Command Pilot's position passed to Grissom, with Flight B astronaut John Young as his Pilot. Wally Schirra was named as Grissom's back-up with Stafford serving as back-up to Young. The three-orbit flight of GT-III took place on 23 March 1965. Grissom became the first person to fly in space twice and the first astronaut to change the trajectory of his spacecraft in orbit. It was the first stage in developing orbital rendezvous, which was vital to the goals of Project Apollo. Grissom and Young served as back-ups to Schirra and Stafford when they were named as the prime crew of GT-VI.

In March 1966 Grissom was named as Commander of the first manned flight of Project Apollo, then planned for launch in October 1966. His crew were Senior Pilot Edward White and Pilot Roger Chaffee. Their back-ups were James McDivitt, David Scott and Russell Schweickart respectively. Apollo Spacecraft 012 was developed by North American Aviation and was intended to fly on the flight designated Saturn Apollo 204 (SA-204). The spacecraft had numerous technical difficulties and the launch was continually delayed until it was finally assigned to the end of February 1967.

On 27 January 1967 Grissom, White and Chaffee participated in a plugs-out Countdown Demonstration Test. An electrical short circuit ignited the pure oxygen in the spacecraft Crew Compartment and the three astronauts perished in the resulting flash fire.

Deke Slayton has stated in writing that had the SA-204 fire not occurred, Grissom was in line to command the first manned lunar landing flight and would therefore have become the first person to walk on the lunar surface. When the fire occurred all Project Apollo crew assignments were cancelled. When manned flights were resumed, new crew assignments were announced but a number of random events interfered with those assignments before Neil Armstrong took his 'one small step' at Tranquillity Base.

A.5 WALTER MARTY SCHIRRA

Wally Schirra was born on 12 March 1923, in Hackensack, New Jersey. He graduated from Dwight D. Murrow High School in June 1940 and spent the following two years studying aeronautical engineering at Newark College of Engineering. In 1942 he was appointed to the US Naval Academy and received his Bachelor of Science degree in June 1945. On graduation he received a commission in the Navy as an Ensign and joined the battle cruiser *Alaska*. World War II ended before he saw combat.

On 23 February 1946 Schirra married Josephine Fraser. The couple would have two children, Walter and Suzanne. In the same year he was assigned to the 7th Fleet. Following his completion of flight training at Pensacola Naval Air Station he was stationed with Fighter Squadron 71. During the Korean War he flew 90 combat missions as an exchange pilot with 154th Fighter-Bomber Squadron and shot down one MIG and possibly a second.

From 1952 to 1954 he served as a test pilot at the Naval Ordnance Training Station, China Lake, California. The following two years saw him serve as project pilot and instructor on the F7U-3. From 1956 to 1957 he flew F3H-2Ns as operations officer of Fighter Squadron 124, on the USS *Lexington*. In 1957 he attended the Naval Air Safety Officer School, before completing test pilot training at the Naval Air Test Centre, Patuxent River, over 1958–59.

Schirra was serving at Patuxent River when he was instructed to attend the briefing on Project Mercury in Washington, DC. He subsequently agreed to undergo the selection process to become a Research Astronaut Candidate. During the selection process he was subjected to surgery to remove a polyp from his larynx. He was named as one of the Flight A astronauts on 9 April 1959 and was given the spacecraft ECS and the astronauts' pressure suit as his areas of responsibility.

He was originally named as back-up pilot to Deke Slayton on MA-7. When Slayton was medically grounded the flight went to Scott Carpenter and Schirra served as his back-up too. On 3 October 1962 he became the third American to orbit Earth, on MA-8. He named his spacecraft for the six-orbit flight Sigma 7.

Schirra then began work on Project Gemini, serving as Gus Grissom's back-up on GT-III, the first manned Gemini flight. His next position was as Command Pilot on the prime crew of GTA-VI with Flight B astronaut Thomas Stafford as his Pilot. When the Gemini Agena Target Vehicle was lost on launch, Schirra and Stafford used the Gemini VII spacecraft as the target for the first rendezvous in orbit of two manned spacecraft, on 15 December 1965. The flight of GT-VIA lasted 25 h 51 min 24 s.

Schirra's next crew assignment was as Commander on the second manned flight in Project Apollo. His crew were Senior Pilot Walter Cunningham and Pilot Don Eisele. After the flight was cancelled Schirra's crew became the back-up crew for the first manned Apollo flight, crewed by Gus Grissom, Edward White and Roger Chaffee. When Grissom's crew died in the fire at Launch Complex 34 on 27 January 1967, all crew assignments were cancelled.

Schirra's crew was named as the prime crew for the first flight of the redesigned Apollo Spacecraft on 20 November 1967. Apollo VII was launched on 11 October 1968 and was an Earth orbital flight-test of the Command and Service Module. The flight lasted almost 11 days.

Having become the only person to fly the first three generations of American manned spacecraft, Schirra resigned from the US Navy and NASA on 1 July 1969. He held the service rank of Captain. Since then he has held senior management positions with a number of companies.

A.6 ALAN BARTLETT SHEPARD

Al Shepard was born in East Derry, New Hampshire, on 18 November 1923. He graduated from Pinkerton Academy, East Derry, in 1940. After one year at Admiral Farragut Academy he entered the United States Naval Academy, Annapolis, Maryland, and graduated with a Bachelor of Science Degree in June 1944. He served the final year of World War II as an Ensign on board the destroyer USS *Cogswell.* On 3 March 1945 he married Louise Brewer. The couple subsequently had two daughters, Laura and Julie. They also raised their niece, Alice.

Shepard received flight training at Corpus Christi Naval Air Station, Texas, and Pensacola Naval Air Station, Florida, and gained his wings as a Naval aviator in March 1947. He was assigned to Fighter Squadron 42 at Norfolk, Virginia, and served several tours on board aircraft carriers with the squadron. In 1950 he attended the US Naval Test Pilot School, Patuxent River, and followed his graduation with two tours of flight-test at the school. Between the two tours he served with Fighter Squadron 193 at Moffett Field, California, and completed two tours with the squadron on the aircraft carrier USS *Oriskany*. Following his second tour at Patuxent River he attended the Naval War College, Newport, Rhode Island, and was then assigned to the staff of the Commander-in-Chief, Atlantic Fleet.

Shepard was one of 110 military test pilots qualified to meet the selection criteria for the NASA Research Astronaut Candidate programme. Having attended the briefing on Project Mercury in Washington, DC, he elected to undergo the selection process. On 9 April 1959 he was named as one of the seven Flight A astronauts. He was allocated the specialisation areas of the World-Wide Tracking Network and recovery procedures.

On 21 February 1961 Shepard was named to be the Pilot of MR-3, the first manned launch of Project Mercury. John Glenn would serve as his back-up and Gus Grissom would serve in a supporting role. Shepard made a suborbital flight on 5 May 1961. Shepard then served as Capsule Communicator on the next two Mercury flights, before being named as Gordon Cooper's back-up for Mercury Atlas 9. When it became clear that Cooper would make the flight, on 15–16 May 1963, Shepard began pushing for a three-day MA-10 mission. The flight did not take place.

Shepard was named as Command Pilot for GT-III, the first manned flight of Project Gemini. His Pilot was Thomas Stafford, a Flight B astronaut. In 1964 Shepard contracted Meniere's disease, which manifested itself in the form of fluid building up in the inner ear, making the motion detectors in the inner ear extremely sensitive. The outward symptoms included a ringing in the ears, nausea, dizzy spells and disorientation. He was removed from GT-III and Stafford also lost his place. The new prime crew was named as Gus Grissom and John Young. Grounded by both NASA and the US Navy, Shepard took over Slayton's position as Chief of the Astronaut Office at MSC, Houston, Texas.

In early 1969 Stafford made Shepard aware of a physician who claimed to be able to perform minor surgery to cure Meniere's disease. The operation involved the installation of a small tube to drain fluid from the inner ear to the spinal cavity. Shepard successfully underwent the operation and was returned to full flight status on 7 May 1969.

Slayton promptly assigned Shepard to the Commander's position on the next available Apollo flight – Apollo XIII, but this was the only crew recommendation made by Slayton that NASA Headquarters refused to endorse. Rather, they swapped Shepard's crew to Apollo XIV to give Shepard more time to prepare. The Command Module Pilot on Apollo XIV was Stuart Roosa and the Lunar Module Pilot was Edgar Mitchell. Apollo XIV lifted-off on 31 January 1971 and when Shepard and Mitchell landed their spacecraft at Fra Mauro on 5 February, Shepard had become the only Flight A astronaut to reach the Moon. Shepard is also remembered for hitting two golf balls at the end of his second lunar surface exploration.

After Apollo XIV Shepard resumed his role as Chief of the Astronaut Office in June 1971, before serving as a delegate to the 26th United Nations General Assembly from September to December 1971. He was promoted to Rear Admiral on 1 December 1971. Shepard retired from the Navy and NASA on 31 July 1974 and served in senior management positions in several companies before setting up his own company. He died in hospital on 21 July 1998.

A.7 DONALD KENNETH SLAYTON

Deke Slayton was born on 1 March 1924, in Sparta, Wisconsin. He graduated from Sparta High School in 1942 and enrolled in the Army Air Corps as an aviation cadet on 1 March 1942. He underwent flight training at Vernon and Waco, Texas, and received his wings and a commission in April 1943. He then flew 56 combat missions in Europe, flying B-25 bombers with the 340th Bombardment Group. He returned to America in 1944 and served as a B-25 instructor pilot. In April 1945 he joined the 319th Bombardment group on Okinowa, where he flew a further seven combat missions in A-26s before World War II ended. Slayton served a further year as a B-25 instructor pilot before being discharged from the USAAF with the rank of Captain in 1946.

He doubled up on courses at the University of Minnesota and achieved his Bachelor's Degree in aeronautical engineering in 1949. Boeing Aircraft Company,

Seattle, Washington, employed him as an engineer for two years. In 1951 he was recalled to active service with the Minnesota National Guard. He served with an F-51 squadron, and then spent 18 months at Headquarters 12th Air Force. He was then assigned to 36th Fighter Day Wing stationed in Bitburg, West Germany. While serving in Germany Slayton met Majorie Lunney and married her on 15 May 1955. They subsequently had one son, Kent.

Slayton returned to America in 1955 and attended the USAF Test Pilot School, Edwards AFB, California. On graduation he served as a test pilot at Edwards AFB from January 1956 to April 1959. He was at Edwards AFB when he was invited to Washington for a briefing on Project Mercury. He subsequently underwent selection and was named as one of the Flight A astronauts on 9 April 1959. Slayton was assigned the Mercury Atlas launch vehicle as his area of specialisation.

Originally Slayton was assigned as prime astronaut on MA-7 with Wally Schirra as his back-up. However, in November 1959 Slayton had been diagnosed as having a heart condition called idiopathic atrial fibrillation (an irregular erratic heart beat). Despite numerous doctors stating that Slayton was fit enough to fly in space NASA grounded him on 15 March 1962. The MA-7 flight went to Scott Carpenter with Schirra serving as his back-up.

Slayton assumed the role of Coordinator of Astronaut Activities, including the post of Chief of the Astronaut Office, in September 1962. In November 1963 Major Slayton retired from the USAF and continued his employment with NASA as a civilian. For three years he served as assistant director of flight crew operations. From 1966 he served as director of flight crew operations with flight crew selection as part of his responsibilities. Slayton selected the NASA flight crews for Project Gemini, Project Apollo, Project Skylab and the Apollo–Soyuz Test Project.

In April 1970, during the long hours spent fighting to return the Apollo XIII crew safely to Earth, Slayton began taking vitamin supplements. In July 1970 his heart fibrillation ceased and he began the process of having himself returned to flight status. He succeeded in March 1972 and was assigned to the prime crew of the Apollo–Soyuz Test Project with Commander Thomas Stafford and Command Module Pilot Vance Brand. Slayton served as Docking Module Pilot. The flight was launched on 15 July 1975 and completed the first docking between an American and Soviet manned spacecraft. The Apollo portion landed after 217 h 28 min 24 s.

From December 1975 to November 1977 Slayton was manager of the Approach and Landing Test programme carried out using the Shuttle Orbiter Vehicle 'Enterprise' at Edwards AFB, to prove the landing capabilities of Shuttle Orbiter, prior to manned space flights commencing. From November 1977 to February 1982 he was manager for the Orbital Flight Training Programme for the first four manned Shuttle flights. He was also responsible for the Shuttle Carrier Aircraft/Orbiter ferry programme.

Slayton retired from NASA on 27 February 1982 and subsequently served in a consultant role and in senior management positions in a number of aerospace companies. Following the break-up of his first marriage, he married Bobbie Osborn on 8 October 1983. Slayton died of brain cancer in June 1993.

Appendix B

The 'Mercury 13'

In December 1959 President Eisenhower had insisted that the Astronaut Candidates be qualified test pilots. At that time the only way to become a test pilot was to earn the qualification in the military services. As there were no women pilots in the military services at that time they could not attend test pilot school. With no women test pilots there could be no women astronauts. In 1960 Dr Randolph Lovelace II of the Lovelace Foundation, Albuquerque, agreed to put a group of American women pilots through the same selection process that had been used one year earlier to select the seven Flight A astronauts. The tests were carried out in secret and Jacqueline Cochran, who in 1953 had become the first woman to fly faster than Mach 1, funded the test programme from her private fortune. At no time was there any suggestion that the women who underwent the tests would be selected as NASA astronauts.

Geraldine 'Jerrie' Cobb was the first female pilot to undergo the tests. At the time of her selection she was 29 years old and had 7,000 hours flight experience, 2,000 hours more than John Glenn, the most experienced pilot among the Flight A astronauts. During five days of physical tests Cobb was subjected to the same 87 Phase I tests as the 32 men who had undergone the first phase of the selection process to become a member of NASA's first astronaut selection. With excellent results in all of the Phase I tests Cobb underwent Phase II testing at the Veteran's Administrative Hospital Laboratory, Oklahoma. Finally, she completed Phase III, two days of psychological and medical assessment, at the Naval Air Station, Pensacola, Florida. Cobb passed the whole process with flying colours, performing better on some of the tests than the seven men who had been selected to fly in Project Mercury. Cobb was introduced to NASA Administrator James Webb who assigned her as 'consultant' to his Administration.

Cobb's results led Cochran and Lovelace to select a group of 25 of America's leading women pilots to undergo the selection process, thereby giving the Lovelace clinic a considerable database of information on women's physical and psychological performance under the conditions examined by the selection process, which one year earlier had been recognised as one of the most thorough ever undertaken.

Throughout the summer of 1961 the women underwent Phase I testing in ones and twos, so as not to draw attention to the still secret programme. By the end of that summer twelve women had passed Phase I and Phase II testing. They were:

Myrtle Cagle	Sara Ratley (Gorelick)	Jean Hixon
Marion Dietrich	Geraldine Truhill	Irene Leverton
Jane Hart	Jan Dietrich	Bernice Steadman
Gene Nora Jessen (stumbough)	Mary Funk	Rhea Woltman (Allison)

Together with Jerrie Cobb the group were named the Fellow Lady Astronaut Trainees (FLATs). When their story became public the media quickly christened them the 'Mercury 13', as opposed to the Flight A astronauts, who were frequently referred to as the 'Mercury 7'.

The Lovelace test programme, along with data from other tests carried out on women, had shown them to be smaller and lighter than their male colleagues and they consumed less oxygen and food. All of these factors were important in Project Mercury where every last kilogram that could be removed from the payload the launch vehicle had to boost into space was vital. Women were also shown to be less prone to heart conditions and better able to withstand the extremes of heat and cold, pain, monotony and loneliness than men.

Phase II training was due to take place in the latter part of the 1961 summer and the women were sent train tickets for their journey to Oklahoma. Funk and Woltman completed Phase II training and Funk continued on to complete Phase III in Pensacola. For most of the women, however, their dreams of ever qualifying to become an astronaut died at the close of Phase I. In July 1961 the remainder of the Mercury 13 were told that their Phase II test programme had been cancelled. This despite the fact that Cobb had come out in the top 2 per cent of the 34 men and women who had completed the entire three-phase selection programme. In some tests Funk out-performed John Glenn, the best performer among the Flight A astronauts in the relevant tests.

With the cancellation of Phase II Cobb and Hart lobbied Vice President Johnson in Washington, forcing him to set up a Congressional Review to look into why NASA was not prepared to select women candidates. The review backed NASA and the women were told that there was no position for them in the manned space programme.

One year later the Soviet textile worker Valentina Tereshkova became the first woman to fly in space. Her flight was a political move orchestrated by Nikita Khrushchev and the female cosmonaut programme was discontinued when he was removed from power.

NASA finally selected its first group of female astronauts, as part of Flight H during their January 1978 selection for the Space Shuttle. When NASA announced plans to launch Sally Ride on STS-7, the Soviet Union launched Svetlana Savitskaya on board Soyuz T-7 before they could do so. Soyuz T-7 was launched on 19 August 1982 carrying a long-duration crew to the Salyut 7 space station. Savitskaya remained on Salyut 7 for almost four months, returning to Earth on 10 December 1982. Ride became the first American woman in space onboard STS-7, a solo Shuttle flight launched on 18 April 1983. The flight lasted six days. Over the next few years mixed crews on the Space Shuttle became a regular occurrence and were no longer met with pages of newsprint, as Ride's flight had been.

Following the announcement that Kathryn Sullivan would become the first female American astronaut to make an extravehicular activity (EVA) on STS-13/41G, the Soviets launched Savitskaya to Salyut 7 for the second time. Soyuz T10 was a short duration flight between 17 and 29 July 1983. During the flight Savitskaya became the first woman to fly in space twice and the first woman to make an EVA. Sullivan flew in space on STS-13 between 5 and 13 October 1984 and became the first American woman to make an EVA.

On 3 February 1995 astronaut Eileen Collins invited the 11 surviving members of the Mercury 13 (Marion Dietrich and Jan Hixson had died) to Cape Canaveral. The occasion was the first launch of an American spacecraft commanded by a female astronaut – Collins herself. Seven of the 11 surviving members made the trip as Collins's special guests.

Appendix C

Hardware dimensions

C.1 LITTLE JOE

Prime contractor:	North American Aviation Corporation
Length:	14.6 m
Diameter:	2 m (body)
	7 m (fins)
Engine type:	4 × Thiokol Recruit + 4 × Thiokol Pollux or,
	4 × Thiokol Recruit + 4 × Thiokol Castor
Propellant:	Solid
Thrust (gross):	4 × 154,078 N × 1.5 s (Recruit)
	4 × 222,400 N × 27 s (Pollux)
	4 × 278,356 N × 27 s (Castor)
Launch Complex:	Mobile, NASA Wallops Island Test Range

C.2 MERCURY REDSTONE

Prime contractor:	Chrysler Corporation
Length:	29 m
Diameter:	1.7 m (body)
	3.6 m (fins)
Engine type:	1 × North American A-7
Propellant:	Liquid oxygen / 75 per cent alcohol – 25 per cent water
Thrust:	346,944 N
Launch Complex:	No. 5, Atlantic Missile Range

C.3 MERCURY ATLAS

Prime contractor:	General Dynamics Aerospace Division

Length:	28.7 m
Diameter:	4.8 m (Booster) 3 m (Sustainer)
Engine type:	2 × Rocketdyne LR89-5 (Booster) 1 × Rocketdyne LR105-5 (Sustainer) 2 × Rocketdyne LR101-5 (vernier)
Propellant:	Liquid oxygen/kerosene (RP-1)
Thrust:	2 × 734,057.9 N (Booster) 1 × 253,581.7 N (Sustainer) 2 × 4,452.7 N (vernier)
Launch Complex:	No. 14, Atlantic Missile Range

C.4 MERCURY SCOUT

Stage One

Length:	9.1 m
Diameter:	1.2 m
Engine type:	Algol
Propellant:	Solid
Thrust	458,005 N × 36 s

Stage Two

Length:	6.1 m
Diameter:	78.7 cm
Engine type:	Castor
Propellant:	Solid
Thrust:	278,356 N × 27 s

Stage Three

Length:	2.8 m
Diameter:	78.7 cm
Engine type:	Antares X254
Propellant:	Solid
Thrust:	598,256 N × 39.7 s

Stage Four

Length:	77 cm
Diameter:	50.8 cm
Engine type:	Altair X248
Propellant:	Solid
Thrust:	12,543 N × 41.4 s
Launch Complex:	No. 18, Atlantic Missile Range

Appendix D

How space flight is achieved

D.1 MOTION IN SPACE

Objects travel through space along a curved trajectory called an ellipse. An ellipse is a closed, curved form resembling a squashed circle. Every ellipse has two foci (singular: focus). The centre of the ellipse lies midway between its two foci. The measure of the distance between the centre of an ellipse and either one of the two foci is called *eccentricity*. It is the measure of the shape of an ellipse. The longer the eccentricity the more elongated the ellipse. Alternatively, the closer together the foci, the more circular the ellipse.

The longest straight line that it is possible to draw through an ellipse passes through both foci and the centre and is called the *Major Axis*. The distance from the centre to one end of the Major Axis is called the *Semi-Major Axis*. The *Minor Axis* crosses the Major Axis at 90 degrees at the centre of the ellipse. The distance from the centre, to one end of the Minor Axis, is called the *Semi-Minor Axis*.

In 1609 Johannes Kepler (1571–1630) published two laws of motion. They were originally written to describe the movement of the planets around the Sun. The first law describes the motions of a Mercury Spacecraft in orbit and may be rewritten as:

A Mercury spacecraft orbits Earth in an ellipse.
Earth's centre occupies one focus of the elliptical orbit.
The other focus is empty.

The occupied focus lies slightly to one side of the centre of the spacecraft's elliptical orbit. Therefore, the spacecraft is at the highest and lowest altitudes above Earth's surface at opposite ends of the orbit's Major Axis. The highest point in an orbit around Earth is called *apogee*, while the lowest point is called *perigee*.

Kepler's second law states:

The radius vector sweeps out equal areas in equal time.

It explains the effects of gravitational attraction on the Mercury Spacecraft in orbit. The *radius vector* is an imaginary straight line from Earth's centre to the centre of the spacecraft. The law says that no matter which portion of its orbit the spacecraft travels around in a given period of time, the area swept by the radius vector will always be the same for the same period of time. This law dictates that a Mercury Spacecraft travels around its orbit fastest when close to perigee and slowest when close to apogee.

Sir Isaac Newton (1642–1727) was an English scientist and mathematician who defined three further laws of motion during the year 1665–66. His first law states:

> *A body at rest remains at rest and a body in motion remains in motion, at the same velocity and in a straight line unless acted upon by some external force.*

In this case 'Force' is a push, or pull, that alters an object's velocity and/or direction of travel. In Project Mercury an example of a body at rest is a launch vehicle sitting on the launch pad. It remains at rest until the rocket motors are ignited, at which time it is pushed in the opposite direction to its exhaust gases. When the rocket motors shut down, the launch vehicle and its payload are trying to leave Earth along a straight line and in the direction in which the launch vehicle was travelling when its rocket motors shut down. Gravitational attraction is an example of a pulling force. It acts upon the launch vehicle and its payload, pulling it towards Earth's centre, and changing its trajectory from a straight line to a curve. The second law states that:

> *Any change in a body's motion is proportional to the force acting upon it and takes place in the direction in which that force is acting.*

The various rocket motors used in Project Mercury are excellent examples of this law. Those employed in the Mercury Atlas launch vehicle were powerful enough to push against Earth's gravitational attraction and place the Mercury Spacecraft into orbit. On the other hand, the small thrusters used on the spacecraft were not powerful enough to change the spacecraft's orbit and could only be used to change its attitude (positioning). In all rocket motors the thrust acts in the opposite direction to the escaping exhaust gases. The third law states that:

> *The mutual attraction of two bodies acting upon each other are always equal and directed to contrary parts.*

The third law is often quoted as '*to every action there is an equal and opposite reaction*' and this is the principle behind rocket propulsion. The action in rocket propulsion is the release of pressure through the throat of the combustion chamber, which causes the exhaust gases to be compressed and expand rapidly within the extension nozzle. The reaction is the thrust thus produced against the inside of the combustion chamber and the extension nozzle, which pushes the vehicle in the opposite direction to the exhaust gases.

Newton's discovery of *gravitational attraction* is the subject of a legend, in which he was sitting under an apple tree when an apple fell to the ground nearby. He realised that the apple was at rest when attached to the tree, until an unidentified force (which was later named gravitational attraction) broke its connection to that tree and pulled it towards Earth's centre. When it struck the Earth's crust it could go no further.

Gravitational attraction also pulled the orbiting Mercury Spacecraft towards

Earth's centre, only the spacecraft was trying to fly through space, away from Earth, in a straight line and at a constant high forward velocity, in keeping with Newton's own first law of motion. Gravitational attraction acted on the spacecraft, bending its straight trajectory into a closed elliptical orbit around Earth. That orbit did not intersect the Earth's surface at any point because the surface fell away beneath the spacecraft in a steeper curve than that of the spacecraft's trajectory.

Newton's theory of Universal Gravitational Attraction states:

> *Any two particles of matter attract each other with a force directly proportional to the product of their masses and inversely proportional to the square of the distance between them.*

The Inverse Square Law, on which this theory is based, states '*The intensity of an effect is reduced in inverse proportion to the square of the distance from the source.*' That is to say, gravitational attraction acting on a square metre of space at a distance of 1 unit from the source is 1 ($1 \times 1 = 1$). However, if the distance from the source is doubled (2 units) then gravitational attraction acting on one square metre of space is reduced in inverse proportion to the distance from the source multiplied by itself ($2 \times 2 = 4$). Gravitational attraction is quartered. At a distance of 3 units gravitational attraction is one-ninth ($3 \times 3 = 9$) and so on. It is this falling off of gravitational attraction with increased distance that lies behind Kepler's second law of motion and made the Mercury Spacecraft travel more slowly the greater the distance between it and the centre of Earth's mass.

Because humans have evolved on Earth, we consider Earth's gravitational attraction to be normal. An object in free fall within Earth's gravitational attraction accelerates at a rate of 9.80665 m/s^2. This natural rate of acceleration is called one *gravity*, or 1*g*. If an object is accelerated artificially, beyond that velocity, it feels heavier than normal. If it accelerates at twice that velocity (2*g*) it feels twice as heavy. At 3*g* it feels three times as heavy, and so on.

D.2 MICROGRAVITY

A Mercury Spacecraft remained in orbit because it had established equilibrium between its forward velocity and Earth's gravitational attraction. It was being pulled towards Earth's centre along a trajectory that never intersects that Earth's surface. Earth's gravitational attraction was still present; indeed, it was keeping the spacecraft in orbit and dictating the shape of that orbit. Everything in and on the spacecraft was in a state of free fall, at the same rate. Gravity did not appear to act on objects, although it was always present. Objects floated freely if they were not held in position.

Today this condition is often called *microgravity*. At the time of Project Mercury it was called weightlessness, or zero gravity. Microgravity is defined as $1 \times 10^{-4}g$ (0.0001*g*) or less. Under these conditions, where gravity is present but not acted upon, other natural forces, normally swamped by gravity, become more obvious. In microgravity, these weaker forces can be subjected to practical experiments.

Newton's laws of motion can be observed in their purest form under microgravity conditions.

One NASA educational document defines microgravity as:

The absence of any apparent gravitational attraction on an object.

Up and down are terms that have no meaning in a microgravity environment. Despite this, the Mercury Spacecraft did display a sense of up and down, inside the Crew Compartment. This was achieved by locating the spacecraft's instrument panels and other equipment to suit the astronaut's position strapped in his prone couch.

In microgravity engineers apply a three-axis system, to give a fixed reference by which to gauge a spacecraft's movements. The axes (called X, Y and Z) are fixed in relation to the spacecraft and move with it. The three axes radiate out from the centre of the spacecraft, with each axis at right angles to the other two. Opposite ends of each axis are labelled + (plus) and − (minus). The symbol is written before the axis identification letter ($+Z$, or $-Z$).

For a launch vehicle standing on the launch complex, the X (longitudinal) axis passes along the length of the vehicle. The $+X$ axis points towards the sky while the $-X$ axis points towards the launch pedestal on which the vehicle is standing. In the Mercury Spacecraft the $+X$ axis was the direction that the astronaut faced when strapped in his prone couch. The $-X$ axis was to his back. The $+Y$ axis was to his left and the $-Y$ axis to his right. The $+Z$ axis extended out of the spacecraft above his head with the $-Z$ axis extended beneath his feet.

When manoeuvring, the Mercury Spacecraft rolled about its X axis, like a child's roundabout turning on its axis. The spacecraft pitched up and down (like someone nodding his head) about its Y axis and yawed (like someone shaking his head from side to side) about its Z axis.

The microgravity environment only existed while the spacecraft was in free fall. If the spacecraft's rocket motors were fired, acceleration forces reasserted themselves to the spacecraft's structure and anything firmly connected to it. The spacecraft's structure was then travelling faster than any loose objects inside it, which were still in free fall. The structure would move past those objects, giving them the appearance of falling towards whichever internal surface was opposite the direction of acceleration. When the rocket motors stop firing acceleration forces disappeared and the spacecraft's structure returned to a state of free fall.

D.3 ROCKET PROPULSION

A rocket motor's combustion chamber is a sealed chamber with an exit hole in one direction. When the rocket motor is firing external air pressure is much lower than the pressure inside the combustion chamber. Because of this inequality of pressure the exhaust gases, from the combustion of propellants inside the chamber, are drawn out through the exit hole. They are compressed as they pass through the exit hole, or throat, of the combustion chamber and enter the external bell-shaped exhaust nozzle where they accelerate rapidly to supersonic velocities as they expand to meet the inner wall of the extension nozzle. In expanding against the inner wall of the exhaust nozzle the gases transfer their energy to that wall, in the form of an impressed force called *thrust*. The exhaust gases then escape out of the open end of the exhaust nozzle.

Thrust is transmitted through the rocket motor's mountings to the thrust frame and thus to the vehicle's physical structure. It is the exhaust escaping in one direction that lowers the pressure in that direction and allows the higher pressure (thrust) to impress its force in the opposite direction. All rocket engines use this simple principle, regardless of their size, or the propellants they burn. The escaping exhaust is the action in Newton's third law of motion while the thrust, which impresses its force in the opposite direction, is the reaction.

D.4 MONO-PROPELLANT ROCKET ENGINES

Mono-propellant rocket motors were used for attitude manoeuvring thrusters on the Mercury Spacecraft. Mono-propellant rocket engines use the decay of a chemical component over a catalyst to produce a gas, which is then used as a propellant. These rocket engines employ the opening and closing of valves to release a small amount of gas under pressure, through the extension nozzle. The spacecraft moved in the opposite direction, by reaction to the impressed force.

D.5 SOLID ROCKET MOTORS

Solid rocket motors were used throughout Project Mercury. They propelled the Little Joe launch vehicle and all four stages of the Scout launch vehicle. On the Launch Escape System both the Escape Motor and the Jettison Motor burned solid propellants, as did the Posigrade Motors and Retrograde Motors mounted in the spacecraft's Retrograde Package.

Solid rocket motors provide high thrust with relatively low technology. Their one major drawback is the fact that they are uncontrollable once ignited and only stop firing when their propellant has been totally consumed. The solid rocket propellant used during Project Mercury consisted of a slurry of several chemicals containing everything necessary to ensure that the propellant would ignite when required and burn throughout the required flight time. The rocket casing provided a container to hold the slurry while it set into a rubbery substance called *grain*.

To achieve ignition an igniter flame was flashed along the centre of the grain and, once ignited, the grain always burned at right angles to the flame front. Combustion took place directly within the rocket casing, which served as the rocket motor's combustion chamber. The burning of the grain caused a very rapid build up of pressure within the rocket casing. Lower air pressure, directly outside the exhaust exit hole, drew the combustion gases out of the combustion chamber in that direction and impressed force pushed the vehicle in the opposite direction to the exhaust gases.

D.6 LIQUID-FUELLED ROCKET MOTORS

Both the Mercury Redstone and the Mercury Atlas launch vehicles burned liquid propellants in their rocket motors. In the case of the Mercury Atlas, three main propulsion units and two vernier steering rockets were all fed from one set of propellant tanks in a configuration known as a one and one-half stage launch vehicle.

Liquid propellant rocket motors required the mixing of two liquids – oxidiser and fuel – to form a controlled explosion within a combustion chamber. Separate propellant tanks house the oxidiser and fuel within the body of the rocket. At launch the liquid fuel and a liquid oxidiser mix, ignite and explode. The controlled release of the energy from that explosion, through the extension nozzle, causes the vehicle to move, by impressed force, in the opposite direction to the escaping exhaust.

At the appropriate time in the countdown, mechanical pumps drive the propellants out of their storage tanks alongside the launch pad. They pass through pipes, up the side of the umbilical tower, across the access arms and into the empty tanks in the launch vehicle.

Liquid oxygen (LOX), which was used as an oxidiser in both Mercury launch vehicles, was cryogenic (turned from liquid to gas at extremely low temperatures) and required very complicated and strict handling conditions. A layer of ice formed on the pipes used to transport it and on the exterior of the launch vehicle propellant tanks used to house it. The vibration at lift-off caused this ice to break free of the rocket stage, fall away and melt in the exhaust gases.

During loading the LOX 'boiled off' through relief valves, turning to gas on contact with the much warmer ambient air. To compensate for this boil off, the propellant tanks were continually topped up, until the moment in the countdown when the propellant umbilicals were closed and disconnected, just before lift-off. Simultaneously, the propellant tank relief valves were closed, thus preventing any further loss of propellant.

A turbo pump fed the fuel and LOX to the injector heads in the combustion chamber. At ignition liquid fuel and oxidiser were forced, at extremely high pressure, through separate injector heads, the high pressure forcing the liquids to atomise, forming a very fine spray. Leaving the injectors, the atomised fuel and oxidiser mixed together and an igniter heated and evaporated the atomised propellant mixture, causing the fumes to ignite and burn.

The heat produced inside a working rocket motor's combustion chamber was enough to melt the individual metals in that motor. To overcome this problem the fuel was passed through a series of pipes built into the rocket motor's outer surface. After passing around the exterior of the combustion chamber and exhaust extension nozzle, the fuel passed back to the fuel turbo pump where it was fed to the injector head. In this way the constantly moving fuel carried away much of the heat transferred to the inner wall of the rocket motor. This system is known as a regeneratively cooled rocket motor.

At ignition the first-stage rocket motors build up to maximum thrust. This build-up is called the transition to mainstage. Once the rocket is firing at full thrust (mainstage) the rocket lifted-off the launch pad. Some early Mercury Atlas launch vehicles were held in place on the launch pad by huge mechanical clamps that only released after the three rocket motors had achieved mainstage. During the early stages of lift-off the launch vehicle had a nearly full propellant load and was at its heaviest. It travelled relatively slowly.

As the vehicle gained altitude external atmospheric gases became thinner and air resistance on the outside of the launch vehicle decreased. Gravitational attraction also decreased with the increase in distance from Earth's centre of mass. As a result of these factors the rocket motors' performance increased as the launch vehicle gained altitude.

The rocket motors burned for a predetermined length of time, after which they shut down and became ineffective. Explosive charges inside the stage in place, fired, and the now useless stage separated and fell away under gravitational attraction. As the Mercury Redstone was a single-stage launch vehicle that was not powerful enough to place itself or its payload into orbit, gravitational attraction pulled it back to Earth once it had shut down and been jettisoned.

On the Mercury Atlas only the two outer motors shut down in this manner and were jettisoned. The single Sustainer motor, which had ignited at lift-off, continued to burn and placed the Sustainer stage into orbit along with the Mercury Spacecraft. The Mercury Spacecraft was separated from the Sustainer once it had achieved orbit and shut down.

D.7 GUIDANCE

The combination of the Mercury Spacecraft and any of its three launch vehicles climbing into the sky on a tail of exhaust gases was an extremely unstable object. Instability resulted from the fact that the launch vehicle's centre of pressure no longer lay above its centre of gravity. Instability increased the angle of attack to the oncoming stream of atoms making up the gases in the atmosphere. That in its turn increased the air resistance (drag) acting against the external skin of the climbing combination. An unguided launch vehicle or one in which the guidance system had malfunctioned might tumble until aerodynamic forces caused it to break up.

The Mercury Little Joe had no guidance system. Evenly spaced fins around the base of the rocket stabilised it as it passed through the thick lower atmosphere. The fins functioned in the same manner as the flights on an arrow. In the Mercury Redstone the fins were combined with an active guidance system. Any deviation from the correct flight path displaced the gyroscopes and sent an electronic signal to the rocket's guidance system. The guidance system then commanded carbon vanes acting within the exhaust gases to move, redirecting the exhaust gases from the single fixed rocket motor and causing the Redstone to move in the opposite direction to the new thrust vector.

Mercury Atlas employed gimballed, or movable, rocket motors under the control of an inertial guidance system, which contained a 'map' of the planned trajectory within the rocket's guidance computer. Signals from accelerometers aligned along the vehicle's three axes provided the guidance computer with information on the rocket's attitude, its trajectory and the ground elapsed time since lift-off. The computer converted any deviations between the planned and the actual trajectory into signals, which it sent to the rocket motors. Gimballing the rocket motor corrected the trajectory by redirecting the exhaust gases and causing the rocket to move in the opposite direction to the new thrust vector. Inertial guidance made the fitting of stabilisation fins around the base of a rocket vehicle largely obsolete. Mercury Atlas also had two vernier rocket motors that were used to make minor course correction too small to warrant gimballing the main propulsion units.

D.8 ABORT SENSING SYSTEM

A Mercury launch vehicle's Abort Sensing System measured a series of operating parameters of vital spacecraft and launch vehicle systems. It also used the onboard computers to compare the actual flight trajectory with maximum allowable discrepancies from the planned flight trajectory. If any of the monitored flight conditions exceeded the set parameters, the Abort Sensing System aborted the flight.

An abort may have involved simply stopping the countdown, or shutting off the rocket engines before the moment of launch. Once the launch vehicle was in motion, such an abort might involve the Range Safety Officer sending a signal to destroy the launch vehicle to prevent its falling on a populated area. If a manned launch had had to abort, the Abort Sensing System may have automatically activated the Launch Escape System (LES), which would have carried the Mercury Spacecraft and its astronaut to safety. The LES was only active while on the launch pad, or in the first few minutes of flight.

D.9 LAUNCH ESCAPE SYSTEM

A LES tractor rocket was a relatively small solid rocket motor mounted on the top of the Mercury Spacecraft as it stood on its launch vehicle. The LES extracted that spacecraft from a malfunctioning launch vehicle. The LES fired automatically, under the control of the launch vehicle's Abort Sensing System, if there was a catastrophic launch vehicle failure or if the launch vehicle flew outside its predetermined flight parameters. Alternatively, the LES could be fired under the manual control of the astronaut. However the LES was activated, the Abort Sensing System commanded the separation of the Mercury Spacecraft from the launch vehicle. A fraction of a second later, the LES Escape Motor ignited to pull the spacecraft away from the impending catastrophe.

Following the consumption of its solid rocket propellant, the LES Escape Motor stopped firing. The explosive bolts holding it to the top of the spacecraft exploded, separating the two. The LES Jettison Motors then ignited, to pull the now useless LES clear of the spacecraft. Using its Earth Landing System, the spacecraft descended to earth, as it would at the end of a successful mission.

If no launch abort occurred during the launch the LES was not used. It was jettisoned when the launch vehicle was at an altitude from which the Mercury Spacecraft could make a safe return to Earth on its own. To do this, the LES was separated from the spacecraft and both the Escape and Jettison Motors were ignited at a predetermined time in the launch schedule. This resulted in the LES being pulled clear of the still ascending launch vehicle.

D.10 ACHIEVING ORBIT

The Project Mercury launches into space took place in the same direction that Earth spins about its axis. This imparted Earth's rotational velocity to the launch vehicle even before it left the launch pad. The launch vehicle ascended vertically until clear of the thick, lowest portion of the atmosphere. It then pitched over, towards the east, to fly the final portion of its trajectory parallel to Earth's surface.

The three Mercury launch vehicles were required to impart the correct combination of altitude, velocity and direction to the Mercury Spacecraft, to allow it to complete its mission. During each launch gravitational attraction towards Earth's centre continually reduced the launch vehicle's velocity. Air resistance (drag), the result of molecules of atmospheric gases resisting the launch vehicle as it ascended through the atmosphere, also retarded a launch vehicle's forward velocity. All appreciable air resistance ended at an altitude of 193 km.

To overcome these natural forces the launch vehicle had to impart the required final velocity to the payload when the final rocket motor stop firing. When the launch vehicle's final velocity was less than 7.91 km/s, the spacecraft achieved a closed elliptical orbit, which intersected Earth's surface at some point. This was called a suborbital or ballistic trajectory.

Each Mercury launch vehicle attempted to fly into space in a straight line, as dictated by Newton's first law of motion. Earth's gravitational attraction overcame the launch vehicle's outward velocity and curved its trajectory into a closed ellipse with Earth's centre at one focus. The launch vehicle's final altitude, itself a result of the vehicle's velocity, dictated the curve of the elliptical trajectory.

Forward velocity affected the launch vehicle's resistance to gravitational attraction. The faster it was travelling, the further it would travel before gravitational attraction curved its trajectory, pulling it back towards Earth's centre. This also affected the downrange distance travelled before it intersected Earth's surface. As a general rule of thumb, high velocities resulted in high altitudes and large downrange distances and low velocities resulted in low altitudes and short downrange distances.

It was possible to predict where a Mercury Spacecraft would intersect Earth's crust at the end of its suborbital trajectory and recover it after the flight. This required a suitable Earth Landing System to be installed in the spacecraft and Recovery Forces in the correct recovery zone.

Where the launch vehicle's final velocity lay between 7.91 and 11.5 km/s the curve of the elliptical trajectory would not intersect Earth's surface because that surface fell away in a steeper curve than that of the ellipse forming the launch vehicle's trajectory through space. In keeping with Kepler's first law of motion, Earth's centre lay at one focus of the ellipse, the other focus being empty. The closed elliptical trajectory was called an orbit.

D.11 APOGEE AND PERIGEE

In keeping with Kepler's first law of motion the Mercury Spacecraft's orbits were elliptical, with the centre of the primary body at one focus. One end of the major axis of the ellipse was higher above the Earth's surface than the other end. The highest point in an orbit about Earth is called the apogee and the lowest point is the perigee. Apogee and perigee of an orbit are always directly opposite each other. The Mercury Spacecraft entered orbit at perigee and climbed towards apogee, before falling back towards perigee.

D.12 ALTITUDE

Altitude was a way of stating how high a Mercury Spacecraft was above Earth's surface; it is the length of an imaginary line from Earth's surface to the centre of the spacecraft. Apogee and perigee are measurements of altitude.

D.13 ORBITAL PERIOD

The orbital period was a measure of the time it took the Mercury Spacecraft to complete one orbit. An orbit is a closed elliptical trajectory beginning and ending at a point, fixed in space (usually the point at which the spacecraft entered orbit) after 360 degrees of travel.

D.14 INCLINATION

Inclination was the angle at which the orbital trajectory crossed Earth's equator. Posigrade orbits such as those used by the Mercury Spacecraft have inclinations between zero and 90 degrees. The angle of inclination of the orbit was equal to the maximum latitude north or south of the equator that the orbit reached over Earth's surface. No launch could achieve an orbit that had an inclination (in degrees) less than the latitude (in degrees) of the launch complex.

D.15 GROUND TRACKS

Ground track is the name given to the imaginary line described on Earth's surface directly beneath the orbital trajectory of a Mercury Spacecraft. Shown in two dimensions, the pattern of a ground track depends on the angle of inclination of the spacecraft's orbit. The ground track of an inclined orbit, such as those employed by the Mercury Spacecraft, crossed the equator at opposite sides of the planet. If apogee lay north of the equator then perigee would lie south of the equator, at opposite ends of the major axis of the orbital ellipse.

The Mercury Spacecraft entered orbit at perigee and travelled north, crossing the equator and continuing north until it reached apogee. Now moving south, it crossed the equator once more and continued south until it returned to perigee, where it began moving north once more. In two dimensions an inclined orbit resulted in a wavy line that passed both above and below the straight line representing Earth's equator. The angle of inclination dictated the highest and lowest degrees of latitude above and below the equator reached by the orbit at apogee and perigee. A few degrees separated the lines representing the ground tracks of individual orbits. The number of degrees that Earth turned about its axis, during one orbit of the Mercury Spacecraft, dictated the exact separation between the lines.

D.16 ATTITUDE CONTROL

Attitude was the word used to define a satellite's local position, with reference to its *X*, *Y* and *Z* axes. The ability to manoeuvre the Mercury Spacecraft about its three axes while it was in orbit allowed it to take up the correct attitude for its various in-flight manoeuvres, or to point the communication antennae in the correct direction.

Attitude control required a series of low thrust rocket motors, called *thrusters*, to be installed on the spacecraft. Thrusters worked like any other rocket, by applying thrust to move the Mercury Spacecraft in reaction to impressed force, in the opposite direction to their exhaust gases. Engineers called these thruster firings 'burns'. The two systems of attitude control thrusters installed on the Mercury Spacecraft used mono-propellant rocket motors.

Mercury's thrusters were not powerful enough to significantly add to, or decrease from, the spacecraft's forward velocity, so attitude manoeuvres did not adjust its orbital trajectory. Rather, they pointed the spacecraft in the correct direction for a particular experiment, observation or communications link. In the case of the Mercury Spacecraft they were also used to turn the spacecraft around following separation from its Atlas launch vehicle and to position it in the correct orbital attitude. They were also used to place the spacecraft in the correct attitude for retrofire and re-entry.

Having been established in the correct orbit and stopped all movements and vibrations, the Mercury Spacecraft would maintain the selected position without further manoeuvres. The spacecraft's attitude was allowed to drift out of alignment within set parameters. When it reached the limits of those parameters the Automatic Stabilisation and Control System (ASCS) fired the RCS thrusters and brought it back to the correct attitude.

During Project Mercury some flights conserved limited attitude manoeuvring propellant by employing drifting flight. In that flight mode the astronauts turned off the ASCS and 'caged', or locked, the spacecraft's gyroscopes. With no information coming from the gyroscopes, the ASCS remained 'ignorant' of the spacecraft's attitude and could not order any manoeuvres to correct it. With no attitude control, the spacecraft was free to drift in any attitude while still travelling around its orbit. When the astronauts uncaged the gyroscopes and turned the ASCS back on, it re-established the spacecraft in the correct attitude.

D.17 RE-ENTRY

To push their way through Earth's atmosphere the Mercury Spacecraft/launch vehicles were aerodynamically streamlined and travelled relatively slowly. The launch vehicle gained velocity as it gained altitude and the atmosphere got thinner. These procedures reduced air resistance (drag) to a minimum. Even so, the energy imparted to the launch vehicle and its spacecraft as it accelerated into orbit was sufficient to vaporise any known material. That energy had to be lost before the payload could return to Earth.

A Mercury Spacecraft could only leave orbit and return to Earth in one way – its velocity had to fall below that required to keep it in an orbit, the curve of which did not intercept Earth's surface. Such a loss of velocity could happen naturally, due to air resistance, or it could be deliberate, the result of the retrograde rockets being deliberately fired to end the flight. Retrofire took place against the spacecraft's direction of travel and caused it to lose velocity.

When the spacecraft's velocity fell below that required to remain in orbit, gravitational attraction pulled it along a curved trajectory that intersected Earth's surface. By calculating the spacecraft's velocity after retrofire, it was possible to predict the curve along which it would travel through the atmosphere. That made it possible to have recovery forces waiting for the spacecraft in the predicted landing area.

To return to Earth the Mercury Spacecraft had to pass back through the atmosphere (re-entry). The spacecraft designers at the PARD were able to employ air resistance (drag) to slow the vehicle down and reduce landing loads to within limits that the spacecraft and any living passengers could survive.

American engineers developed the broad-faced, high-drag design used on modern re-entry vehicles in the early 1950s. The leading face was the widest portion of the re-entry vehicle. Air in the bow wave travelled harmlessly around the edges of the leading face. Little or none of that airflow touched the spacecraft's afterbody. The spacecraft's kinetic energy was transferred to heat. Air pushed away in front of the re-entry vehicle became compressed, leading to its becoming heated to many thousands of degrees. A sheath of ionised air molecules formed around the re-entry vehicle, making radio communications impossible.

Air resistance reduced the velocity of the spacecraft. Ninety-nine per cent of the energy built up during lift-off was lost harmlessly to the airflow. Even so, the remaining 0.1 per cent of the energy was still sufficient to vaporise the re-entry vehicle's structure. The remaining energy was expressed as heat, which must be returned to the airflow, before it can reach the vehicle's structure. This was the task of the ablative heat shield.

The Mercury Spacecraft's ablative heat shields employed a variety of advanced plastics to absorb the final 1 per cent of energy, as heat. The heat shield charred, melted, or vaporised, depending on its design, and thus carried the absorbed heat away in the airflow. Sufficient ablative material had to remain, at the end of the heating phase to prevent the heat reaching the structure of the re-entry vehicle.

When a Mercury Spacecraft reached the lower atmosphere, most of its energy had been safely lost. Its velocity entered the subsonic range (less than Mach 1). The bow wave of compressed air in front of the spacecraft no longer formed because the vehicle was now travelling slowly enough to move the air out of its way. With no bow wave, radio communications were re-established. The spacecraft fell through the atmosphere, towards the primary body's centre, under gravitational attraction. During the final few thousand metres of descent, parachutes were employed to achieve final braking.

Following deployment of the parachute(s) two forces combine to counteract the weight of the falling spacecraft. First, the weight of the falling spacecraft pulled the deployed parachute downwards, through the atmospheric gases. This produced air resistance (drag), slowing the descent. Second, atmospheric gases push upwards, against the underside of the parachute, causing lift.

The Mercury Spacecraft employed a two-parachute landing system. A barometric switch, which sensed atmospheric pressure, initiated their deployment. Explosive bolts fired to released the Antenna Canister and deployed the drogue parachute, which stopped the spacecraft oscillating and steadied it for the deployment of the larger main parachute. The Antenna Canister was carried away by the drogue parachute. In separating, the Antenna Canister deployed a lanyard that ran from the Antenna Canister to the main parachute and the lanyard pulled the main parachute from its container.

To reduce deployment loads the main parachute was initially deployed reefed, or partially open. This further reduced the spacecraft's velocity. Cutting the reefing lines allowed the main parachute to deploy fully. The main parachute reduced landing loads to within limits that the spacecraft and its contents could survive.

The Mercury Spacecraft landed in the ocean. This allowed the seawater to absorb the final landing forces, together with shock attenuation systems built into the spacecraft. The water also cooled the extremely hot spacecraft before its recovery by the US Navy.

Appendix E

The human element

E.1 THE EFFECTS OF ALTITUDE

The human body has evolved to function at ground level, at the bottom of Earth's atmosphere. At that altitude the atmosphere consists of 77 per cent nitrogen, 21 per cent oxygen and 2 per cent other gases and trace elements. A gravitational attraction of one-gravity attracts the gases of the atmosphere towards the centre of the planet and they rest on the planet's surface and on the outer surface of the human body, with a pressure of 760 mm of mercury. Under those conditions the average adult human heart beats 72 times per minute and the average person breathes between 16 and 18 times per minute.

When the average person inhales, he sucks in enough atmospheric gases to fully inflate his lungs. The inhaled gases are collected in the lung's alveoli, or air sacs, and oxygen is collected from the inhaled breath by the red cells in the bloodstream. The left side of the heart pumps oxygenated blood around the body within a network of arteries. As the blood moves around the body the arteries divide, continually becoming smaller, until they become capillaries. Capillaries are very thin blood vessels through which the two-way exchange of gases and fluids to and from the body tissues take place. Oxygen in the bloodstream is diffused into the body tissues, where it serves as a fuel for the muscles and the brain.

When their body completes an action, the muscles give off carbon dioxide as a waste product, which enters the deoxygenated bloodstream in the capillaries and is carried away from the body tissues. The capillaries continually join up to become veins, which return the deoxygenated blood to the right side of the heart and thus to the alveoli in the lungs. From the alveoli the carbon dioxide mixes with the inhaled air in the lungs, replacing the oxygen that has entered the bloodstream. The carbon dioxide leaves the body in exhaled breath. Under normal living conditions, close to sea level on Earth, the human body is saturated with oxygen in the air around it. Even so, the body only has a storage capacity for approximately one litre of the gas.

An unprotected body travelling vertically through Earth's atmosphere would feel little or no ill effects until it reached an altitude of approximately 3,000 m. With considerably less oxygen in the air at that altitude to feed the muscles and brain, the body starts to suffer from hypoxia, literally a lack of oxygen.

Around 9,150 m the body tissues and fluids begin out-gassing nitrogen and other gases because of the low pressure. This painful condition is called dysbarism, or air embolism. Breathing 100 per cent oxygen overcomes this condition. Above 12,000 m pure oxygen at increased pressure is necessary, to avoid dysbarism. Even pure oxygen at increased pressure will not prevent the condition above 13,000 m.

During the breathing process the lungs produce carbon dioxide and water vapour under a constant pressure, while air is inhaled at the surrounding external pressure. As the unprotected body rises through the atmosphere, the surrounding external air pressure decreases and carbon dioxide and water vapour occupy more of the interior volume of the lungs, leaving less room for inhaled air.

The internal temperature of the alveoli in the human lungs is 37 °C. At an altitude of 15,000 m the surrounding air pressure is 87 mm of mercury, equal to the combined pressures of carbon dioxide and water vapour at 37 °C. At that altitude the unprotected lungs are full of carbon dioxide and water vapour. There is no internal volume left for the inhalation of external air. This condition is called anoxia, or oxygen starvation.

At an altitude of 19,000 m the surrounding air pressure is down to 47 mm of mercury. At that pressure the boiling point of the fluids in an unprotected human body is 37 °C. This also happens to be the temperature of a healthy human body. Therefore, at 19,000 m the fluids in an unprotected human body boil, even if that body is at rest. The use of an artificially induced higher surrounding pressure increases the temperature at which the body fluids boil and overcomes anoxia. This was the function served by the pressurised Crew Compartment in the Mercury Spacecraft and the astronaut's pressure suit, with their artificial Life Support Systems.

Anything ascending to 24,000 m in Earth's atmosphere passes through the ozone layer. Ozone is toxic to the human body, therefore a person wanting to pass through the ozone layer requires a fully sealed unit, with a self-contained Life Support System. At approximately 40,000 m the effects from Cosmic Radiation increase and the human body requires protection. Solar Ultra Violet Radiation increases dramatically at an altitude of 45,000 m. The human body requires further protection against the effects of this increase.

E.2 THE EFFECTS OF ACCELERATION

Acceleration is defined as: *Change of velocity per unit of time squared.* There are three forms of acceleration forces:

Linear	an increase, or decrease, in velocity.
Radial	changes in the direction the movement.
Angular	combination of both of the above.

These forces occur against the following four criteria:

Acceleration

An object in free fall in Earth's atmosphere accelerates at a velocity of 9.75 m/s under gravitational attraction. This rate of acceleration (one gravity or 1*g*) is the principal measure of acceleration forces on Earth. An object travelling at twice this speed is accelerating at 2*g*, and so on.

Duration

The type of acceleration forces experienced depends on the length of time (duration) for which the subject experiences those forces, in seconds. The definitions are as follows:

Less than 0.2 second	Impact acceleration
0.2 to 2 seconds	Abrupt acceleration
2 to 10 seconds	Brief acceleration
10 to 60 seconds	Long-term acceleration
over 60 seconds	Prolonged acceleration

Onset

This is the increase in acceleration (*g* forces) per second.

Direction

When acceleration occurs in a given direction, acceleration forces act in the opposite direction. The direction of the acceleration forces are defined using the same three-axis system that defines the spacecraft's attitude in space. Definition of the forces is in relation to the direction in which they displace the astronaut's body organs.

If the astronaut fires the rocket engines behind his back, in a posigrade burn giving an acceleration of 1*g*, he will receive acceleration forces of 1*g* along the *X* axis, from his chest to his spine. Those forces will displace his body organs towards his spine. He receives an acceleration force of $+1g\ X$ (1*g* acceleration (posigrade)) along the *X* axis. If the manoeuvre had been a retrograde burn, the deceleration forces occur in a spine to chest direction, displacing the organs towards the ribs. The forces concerned are $-1g\ X$ (1*g* deceleration (retrograde)) along the *X* axis.

The highest levels of acceleration forces experienced during the manned space flights of Project Mercury took place along the spacecraft's *X* axis. The astronauts' couches were placed so that they experienced those forces in a transverse, or chest to back position. In that position the acceleration forces acted across the main blood vessels. Under high levels of transverse acceleration forces the subjects experienced considerable difficulty in breathing. This is the result of the increased difficulty experienced in the lifting of the anterior wall of the thorax. An astronaut's couch held its occupant in the correct position to receive transverse acceleration forces with the minimum of discomfort.

E.3 THE EFFECTS OF MICROGRAVITY

On entering microgravity the human body undergoes a series of physiological changes. With no apparent gravitational attraction to pull body fluids into the lower body, those fluids pool in the upper body. This causes the girth of the legs to shrink, while causing a red, puffy appearance in the face. All astronauts experience distended veins, puffed eyelids and nasal congestion. More blood enters the right ventricle of the heart, while the left ventricle decreases in size. Because the heart is pumping more blood with each beat, it needs fewer beats per minute to maintain the same internal body pressure. This collection of body fluids in the upper body sets in motion a series of natural reactions. These reactions are the body's attempts to purge what it sees as excess fluid, through urination.

By altering its hormone balance, the body produces fewer red blood cells, less plasma and less potassium. These changes lead to a loss of calcium, hydroxyproline (an amino acid associated with the metabolic turnover of bone) and nitrogen, in the astronaut's urine. High calcium levels in the urine leave astronauts on long-duration flights open to the possibility of developing kidney stones.

Microgravity can cause calcium to be lost from the bones. Structural material in particular is lost from load-bearing bones, normally compressed under gravitational attraction. Bone mineral loss is not substantial, but it does lead to the bones becoming spongy and brittle. Nitrogen loss suggests a loss of muscle mass, and this atrophy is the result of the fact that many muscles do not have to work as hard, if at all, under microgravity conditions. Leg muscles in particular receive very little use during periods of microgravity. All astronauts that have flown in space have readapted to their normal physical condition within weeks of returning to Earth.

Almost half of the astronauts who have flown in space have suffered from Space Adaptation Syndrome, a form of motion sickness. The symptoms are nausea, dizziness, sweating, fatigue and vomiting. These symptoms may persist for several days, making an astronaut unable to complete the heavy work schedule in the early portion of the flight. The problem is caused by conflicting signals reaching the brain from the body's sensory organs. The otolists, a natural orientation device inside the inner ear, are no longer subject to the effects of gravitational attraction, and this prevents them from sending accurate information on the body's orientation to the brain. The inaccurate information received from the otolists is thought to contradict the information received from the eyes, thus causing the astronaut to become disoriented. That disorientation may cause the brain to release a substance that the body identifies as toxic, making the astronaut vomit, to purge his body of that substance.

With no gravitational attraction compressing the vertebrae of the spine, they are free to expand. An astronaut's spine may lengthen by up to 7.4 mm while it is subject to a microgravity environment. This can cause back pain and a partial loss of the sense of touch, due to blocked nerves. The spine returns to its normal length when the astronaut returns to Earth. This frees any blocked nerves and the full sense of touch returns.

E.4 LIVING IN SPACE

If the human body is to function at optimum levels it requires the five environmental conditions detailed below.

Pressure

Depending on the amount of available oxygen, the body can function for varying lengths of time, at pressures between 175.25 and 2812.4 g/cm^2. The correct external pressure on the body prevents a series of extremely painful internal effects. Uncorrected, these conditions would ultimately result in the death of the subject.

Inside the Mercury Spacecraft the pressure vessel, the Crew Compartment, held the artificial atmosphere under the correct conditions. The spacecraft's Life Support System produced those conditions. If the spacecraft pressure vessel failed the supply of environmental gases to the Crew Compartment could be turned off and the astronaut could return to Earth using his pressure suit. The pressure suit worn by astronauts was a pressure vessel of the smallest possible dimensions, which received its oxygen through short hoses called umbilicals from the spacecraft's Life Support System.

During Project Mercury the astronauts wore their pressure suits for the full duration of the flight. This was because the spacecraft lacked the room for them to remove their suits. The suit was only pressurised during the most dangerous portions of the flight. Throughout the remainder of the flight the pressure suit was worn with the helmet visor open. At such times, oxygen passed through the pressure suit to keep the astronaut as cool and as comfortable as possible.

Temperature

The human body only gives optimum performance at temperatures close to 37 °C. Too hot, or too cold and the body will die. Both the astronaut's body and the Mercury Spacecraft's electrical systems gave off heat as a by-product of their correct functioning. Removal of that heat was necessary to prevent overheating and malfunction of both the astronaut and his machinery. Excessive overheating would have caused malfunctions of the spacecraft equipment and incapacitated the astronaut.

Oxygen

To function at optimum performance the human body also requires a steady supply of oxygen. Three minutes without oxygen causes serious brain damage. Five minutes without oxygen and the body will die.

Oxygen was stored in the Mercury Spacecraft in high-pressure bottles. It was fed through two separate channels to the Crew Compartment and the astronaut's pressure suit. The oxygen carried away heat from the astronaut's body and the spacecraft's equipment, along with exhaled carbon dioxide and moisture. The used oxygen was chemically scrubbed to remove carbon dioxide and other impurities. Exhaled moisture was caught in a condensation trap and heat was removed by means of a heat exchanger. Fresh oxygen was added to the cleaned oxygen, which was then recycled.

Water

Water is the most vital of the listed requirements. Even if an astronaut receives an adequate supply of each of the other four, death will occur after two days without water. The Mercury Spacecraft therefore carried high-pressure storage tanks containing potable water. The astronaut consumed water throughout his flight and his body rejected water in exhaled breath, perspiration, urine and solid body waste (faeces). Perspiration entered the Crew Compartment atmosphere through evaporation. Along with exhaled water vapour it was separated and processed in the oxygen regeneration cycle.

Food

The human body requires food to keep functioning and to maintain the correct internal temperature. The Mercury astronauts ate purėed food from tubes and bite-size snacks held in sealed bags until they were required. All food consumed was recorded to assist the flight surgeons on Earth in ensuring correct results from their post-flight examination of the astronauts.

E.5 DEALING WITH HUMAN BODY WASTE

During its normal functions the human body produces five forms of natural waste: exhaled breath, heat, hair, urine and faeces. These must all be adequately dealt with if the astronauts are not to suffer intolerable conditions onboard their spacecraft during flight.

Exhaled breath

Exhaled breath contains unused oxygen, carbon dioxide and water vapour. It entered the atmosphere of the Crew Compartment or the astronaut's pressure suit. The exhaled breath was processed as part of the Mercury Spacecraft's oxygen revitalisation process.

A bed of lithium hydroxide removed carbon dioxide from the used atmosphere and a separate bed of activated charcoal removed unwanted odours. A water separator then removed the condensed water vapour. The cleaned oxygen was mixed with unused oxygen and warmed in a heat exchanger before being re-introduced into the Crew Compartment or pressure suit.

Body heat

Body heat was carried away in the oxygen circulating through the astronaut's pressure suit. It was dealt with as part of the oxygen revitalisation process.

Urine

On early space flights during Project Mercury toilet facilities were extremely basic. Alan Shepard, the first American astronaut to fly in Project Mercury, was only expected to be in flight for 15 minutes during his suborbital flight. With this fact in mind no consideration was given to his having to relieve himself in flight. Long delays in launching the flight led to the astronaut lying on his back, on top of his launch vehicle for many hours. Ultimately, he felt the need to urinate. With no solution to his problem forthcoming from the men on the ground the astronaut took it upon himself to urinate directly into his pressure suit. With the technical problems overcome, the flight took place as planned. The problem of human body waste would be addressed before the next flight.

Thereafter, while wearing his pressure suit, the American astronaut wore a condom-like device on his penis. This attached, via a tube, to the Urine Collection Bag inside his pressure suit. The need to defecate was minimised by the astronaut's strictly controlled diet in the days leading up to launch.

E.6 EXPOSURE TO VACUUM

If astronauts were to function in space, their bodies had be protected from the vacuum of space. Under normal conditions, the spacecraft's Crew Compartment contained the correct gases (atmosphere), supplied by the Life Support System, at the correct temperature and pressure to allow the human body to function correctly. However, there was always a danger of decompression.

Exposure to vacuum is not instantly fatal and does not result in any of the spectacular effects shown in the popular films of recent years. In a failed pressure suit test a NASA technician was exposed to vacuum in a vacuum chamber for 30 seconds and survived with no long-term effects.

In 1971 the crew of Soyuz 11 were exposed to the vacuum for 5 minutes, when a pressure equalisation valve on their spacecraft jammed in the open position. Their spacecraft returned to Earth using automatic recovery systems and landed safely, but the three cosmonauts were found dead in their couches. Film of the recovery crew attempting to revive one of the cosmonauts shows his body lying on the steppe alongside his spacecraft. He appears externally normal.

During World War II the German Nazi regime used Jewish prisoners in their concentration camps as live subjects in decompression tests. The data obtained was passed to the Luftwaffe. After the war the German data was taken to America. Early post-war animal tests of 'explosive (very rapid) decompression' resulted in the loss of most, but not all, of the test subjects. Further tests, to define exactly what happened to the human body during decompression, showed that the rate at which the pressure change takes place and the difference between the starting pressure and the lowest pressure achieved were both pertinent factors.

If astronauts had been exposed to the vacuum of space through decompression, a series of internal biological effects would have taken place. At a pressure of 47 mm of mercury the liquids in their body would have turned to gas, extracting body heat from the muscles and organs in the process and a deep cold would set into the body. After 6 seconds the vaporised liquids would have left the heart and lungs, causing them to collapse. Blood pressure would drop, causing circulatory interruption, anoxia (oxygen starvation), convulsions and relaxation of the bowel muscles and a final, large-scale evaporation of the remaining body fluids. A complete interruption of the blood supply for 13 seconds or more would have led to mental confusion and the inability to make decisions. A 20-second interruption of the blood supply would have resulted in loss of consciousness. Their hearts would have stopped beating after some 2 minutes and they would have died.

The Mercury astronauts wore pressure suits to serve as a secondary line of defence against a loss of pressure in the spacecraft's Crew Compartment. None of the astronauts that flew during Project Mercury experienced an unplanned decompression of their spacecraft.

Appendix F

Tables

Table F.1 Mercury Little Joe flight programme, January 1959

Flight	Launch vehicle	Spacecraft	Launch date	Test subject	Remarks
–	1	–		–	Manufacturer's prototype
LJ-1	2	Prototype	Jul. 1959	–	LES max-*q* qualification
LJ-2	3	Prototype	Jul. 1959	–	Spacecraft re-entry dynamics development
LJ-3	4	Prototype	Aug. 1959	–	Spacecraft re-entry dynamics development
LJ-4	5	Prototype	Sep. 1959	–	Spacecraft re-entry dynamics development
LJ-5	6	SC-3	Nov. 1959	Chimpanzee	Production SC LES max-*q* qualification
LJ-6	7	Not designated	Dec. 1959	–	Back-up to be flown if necessary

Table F.2(a) Mercury Little Joe flight programme, January 1959–61

Flight	Launch vehicle	Spacecraft	Launch date	Test subject	Remarks
LJ-1	2	Prototype	21 Aug. 59	–	LES max-*q* qualification
LJ-2	3	Prototype	4 Nov. 59	–	LES max-*q* qualification
LJ-3	4	Prototype	4 Dec. 59	Rhesus monkey	LES max-*q* qualification
LJ-4	5	Prototype	21 Jan. 60	Rhesus monkey	High altitude LES qualification
LJ-5	6	SC-3	8 Nov. 60	–	LES max-*q* qualification
LJ-6	7	SC-14	18 Mar. 61	–	LES max-*q* qualification
LJ-7	1	SC-14A	28 Apr. 61	–	LES max-*q* qualification

Table F.2(b) Mercury Little Joe flight programme 1959–61

Flight	Launch vehicle	Spacecraft	Launch date	Test subject	Apogee (km)	Downrange (km)	Duration h.min.s.
LJ-1	2	Prototype	21 Aug. 59	–	–	–	–
LJ-6	2	Boilerplate	4 Oct. 59	–	59.72	127.75	00.05.10
LJ-1A	3	Prototype	4 Nov. 59	–	14.48	18.50	00.08.11
LJ-2	4	Prototype	4 Dec. 59	Rhesus monkey SAM	85.32	312.78	00.11.06
LJ-1B	5	Prototype	21 Jan. 60	Rhesus monkey Miss SAM	14.96	18.82	00.08.35
LJ-5	6	SC-3	8 Nov. 60	–	16.25	21.88	00.02.22
LJ-5A	7	SC-14	18 Mar. 61	–	12.43	31.85	00.23.48
LJ-5B	1	SC-14A	28 Apr. 61	–	04.45	14.48	00.05.25

Table F.3 Mercury Redstone flight programme, January 1959

Flight	Launch date	Test subject	Remarks
MR-1	November 1959	–	Spacecraft and Redstone qualification
MR-2	December 1959	–	Spacecraft and Redstone qualification
MR-3	January 1960	Manned	Suborbital training flight
MR-4	February 1960	Manned	Suborbital training flight
MR-5	March 1960	Manned	Suborbital training flight
MR-6	April 1960	Manned	Suborbital training flight
MR-7	May 1960	Manned	Suborbital training flight
MR-8	June 1960	Manned	Suborbital training flight

Table F.4 Mercury Redstone flight programme 1960–61

Flight	Launch vehicle	Spacecraft	Pilot	Launch date	Launch time	Apogee (km)	Down range	Duration (h.min.s)
MR-1	1	SC-2	Mechanical man	21.11.60	1400	–	–	–
MR-1A	3	SC-2	Mechanical man	19.12.60	1645	210.00	377.79	00.15.45
MR-2	2	SC-5	Chimpanzee HAM	31.01.61	1655	252.61	672.56	00.16.39
MR-BD	5	Prototype	–	24.03.61	1730	182.62	494.60	00.08.23
MR-3	7	SC-7	Shepard	05.05.61	1434	187.44	487.20	00.15.22
MR-4	8	SC-11	Grissom	21.07.61	1220	190.28	486.07	00.15.37

Table F.5 Mercury Atlas flight programme 1959–63

Flight	Launch vehicle	Space-craft	Pilot/ test subject	Launch date	Launch time	Apogee (km)	Perigee (km)	Inclination (degrees)	No. of orbits	Landing date	Landing time	Duration (h.min.s)
BJ	10D	Prototype	–	09.09.59	0819	152.8	–	–	0	09.09.59	0832	00.13.00
MA-1	50D	SC-4	–	29.07.60	0913	3.3	–	–	0	29.07.60	0916	00.03.30
MA-2	67D	SC-6	–	21.02.61	0910	183.5	–	–	0	21.02.61	0928	00.17.56
MA-3	100D	SC-8	–	25.04.61	1115	07.24	–	–	0	25.04.61	1122	00.07.07
MA-4	88D	SC-8A	–	13.09.61	0904	228.6	159.1	32.8	1	13.09.61	1055	01.49.20
MA-5	93D	SC-9	Chimpanzee Enos	29.11.61	1008	237.1	160.9	32.6	2	29.11.61	1311	03.02.59
MA-6	109D	SC-13	Glenn	20.02.62	0947	260.9	160.9	32.5	3	20.02.61	1442	04.55.23
MA7	107D	SC-18	Carpenter	24.05.62	0745	268.4	160.7	32.5	3	24.05.62	1241	04.56.05
MA-8	113D	SC-16	Schirra	03.10.62	0715	289.9	161.3	32.5	6	03.10.62	1628	09.13.00
MA-9	130D	SC-20	Cooper	15.05.63	0804	269.9	161.3	32.5	22	16.10.63	1823	34.19.19

Table F.6 Vostock flight programme 1960–63

Flight	Launch vehicle	Pilot/ test subjects	Launch date	Launch time (MT)*	Perigee (km)	Apogee (km)	Inclination (degrees)	No. of orbits	Landing date	Landing time
T-1	R-7	–	00.01.60†		–	1	–	0	00.01.60†	
T-2	R-7	–	30.01.60			–	–	0	31.01.60	
KB-1	R-7	–	15.05.60		312	368	65		15.10.65	No recovery
KB	R-7	Chiaka Lisichka	28.07.60		–	–	–	–	–	–
KB-2	R-7	Strelka Belka	19.08.00	1138	339	360	65	18	20.08.60	1402
KB-3	R-7	Pchelka Muska	01.12.60	1026	166	232	65	17	02.12.60	No recovery
KB	R-7	Shutka Kometa	22.12.60		–	8	–	0	22.12.00	Aborted orbital flight
KB-4	R-7	Chenushka Manequin	09.03.61	0929	173	239	65	1	09.03.61	1116
KB-5	R-7	Zvedochka Manequin	25.03.61	0900	164	230	65	1	25.03.61	1047
V-I	R-7	Gagarin	12.04.61	0907	169	315	65	1	12.04.61	1055
V-II	R-7	Titov	06.08.61	0900	111	160	64.9	17	06.08.61	1018
V-III	R-7	Nikolayev	11.08.62	1130	166	218	65	64	15.08.62	0954
V-IV	R-7	Popovich	12.08.62	1102	169	222	65	48	15.08.62	1009
V-V	R-7	Bykovsky	14.06.63	1500	175	221	65	81	19.06.63	1400
V-VI	R-7	Terishkova	16.06.63	1230	181	231	65	48	19.06.63	1120

* Moscow time.
†Date not released.

Bibliography and sources

Contemporary books covering Project Mercury are difficult to acquire, although some are still to be found in second-hand book shops. Many of the contemporary NASA publications are also out of print although some are available through the various NASA History Offices or their Internet sites. Several of the astronauts involved in Project Mercury have published their autobiographies and John Glenn's return to space on the Space Shuttle resulted in a number of new books covering his career.

The following chronological list is intended to point the reader in the direction of other books and publications containing coverage of Project Mercury.

Seven Into Space. Joseph N. Bell. Ebury Press, London, 1960.

John H. Glenn Astronaut. Lt Col. Philip Pierce and Karl Schuon. George C. Harrap & Co. Ltd, London, 1962.

Mercury Project Summary: *Including the Results of the Fourth Manned Orbital Flight 15 and 16 May 1963*. NASA SP-45, October 1963.

Project Mercury: A Chronology. James M. Grimwood. NASA SP-4001, 1963.

Word-Wide Network Support. Paper 8, Mercury Project Summary (NASA SP-45). NASA 1963.

Flight Control Operations. Paper 15, Mercury Project Summary (NASA SP-45). NASA 1963.

Astronaut Training. Paper 10, Mercury Project Summary (NASA SP-45). NASA 1963.

This New Ocean: A History of Project Mercury. Loyd S. Swenson Jr, James M. Grimwood and Charles C. Alexander. NASA 1966.

A New Dimension: Wallops Island Flight Range: The First Fifteen Years. Joseph A. Shortal. NASA Reference Publication 1028, 1978.

America's Astronauts and their Indestructible Spirit. Dr F. Kelly. Tab Books Inc., 1986 (ISBN 0-8306-8396-8).

An Illustrated History of Space Shuttle: US Winged Spacecraft: X-15 to Orbiter. Melvyn Smith. Haynes Publishing Group, 1986 (ISBN 0-85429-480-5).

Schirra's Space. Walter M. Schirra and Richard N. Billings. Quilan Press, Boston, 1988.

DEKE! U.S. Manned Space: From Mercury to the Shuttle. Donald K. Slayton and Michael Cassutt. Forge Books, 1993 (ISBN 0-312-85503-6).

Moonshot: The Inside Story of America's Race to the Moon. Alan Shepard, Deke Slayton with Jay Barbee and Howard Benedict. Virgin Publishing Ltd, London, 1994 (ISBN 1-85227-498-0).

The Pre-Astronauts: Manned Ballooning on the Threshold of Space. Craig Ryan. Naval Institute Press (ISBN: 1-55750-732-5).

40th Anniversary of Mercury 7: M. Scott Carpenter. Tara Gray. NASA 1999.

40th Anniversary of Mercury 7: L. Gordon Cooper, Jr. Tara Gray. NASA 1999.

40th Anniversary of Mercury 7: John H. Glenn, Jr. Tara Gray. NASA 1999.

40th Anniversary of Mercury 7: Virgil Ivan 'Gus' Grissom. Mary C. Zorinio. NASA 1999.

40th Anniversary of Mercury 7: Walter M. Schirra, Jr. Tara Gray. NASA 1999.

40th Anniversary of Mercury 7: Alan B. Shepard, Jr. Tara Gray. NASA 1999.

40th Anniversary of Mercury 7: Donald K. 'Deke' Slayton. Tara Gray. NASA 1999.

First Up? Tony Reichhardt. Air & Space, Smithsonian, Aug./Sept. 2000.

SOURCES

Every effort has been made to seek permission to quote from each of the following sources. Although, in some instances, permission could not be obtained as the publisher was no longer at the address printed in the original, the source is nevertheless acknowledged.

This New Ocean: A History of Project Mercury. Loyd S. Swenson Jr, James M. Grimwood and Charles C. Alexander. NASA, 1966.

World-Wide Network Support. Paper 8, Mercury Project Summary. NASA SP-45, 1963.

Flight Control Operations. Paper 15, Mercury Project Summary. NASA SP-45, 1963.

Astronaut Training. Paper 10, Mercury Project Summary. NASA SP-45, 1963.

Mercury Project Summary. Including the Results of the Fourth Manned Orbital Flight May 15 and 16, 1963. NASA SP-45, October 1963.

Project Mercury: A Chronology. James M. Grimwood. NASA SP-4001, 1963.

Results of the Second United States Manned Orbital Space Flight. NASA SP-6, 1962.

Results of the Third United States Manned Orbital Space Flight. NASA SP-12, December 1962.

A New Dimension: Wallops Island Flight Range: The First Fifteen Years. Joseph A. Shortal. NASA Reference Publication 1028, 1978.

On the Shoulders of Titans. A History of Project Gemini. Barton Hacker and James Grimwood. NASA SP-4203, 1977.

Chariots for Apollo. A History of Manned Lunar Spacecraft. Courtney Brooks, James Grimwood and Loyd Swenson Jr. NASA SP-4205, 1979.

John H. Glenn Astronaut. Lt Col. Philip Pierce and Karl Schuon. George C. Harrap & Co. Ltd, London, 1962.

Schirra's Space. Walter M. Schirra and Richard N. Billings. Quilan Press, Boston, 1988.

America's Astronauts and Their Indestructible Spirit. Dr F. Kelly. Tab Books Inc., 1986 (ISBN 0-8306-8396-8).

DEKE! U.S. Manned Space: From Mercury to the Shuttle. Donald K. Slayton and Michael Cassutt. Forge Books, 1993 (ISBN 0-312-85503-6).

Disasters and Accidents in Manned Spaceflight. David Shayler. Springer-Praxis, 2000 (ISBN 1-852-332-255).

An Illustrated History of Space Shuttle. U.S. Winged Spacecraft: X-15 to Orbiter. Melvyn Smith. Haynes Publishing Group, 1986 (ISBN 0-854 29-480-5).

The New Russian Space Programme. From Competition to Collaboration. Brian Harvey. Wiley-Praxis, 1996 (ISBN 0-471-96014-4).

German Secret Weapons of World War Two: The Missiles, Rockets, Weapons and New Technology of the Third Reich. Ivan V. Hogg. Greenhill Books, 1999 (ISBN 1-85367-325-0).

The 20th Century: The Pictorial History. Hamlyn Publishing Group, 1990 (ISBN 0-600-50353-4).

The Cold War. Jeremy Isaacs and Taylor Downing. Bantam Press, 1998 (ISBN 0593-04309X). All information extracted from The *Cold War* is copyright Jeremy Isaacs Productions and Turner Original Productions, 1998. (*The Cold War* is published by Bantam Press, a division of Transworld Publishers. All rights reserved.)

High Road To The Moon. Bob Parkinson. British Interplanetary Society, 1979 (no ISBN shown).

Spaceflight – The Basic Principles I-II. John Catchpole. Quest for Knowledge Magazine, Top Events Publishing, 1998.

Interview with Malcolm Scott Carpenter. Nigel MacKnight. Space Flight News, 1986.

Index